Algal Biorefineries and the Circular Bioeconomy

Algal Biorefineries and the Circular Bioeconomy
Algal Products and Processes

Edited by
Sanjeet Mehariya, Shashi Kant Bhatia, and
Obulisamy Parthiba Karthikeyan

CRC Press
Taylor & Francis Group
Boca Raton London New York

CRC Press is an imprint of the
Taylor & Francis Group, an **Informa** business

First edition published 2022
by CRC Press
6000 Broken Sound Parkway NW, Suite 300, Boca Raton, FL 33487-2742

and by CRC Press
4 Park Square, Milton Park, Abingdon, Oxon, OX14 4RN

CRC Press is an imprint of Taylor & Francis Group, LLC

ISBN: 978-1-032-04891-8 (hbk)
ISBN: 978-1-032-04972-4 (pbk)
ISBN: 978-1-003-19540-5 (ebk)

DOI: 10.1201/9781003195405

Typeset in Times
by MPS Limited, Dehradun

To all contributors and future algal biotechnologists for their contributions to sustainable development of algal biorefinery.

Contents

Preface...ix
Acknowledgments...xiii
Editors...xv
Contributors xvii

Chapter 1   Marine Macroalgal Biorefinery: Recent Developments and Future
Perspectives 1

                    Nitin Trivedi, Dhanashree Mone, Arijit Sankar Mondal, and
Ritu Sharma

Chapter 2   Valorization of Algal Spent Biomass into Valuable Biochemicals
and Energy Resource..37

                    Saravanan Vasanthakumar, K. Greeshma, and
Muthu Arumugam

Chapter 3   Algal Role in Microbial Fuel Cells 71

                    Ghadir Aly El-Chaghaby, Sayed Rashad, Meenakshi Singh,
K. Chandrasekhar, and Murthy Chavali

Chapter 4   Potential of Microalgae for Protein Production 91

                    Elena M. Rojo, Silvia Bolado, Alejandro Filipigh,
David Moldes, and Marisol Vega

Chapter 5   Microalgae for Pigments and Cosmetics........................133

                    Nídia S. Caetano, Priscila S. Corrêa, Wilson G. de Morais
Júnior, Gisela M. Oliveira, António A.A. Martins, Monique
Branco-Vieira, and Teresa M. Mata

Chapter 6   Microalgae for Animal and Fish Feed 177

                    Margarida Costa, Joana G. Fonseca, Joana L. Silva, Jean-Yves
Berthon, and Edith Filaire

Chapter 7   Algal-Sourced Biostimulants and Biofertilizer for Sustainable
Agriculture and Soil Enrichment: Algae for Fertilizers and Soil
Conditioners...211

                    P. Muthukumaran, J. Arvind, M. Kamaraj, and A. Manikandan

Chapter 8 Recent Trends in Microalgal Refinery for Sustainable
              Biopolymer Production 237

              *Menghour Huy, Ann Kristin Vatland, Reza Zarei, and
              Gopalakrishnan Kumar*

Chapter 9 Algae as a Source of Polysaccharides and Potential
              Applications 257

              Sonal Tiwari, E. Amala Claret, and Vikas S. Chauhan

Chapter 10 Algae as Food and Nutraceuticals 295

              *Chetan Aware, Rahul Jadhav, Jyoti Jadhav,
              Virdhaval Nalavade, Shashi Kant Bhatia,
              Yung-Hun Yang, and Ranjit Gurav*

Chapter 11 Cyanobacterial Phycobiliproteins – Biochemical Strategies to
              Improve the Production and Its Bio Application 339

              Khushbu Bhayania, Sandhya Mishra, and Imran Pancha

Index 369

Preface

In recent years, the algal biorefinery has been seen as a promising alternative to fossil-derived products that reduce environmental pollution, reduce product costs, and support circular bioeconomy. However, upstream algal cultivation and downstream processing are energy-intensive processes and are considered as bottlenecks in promoting algal biorefinery. Improving the biomass productivity and bioproduct developments are still underway, while number of novel bioprocess and bioreactor engineering technologies were developed recently. Therefore, this book will offer complete coverage of algal bioproducts process, including biotechnological applications and environmental effects of microalgae cultivation. With contributions from world experts, this book focuses on microalgae from an organism perspective to offer a complete picture from evolution to bioproducts. In addition, it proposes an integrated microalgal refinery for the advancement of existing technologies. Also, it summarizes the strategies and future perspectives of the microalgal refinery to integrate with circular bioeconomy concepts. The book provides a concise introduction to the science, biology, technology, and application of algae. It covers downstream and upstream steps of the algal refinery for the production of algal biomass, which has several social benefits.

In today's world, marine macroalgae (or seaweeds) are emerging as the most promising source of distinct valuable products. Chapter 1 discusses how marine macroalgae have gained the spotlight as the world's third-generation feedstock. Traditionally, macroalgae have been widely utilized for the direct extraction of value-added bioactive components. However, the vision with the macroalgae has been stretched with a sequential extraction process by introducing the "macroalgal biorefinery concept". Therefore, in this chapter, the author summarizes and discusses the different zero waste management strategies to utilize the complete macroalgal biomass (from fresh to residue) by producing various primary and secondary products, along with an improved cultivation system to uplift the biorefinery concept, thereby fulfilling the supply chain demand of value products.

Chapter 2 explores how algae are the largest primary producers and are found in every ecological niche conceivable. They accumulate numerous bioactive compounds that are gaining more interest progressively since they can be utilized in several ways. However, algae production facilities should run in a closed-loop biorefinery approach in order to produce a maximum number of possible products, which have a key role in launching sustainable bioprocess technologies, industrial translation, and circular economy. In this chapter, the author discusses the cost-effective optimization of different extraction methods, valorization of each residual biomass fraction obtained after extraction, composition, and different industrial utilities for the multiple bio-based products.

Chapter 3 describes how microbial fuel cells offer new opportunities for the sustainable production of energy from biodegradable, reduced compounds – a novel biotechnology for energy production. The chapter discusses the role of algae in microbial fuel cells (MFCs), gives a brief introduction about the importance of energy sources as well as the drawbacks of traditional sources, and explains the importance of finding sustainable alternatives. Also, this chapter presents algae, their definition, classification, and importance in bio-energy production. The chapter covers the different algal substrates as an autotrophic and heterotrophic mode of MFCs operations and their benefits. Further, a few challenges of algal MFC and its future perspectives are also discussed.

Chapter 4 explores how microalgae are an important source of functional food, especially due to their high protein content. The production of protein depends on multiple factors – from their location within the microalgae to their future uses. Once microalgae with a determined protein content have been cultivated and harvested, it is necessary to extract and separate the proteins by different methods aimed at breaking the cell wall (extraction process), and isolate proteins from other undesired compounds (separation and purification process). Thus, this chapter discusses the wide range of potential applications for microalgae-extracted proteins in the food science field or in products with techno-functional properties, especially feeding and nutraceuticals.

Chapter 5 explains how microalgae are among the most promising cell factories of the near future. Their renewable nature, and ability to rely on photosynthesis to use CO_2 or nutrients from wastewater to grow and multiply, make them an emergent source of valuable natural compounds. Although in the past, a few of these microalgae have been known for their value as a source of proteins, carbohydrates, exopolysaccharides, polyunsaturated fatty acids, omega 3, and omega 6 fatty acids, recently they have also been found increasingly important as sources of more valuable compounds such as carotenoids, of which astaxanthin, lutein, and β-carotene are of extreme importance in food, feed, and cosmeceutical industries, and phycobiliproteins and chlorophylls are finding their place in the commercial market. In this chapter, the author presents and discusses the potential microalgae sources for these value compounds, potential applications, as well as the major ongoing research projects, and their contribution driving the blue bioeconomy.

Chapter 6 discusses how microalgae are a group of microorganisms with a nutrient-rich biochemical composition, including several biologically active compounds. These organisms represent not only alternative sources to animal proteins, but also to fish sources of omega-3. Furthermore, microalgal color pigments can act as antioxidants, along with several other health benefits. They can also produce molecules that are of great interest, having applications in many fields, including astaxanthin, lycopene, chlorophylls, phycobiliproteins, and exopolysaccharides. These compounds mean that microalgae can be used as an attractive natural resource and applied as a functional ingredient for food and feed. This chapter aims to give an overview of microalgae properties and review their potential to be used as feed products.

Chapter 7 describes how algae are achieving greater traction for agro-allied industries and agriculturists, especially for their contribution as biostimulants or biofertilizers to impact agricultural sustainability. Algae represent a potential sustainable replacement for the meliorism and preservation of crops and plants. Algal cellular extracts via biomass could be utilized as a seed galvanizing agent;

foliar spray, and biofertilizer, or a plant growth advocate; they can be used as a fruit production enhancer, also to promote rapid germination of seeds, etc. This chapter targets the recent research advancement in algae-sourced based plant biofertilizers and biostimulants in agricultural applications and soil attribute enhancement, and the contingencies and threats to commercializing these biofertilizers and biostimulants are also deliberated.

Chapter 8 explains how microalgae biomass is packed with valuable compounds and can accumulate highly targeted composition with components such as proteins, carbohydrates, and lipid for the final refinery of the products. Polyhydroxyalkanoate (PHA), for instance, is known to be obtained from microalgae cells synthesized from lipid in microalgae biomass or derived from other microbial sugar fermentation using microalgae biomass as feedstock. Therefore, sugar fermentation is predicted to be the largest production method in the PHAs market between 2020 and 2025. Therefore, this chapter discusses the different routes for bioplastic production from algal biomass.

Chapter 9 explores how marine macroalgae are a major source of natural polysaccharides, which are a large chain of covalently linked simple sugars. Their structure, function, and bio-active properties are unique to each species. A wide range of biological activities is aided by extracellular polysaccharides (EPSs) generated by microalgae and cyanobacteria. The polysaccharides derived from algae are emerging as potential biomolecules for innovative therapeutic applications, such as wound dressing, drug delivery, etc. Further, algal polysaccharides are widely utilized for commercial applications in cosmeceuticals, nutraceuticals, pharmaceuticals, functional foods, food packaging, bio-plastics, and plant biostimulants. This chapter provides an insight into the recent developments in research on algal polysaccharides, their source, structural properties, bioactive potential, and applications.

Chapter 10 describes a market survey and trend to increase the global demand and consumption for macroalgal and microalgal foods owing to their functional benefits above their traditional uses and considerations. Therefore, this chapter emphasizes the developing fields of algal science with a specific focus on algal diversity, digestibility, bioavailability, active bioactive components, and applications that are required for better assessment of the health benefits of these algae or algal by-products. Furthermore, vast opportunities exist for phycologists, other researchers, and people interested in developing new products and their industrial applications in this emerging field.

Chapter 11 explores how cyanobacteria are gram-negative prokaryotes with oxygenic photosynthesis on earth and prolonged evolutionary history. The application of photosynthetic cyanobacteria in directly converting carbon dioxide to food, feed, and fuel is an emerging area of interest. With simple growth requirements, their ease of genetic modifications allows them to serve as a host for various biotechnological applications. Phycobiliproteins, carotenoids, scytonemin, chlorophylls, phenolics, polysaccharides, and phytohormones are some of the important compounds obtained from cyanobacteria. However, due to less yield in biomass and products, the commercialization of cyanobacteria for bioproducts is a major challenge. Knowledge of factors regulating biomass with bioproducts accumulation and optimization of downstream processing in cyanobacteria may help improve the overall yield and lower the cost of production. In this chapter, the authors have insight into phycobiliproteins from cyanobacteria along with its production, purification, and potential biotechnological applications.

Sanjeet Mehariya
Shashi Kant Bhatia
Obulisamy Parthiba Karthikeyan

Acknowledgments

First and foremost, we would like to express our sincere gratitude to all the distinguished authors for their thoughtful contribution for making this project successful. We really appreciate their patience and diligence in revising the first draft of the chapters after assimilating the suggestions and comments. I would like to thank and acknowledge the solicitous contributions of all the reviewers who spent their valuable time in providing constructive comments to improve the quality of the chapters. We are also grateful to the staff at CRC Press, particularly Nick Mould and Marc Gutierrez, who supported us tremendously throughout the project and for their immediate response to allow us to publish quickly.

Finally, we would like to acknowledge the support of our mentors, family members, friends, and colleagues for their love and encouragement; this work is dedicated to their smiles.

Editors

Sanjeet Mehariya, PhD, is a Postdoctoral Researcher at the Department of Chemistry, Umeå University, Umeå, Sweden. He earned a PhD in engineering at the University of Campania "Luigi Vanvitelli", Italy. Dr. Mehariya's research expertise includes the development of blue-bioeconomy through the algal biorefinery, the production of value-added chemical and the formulation of functional food from algal biomass, and agro-industrial waste-based resource recovery. He has collaborated with the national and international stakeholders, including policy makers, industries, and prominent RTD institutes in the field of biobased economy. He has worked at CSIR-Institute of Genomics and Integrative Biology, Delhi, India; Konkuk University, Seoul, South Korea; ENEA-Italian National Agency for New Technologies, Energy and Sustainable Economic Development, Rome, Italy; Hong Kong Baptist University, Hong Kong; University of Campania "Luigi Vanvitelli", Italy; and Sapienza - University of Rome, Italy.

Shashi Kant Bhatia, PhD, is an Associate Professor in the Department of Biological Engineering at Konkuk University, Seoul, South Korea, and has more than ten years' experience in the area of biowastes valorization into bioenergy, biochemicals, and biomaterials. Dr. Bhatia has contributed extensively to the industrial press and served as an editorial board member of Sustainability and Energies journal, and as Associate Editor of Frontier of Microbiology, Microbial Cell Factories, and PLOS One journal. He has published more than 140 research and review articles on industrial biotechnology, bioenergy production, biomaterial, biotransformation, microbial fermentation, and enzyme technology in international scientific peer-reviewed journals and holds 12 international patents. He was selected to the Top 2% Scientist List of Stanford University. He earned an MSc and a PhD in biotechnology at Himachal Pradesh University (India). He has qualified JNU combined biotechnology entrance, DBT-JRF, GATE, CSIR exams. Dr. Bhatia has worked as a Brain Pool Post Doc Fellow at Konkuk University (2014–2016), South Korea.

Obulisamy Parthiba Karthikeyan, PhD, is a Research Scientist and Lecturer (Adjunct) at the Department of Civil and Environmental Engineering, South Dakota School of Mines and Technology, Rapid City, South Dakota, USA. He has over 12 years of international research experience in the field of environmental microbiology and biotechnology. Dr. OPK serves as a peer review member assessor for the U.S. Department of Agriculture (USDA) – USA, Agriculture and

Agri-Food Canada, Horizon H2020 – Marie Sklodowska – Curie Individual Fellowships – Europe, Australian Research Council – Australia, and the Research Foundation – Flanders (FWO), Belgium. He is an Associate Editor for the Frontiers of Microbiotechnology and guest edited some special issues for publishers such as Elsevier, Springer, and MDPI. He has published more than 60 research and review articles in Scopus-indexed journals, 2 books, 10 chapters, and 2 editorials. He has received meritorious awards such as Australia–Thailand Early Career Research Exchange Award, 2014; Sêr Cymru National Research Network for Low Carbon Energy and Environment Writing Fellowship and Award, 2018 (UK); and Ramalingaswamy Re-Entry Fellowship, 2018 (India). His research interests include microbial engineering, omics for process understanding, greenhouse gas fermentation into value products, algal biorefinery, bioreactor engineering, and bioleaching of metals.

Contributors

Muthu Arumugam
Microbial Processes and Technology
    Division
Council of Scientific and Industrial
    Research – National Institute for
    Interdisciplinary Science and
    Technology
Thiruvananthapuram, India
and
Academy of Scientific and Innovative
    Research
Ghaziabad, India

J. Arvind
Dhirajlal Gandhi College of
    Technology
Omalur, Tamil Nadu, India

Chetan Aware
Department of Biotechnology
Shivaji University
Maharashtra, India

Jean-Yves Berthon
Greentech SA
Rue Michel Renaud
Saint Beauzire, France

Shashi Kant Bhatia
Department of Biological Engineering,
    College of Engineering
Konkuk University
Seoul, South Korea

Khushbu Bhayani
CSIR-Central Salt and Marine
    Chemicals Research Institute
Council of Scientific and Industrial
    Research
Gujarat, India

Silvia Bolado
Department of Chemical Engineering
    and Environmental Technology
School of Industrial Engineering
and
Institute of Sustainable Processes
University of Valladolid
Valladolid, Spain

Monique Branco-Vieira
LEPABE-Laboratory for Process
    Engineering, Environment,
    Biotechnology and Energy
Faculty of Engineering
University of Porto (FEUP)
Porto, Portugal

Nídia S. Caetano
LEPABE-Laboratory for Process
    Engineering Environment,
    Biotechnology and Energy
Faculty of Engineering
University of Porto (FEUP)
and
Porto, Portugal
CIETI/ISEP – Centre of Innovation on
    Engineering and Industrial
    Technology/IPP-ISEP School of
    Engineering
Porto, Portugal

K. Chandrasekhar
School of Civil and Environmental
    Engineering
Yonsei University
Seoul, Republic of Korea
and
Alliance University (Central Campus)
Bengaluru, Karnataka, India
and
NTRC-MCETRC and Aarshanano
    Composite Technologies Pvt. Ltd.
Andhra Pradesh, India

Vikas S. Chauhan
Plant Cell Biotechnology
    Department (PCBT)
CSIR-Central Food Technological
    Research Institute (CFTRI)
Mysuru, Karnataka, India
and
Academy of Scientific and Innovative
    Research (AcSIR)
Ghaziabad, India

Murthy Chavali
Office of the Dean (Research) and
    Department of Chemistry
Alliance College of Engineering and
    Design
Faculty of Science and Technology
Karnataka, India

E. Amala Claret
Plant Cell Biotechnology
    Department (PCBT)
CSIR-Central Food Technological
    Research Institute (CFTRI)
Mysuru, Karnataka, India
and
Academy of Scientific and Innovative
    Research (AcSIR)
Ghaziabad, India

Priscila S. Corrêa
CIETI/ISEP – Centre of Innovation on
    Engineering and Industrial
    Technology/IPP-ISEP
School of Engineering
Porto, Portugal

Margarida Costa
Allmicroalgae – Natural Products
    Avenida
Pataias, Portugal

Wilson G. de Morais Júnior
CIETI/ISEP – Centre of Innovation on
    Engineering and Industrial
    Technology/IPP-ISEP
School of Engineering
Porto, Portugal

Ghadir Aly El-Chaghaby
Bioanalysis Laboratory
Regional Center for Food and
    Feed (RCFF)
Agricultural Research Center (ARC)
Giza, Egypt

Edith Filaire
University Clermont Auvergne
INRA-UcA
UNH (Human Nutrition Unity)
ECREIN Team
Clermont-Ferrand, France

Alejandro Filipigh
Department of Chemical Engineering
    and Environmental Technology
School of Industrial Engineering
University of Valladolid
Valladolid, Spain

Joana G. Fonseca
Allmicroalgae – Natural Products
Avenida
Abril, Pataias, Portugal

K. Greeshma
Microbial Processes and Technology
    Division
Council of Scientific and Industrial
    Research – National Institute for
    Interdisciplinary Science and
    Technology
Thiruvananthapuram, India

Ranjit Gurav
Department of Biological Engineering
College of Engineering
Konkuk University
Seoul, South Korea

Menghour Huy
Department of Chemistry Bioscience,
    and Environmental Engineering
Faculty of Science and Technology
University of Stavanger
Stavanger, Norway

Jyoti Jadhav
Department of Biotechnology
Shivaji University
Maharashtra, India

Rahul Jadhav
Department of Biotechnology
Shivaji University
Kolhapur, Maharashtra, India

M. Kamaraj
Department of Biotechnology
College of Biological and Chemical
    Engineering
Addis Ababa Science and Technology
    University
Addis Ababa, Ethiopia

Gopalakrishnan Kumar
Department of Chemistry Engineering,
    Environment, Biotechnology and
    Energy
Faculty of Science and Technology
University of Stavanger
Stavanger, Norway

A. Manikandan
Department of Industrial Biotechnology
Bharath Institute of Higher Education
    and Research
Chennai, Tamilnadu, India

António A.A. Martins
LEPABE-Laboratory for Process
    Engineering, Environment
    Biotechnology and Energy
Faculty of Engineering
University of Porto (FEUP)
Porto, Portugal

Teresa M. Mata
INEGI-Institute of Science and
    Innovation in Mechanical and
    Industrial Engineering
R. Dr. Roberto Frias
Porto, Portugal

Sandhya Mishra
CSIR-Central Salt and Marine
    Chemicals Research Institute
Council of Scientific and Industrial
    Research
Bhavnagar, Gujarat, India

David Moldes
Department of Analytical Chemistry
Faculty of Sciences
University of Valladolid
Campus Miguel Delibes
Valladolid, Spain

Arijit Sankar Mondal
Department of Microbiology
Guru Nanak Institute of Pharmaceutical
    Science and Technology (Life
    Science)
Kolkata, India

Dhanashree Mone
DBT-ICT Centre for Energy
    Biosciences
Institute of Chemical Technology
Matunga, Mumbai, India

P. Muthukumaran
Department of Biotechnology
Kumaraguru College of Technology
Coimbatore, Tamilnadu, India

Virdhaval Nalavade
Department of Biotechnology
Yashavantrao Chavan Institute of Science
Satara, Maharashtra, India

Gisela M. Oliveira
UFP Energy, Environment and Health
    Research Unit
University Fernando Pessoa
Porto, Portugal

Imran Pancha
Department of Biology
SRM University-AP
Amaravati, Andhra Pradesh, India

Sayed Rashad
El-Fostat Laboratory
Cairo Water Company
Cairo, Egypt

Elena M. Rojo
Department of Chemical Engineering
    and Environmental Technology
School of Industrial Engineering
and
Institute of Sustainable Processes
University of Valladolid
Valladolid, Spain

Ritu Sharma
Department of Microbiology
Guru Nanak Institute of Pharmaceutical
    Science and Technology (Life Science)
Kolkata, India

Joana Laranjeira Silva
Allmicroalgae – Natural Products
Pataias, Portugal

Meenakshi Singh
Department of Botany
The M.S. University of Baroda
Gujarat, India

Sonal Tiwari
Plant Cell Biotechnology
    Department (PCBT)
CSIR-Central Food Technological
    Research Institute (CFTRI)
Mysuru, Karnataka

Nitin Trivedi
DBT-ICT Centre for Energy
    Biosciences
Institute of Chemical Technology
Matunga, Mumbai, India

Saravanan Vasanthakumar
Microbial Processes and Technology
    Division
Council of Scientific and Industrial
    Research – National Institute for
    Interdisciplinary Science and
    Technology
 Thiruvananthapuram, India

Ann Kristin Vatland
Department of Chemistry, Bioscience,
    and Environmental Engineering
Faculty of Science and Technology
University of Stavanger
Stavanger, Norway

Marisol Vega
Institute of Sustainable Processes
and
Department of Analytical Chemistry
Faculty of Sciences
University of Valladolid
Valladolid, Spain

Yung-Hun Yang
Department of Biological Engineering
College of Engineering
Konkuk University
Seoul, South Korea

Reza Zarei
Department of Chemistry, Bioscience,
    and Environmental Engineering
Faculty of Science and Technology
University of Stavanger
Stavanger, Norway

1 Marine Macroalgal Biorefinery
Recent Developments and Future Perspectives

Nitin Trivedi and Dhanashree Mone
DBT-ICT Centre for Energy Biosciences, Institute of
Chemical Technology, Matunga, Mumbai, India

Arijit Sankar Mondal and Ritu Sharma
Department of Microbiology, Guru Nanak Institute of
Pharmaceutical Science and Technology (Life Science),
Kolkata, India

CONTENTS

1.1 Introduction...2
1.2 Green Macroalgae 3
    1.2.1 Advanced Green Macroalgae Cultivation Strategies to Uplift the
        Biorefinery System 9
    1.2.2 Essential Components in Green Macroalgae Important for
        Sequential Biorefinery Processes 10
    1.2.3 Green Macroalgae Biorefinery Protocol with the Advanced
        Extraction Process 10
    1.2.4 Potential Primary Products Obtained during the Biorefinery
        Process 11
    1.2.5 Potential Secondary Products Obtained during the Biorefinery
        Process 13
        1.2.5.1 Bioethanol 13
        1.2.5.2 Biohydrogen...14
        1.2.5.3 Biogas 14
        1.2.5.4 Biodiesel 15
        1.2.5.5 Other Miscellaneous Applications 15
1.3 Red Macroalgae...16
    1.3.1 Cultivation of Red Macroalgae for a Biorefinery Approach............16

DOI: 10.1201/9781003195405-1

1

1.3.2   Bioactive Components of Red Macroalgae for a Biorefinery
        Approach...17
1.3.3   Primary Products Obtained during the Biorefinery Process............18
1.3.4   Secondary Products Obtained during the Biorefinery Process.........19
        1.3.4.1   Bioethanol...19
        1.3.4.2   Biodiesel...20
        1.3.4.3   Biohydrogen and Biogas.......................................21
        1.3.4.4   Other Miscellaneous Products................................21
1.4   Brown Macroalgae 23
        1.4.1   Cultivation of Brown Macroalgae for a Biorefinery Approach 23
        1.4.2   Bioactive Components of Brown Macroalgae for a Biorefinery
                Approach...24
        1.4.3   Primary Products Obtained during Biorefinery Process 25
        1.4.4   Secondary Products Obtained during the Biorefinery Process.........27
                1.4.4.1   Biogas 27
                1.4.4.2   Bioethanol..27
                1.4.4.3   Biohydrogen and Biomethane................................28
                1.4.4.4   Other Miscellaneous Applications 28
1.5   Future Perspectives..29
1.6   Conclusion 29
Acknowledgments...29
References...30

1.1   INTRODUCTION

Marine macroalgae are one of the precious gems of the marine ecosystem. Based on the morphology and pigmentation, macroalgae have been classified into three types: red (Rhodophyceae), green (Chlorophyceae), and brown (Phaeophyceae). Macroalgae are a valuable bioresource with several added advantages over terrestrial crops, such as (a) no fertile land and freshwater requirement, (b) no pesticide and fertilizer requirement, (c) no competition with food, feed, and industrial crops, (d) speedy biomass production in several macroalgal species such as *Ulva meridionalis* (Tsubaki et al., 2020) and *Saccharina latissima* (Venolia et al., 2020), and (e) CO_2 sequestration and in-turn production of O_2 by the process of photosynthesis (Kraan, 2013).

Macroalgae are abundant sources of bioactive components, namely polysaccharides, minerals, fatty acids, vitamins, pigments, etc., which are useful in various industrial applications (cosmeceutical, pharmaceutical, food, feed, paint, fertilizer, and energy source). In addition, macroalgae are effective against numerous health conditions, such as diabetes, obesity, hypertension, inflammation, and viral infections (Ganesan et al., 2019; Kılınç et al., 2013). Although the global macroalgae production is massive, approximately, 32386.2 thousand tons wet weight (in 2018) (FAO (Food and Agricultural Organization), 2018), still there is a need for the efficient utilization of the biomass to fulfill the rising demands of macroalgae-based bioproducts. Earlier macroalgae were used to extract single components for specific applications, e.g., only carrageenan extraction from red macroalgae (Naseri et al., 2019). However, in the past decade, researchers across the globe have designed strategies to extract multiple value-added products in a sequential biorefinery process.

Biorefinery can be defined as the sustainable utilization of biomass for producing a wide range of biological products viz. food, feed, and chemicals as well as biofuel. The benefit of biorefinery includes the reduction in the energy and chemical requirements during the product extraction process, cost-effectiveness, and environmental friendliness, along with the production of multiple products within a cascading manner. Efficient utilization and production of the spectrum of products result in minimal waste generation and effortless effluent management (Mehariya et al., 2021a, 2021b, 2021c).

Over the past decades, increasing demand for macroalgae products has resulted in growing macroalgae waste and residual biomass; for example, before processing, feedstock gets rejected when quality standards are not met, or after processing, residual generation leads to the wastage of feedstock. To effectively utilize the macroalgae biomass, the residue from the first process could be taken into account as an input for another process. In this cascading approach, biomass is first used for producing value-added primary products, followed by secondary products. The optimization of the process for the re-utilization of the biomass yields multiple valuable products, such as pharmaceuticals and nutraceuticals, besides focusing on the traditional applications, such as phycocolloids. Further investigating the process for the effective utilization of the biomass can lead to zero waste generation (Torres et al., 2019b; Balina et al., 2017). Figure 1.1 illustrates the schematic flowchart of a complete sequential macroalgae biorefinery workflow.

The circular bioeconomy focuses on economic development by minimizing resource consumption along with the reduction of waste generation. This process uses the circular approach of reuse, remanufacture, and recycle rather than the linear approach of make-use-waste. The biorefinery concept uplifts the circular economy resulting in the valorization of macroalgae biomass by drawing multiple value-added products from the same biomass and reducing waste generation (Ubando et al., 2020). Figure 1.2 demonstrates the possible model for circular bioeconomy through macroalgae biorefinery.

In the last decade, tremendous progress has been made in the area of macroalgal growth engineering for higher and quality biomass production, followed by process optimization for sequential extraction of multiple value-added products via biorefinery approach. However, demonstration of biorefinery process is still in progress. Biorefinery studies of green, red, and brown macroalgae, including different extraction methods and products obtained, are shown in Table 1.1.

1.2 GREEN MACROALGAE

Green macroalgae (Chlorophyceae) are similar to terrestrial plants regarding biochemical composition and chlorophyll (a and b content) (del Río et al., 2020). There are around 4500 Chlorophyta species that are broadly classified into fresh and seawater macroalgae (Jung et al., 2013). In contrast to terrestrial sources, the higher growth rate and development of green macroalgae merely or does not depend on the geographical and ecological conditions and external nutrient (fertilizer) input

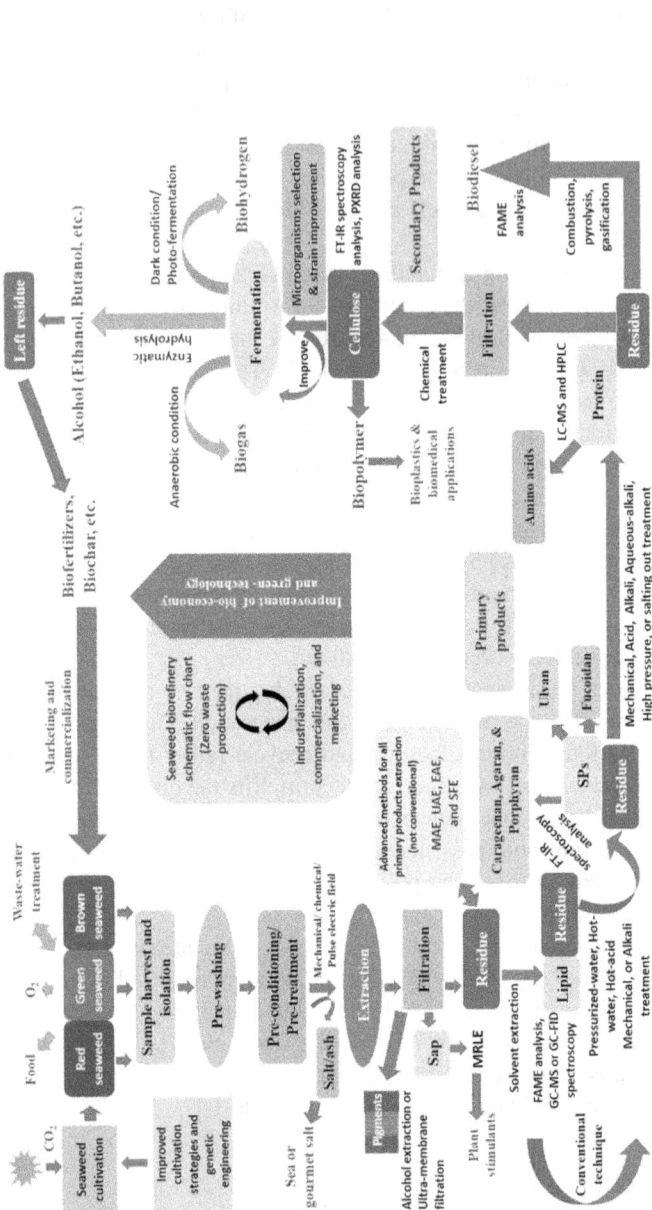

FIGURE 1.1 The schematic flowchart of a possible sequential macroalgal biorefinery. CO_2, carbon dioxide; O_2, oxygen; MRLE, mineral-rich liquid extract; FAME analysis, fatty acid methyl ester analysis; GC-MS, gas chromatography-mass spectrometry; GC-FID, gas chromatography-flame ionization detector; MAE, microwave-assisted extraction; UAE, ultrasound-assisted extraction; EAE, enzyme-assisted extraction; SFE, supercritical fluid extraction; FT-IR, Fourier-transform infrared spectroscopy; SPs, sulfated polysaccharides; LC-MS, liquid chromatography-mass spectrometry; HPLC, high-performance liquid chromatography; PXRD, powder X-ray diffraction.

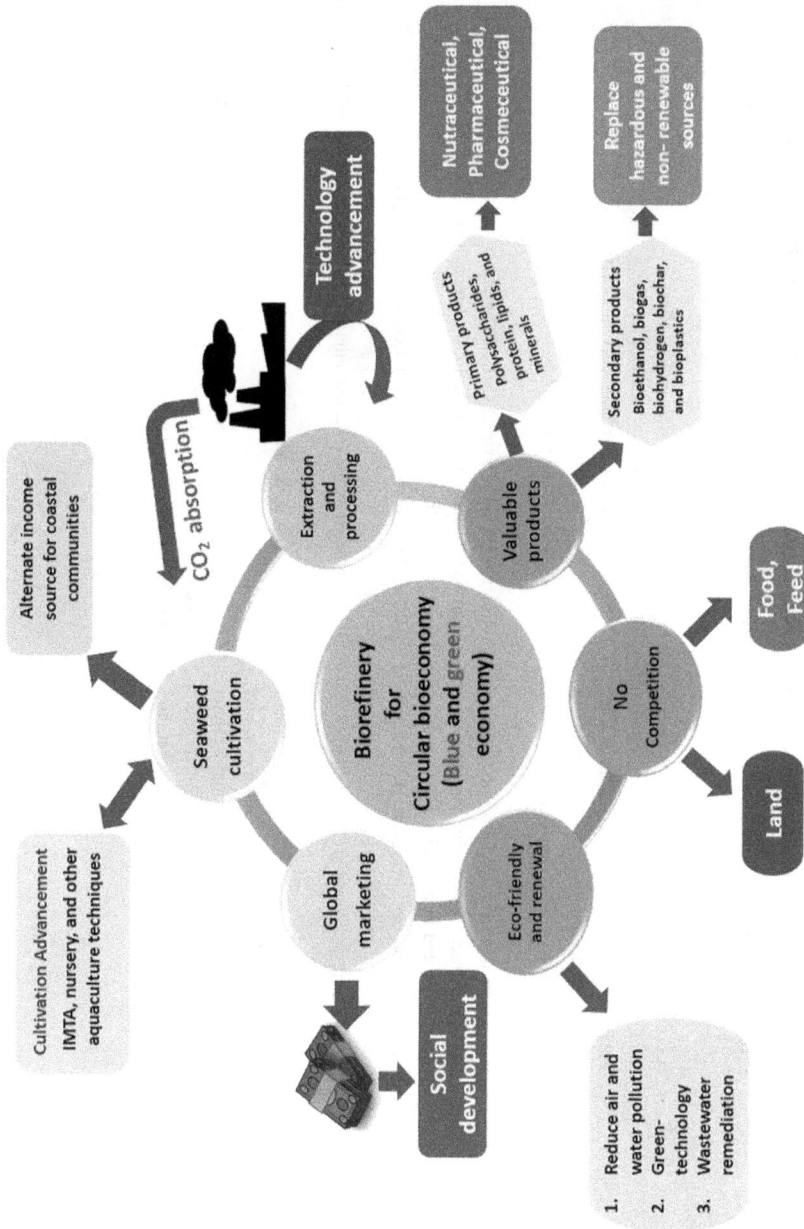

FIGURE 1.2 Macroalgal biorefinery as an important source for circular bioeconomy.

TABLE 1.1

Examples of Extraction Methods/Processes for Obtaining Several Biorefinery Products from Different Macroalgae

Macroalgae	Extraction Method/Process	Biorefinery Products Obtained	References
Green Macroalgae			
Ulva rigida	Pulse electric field (PEF), microwave, and ultrasound treatments; acid and hydrothermal hydrolysis extraction; pyrolysis; anaerobic digestion; saccharification; and fermentation	Protein, ash (salt) removal, polysaccharides (ulvan, cellulose, etc.), feedstock, biofuel (bioethanol), and other bioenergy products	(Zollmann et al., 2019)
Ulva ohnoi	Aqueous, thermal, and chemical (acid/alkaline) treatment	Salt, pigment, ulvan, and protein	(Glasson et al., 2017)
Ulva fasciata	Mechanical treatment (grinding), aqueous and solvent extraction treatment, enzymatic hydrolysis, and fermentation	Mineral-rich liquid extract (MRLE) as liquid fertilizer, cellulose, ulvan, lipid, feedstock, and biofuel (bioethanol)	(Trivedi et al., 2016)
Ulva lactuca	Mechanical and heat treatment, solvent, alkali, and chemical extraction	MRLE, lipid, ulvan, protein, and cellulose	(Gajaria et al., 2017)
Ulva lactuca	Pretreatment, enzymatic hydrolysis, and fermentation	Biofuel (acetone, butanol, and ethanol)	(van der Wal et al., 2013)
Ulva lactuca	Aqueous, thermal, and chemical (acid/alkaline) treatment and anaerobic digestion	Sap, ulvan, protein, and biogas	(Mhatre et al., 2019)
Ulva lactuca	Mechanical, aqueous, and chemical treatment, hydrothermal, thermochemical (acid) and enzymatic hydrolysis, saccharification, and fermentation	Proteins, lipids, ulvan, cellulose, pigments, dietary food, feedstocks, and biofuel (biodiesel, bio-oil, bioethanol, acetone, and biogas	(Dominguez and Loret 2019)

(Continued)

TABLE 1.1 (Continued)
Examples of Extraction Methods/Processes for Obtaining Several Biorefinery Products from Different Macroalgae

Macroalgae	Extraction Method/Process	Biorefinery Products Obtained	References
Red Macroalgae			
Eucheuma cottonii	Enzymatic hydrolysis, saccharification, and fermentation	K-carrageenan and biofuel (bioethanol)	(Tan and Lee 2014)
Gracilaria gracilis	Chemical (acid) treatment and pyrolysis	Polysaccharides, phycobiliproteins, and biofuel (bio-oil and biochar)	(Francavilla et al., 2015)
Glacilaria corticata	Chemical (acid) treatment, mechanical treatment (centrifugation), solvent extraction, aqueous extraction, saccharification, and fermentation	Protein (phycobiliproteins), phenol, polysaccharides (Agar and cellulose), lipids, mineral-rich water, pigments, biofertilizer, and biofuel (bioethanol)	(Sudhakar et al., 2016)
Kappaphycus alvarezii	Chemical treatment, mechanical treatment, fermentation, anaerobic digestion, and acid treatment	Carrageenan, fertilizer, biofuel (bioethanol and biogas), and organic acids	(Shanmugam and Seth 2018)
Solieria filiformis (Gigartinales)	Enzymatic and microwave-assisted extraction	Carrageenan, lipid, sulfated polysaccharide, and protein	(Peñuela et al., 2018)
Asparagopsis taxiformis	Solvent extraction, mechanical treatment, and enzymatic hydrolysis	Lipids, carrageenan, cellulose, phenolics, carotenoids, and biofuel (bioethanol)	(Nunes et al., 2018)
Gelidium amansii	Chemical (acid) treatment and fermentation	Polysaccharides (Agar and cellulose), feedstock, D-galactonic acid, and biofuel (bioethanol)	(Liu et al., 2020)
Brown Macroalgae			
Laminaria digitata	Acid hydrolysis, Soxhlet extraction, enzymatic hydrolysis, and fermentation	Bioethanol and protein	(Hou et al., 2015)
Carpophyllum flexuosum, Carpophyllum plumosum,	Solvent extraction, microwave-assisted extraction, and centrifugation	Pigments, mannitol, alginates, carbohydrates, and residue	(Zhang et al., 2020)

(Continued)

TABLE 1.1 (Continued)
Examples of Extraction Methods/Processes for Obtaining Several Biorefinery Products from Different Macroalgae

Macroalgae	Extraction Method/Process	Biorefinery Products Obtained	References
Ecklonia radiata and *Undaria pinnatifida*			
Ascophyllum nodosum	Microwave-assisted extraction and hydrothermal treatment	Fucoidan, alginate, mono and polysaccharides, feedstock, and biofuel (biochar)	(Yuan and Macquarrie 2015)
Saccharina latissima	Acid hydrolysis, combustion, Soxhlet extraction, and chromatography	Biofuel, mannitol, fertilizer, and other value-added products	(López-Contreras et al., 2014)
Laminaria digitata	Enzymatic hydrolysis and fermentation	Sugars, biofuel (acetone, biobutanol, bioethanol), acetic acid, lactic acid, and other bioactive components	(Hou et al., 2017)
Saccharina latissima	Mechanical treatment (grinding and centrifugation), aqueous extraction, osmotic shock, alkaline extraction, isoelectric precipitation	Protein, food, alginate, and pigments	(Vilg and Undeland 2017)
Laminaria japonica	Chemical (acid and alkali) and hydrothermal treatment, centrifugation, and fermentation	Alginate, proteins, lipid, feedstocks, and biofuel (bioethanol and biodiesel)	(Kim et al., 2019)

requirements (Lakshmi et al., 2020). Therefore, it can be well understood that green macroalgae causing marine environment eutrophication are commonly known as "green tides". However, to overcome this crucial issue (green tides), biorefinery strategies in recent decades can be utilized with a dual intention of tackling marine eutrophication as well as uplifting the blue bio-economy of the global coastal areas. Moreover, the rapid development of green macroalgal species exceeds one-time harvest every year, which will uplift the development of the macroalgal industries, thereby fulfilling people's demand globally (Lakshmi et al., 2020).

1.2.1 ADVANCED GREEN MACROALGAE CULTIVATION STRATEGIES TO UPLIFT THE BIOREFINERY SYSTEM

In the recent era, the green macroalgae biorefinery has been selected as a huge source for low lignin (Lakshmi et al., 2020) sustainable bioeconomy. Therefore, cultivation strategy implementation plays a vital role that needs a huge cultivation site along with innovative cultivation techniques. Additionally, several growth parameters (temperature (T), salinity (S), and light intensity (I)) are also essential for maintaining algal growth rate (μ) under controlled cultivation conditions (Buschmann et al., 2017). An offshore cultivation system, along with new technique installations (aquaculture rafts (Zollmann et al., 2019), flexible and submersible aquaculture (SUBFLEX) (Drimer 2016), offshore wind farms, and distance offshore cultivation from coast), has become the new option that replaced the onshore cultivation system due to its growing competition with terrestrial plants, further uplifting mass production. Apart from new technique installations, the traditional methods (rafts, ropes, and cages) are still used in certain developing countries due to their cost efficiency and easy installation (Zollmann et al., 2019). Nevertheless, various other techniques, such as strain development via genetic engineering (Lin and Qin 2014), hatchery and protoplast cultivation (Gupta et al., 2018), macroalgae growth-promoting bacteria (*Ulva mutabilis-Roseovarius-Maribacter*) system (Ghaderiardakani et al., 2019), integrated multi-trophic aquaculture (IMTA) with other aquatic organisms, free-floating enclosure cultivation, and photobioreactor installment can be used for enhancing cultivation with minor limitations (Zollmann et al., 2019). Among all the techniques, the IMTA system has been widely utilized in the present day for the synergistic cultivation of macroalgae coupled with fishery cultivation. The inorganic nutrients (nitrate, ammonia, phosphate, and carbon dioxide) produced from aquatic animals are completely used as the nutrient source for macroalgal growth. For example, a marine-based company, ALGAplus from Portugal, has introduced an innovative IMTA technique where *Ulva rigida* was co-cultured with fish both in tanks and in lagoons, along with the natural activity of algal growth-morphogenesis promoting factors (AGMPFs), restoring macroalgae growth under anexic conditions. The AGMPFs are obtained from morphogenesis, inducing epiphytic bacteria (*Maribacter mutabilis*, *Roseovarius* sp. (*Sulfitobacter* spp.)) and also inducing morphogen production, which is further supplemented with fish excreted nutrients year-round with a positive effect on macroalgae growth (Ghaderiardakani et al., 2019).

However, ocean currents certainly affect offshore cultivation, imposing cultivation loss, which has been identified during free-floating cultivation (Zollmann et al., 2019). The above-mentioned cultivation techniques and approaches are important for the biorefinery approach at an industrial scale and can be implemented for global commercial profit, which will uplift the bioeconomy. Moreover, the biorefinery concept is a cheap, renewable alternative toward non-renewable sources, along with tackling worldwide food security issues.

1.2.2 Essential Components in Green Macroalgae Important for Sequential Biorefinery Processes

Macroalgae are mostly known for their polysaccharide contents apart from protein and other biochemical components. Ulvan (comprised of glucuronic acid, xylose, sulfated rhamnose, and iduronic acid) is the unique sulfated polysaccharide that is mainly found in green macroalgae of genus *Ulva*. Few other polysaccharides, such as xyloglucan, glucuronan, and cellulose, are also present but limited in the cell wall (Lakshmi et al., 2020; Kidgell et al., 2019). The presence of these polysaccharides (mostly ulvan) upholds the green macroalgae as an attractive source for the feedstock production of high to low-grade biorefinery products. These biorefinery products obtained from the residual biomass via combined treatment processes (physical, chemical, thermochemical, and biochemical) comprise bio-based valuable feed, food, and non-food products (Sadhukhan et al., 2019). Apart from ulvan, various other biochemical components of green macroalgae, such as protein, lipids, fatty acids, pigments, etc., along with the dietary fibers, are also used for producing distinct primary value-added products, followed by the production of secondary products: animal feed, biorefinery feedstock, bioenergy, and other miscellaneous products (bioadsorbents, biomaterials, and bioelectronics) from the leftover biomass (Zollmann et al., 2019; Shannon and Abu-Ghannam 2019; Lakshmi et al., 2020). The above applications highlight the complete utilization of the macroalgae with zero-waste production, which strictly obeys the protocols of the macroalgal biorefinery approach.

1.2.3 Green Macroalgae Biorefinery Protocol with the Advanced Extraction Process

The designing of an efficient biorefinery protocol is very crucial for the extraction of various high-to-low and sensitive-to-resistive products in terms of quality and physicochemical nature, respectively. The high-grade and sensitive products (bioactive molecules) are primarily extracted, followed by low-grade and resistive products (biomaterials and bioenergy products) (Balina et al., 2017). The extraction process varies for both these products. Further, the downstream yield of products is greatly improved via sequential extraction over the direct extraction process. Each step in sequential extraction reduces the demand of biomass quality, resulting in bio-energy productions, even from the lowest grade leftover biomass (Zollmann et al., 2019). The sensitive bioactive components are extracted through various advanced techniques: ultrasound-assisted extraction (UAE), microwave-assisted extraction (MAE), enzyme-assisted extraction (EAE), pressurized-liquid extraction (PLE), supercritical-fluid extraction (SFE), and soxhlet extraction (Gullón et al., 2020). However, the processing of low-grade biorefinery products includes the conversion of either dried or wet feedstock through various treatments. Dried feedstock undergoes thermochemical treatments (pyrolysis, combustion, liquefaction, gasification, and trans-esterification) for bio-oil/biodiesel and biofuel production, but alcohol, acetone, methane, and biohydrogen production requires wet biomass via biochemical treatments (anaerobic digestion, hydrothermal treatments, and fermentation) from dried biomass (Torres et al., 2019a). Most of the extraction methods of biocomponents from all three macroalgae are similar. Figure 1.3 shows the different extraction methods (direct and sequential) generally used to extract multiple products from red, green, and brown macroalgal biomass with possible advantages and disadvantages.

1.2.4 POTENTIAL PRIMARY PRODUCTS OBTAINED DURING THE BIOREFINERY PROCESS

Based on the integrated sequential extraction procedures, a well-designed biorefinery protocol must extract the high-grade macroalgal components (bioactive components) first, followed by the low-grade product extraction from the residual

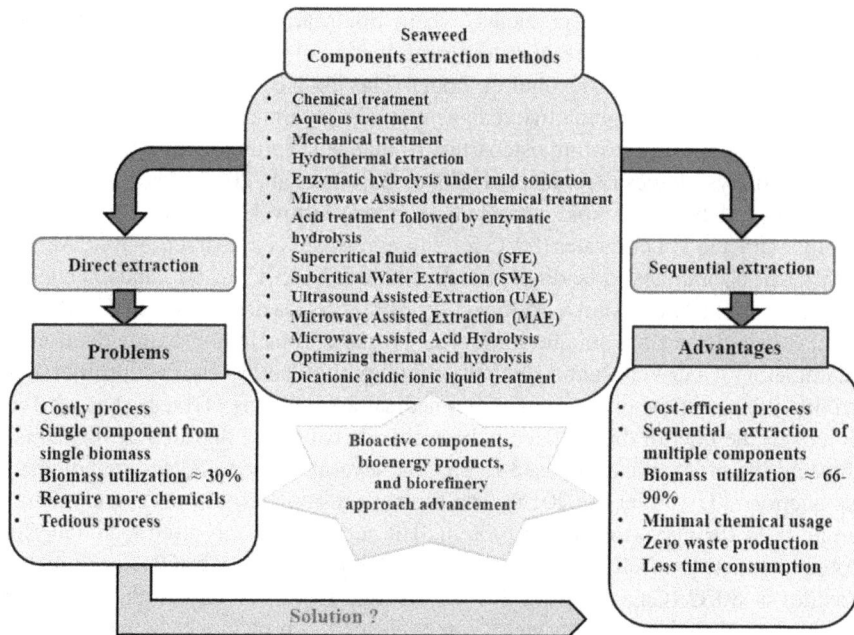

FIGURE 1.3 Seaweed component extraction process along with the advantages and disadvantages of the sequential and direct extraction process.

biomass of the former. The primary step of the biorefinery process initiates with salt (ash) removal to improve ulvan extraction either by dissolving the dry biomass in distilled water at 1:10 (biomass: water) ratio for 30 minutes at 40°C (Glasson et al., 2017) or by pulse electric field (PEF) extraction process (Robin et al., 2018). However, either increasing the temperature or time in the former process results in excess salt yield. The salt must be extracted and removed from the biomass in the preliminary step as fertilizers because biostimulants with high salt concentrations may cause saline toxicity in plants. However, the extracted salt has a huge significance in reducing the consumption of excess sodium (Na)-treated processed food by the gourmet salt as a substitute due to the presence of balanced mineral composition, such as magnesium (Mg), potassium (K), and sodium (Na). The presence of these natural mineral salts prevents the risks of cardiovascular diseases by reducing the chances of high blood pressure (Magnusson et al., 2016). The aforementioned step can be skipped by using fresh macroalgal stock followed by sap extraction via a mechanical grinding process under the aqueous condition, which results in mineral-rich liquid extract (MRLE) isolation containing essential nutrients (carbon (C), nitrogen (S), sulfur (S), etc.) and phytohormones that are essential for plant growth as liquid biofertilizers (Trivedi et al., 2016). Therefore, the MRLE extraction describes the advanced biorefinery strategy, where the sap along with the last leftover residue can act as fertilizer in liquid or solid form, respectively. The biomass residue left after either of these two processes is utilized for producing high-grade bioactive components, which include polysaccharides, proteins, and lipids. The primary components produced from the residual biomass after MRLE are lipids and pigments, where lipids are extracted with hexane or ethanol or by using a mixture of chloroform: methanol (1:2, v/v) and pigments via dissolving the biomass in alcohol (1 hour). The lipid content extracted from the fractionated biomass of green macroalgae contains essential $\omega 3$ and $\omega 6$ fatty acids, which are essential for treating osteoarthritis, diabetes, cardiovascular diseases, and other pathophysiologies (Trivedi et al., 2016; Gajaria et al., 2017). Thereafter, ulvan is fractionated from the lipid-extracted (defatted) leftover biomass. The ulvan is usually extracted via hot water (90°C for 2 hours) or hot acid extraction (0.05M HCl at 85°C for 60 minutes) or sodium oxalate extraction (85°C for 60 minutes) chelate the calcium (Ca^{2+}) crosslinks, followed by filtration (Gajaria et al., 2017; Glasson et al., 2017). Ulvan has unique biological properties, including antioxidant, anti-inflammatory, antiviral, anticoagulant, antihyperlipidemic, and antiproliferative activity along with its diverse use in biomedical applications (Trivedi et al., 2016). Moreover, the fiber of this sulfated polysaccharide cannot be digested by the human stomach; therefore, it can be used to prevent several intestine-related pathological dysfunctions (Trivedi et al., 2016). The biomass residue left after ulvan extraction can be used for the extraction of protein. The most commonly used technique for protein extraction is alkaline extraction treatment using 1N NaOH (sodium hydroxide) at 80°C (Gajaria et al., 2017). Alkaline treatment fastens the extraction process but gives low protein yield. The yield of the protein can be increased by using aqueous-alkaline treatment. The protein from the green macroalgae contains essential amino acids, which are comparatively higher than fish and soybean meal. In addition, this protein with fiber content can be used as animal feedstock. Finally, the last fractionation is performed for the extraction of the cellulose via buffer and alkaline treatment, which further undergoes enzymatic hydrolysis followed by fermentation or anaerobic digestion, producing numerous low-grade bioenergy products, discussed below (Zollmann et al., 2019).

1.2.5 POTENTIAL SECONDARY PRODUCTS OBTAINED DURING THE BIOREFINERY PROCESS

1.2.5.1 Bioethanol

Green macroalgae, in the recent era, have been designated as an attractive source for alcohol production. Bioethanol production from macroalgae includes successive stages, such as pretreatment, hydrolysis, fermentation, and recovery of ethanol. Cellulase treatment is a crucial parameter for digesting the seaweed cell wall. Thus, novel cellulase has been isolated from various microbes (*Bacillus, Clostridium, Ruminococcus, Streptomyces, Acetovibrio, Penicillium, Humicola, Schizophillum, Fusarium*, etc.) (Offei et al., 2018). The fermentative microbes come next, which helps in the conversion of these saccharides to ethanol, such as *Escherichia coli, Saccharomyces cerevisiae*, and *Pichia angophorae* (Offei et al., 2018). Yoza and Masutani (2013) reported that *Ulva reticulata* exhibits high biomass (20%) and soluble carbohydrate (18%) composition that has been used for the effective production of bioethanol of about 90 L/tonne of dried macroalgae via enzymatic hydrolysis using cellulase from *Trichoderma reesei* followed by fermentation with *Saccharomyces cerevisiae* WLP099. Trivedi et al., (2016) have demonstrated the integrated sequential extraction in comparison to direct extraction of macroalgal biomass exploitation by extracting high-grade bioactive components (ulvan, lipid, and cellulose) along with MRLE to low-grade compounds (bioethanol) from *Ulva fasciata*. The MRLE extract exhibits a high potential to act as liquid fertilizer. High yields of ulvan extracted can be widely used in nutraceutical applications. In contrast to red and brown, green macroalgae have high cellulose yields. Genetically modified *Escherichia coli* can be utilized as an alternative approach for enzyme hydrolysis and fermentation of non-glucose monosaccharide (rhamnose and glucuronic acid) present in *Ulva* sp. to overcome the challenges while fermenting with *Saccharomyces cerevisiae*. However, non-glucose monosaccharide exhibits very limited production of bioethanol. Moreover, *Saccharomyces cerevisiae* is sensitive to the saline environment, which makes it challenging to ferment residual biomass if the prewashing step is skipped. However, the introduction of halotolerant seaweeds through genetic engineering can overcome this limitation (Zollmann et al., 2019). Hamouda et al. (2016) classified a dual saccharification process of *Ulva fasciata* by both biological (*Bacillus subtilis* SH04 and *Bacillus cereus* SH06 (B6)) and chemical (acid and base) hydrolysis methods for determining the difference in the sugar yield and bioethanol production. After sugar yield, both hydrolysates were fermented with *Saccharomyces cerevisiae* SH02 to analyze the ethanol yield. The biologically treated hydrolysate with yeast produced 55.9% ± 5.2 ethanol, further utilizing the same hydrolysate with yeast along with 5% acid hydrolyzed sugar concentration produced 78.3% ± 6 ethanol. In addition, few species and strains of *Clostridium,* such as *Clostridium acetobutylicum* ATCC 824, *Clostridium beijerinckii* NCIMB 8052 and ATCC 55025, and *Clostridium saccharoperbutylacetonicum* ATTC

27021, exhibit enormous potential to produce ABE (acetone: butanol: ethanol) from *Ulva lactuca* via anaerobic fermentation (ABE fermentation) of pentose and hexose sugar molecules. Interestingly, *Clostridium acetobutylicum, Clostridium beijerinckii,* and *Clostridium saccharoperbutylacetonicum* have demonstrated high amounts of organic acids, 1,2-propanediol from rhamnose, and butanol production, respectively (van der Wal et al., 2013; Potts et al., 2011).

Macroalgal sugar hydrolyzation and fermentation is the crucial step for bioethanol and acetone production. However, there are various challenges and limitations that restrict the efficient production of bioethanol. Moreover, the pretreatment process must be optimized with further innovative techniques to overcome the toxic materials generated due to chemical treatments, which may interfere with bioethanol yield.

Nevertheless, bioethanol production from macroalgae not only depends upon highly efficient and cheap cellulase enzymes but also on the participation of a modified fermentative microbial strain. According to the aforementioned experiments, it is well understood that besides macroalgae, strain selection also plays a pivotal role in better product yield, which should not be overlooked.

1.2.5.2   Biohydrogen

Green macroalgae can be a potential producer of biohydrogen, which can be an alternative to fossil fuel. Hydrogen is commonly known as "clean fuel", as it does not cause any environmental pollution in contrast to non-renewable fuels. Several other sugars and lignocellulosic-containing terrestrial crops have been utilized for biofuel production. However, the higher downstream cost and food and feed competition become the major disadvantage that can be mitigated by producing feedstock from green macroalgae. As previously discussed, the high growth rate and low lignin content of green macroalgae satisfy the aforementioned statement. Various hydrogen-producing mesophilic (*Clostridium butylicum*) and thermophilic (*Thermotoga raeoplantia*) microorganisms can be utilized for biohydrogen production via dark fermentation, whereas purple non-sulfur bacteria (PNS) is used via photo fermentation. In addition, various other organic acids (acetic, butyric, and lactic acid) of industrial importance are also produced simultaneously during the process (Sharma and Arya 2017). Nevertheless, mostly green microalgae have been introduced toward biohydrogen production by various processes such as photo fermentation, and direct and indirect biophotolysis. Therefore, very limited data are available on macroalgal biohydrogen production.

1.2.5.3   Biogas

Besides biohydrogen production, the cultivation of green macroalgae for biogas and biofertilizer production has attained a good platform in the biorefinery strategy. The macroalgae are used as a co-substrate to increase the efficiency of anaerobic digestion due to their enormous carbohydrate and moisture contents, which further improves the production and quality of biogas and biofertilizer (Milledge et al., 2019). Mhatre et al. (2019) studied biomethane potential (BMP) of *Ulva lactuca* via anaerobic digestion from the residue collected after the primary product (sap, ulvan, and protein) extraction. The untreated macroalgal biomass produced only 210.08 ± 6.15 mlg^{-1} methane due to the interference of inhibitory factors (nitrogen and sulfur). However, the treated biomass, which is free of nitrogen, resulted in up to a two-fold increase in methane production. In addition, high-value-added products extracted before biogas production have enormous application in the development of marine-based products.

1.2.5.4 Biodiesel

The macroalgal biomass exhibits huge potential to be an eco-friendly source of biofuel. Besides bioethanol, biodiesel is the next liquefied form of renewable bio-fuel sustainably obtained from macroalgae. Like red and brown, green macroalgae can also be an efficient producer of biodiesel from their leftover residue. The biodiesel production from macroalgae follows many processes, such as drying/dewatering, direct combustion, co-combustion, pyrolysis, and gasification. However, the most important step, i.e., "trans-esterification", results in the production of biodiesel from algal long-chain mono alkyl fatty acid. Moreover, the lipid content of the macroalgae (0.3–6%) is very low compared to microalgae (>70%). Therefore, in comparison to microalgae, macroalgae are less suitable for biodiesel production (Lakshmi et al., 2020; Milledge et al., 2014). Hence, with the increasing information about macroalgal genomes, genetic engineering can help in increasing the lipid contents in macroalgae.

1.2.5.5 Other Miscellaneous Applications

Apart from value-added products, biofuel, and bioenergy, using a biorefinery approach, green macroalgae can be utilized for other applications as well such as wastewater treatment, adsorbents, biomaterials, and feed.

Green macroalgae are used in bioremediation due to their rapid growth and carbon sequestration capability. The wastewater from industries, domestics, and other sectors, including agricultural lands, is loaded with an enormous amount of organic (benzene, toluene, ethylbenzene, xylene (BTEX), chemicals, and dyes) and inorganic (arsenides, heavy metals, and fluorides) pollutants. Pollutants with hazardous and xenobiotic compounds may or may not result in eutrophication but are passed through the food chain, affecting higher trophic levels, if not treated. Bioabsorption of these hazardous compounds using macroalgae is cost-effective and less labor-intensive. Mostly red and brown macroalgae have been used in treating wastewater (Arumugam et al., 2018) among macroalgae, with limited use of green macroalgae. *Ulva lactuca, Ulva rigida, Caulerpa recemosa var. cylindracea*, and *Caulerpa lentillifera* are a few green macroalgae that are used for bioremediating different wastewater (Arumugam et al., 2018). Further, Ulva spp. (*Ulva ohnoi* and *Ulva lactuca*) are industrially important sources of biochar that help in soil amelioration and carbon sequestration (Lakshmi et al., 2020; Arumugam et al., 2018). Green macroalgae can also be utilized as cost-efficient absorbents for the remediation of various metal ions (copper, lead, cadmium, and chromium). Various green macroalgae, such as *Ulva fasciata, Ulva lactuca, Cladophora sericioides, Cladophora glomerata*, and *Caulerpa Serrulata,* exhibits potent bioabsorbent capabilities. However, *Caulerpa Serrulata* (wild type), among all the aforementioned species, has been reported to act as an efficient multi-bioabsorption agent, followed by improved bioabsorption after modifying with ethylenediamine (Lakshmi et al., 2020). Besides these applications, various biochemical components (sap, lipid, protein, carbohydrates, MRLE, cellulose nanocrystals (CNCs), pigments, antioxidants, and others) obtained from green macroalgae can be utilized in nutraceutical, pharmaceutical, cosmeceutical, food, and bioenergy production. Ulvan-based biomaterials are now widely utilized as natural polymers, i.e., hydrogel (for biomedical, tissue engineering, cell encapsulation, and drug delivery) from *Ulva lactuca* and *Ulva armonicana*, membrane and films (for packaging and biomedical application) from *Ulva lactuca*, *Ulva rigida*, *Ulva fasciata*, and Ulvan/chitosan complex, nanofibers from *Ulva fasciata*/PEO and PCL, 3D-porous scaffolds (for tissue engineering, bone graft substitutes, and bone tissue engineering) from *Ulva armonicana* (Lakshmi et al., 2020).

1.3   RED MACROALGAE

Red macroalgae (Rhodophyceae) comprise 61% of total macroalgal production worldwide, with around 6000 species, containing polysaccharides (carrageenans, agarans, floridean starch, cellulose), lipids, proteins, minerals, and other valuable compounds like polyphenols as major components (Torres et al., 2019b; Álvarez-Viñas et al., 2019; Torres et al., 2019a). Depending upon the species and the environmental conditions, the use of raw materials obtained from red macroalgae is an excellent source of biorefinery to produce valuable products.

Red macroalgae can grow in a wide range of temperatures depending upon the species and its geographical locations. Depending upon the species, the polysaccharide component of red macroalgae is classified as carrageenan with~75% dry weight composed of α- and β-D-galactopyranose and agaran with~52% dry weight comprised of α-L-galactose and β-D-galactose. The presence of enormous bioactive compounds has primarily commercialized the red macroalgae for agar and carrageenan production owning to numerous food, pharmaceutical, and nutraceutical applications. Also, the presence of floridean starch and floridoside makes them stand out from green and brown macroalgae (Jung et al., 2013). Furthermore, the quantity of compounds existing in red macroalgae contains carbohydrates with~83.6% dry weight, proteins up to 45% dry weight, and lipids with~0.9% dry weight. Besides the above compounds, certain minerals, trace elements, and ash are also present in red macroalgae. Although the polysaccharide abundance has been identified in all macroalgae, red macroalgae exhibit more protein content compared to other green and brown macroalgae, which uplifts their biorefinery choice in terms of biofuel, feedstocks, and biofertilizers (Álvarez-Viñas et al., 2019; Chen et al., 2015).

1.3.1   CULTIVATION OF RED MACROALGAE FOR A BIOREFINERY APPROACH

Besides innovation in extraction procedures, the cultivation of macroalgae also plays a major role in macroalgal biorefinery development. *Kappaphycus* is the most common red macroalgae with a broad number of applications. There is an impetus to produce biorefinery products from sequential extraction methods apart from direct extraction to meet the increasing consumption/utilization demand. Accordingly, to fulfill the macroalgae demand for value-added products, there is a need to upscale cultivation of red macroalgae under controlled process preventing macroalgal eutrophication, seasonal variation in product yields, environmental variations, grazing, predation, sometimes pests and disease, etc., which should be monitored for sustainable production of macroalgae at commercial scale (Hurtado et al., 2014). The integrated multi-trophic aquaculture (IMTA) technique is one of the innovative processes to cultivate red macroalgae to reduce environmental issues like eutrophication. For example, a preindustrial study was recently conducted in the Gulf of Taranto (northern Ionian Sea) under the EU Remedia Life Project, where an IMTA system co-cultured with *Sabella spallanzanii* (polychaete), *Sarcotragus spinosulus* (sponge), and *Gracilaria bursa-pastoris* (macroalgae) was utilized to improve the environmental quality, thereby bioremediating the excess nutrients (Giangrande et al., 2020). Few offshore cultivation techniques such as deeper water (Triangular, spiderweb, hanging long line, hanging basket, free-swing, single and multiple rafts long-line) and shallow water (fixed off-bottom) have been introduced in Indonesia, the Philippines, and Malaysia (Hurtado et al., 2014).

1.3.2 BIOACTIVE COMPONENTS OF RED MACROALGAE FOR A BIOREFINERY APPROACH

The extraction of polysaccharides, agar, agarose, and carrageenan is one of the fundamental goals for macroalgal industrial applications. Macroalgal proteins can be utilized as an inexpensive protein food source for malnourished world populations, apart from acting as a good source of bioactive amino acids for nutraceutical and pharmaceutical industries (Sadhukhan et al., 2019). The biorefinery strategy applied to the residual wastes of agarophytes and carragenophytes obtained after agar and carrageenan extraction could be exploited to produce low-molecular-weight carbohydrates, agricultural biostimulants, ethanol, nutraceutical supplements, biochar, methane, lipid-rich in PUFA, biofertilizer, and iron-based sorbents. Residues of porphyran-rich macroalgae could be used to generate edible films, hydrolysate with low molecular weight peptides owning to antioxidant and anti-hypertensive properties, and animal feed (Álvarez-Viñas et al., 2019). Beginning with the agarophytes, the residues (15–40%) of agar extraction collected from commercial exploitation of *Gracilaria, Gelidium,* and *Gelidiella* comprise 30–70% of carbohydrates, of which *Gelidium sesquipedale* treatment for agar extraction generated low-molecular-weight carbohydrate residues that exhibited beneficial effects on human health and plant growth (Lebbar et al., 2018). Lipid extraction from the residual biomass of *Gracilaria cortica*, utilized for food and nutraceutical applications, retained 65.6% of the PUFA fraction of total fatty acids (Baghel et al., 2016). Carrageenan having the highest residual fraction (30%) among all hydrocolloids is essential to be used for biorefinery purposes to minimize waste production (Álvarez-Viñas et al., 2019). Recently, κ-carrageenan and the residual fraction of *Kappaphycus alvarezii* showed activity against obesity and related metabolic disorders in a pilot study and thus can be of pharmacological significance as well (Chin et al., 2019). During phycocolloid extraction from *Porphyra columbina*, the phycobiliproteins present in the discarded extract conferred antioxidant activity, enabling it to be involved in preparing edible films (Cian et al., 2014). Apart from the immunosuppressive effect on rat splenocytes, some antihypertensive and antioxidant activity was also manifested by the hydrolysates with low molecular-weight peptides rich in threonine, leucine, proline, aspartic, glutamic, and alanine, obtained from the enzymatic hydrolysis of the discarded extract (Cian et al., 2012). Another hydrolysate collected from the enzymatic hydrolysis of the residual cake of phycocolloid extraction of *Porphyra columbina* had higher protein digestibility to be utilized as animal feed, along with copper-chelating, radical scavenging, and ACE inhibition activity (Cian et al., 2013).

1.3.3 PRIMARY PRODUCTS OBTAINED DURING THE BIOREFINERY PROCESS

The biorefinery concept of utilization of macroalgae by cascading approach is widely emphasized for the extraction of products with higher values first followed by the sequential extraction of lower value products. The massive quantity of algal biomass generated during the processing of value-added products from red macroalgae is usually disposed of, thereby increasing the environmental burden. Co-production of value-added compounds by integral utilization of biomass with maximum efficiency can generate a zero-waste production strategy and a sustainable bioeconomy (Balina et al., 2017).

The extraction of products with high graded value involves the production of agar, carrageenan, pigments, proteins, and lipids apart from vitamins and minerals, obtained from red macroalgae. Initially, the sequential extraction process begins with the homogenization of the collected fresh macroalgal sample in distilled water, incubated at 4°C for 12 hours. The homogenized sample was filtered to collect pigments containing liquid fraction and leftover biomass. The following step could be recurrently used for more pigment extraction. The leftover biomass from pigment extraction was utilized for extracting other products while the filtrates were used for segregating the pigment extract and mineral-rich permeate via ultra-membrane filtration (UMF). The separated pigment could then be used to analyze the phycobiliproteins (phycoerythrin and phycocyanin). The phycobiliproteins are commercially applied to food, nutraceuticals, and cosmetics as a natural colourant and possess numerous therapeutic values, such as antioxidant, anti-inflammatory, immunomodulating, antimicrobial, anticarcinogenic, and hepatoprotective properties. Additionally, the phycobiliproteins can act as fluorescent markers and as photosensitizers in immunological and clinical diagnostics (Torres et al., 2019a; Baghel et al., 2016). For at least two batches of fresh samples, the mineral-rich permeate was reused for pigment recovery. Subsequently, the filtrate was analyzed for the presence of plant nutrients (micro and macro) so that it could be employed as a biostimulant. Currently, the presence of minerals (Ca, K, Fe, Mg) and several other minor elements could also be analyzed to be used in low-sodium diets. It has been reported to function as anti-genotoxic, anti-hematotoxic, antioxidant, and retains osteomineral metabolism and bone histo-architecture in adult rats (Jaballi et al., 2018). Extraction of the pigment-discharged residues (dried) were employed for the lipid extraction by dissolving in chloroform: methanol (1:2 v/v) solvent. The incubation of the mixture was performed for 30 minutes at room temperature, thus filtering and obtaining the lipid filtrate. To make the organic layer colorless, the residue was treated several times with the same solvent as above, and all the lipid fractions were eventually mixed to filter through 44 μm Whatman filter paper. Consequently, the bilayer suspension was obtained via aqueous extraction. The upper layer was assigned as the aqueous layer, while the lower layer was the lipid layer. The lipid layer was processed for drying using a rotary evaporator, which was also accountable for the recovery of solvents (chloroform in the lower layer and methanol in the upper layer). The recovered solvents could then be reused for further lipid extraction, attentive to the fact that the volume of the recovered solvent reduced after each extraction cycle. Most of the red macroalgae have a beneficial ω6/ω3 ratio with high C20:4 and C20:5 contents. The beneficial ω3 polyunsaturated fatty acids (PUFAs) can only be acquired through dietary sources as they are not synthesized by humans. Hence, they have substantial therapeutic, nutraceutical, and food applications. Upon lipid extraction, the dried residual mass obtained is exploited for agar or carrageenan extraction. Agar is commonly known for its thickening, gelling, and stabilizing properties and is also used as cryoprotectants in the food, pharmaceutical, and cosmeceutical industries. Apart from the aforementioned statement, the microbiological media preparations and chromatographic techniques are the other applications of agar. Carrageenans have several food-grade applications, including food additives. They are implemented in nutraceutical and pharmaceutical industries based on anti-inflammatory, antioxidant, anticoagulant, neuroprotective, antiviral, and other activities (Torres et al., 2019a; Baghel et al., 2016). The residual after the agar/carrageenan extraction (residual pulp) is dried for saccharification and fermentation to produce biofuels. The fermentation residual can be sun-dried and used as a soil conditioner or biofertilizer. Additionally, proteins extracted from the residual biomass are rich in essential amino acids making them suitable to be used as animal feed (Baghel et al., 2016).

1.3.4 Secondary Products Obtained during the Biorefinery Process

1.3.4.1 Bioethanol

Red macroalgae can be widely used as a source of bioethanol. The higher carbohydrate content compared to protein and lipids makes them suitable for fermentation. Various fermentative microorganisms are discussed below to facilitate the fermentation process via enzymatic hydrolyzation. However, *Saccharomyces cerevisiae* has been widely utilized as fermentative microbe for the production of bioethanol from red macroalgae by cellulase treatment. Bioethanol processed from algal biomass is considered to be a third-generation (3G) biofuel. For example, in a biorefinery approach using *Gracilaria corticata* residual pulp, the reducing sugars obtained were 269 ± 2.6 per g residue via enzymatic hydrolysis. The hydrolysate on fermenting with *S. cerevisiae* (MTCC 180) yielded bioethanol of about 472 ± 4.9 mg/g reducing sugar along with 12.72% bioethanol production from residual biomass (Baghel et al., 2016). Solid wastes generated after κ-carrageenan extraction from *Eucheuma cottonii* can be either processed by simultaneous saccharification and fermentation (SSF) or separate hydrolysis and fermentation (SHF) process, using baker's yeast. In the SHF process, the glucose yield and bioethanol yield were found to be 99.8% and 55.9%, respectively. Thereafter, SSF showed the highest bioethanol yield of 90.9%, proving it to be an efficient process of bioethanol production (Tan and Lee 2014). The leftover waste of *Kappaphycus alvarezii* from carrageenan extraction generated a high ethanol yield (concentration) of 13.78 g/L with a 69% theoretical yield. In this study, the hydrolysate was hydrolyzed enzymatically and bioconverted using *S. cerevisiae* (ATCC 200062) to produce bioethanol (Meinita et al., 2019). After the agar extraction from *Gracilaria verrucosa*, the leftover pulp went for enzymatic hydrolysis to yield 0.87 g sugars/g cellulose, which produced an ethanol yield of 0.43 g/g sugars by fermenting with *S. cerevisiae* (HAU strain) (Kumar et al., 2013). Biomass of *Gracilaria* sp. obtained by *in situ* sequential hydrolysis (acid and enzyme) followed by *S. cerevisiae* (Wu-Y2) fermentation produces ethanol yield of 4.72 g/L (0.48 g/g sugar consumed with a conversion rate of 94%) (Wu et al., 2014). A work done by Baghel et al. (2015) involving hydrolysis and fermentation of *Gelidiella acerosa*, *Gelidium pusillum*, and *Gracilaria dura* biomass using *S. cerevisiae* (MTCC 180) gave the ethanol yield of 418 mg/g sugars for each red macroalgae.

1.3.4.2   Biodiesel

Biodiesel is mainly derived from animal fats, vegetables, and plant oil on the transesterification of triglycerides. There are scanty data on biodiesel production in red macroalgae compared to other micro- and macroalgae. As the lipid and fatty acid contents are very low in most of the red macroalgae, preparing distinct oil-derived biorefinery products is problematic (Torres et al., 2019b). Nevertheless, here we describe some recent advances in biodiesel production from red macroalgae and their advantageous properties. Sharmila et al. (2012) reported better production of biodiesel with *Gracilaria corticata* when the hexane-ether solvent system   (2ml/10gm) was adopted compared to the benzene solvent system (1.8ml/10gm). Biomass of *Kappaphycus alvarezii* can be used as raw material (carbon source) for biodiesel production by hydrolyzing with *Pichia kudriavzevii* (Sulfahri et al., 2019). An experiment was conducted where biodiesel production from red macroalgae (*Kappaphycus alvarezii* and *Gracilaria* sp.) and waste tuna oil (*Thunnus* sp.) were observed by considering the viscosity test, flash, and freezing point as essential test parameters to determine the biodiesel efficiency. *Glacilaria* sp. exhibited the highest flash point compared to *K. alvarezii*. However, when compared to biosolar, the flashpoint of biodiesel was higher with lower emission of greenhouse gases and hydrocarbons and minimal toxicity. In addition, the lasting and quality of flame was higher for a longer duration of time, which describes biodiesel to be the most eco-friendly fuel alternative to non-renewable fuels. Moreover, an increase in the freezing point lowers the biosolar comparison, which was well observed in the macroalgae and tuna oil, and viscosity of the biodiesel was more than biosolar, making it a cheap source of biofuel compared to waste tuna oil. Therefore, *K. alvarezii* is the most potent red macroalgae for biodiesel production as it contains the flashpoint (5.0) a little less than *Gracilaria* sp. (P5). Nevertheless, the freezing point (−2.6°C) and viscosity score (3.75 cSt) demonstrate better quality biodiesel from red algae (*K. alvarezii*) (Alamsjah et al., 2017). Biodiesel production from *Eucheuma cottonii* residual biomass was performed by transesterification using sodium methoxide as a catalyst mixed with ethanol. The quality of biodiesel obtained was compared to the crude diesel. In contrast to diesel, Total Fuel Consumption (TFC) and Specific Fuel Consumption (SFC) was less in biodiesel with lesser emission of carbon monoxide (CO), carbon dioxide (CO_2), and smoke. The aforementioned statement suggests biodiesel as an alternative eco-friendly, renewable fuel compared to diesel (Selvan et al., 2014).

1.3.4.3   Biohydrogen and Biogas

Although macroalgae are efficient biohydrogen producers, sequential extraction followed by biohydrogen production in red macroalgae is still under research. Therefore, further investigations have to be undertaken to bloom this field in future research.

Apart from previously described biohydrogen production, biogas production can be the added advantage obtained from red macroalgae. The leftover biomass can be utilized as biofertilizers, ameliorating soil fertility. Although very scanty research has been identified in biogas production. The biomethanation efficiency of *Gracilaria* spp. was determined by preparing a triplicate batch assay of the various percentage of Total Solids (TS) at a mesophilic temperature of 37°C, with hydrolysis being the limiting step. Following the anaerobic biodegradability test, biogas production was optimized via activated sludge co-digestion. The biomethane potential (BMP) of *Gracilaria* sp. was found to be 182 ± 23 L CH_4/kg VS at 2.5% TS. The macroalgal (15%) co-digestion with activated sludge (85%) generates an efficient methane production rate of 26%, which is comparatively higher than activated sludge alone (Costa et al., 2012). A study on pretreated (washing and maceration) *Gracilaria vermiculophylla* demonstrated a BMP of 481 ± 9 L CH_4/kg VS. However when *G. vermiculophylla* is co-digested with glycerol (2%), the production of methane increased to 599 ± 25 L CH_4/kg VS. Moreover, co-digesting with sewage sludge has led to the highest methane production of 605 ± 4 L CH_4/kg VS (Oliveira et al., 2014). Likewise, after agar extraction, the industrial biomass of *Gracilaria manilaensis* and *Gracilariopsis persica* were acid pretreated before anaerobic fermentation, yielding BMPs at 70% and 62%, which was compared to the whole plant biomass of each macroalga with only 48% and 46% BMPs, respectively. The methane yield was enhanced by pretreatment up to 47% and 77%, respectively, compared to the untreated residue (whole plant biomass) (Hessami et al., 2019).

1.3.4.4   Other Miscellaneous Products

Besides the above applications of red macroalgal biomass, several other processes and products are under investigation to be used in biorefinery systems to increase the circular economy. Like green macroalgae, biochar and bio-oil obtained from red algae are simply the generations of useful products from algal waste conversion by the most effective thermochemical process (pyrolysis). The heating value (HV) indicates the amount of energy present in the fuel. A good bio-oil must exhibit a HV of 26 MJ/Kg approximately lower than conventional fossil fuels (42–44 MJ/Kg)

(Sadaka and Boateng 2009). For example, *Porphyra tenera* presented the yield and HV of bio-oil to be 47.4% and 29.7 MJ/kg, respectively. The bio-oil and biochar yielded via thermochemical treatment (pyrolysis) were found to be 65.0% and 28.0%, respectively, while the HV of biochar was 13.6 MJ/kg for *Gracilaria gracilis* (Torres et al., 2019b). *Gelidium sesquipedale* undergoes hydrothermal carbonization, thereby producing hydrochars with 60% yield (Méndez et al., 2019). Similarly, *Gracilaria edulis* and *Eucheuma denticulatum* can also be used to produce biochars for soil amelioration (Álvarez-Viñas et al., 2019).

Biosorbents are primarily used to remove toxic compounds from industrial effluents. They can be used to eliminate pollutants in wastewater and are good bioremediating agents for hazardous metal ions. Treatment of aquaculture water by *Gracilaria lemaneiformis, Gracilaria chouae, Gracilaria vermuculophylla,* and *Gracilaria caudate* removed the maximum amount of nitrogen and phosphorous from them (Arumugam et al., 2018). Moreover, bioabsorption of nutrients (nitrogen and phosphorous) to reduce eutrophication in shrimp wastewater can be efficiently implemented by *Gracilaria tikvahiae, Gracilaria birdiae,* and *Gracilaria caudate J. Agardh* (Torres et al., 2019a). Dyes like methylene blue can be removed by *Gelidiella acerosa* biochar and *Gracilaria parvispora*. In addition, *Gracilaria corticata* is responsible for the biosorption of malachite green. Adsorption of cadmium (Cd), lead (Pb), and copper (Cu) could be carried out by the leftover slurry of *Gracilaria verrucosa, Gelidium amansii, Eucheuma denticulatum,* and *K. alvarezii* (Sunwoo et al., 2016). The biochar produced by *Gracilaria* sp. can eliminate molybdenum (Mo), arsenic (As(V)), selenium (Se(VI)), iron (Fe), nickel (Ni), and cadmium (Cd) heavy metals from the landfill leachate. Another red macroalga, *Osmundea pinnatifida*, showed maximum biosorbent capacity against Cd and Cu, while *Ceramium virgatum* displayed the biosorbent capacity only against Cd. *Chondracanthus chamissoi* helped remove Pb(II) and Cd(II). The waste biomass of *Gracilaria edulis* post-extraction of agar generated iron-based sorbents beneficial in the removal of iron compounds (Torres et al., 2019b; Álvarez-Viñas et al., 2019; Torres et al., 2019a).

The extracts obtained after the production of certain value-added compounds can also be used as biostimulants, soil conditioners, and biofertilizers. For example, *Gracilaria,* along with a recommended dose of fertilizer, was applied to *Vigna mungo* to increase the yield (Mahajan et al., 2016). Similarly, *Kappaphycus alvarezii* and *Gracilaria edulis* when applied to *Oryza sativa* promoted morphophysiological parameters, growth, germination, nutritional value, productivity, and yield (Singh et al., 2015; Layek et al., 2017). Likewise, *Hypnea valentiae* enhanced the growth and biochemical characteristics of *Dolichos biflorus* (Renakabai et al., 2013). The mineral-rich liquid extracts of macroalgae could also be used in foliar applications as plant stimulants (Torres et al., 2019b).

Biorefining of macroalgal biomass resulted in the removal of inorganic elements and heavy metals and hence could be used in feed application. As the biomass of *Asparagopsis taxiformis* was sequentially extracted, the extracts later obtained were rich in iodine so that they could be used as nutraceutical supplements. Having the highest protein content among all, red macroalgae could be used to extract the protein to be used as flavor enhancers, food additives, and animal feed, adapting to the possibility of circular bioeconomy. The beneficial effects of macroalgae in animal diets comprise an increase in milk production in cows, retaining of sheep weight during winter for increased wool production, and good iodine content in eggs involved in the poultry diet (Torres et al., 2019b; Álvarez-Viñas et al., 2019).

Renewable seaweed-based resources, such as polysaccharides from red macroalgae, can be a potent source of macroalgae-based eco-friendly bioplastics. Polysaccharides from red macroalgae are highly colloidal with good oxygen and moisture permeability. Recently, *Kappaphycus alvarezii* (3% w/v) was used with plasticizer polyethylene glycol (PEG)-3000 to produce bioplastic film with superior physical and mechanical properties. This paved the way for a better source of alternatives to non-degradable hazardous plastics for a better future (Sudhakar et al., 2020).

1.4 BROWN MACROALGAE

Brown macroalgae (Ochrophyta) have been classified into 13 orders and 300 genera, which contain 1836 identified species. Among brown macroalgae, only three orders – laminarales, dictyotales, and fucales – have been extensively investigated for bioactive natural products (Arumugam et al., 2018). In the coastal region, brown macroalgae are the largest biomass producers (Habeebullah et al., 2020). Phaeophyta includes many alginophytes, such as *Sargassum*, *Laminaria*, *Ascophyllum*, *Macrocystis*, etc. The temperate region waters comprise *Laminaria* as the important source of alginate. In tropical countries, such as India, *Sargassum* is the major alginate source. *Sargassum* sp. was recorded at the Gulf of Mexico and the West African coasts for the golden tide (Baghel et al., 2020).

Brown macroalgae are not only useful in minimizing the eutrophication but also in the absorption of heavy metals, making them a suitable bioindicator for heavy metal contamination (Ditchburn and Carballeira 2019). Along with this, brown macroalgae contain numerous bioactive components. As a result, brown macroalgae were studied for numerous applications, including food, feed, bioenergy, bioremediation, pharmaceutical, cosmeceutical, nutraceutical, functional food, prebiotic, and fertilizers (Arumugam et al., 2018; Wijesinghe and Jeon 2012; Wijesinghe and Jeon 2011). The sequential extraction of these components makes it a sustainable option for valorizing the brown macroalgae feedstock. This exemplifies the biorefinery perspective of the brown macroalgae and consecutively enhances the circular bioeconomy and also the blue economy.

1.4.1 CULTIVATION OF BROWN MACROALGAE FOR A BIOREFINERY APPROACH

The cultivation of brown macroalgae *Sargassum siliquosum* is grounded on sexual reproduction, where a huge amount of seedlings were produced in the hatchery. Among the clay, limestone, and nylon, the clay was found to be the suitable substrate for the zygotes in the seeding frames. Seedling takes 1.5 years in the tank and a further six to seven months in the open sea to get to harvestable size. For cultivation of *Sargassum siliquosum* in open sea, suspension cultivation using a substrate panel was found to be a useful technique (Largo et al., 2020). Another approach for the cultivation of brown macroalgae has been reported by Purcell-Meyerink et al. (2021). It suggested the cultivation of *Laminaria* sp. on the rope of 1060 m on either end of salmon cage yields 1600 kg of dry weight. Alternative species, such as *Macrocystis* and *Nereocystis,* can also be cultivated by this method (Purcell-Meyerink et al., 2021). To commercialize the macroalgae industry, it is essential to optimize the cultivation of the macroalgae. To improve the quality and chances of cultivation, the effect of the seeding technique and hatchery period was studied for the out-planting of the *Saccharina latissima*. It showed that the precultivated meiospores with seedling in the hatchery for 42 days before the distribution results in a higher biomass yield of 7.2 ± 0.1 kg m−1 and longer fronds of 77.0 ± 6.7 cm (Forbord et al., 2020). As mention above, the most common cultivation technique of brown algae is suspension culti-vation, but many other factors are also responsible for the growth. Further, every component of the cultivation, e.g., seedling pretreatment, the substrate used, etc. can be standardized for the commercialization of the macroalgae industry.

Another beneficial approach for the cultivation of macroalgae is by using an integrated multi-trophic system (IMTA). This process of farming includes the cultivation of species at different trophic levels, where a synergistic effect has been seen among the species. Here, the first species leftover feed, nutrients, wastes, and by-products have been utilized by the other species and transformed into fertilizer, energy, and feed along with biomitigation (Alemañ et al., 2019). For example, *Macrocystis pyrifera* (giant kelp) was found to be complementary with salmon farming, where it lowers down nutrient load with the production of the macroalgae (Hwang and Park 2020).

To improve macroalgae production, the use of a productive strain is also crucial. To deal with the species with lower adaptation to the environment, low value-added product formation, and disease-prone macroalgae cultures, the solution for this could be the transplantation of other non-native species. Also, methods such as artificial seedling and cultivation have been implemented to maximize the pro-duction of macroalgae biomass (Habeebullah et al., 2020).

1.4.2 Bioactive Components of Brown Macroalgae for a Biorefinery Approach

Brown macroalgae have a complex cell wall polysaccharides layer that restricts the extraction of bioactive components. Thus, techniques such as enzyme-assisted extraction could be the smart option for the extraction of bioactives from brown macroalgae (Alemañ et al., 2019). Brown algae possess numerous bioactive com-ponents such as carotenoids (fucoxanthin), polysaccharides, peptides, polyphenols, laminarin, fucoidan, alginate, vitamins, and minerals. Brown macroalgae are gen-erally known for the presence of bioactive polyphenolic compounds called phlor-otannins, which are exclusively present in marine brown macroalgae with 0.2–14% of the dry weight of brown macroalgae. Phlorotannins were assembled by poly-merization of phloroglucinol and show lower structural complexity than that of tannins present in the terrestrial plants. These bioactive components show properties such as antioxidant activities, anti-inflammatory properties, antidiabetic properties, antimicrobial and bactericidal activity, and radioprotection (Miyashita et al., 2013, Li et al., 2011, Holdt and Kraan 2011, Kadam et al., 2015, Xing et al., 2020).

Brown macroalgae also contain lipids that are affluent in functional long-chain omega-3 PUFAs, such as EPA (20:5n−3) and SDA (18:4n−3), and a carotenoid, fucoxanthin (Balina et al., 2017). The brown color of the macroalgae is because of the abundance of the carotenoid fucoxanthin (Ali et al. 2017). Carotenoid contributes about 0.08% of the dry weight in *F. serratus* − 70% of which is fucoxanthin. It is reported for its anti-cancer, anti-inflammatory, anti-obesity, and anticancer properties (Li et al., 2011; Kadam et al., 2015).

Brown macroalgae mainly contain laminarin, fucoidan, and alginate as major polysaccharides (Arumugam et al., 2018). Giant bull kelp (*Durvillaea potatorum*) contains 43.57% (w/w) total polysaccharides, from which Fucoidan/Laminarin comprises 7.02% ± 0.59, acid and alkaline extracted alginate comprises 13.32% ± 0.17, and 23.23% ± 0.80 respectively. Fucoidan is the fucose-rich sulphated polysaccharides that fascinate with its biological active properties resulting in the application of cosmeceuticals, as nutrient-rich food, and dietary supplements (Li et al., 2011). Laminarin is composed of β-glucan which is linked with 1–3 glycosidic bonds and a few 1–6 intrachain glycosidic bonds. Laminarin has been a biologically active component with properties such as anti-inflammatory, anticoagulant, antioxidant, anti-tumor, and anti-apoptotic activities. Alginate is a linear negatively charged polysaccharide, which consists of the β-D-mannuronic acid and α-L-guluronic acid, which are linked by β-1, 4-glycosidic bonds. Because of its biocompatibility, inexpensiveness, reduced toxicity, and capability of making a gel along with the divalent cations such as Ca^{2+}, it shows application in food as well as medical industries (Xing et al., 2020). In addition to this, alginate of brown macroalgae is used for dressing wounds, in tablet production, as an industrial molding material, and as a recombinant biocatalyst. It is also utilized for its adhesive effect and with paper to prepare a biodegradable compound that is able to manufacture furniture. Or it is also used with calcium fibers to make fabric (Ditchburn and Carballeira 2019). Nevertheless, its applications have been restricted because of its high molecular weight, which could be resolved by the degradation of the polymer into the low molecular weight oligosaccharides. These alginate oligosaccharides (AOS) exhibit bioactivities in the food, pharmaceutical, and agricultural fields. Along with this, it possesses anti-tumor, anti-inflammatory, antibacterial, neuroprotection, and immunomodulatory properties, and it reduces obesity, blood sugar level, and hypertension; it also promotes the cell proliferation and plant growth regulator (Xing et al., 2020).

Brown macroalgae also comprises a high content of vitamins such as vitamin C. It also contains an adequate amount of vitamin B1 and B2, and a good amount of α-tocopherol along with β-, γ-tocopherols than that of red as well as green macroalgae, which contains α-tocopherol solely (Škrovánková 2011). Besides this macroalgae also contains macro- (K, Na, S, Ca, and Mg) and micronutrients (Zn, B, Mo, V, Se, Cl, P, and I), it has fertilizer potential (Soares et al., 2020).

1.4.3 PRIMARY PRODUCTS OBTAINED DURING BIOREFINERY PROCESS

Brown macroalgae possess a vast range of bioactive components, which attract interest for the manufacture of various value-added products. The individual extraction of these compounds from the macroalgae feedstock results in the wastage of the macroalgae biomass; it could be prevented by the extraction of these products sequentially, preventing the wastage and its reutilization. Conventionally, brown macroalgae, such as giant bull kelp (*Durvillaea potatorum*), was used for the extraction of a single product, such as soluble sodium alginate by the alkaline extraction method, but a recent study has shown that the optimized process of acidic extraction before the alkaline extraction leads to the production of three polysaccharides, i.e., alginate, fucoidan, and laminarin, instead of one (Abraham et al., 2019).

The integrated process of bioethanol and protein extraction from the *Laminaria digitata* was found to be a good example of a brown macroalgae biorefinery. The study carried out by Hou et al. (2015) reported that a simple pretreatment of milling on macroalgae biomass, followed by enzymatic hydrolysis and fermentation, results in an enhanced yield of 77.7% bioethanol. The residue after bioethanol production shows a 2.7-fold higher protein content with an abundance of glutamic acid and aspartic acid amino acid (10% each of crude proteins) (Hou et al., 2015).

The recently reported study on the brown macroalgae biorefinery showed promising results of 93% of utilization of *Sargassum tenerrimum* biomass. The one-kilogram batch produced sap, alginic acid, protein concentrate, and cellulose, with 541.33 ± 5.50 mL, 32 ± 1.5 g, 3.8 ± 0.2 g, and 10 ± 0.5 g weight, respectively, and salt from the residue with 115 ± 5 g weight. The first isolation of sap contents was carried out by grinding fresh macroalgae biomass with the help of a mixer grinder followed by compressing in muslin cloth. Sap comprises macro-mineral (K, Ca, Na, Mg) and micronutrients (Fe, Zn, B, Cu), along with a fair amount of protein and carbohydrates. This extract has a promising potential in seed germination, plant growth stimulator, nutrient absorption in flowering plants, and fighting back against biotic and abiotic factors. This step of sap extraction saves the drying of biomass and, in exchange, gives mineral-rich sap (Baghel et al., 2020).

The residue left after the sap extraction has been further utilized to produce alginic acid. It is one of the most important products obtained from macroalgae biomass with applications in food, textile, paper coating, printing, and biomedical fields such as tissue engineering, hydrogel, and scaffold. The residue was further bleached with $NaClO_2$, hydrolyzed with 1% $NaCO_3$, and filtered out through the muslin cloth. After the filtration, the filtrate was further used for the production of alginic acid, which is recovered by subsequent precipitation by using HCl, and the residue was treated with NaOH and then HCl precipitation for protein extraction. This macroalgal protein contains all essential amino acids. In the digestibility studies, it was found appropriate for the food and feed sources. The residue after the protein extraction was later suspended in 5% HCl and afterward, gives cellulose, which shows vast applications from food, to pharmaceutical, textile, and the paper industry as well as in the production of ethanol. Prior extraction of sap, alginate, and protein concentrate leads to fewer chemical requirements, which could have been more for the pretreatment if single cellulose has been extracted. The mixture of all liquid effluent obtained was sun-dried after neutralization with NaOH, yielding salts. Salt mainly contains sodium chloride with macro- as well as microminerals, such as K, Fe, Cu, Mg, Ca, Co, etc. These salts can be used as a feed source (Baghel et al., 2020).

This study explains the biorefinery of brown macroalgae biomass at its best. At each step, there is a lesser chemical requirement because of prior extraction and processing. It helps in cutting down the pretreatment chemicals and energy requirement to a large extent, which makes the process cost-effective. This step-by-step product extraction from the same biomass with reutilization of the residue results in minimal waste generation, or zero waste formation, which could also be achieved by further optimization. All these advantages of biorefinery make it environmentally friendly and cost-effective.

1.4.4 SECONDARY PRODUCTS OBTAINED DURING THE BIOREFINERY PROCESS

1.4.4.1 Biogas

One of the aspects of biofuel generation from macroalgae is biogas production, which could be put together in an economical model by complementing it with the extraction of other bioproducts (Tedesco and Daniels 2018). Despite all the advantages of macroalgae as a source of bioenergy, the energy required for production is more than the produced bioenergy. To reverse this energy scenario, the biorefinery method was found to be the smart alternative, where various value-added products are extracted in a cascading manner along with biofuel production. According to the IEA (International Energy Agency) report, the gross bioenergy potential of macroalgae in biogas form is over 300 GJ/ha/yr, which is more than the other land crop plants viz. maize, grass, and fodder beet (Tabassum et al., 2017).

It has been reported that the fermentation of the macroalgal residue from the local bioindustry after the extraction of the bioproducts, such as fucoxanthin, alginic acid, laminarin, fucoidan, mannitol, and proteins, yields biogas. In the present study, biogas generation has been studied by anaerobic digestion for the whole year to investigate the seasonal changes and location, on the macroalgae an intern on the biogas production. Among studied *Laminaria* species, *Laminaria digitata* and *L. saccharina* reported CH_4 production of 523 mL CH_4 gVS-1 with acclimatized sludge and 535 mL CH_4 gVS-1 with non-acclimatized sludge sequentially (Tedesco and Daniels 2018).

1.4.4.2 Bioethanol

As mentioned earlier, the brown macroalgae contain a large amount of polysaccharides. These fermentable sugars make brown macroalgae a potential source for the production of bioethanol. The study carried out by Lamb et al. (2018), demonstrated the production of bioethanol from brown macroalgae *Saccharina latissima* collected from the Norwegian Coast, Norway. The collected macroalgae on enzymatic hydrolysis contain 31.31 ± 1.73 g of reducing sugars per 100 g of dry biomass, which on fermentation with *Saccharomyces cerevisiae* yields 0.42 g of ethanol/reducing sugar (g), resulting in an efficiency of 84% (Lamb et al., 2018).

The biorefinery concept can be applied in the extraction of bioethanol, the brown macroalgae such as *Saccharina japonica* used in the biorefinery process, which will lead to the extraction of multiple value-added products like ethanol, mixed alcohol, biodiesel, biogas, and electricity. It can be achieved by the process of anaerobic digestion and pyrolysis, heat, and power (PHP) (Brigljević et al., 2018).

1.4.4.3 Biohydrogen and Biomethane

Brown macroalgae are the sustainable option for bioenergy production. In the study carried out by Ding et al., (2020), *L. digitata* biomass was collected from shallow waters of the West Cork Coast in Ireland, which was followed by drying at 105 °C and cryopreservation at −20 °C. Biomass was further utilized as the potential source to produce biohydrogen and biomethane. As per many previously reported studies, such as the pretreatment of heating at 121 °C for 30 min increases biohydrogen production in the case of *Laminaria japonica* (Liu and Wang 2014), various pretreatments were applied for the depolymerization of *Laminaria digitata*. Among all studied pretreatments, HTP (hydrothermal pretreatment) was found to be the most suitable technique. The use of the H_2SO_4 in HTP increases the solubilization of carbohydrates to monomers. There was a 3.5-fold increase in carbohydrate monomers than that of raw biomass after hydrothermal dilute acid pretreatment (HTDAP), which in turn enhanced the biohydrogen production (57.4 mL/gVS) in the first dark stage of fermentation by 60.8%. The sludge from swine slurry digester was utilized as an inoculum in this dark fermentation.

The by-product of the first step has hydroxymethylfurfural, which damages the second stage of the anaerobic digestion from the leftover residue of the first process. This results in the biomethane yield of 25.9%. The pretreatment of HTP increases the efficiency up to 26.7% versus the untreated biomass of *L. digitata* (Ding et al., 2020).

1.4.4.4 Other Miscellaneous Applications

The study conducted by Zhang et al., (2018) showed that among three brown macroalgae investigated, *Carpophyllum flexuosum* showed the highest phlorotannins production by using MAE when extracted as a single product. To make the extraction cost-effective, it should be complemented with the extraction of other value-added products. Further, studies have been carried out on macroalgae biomass for the sequential extraction of the bioactives. It results in the production of pigments (3.4–9.8%), mannitol (22.2–30.7%), alginates (5.2–15.5%), carbohydrates (12.2–18.5%), and residue (13.5–19.5% of dry weight). The following are uses of these compounds: (a) Brown macroalgae pigments like coxanthin have anti-inflammatory, antioxidant, and anti-obesity properties. (b) Mannitol and laminarin can be further used for food additives, sweetening agents, and bioenergy production, such as ethanol, and butanol by using fermentation. (c) Alginates and other carbohydrates have been used as a gelling agent in food, pharmaceuticals, cosmeceuticals, and for the production of biofuels such as butanol and ethanol. (d) Residual macroalgae can also be converted into biochar, and it has the potential for agricultural application (Zhang et al., 2020).

Similarly, another study reported by Yuan and Macquarrie (2015) for biorefinery shows the sequential extraction of brown macroalgae *Ascophyllum nodosum* by MAE, resulting in value-added products, such as fucoidan, alginates, sugars, and also biochar from the residue (Yuan and Macquarrie 2015).

1.5   FUTURE PERSPECTIVES

* The high carbohydrate content of macroalgae can be used for producing various macroalgal-based polymers/biomaterials.
* The introduction of novel chemical-free pretreatment techniques (ultrasonication) can prevent the production of harmful chemicals (act as inhibitors) that lowers the product yield.
* The utilization of engineered/wild-type strains for crude cellulase isolation can simultaneously increase bioethanol production.
* The future of macroalgal biorefinery as a "green technology" mostly depends on its advanced cultivation and extraction strategies. Therefore, innovative and novel techniques must be applied integrated with efficient technological installment for industrialization.
* The minimal investment in the production of high- to low-grade products with huge global turnover can uplift the advanced seaweed-derived economy "circular bioeconomy" in the near future.
* Finally, the introduction of genetic modulation techniques (seaweed and microbes) can revolutionize the macroalgal biorefinery industry along with maximizing the profits in the near future.

1.6   CONCLUSION

The development of macroalgal biorefinery completely depends on residual biomass conversion into various sustainable products. The entire biorefinery approach is a huge strategy that begins from the feedstock species selection and ends up in global marketing of value-added products, fulfilling people's demand. The aforementioned applications in the development of the biorefinery strategy depict zero waste management, which will uplift both the circular bioeconomy and socioeconomy. The advantages of selecting macroalgae as a biorefinery feedstock relies on its growth rate, high-quality bio components content, and popularization of macroalgal products in western countries. Nevertheless, the biorefinery strategy is a very nascent futuristic commercialization approach encompassing various obstacles and challenges, which can be tackled by developing innovative technologies. Moreover, the improvement in strain selection, advanced cultivation and harvesting strategies, annual feedstock production, involvement of seaweed farmers, waste recovery management, and bioenergy production, along with technology and methods advancement, will boost the development of industrial seaweed commercialization.

ACKNOWLEDGMENTS

Author (NT) would like to thank the Department of Science & Technology, Delhi, India, for the DST INSPIRE faculty award. All the authors have contributed equally to the chapter, with no conflict of interest.

REFERENCES

Abraham, R.E., Su, P., Puri, M., Raston, C.L., Zhang, W., 2019. Optimisation of biorefinery production of alginate, fucoidan and laminarin from brown seaweed. *Durvillaea potatorum*. *Algal Res.* 38, 101389. 10.1016/j.algal.2018.101389

Alamsjah, M.A., Abdillah, A.A., Mustikawati, H., Atari, S.D.P., 2017. Screening of biodiesel production from waste tuna oil (*Thunnus* sp.), seaweed. *Kappaphycus alvarezii* and *Gracilaria* sp. AIP Conference Proceedings 1888, 020009. 10.1063/1.5004286

Alemañ, A.E., Robledo, D., Hayashi, L., 2019. Development of seaweed cultivation in Latin America: Current trends and future prospects. *Phycologia* 58, 462–471.

Ali, A.Y.A., Idris, A.M., Ebrahim, A.M., Eltayeb, M.A.H., 2017. Brown algae (Phaeophyta) for monitoring heavy metals at the Sudanese Red Sea coast. *Appl. Water Sci.* 7, 3817–3824.

Álvarez-Viñas, M., Flórez-Fernández, N., Torres, M.D., Domínguez, H., 2019. Successful approaches for a red seaweed biorefinery. *Mar. Drugs* 17, 620. 10.3390/md17110620

Arumugam, N., Chelliapan, S., Kamyab, H., Thirugnana, S., Othman, N., Nasri, N.S., 2018. Treatment of wastewater using seaweed: A review. *Int. J. Environ. Res. Public Health* 15, 2851. 10.3390/ijerph15122851

Baghel, R.S., Suthar, P., Gajaria, T.K., Bhattacharya, S., Anil, A., Reddy, C.R.K., 2020. Seaweed biorefinery: A sustainable process for valorising the biomass of brown seaweed. *J. Clean. Prod.* 263, 121359. 10.1016/j.jclepro.2020.121359

Baghel, R.S., Trivedi, N., Gupta, V., Neori, A., Reddy, C.R.K., Lali, A., Jha, B., 2015. Biorefining of marine macroalgal biomass for production of biofuel and commodity chemicals. *Green Chem.* 17, 2436–2443.

Baghel, R.S., Trivedi, N., Reddy, C.R.K., 2016. A simple process for recovery of a stream of products from marine macroalgal biomass. *Bioresour. Technol.* 203, 160–165.

Balina, K., Romagnoli, F., Blumberga, D., 2017. Seaweed biorefinery concept for sustainable use of marine resources. *Energy Procedia* 128, 504–511.

Brigljević, B., Fasahati, P., Liu, J.J., 2018. Integrated bio-refinery utilizing brown macroalgae: Process design, simulation and techno-economical assessment, in: Eden, M.R., Lerapetritou, M.G., Towler, G.P. (Eds.), *Computer Aided Chemical Engineering*. Elsevier, pp. 337–342.

Buschmann, A.H., Camus, C., Infante, J., Neori, A., Israel, A., Hernández-González, M.C., Pereda, S.V., Gomez-Pinchetti, J.L., Golberg, A., Tadmor-Shalev, N., Critchley, A.T., 2017. Seaweed production: Overview of the global state of exploitation, farming and emerging research activity. *Eur. J. Phycol.* 52, 391–406.

Chen, H., Zhou, D., Luo, G., Zhang, S., Chen, J., 2015. Macroalgae for biofuels production: Progress and perspectives. *Renew. Sustain. Energy Rev.* 47, 427–437.

Chin, Y.X., Mi, Y., Cao, W.X., Lim, P.E., Xue, C.H., Tang, Q.J., 2019. A pilot study on anti-obesity mechanisms of *Kappaphycus alvarezii*: The role of native κ-carrageenan and the leftover sans-carrageenan fraction. *Nutrients* 11, 1133. 10.3390/nu11051133

Cian, R.E., Alaiz, M., Vioque, J., Drago, S.R., 2013. Enzyme proteolysis enhanced extraction of ACE inhibitory and antioxidant compounds (peptides and polyphenols) from *Porphyra columbina* residual cake. *J. Appl. Phycol.* 25, 1197–1206.

Cian, R.E., Martínez-Augustin, O., Drago, S.R., 2012. Bioactive properties of peptides obtained by enzymatic hydrolysis from protein byproducts of *Porphyra columbina*. *Food Res. Int.* 49, 364–372.

Cian, R.E., Salgado, P.R., Drago, S.R., González, R.J., Mauri, A.N., 2014. Development of naturally activated edible films with antioxidant properties prepared from red seaweed *Porphyra columbina* biopolymers. *Food Chem.* 146, 6–14.

Costa, J.C., Gonçalves, P.R., Nobre, A., Alves, M.M., 2012. Biomethanation potential of macroalgae *Ulva* spp. and *Gracilaria* spp. and in co-digestion with waste activated sludge. *Bioresour. Technol.* 114, 320–326.

Ding, L., Cheng, J., Lin, R., Deng, C., Zhou, J., Murphy, J.D., 2020. Improving biohydrogen and biomethane co-production via two-stage dark fermentation and anaerobic digestion of the pretreated seaweed *Laminaria digitata*. *J. Clean. Prod.* 251, 119666. 10.1016/j.jclepro.2019.119666

Ditchburn, J.L., Carballeira, C.B., 2019. Versatility of the humble seaweed in biomanufacturing. *Procedia Manuf.* 32, 87–94.

Dominguez, H., Loret, E.P., 2019. *Ulva lactuca*, a source of troubles and potential riches. *Mar. Drugs* 17, 357. 10.3390/md17060357

Drimer, N., 2016. Offshore structures first principle approach to the design of an open sea aquaculture system. *Ships Offshore Struct.* 14, 384–395.

FAO (Food and Agricultural Organization), 2018. The global status of seaweed production, trade and utilization. *Globefish Research Programme*. Available from: http://www.fao.org/3/CA1121EN/ca1121en.pdf

Forbord, S., Steinhovden, K.B., Solvang, T., Handå, A., Skjermo, J., 2020. Effect of seeding methods and hatchery periods on sea cultivation of *Saccharina latissima* (Phaeophyceae): A Norwegian case study. *J. Appl. Phycol.* 32, 2201–2212.

Francavilla, M., Manara, P., Kamaterou, P., Monteleone, M., Zabaniotou, A., 2015. Cascade approach of red macroalgae *Gracilaria gracilis* sustainable valorization by extraction of phycobiliproteins and pyrolysis of residue. *Bioresour. Technol.* 184, 305–313.

Gajaria, T.K., Suthar, P., Baghel, R.S., Balar, N.B., Sharnagat, P., Mantri, V.A., Reddy, C.R.K., 2017. Integration of protein extraction with a stream of byproducts from marine macroalgae: A model forms the basis for marine bioeconomy. *Bioresour. Technol.* 243, 867–873.

Ganesan, A.R., Tiwari, U., Rajauria, G., 2019. Seaweed nutraceuticals and their therapeutic role in disease prevention. *Food Sci. Hum. Well.* 8, 252–263.

Ghaderiardakani, F., Califano, G., Mohr, J.F., Abreu, M.H., Coates, J.C., Wichard, T., 2019. Analysis of algal growth- and morphogenesis-promoting factors in an integrated multitrophic aquaculture system for farming *Ulva* spp. *Aquacult. Environ. Interact.* 11, 375–391.

Giangrande, A., Pierri, C., Arduini, D., Borghese, J., Licciano, M., Trani, R., Corriero, G., Basile, G., Cecere, E., Petrocelli, A., Stabili, L., Longo, C., 2020. An innovative IMTA system: Polychaetes, sponges and macroalgae co-cultured in a Southern Italian inshore mariculture plant (Ionian Sea). *J. Mar. Sci. Eng.* 8, 733 10.3390/jmse8100733

Glasson, C.R.K., Sims, I.M., Carnachan, S.M., de Nys, R., Magnusson, M., 2017. A cascading biorefinery process targeting sulfated polysaccharides (ulvan) from *Ulva ohnoi*. *Agal Res.* 27, 383–391.

Gullón, B., Gagaoua, M., Barba, F., Gullón, P., Zhang, W., Lorenzo, J.M., 2020. Seaweeds as promising resource of bioactive compounds: Overview of novel extraction strategies and design of tailored meat products. *Trends Food Sci. Technol.* 100, 1–18.

Gupta, V., Trivedi, N., Simoni, S., Reddy, C.R.K., 2018. Marine macroalgal nursery: A model for sustainable production of seedlings for large scale farming. *Algal Res.* 31, 463–468.

Habeebullah, S.F.K., Alagarsamy, S., Sattari, Z., Al-Haddad, S., Fakhraldeen, S., Al-Ghunaim, A., Al-Yamani, F., 2020. Enzyme-assisted extraction of bioactive compounds from brown seaweeds and characterization. *J. Appl. Phycol.* 32, 615–629.

Hamouda, R.A., Sherif, S.A., Dawoud, G.T.M., Ghareeb, M.M., 2016. Enhancement of bioethanol production from *Ulva fasciata* by biological and chemical saccharification. *Rend. Fis. Acc. Lincei.* 27, 665–672.

Hessami, M.J., Phang, S.M., Sohrabipoor, J., Zafar, F.F., Aslanzadeh, S., 2019. The biomethane potential of whole plant and solid residues of two species of red seaweeds: *Gracilaria manilaensis* and *Gracilariopsis persica*. *Algal Res.* 42, 101581. 10.1016/j.algal.2019.101581

Holdt, S.L., Kraan, S., 2011. Bioactive compounds in seaweed: Functional food applications and legislation. *J. Appl. Phycol.* 23, 543–597.

Hou, X., From, N., Angelidaki, I., Huijgen, W.J.J., Bjerre, A.B., 2017. Butanol fermentation of the brown seaweed *Laminaria digitata* by *Clostridium beijerinckii* DSM-6422. *Bioresour. Technol.* 238, 16–21.

Hou, X., Hansen, J.H., Bjerre, A.B., 2015. Integrated bioethanol and protein production from brown seaweed *Laminaria digitata*. *Bioresour. Technol.* 197, 310–317.

Hurtado, A.Q., Gerung, G.S., Yasir, S., Critchley, A.T., 2014. Cultivation of tropical red seaweeds in the BIMP-EAGA region. *J. Appl. Phycol.* 26, 707–718.

Hwang, E.K., Park, C.S., 2020. Seaweed cultivation and utilization of Korea. *Algae* 35, 107–121. 10.4490/algae.2020.35.5.15

Jaballi, I., Saad, H.B., Bkhairia, I., Cherif, B., Kallel, C., Boudawara, O., Droguet, M., Magné, C., Hakim, A., Amara, I.B., 2018. Cytoprotective effects of the red marine alga *Chondrus canaliculatus* against Maneb-induced hematotoxicity and bone oxidative damages in adult rats. *Biol. Trace Elem. Res.* 184, 99–113.

Jung, K.A., Lim, S.R., Kim, Y., Park, J.M., 2013. Potentials of macroalgae as feedstocks for biorefinery. *Bioresourc. Technol.* 135, 182–190.

Kadam, S.U., Tiwari, B.K., O'Donnell, C.P., 2015. Extraction, structure and biofunctional activities of laminarin from brown algae. *Int. J. Food Sci. Technol.* 50, 24–31.

Kidgell, J.T., Magnusson, M., de Nys, R., Glasson, C.R.K., 2019. Ulvan: A systematic review of extraction, composition and function. *Algal Res.* 39, 101422. 10.1016/j.algal.2019.101422

Kılınç, B., Cirik, S., Turan, G., Tekogul, H., Koru, E., 2013. Seaweeds for food and industrial applications, in: Muzzalupo, I. (Ed.), *Food Industry*. Intech Open Ltd., London, pp. 735–748.

Kim, G.Y., Seo, Y.H., Kim, I., Han, J.I., 2019. Co-production of biodiesel and alginate from *Laminaria japonica*. *Sci. Total Environ.* 673, 750–755.

Kraan, S., 2013. Mass-cultivation of carbohydrate rich macroalgae, a possible solution for sustainable biofuel production. *Mitig. Adapt. Strateg. Glob. Change.* 18, 27–46.

Kumar, S., Gupta, R., Kumar, G., Sahoo, D., Kuhad, R.C., 2013. Bioethanol production from *Gracilaria verrucosa*, a red alga, in a biorefinery approach. *Bioresour. Technol.* 135, 150–156.

Lakshmi, D.S., Sankaranarayanan, S., Gajaria, T.K., Li, G., Kujawski, W., Kujawa, J., Navia, R., 2020. A short review on the valorization of green seaweeds and Ulvan: FEEDSTOCK for chemicals and biomaterials. *Biomolecules* 10, 991. 10.3390/biom10070991

Lamb, J.J., Sarker, S., Hjelme, D.R., Lien, K.M., 2018. Fermentative bioethanol production using enzymatically hydrolysed *Saccharina latissima*. *Adv. Microbiol.* 8, 378.

Largo, D.B., Diola, A.G., Rance, G.M.S., 2020. Culture of the brown seaweed *Sargassum siliquosum* J. Agardh (Phaeophyceae, Ochrophyta): From hatchery to out-planting. *J. Appl. Phycol.* 32, 4081–4098.

Layek, J., Das, A., Ghosh, A., Sarkar, D., Idapuganti, R.G., Boragohain, J., Yadav, G.S., Lal, R., 2017. Foliar application of seaweed sap enhances growth, yield and quality of maize in eastern Himalayas. *Proc. Natl. Acad. Sci. India Sect. B: Biol. Sci.* 89, 221–229. 10.1007/s40011-017-0929-x

Lebbar, S., Fanuel, M., Le Gall, S., Falourd, X., Ropartz, D., Bressollier, P., Gloaguen, V., Faugeron-Girard, C., 2018. Agar extraction by-products from *Gelidium sesquipedale* as a source of glycerol-galactosides. *Molecules* 23, 3364. 10.3390/molecules23123364

Li, Y.X., Wijesekara, I., Li, Y., Kim, S.K., 2011. Phlorotannins as bioactive agents from brown algae. *Process Biochem.* 46, 2219–2224.

Lin, H., Qin, S., 2014. Tipping points in seaweed genetic engineering: scaling up opportunities in the next decade. *Mar. Drugs* 12, 3025–3045. 10.3390/md12053025

Liu, H., Wang, G., 2014. Fermentative hydrogen production from macro-algae *Laminaria japonica* using anaerobic mixed bacteria. *Int. J. Hydrogen Energy* 39, 9012–9017.

Liu, P., Xie, J., Tan, H., Zhou, F., Zou, L., Ouyang, J., 2020. Valorization of *Gelidium amansii* for dual production of D-galactonic acid and 5-hydroxymethyl-2-furancarboxylic acid by chemo-biological approach. *Microb. Cell Fact.* 19, 104. 10.11 86/s12934-020-01357-6

López-Contreras, A.M., Harmsen, P.F.H., Blaauw, R., Houweling-Tan, B., van der Wal, H., Huijgen, W.J.J., van Hal, J.W., 2014. Biorefinery of the brown seaweed *Saccharina latissima* for fuels and chemicals. In: Proceedings of the Mie Bioforum on lignocellulose degradation and biorefinery, Nemunosato Resort, Mie Prefecture, Japan. https://edepot.wur.nl/382063

Magnusson, M., Carl, C., Mata, L., de Nys, R., Paul, N.A., 2016. Seaweed salt from Ulva: A novel first step in a cascading biorefinery model. *Algal Res.* 16, 308–316.

Mahajan, R.V., Bhale, V.M., Deshmukh, J.P., Patil, S.P., Shingrup, P.V., 2016. Utilization of seaweed extract as bio-regulator for enhancement of morpho-physiological biochemical traits and yield of blackgram. *Ecol. Environ. Conserv.* 22, S269–S274.

Mehariya, S., Fratini, F., Lavecchia, R., Zuorro, A., 2021a. Green extraction of value-added compounds form microalgae: A short review on natural deep eutectic solvents (NaDES) and related pre-treatments. *J. Environ. Chem. Eng.* 9, 105989. 10.1016/j.jece.2021.105989

Mehariya, S., Goswami, R.K., Karthikeysan, O.P., Verma, P., 2021b. Microalgae for high-value products: A way towards green nutraceutical and pharmaceutical compounds. *Chemosphere* 130553. 10.1016/j.chemosphere.2021.130553

Mehariya, S., Goswami, R.K., Verma, P., Lavecchia, R., Zuorro, A., 2021c. Integrated Approach for Wastewater Treatment and Biofuel Production in Microalgae Biorefineries. *Energies* 14, 2282. 10.3390/en14082282

Meinita, M.D.N., Marhaeni, B., Jeong, G.T., Hong, Y.K., 2019. Sequential acid and enzymatic hydrolysis of carrageenan solid waste for bioethanol production: A biorefinery approach. *J. Appl. Phycol.* 31, 2507–2515.

Méndez, A., Gascó, G., Ruiz, B., Fuente, E., 2019. Hydrochars from industrial macroalgae *Gelidium sesquipedale* biomass wastes. *Bioresour. Technol.* 275, 386–393.

Mhatre, A., Gore, S., Mhatre, A., Trivedi, N., Sharma, M., Pandit, R., Anil, A., Lali, A., 2019. Effect of multiple product extractions on bio-methane potential of marine macrophytic green alga *Ulva lactuca*. *Renew. Energy.* 132, 742–751.

Milledge, J.J., Nielsen, B.V., Maneein, S., Harvey, P.J., 2019. A brief review of anaerobic digestion of algae for bioenergy. *Energies.* 12, 1166. 10.3390/en12061166

Milledge, J.J., Smith, B., Dyer, P.W., Harvey, P., 2014. Macroalgae-derived biofuel: A review of methods of energy extraction from seaweed biomass. *Energies.* 7, 7194–7222. 10.3390/en7117194

Miyashita, K., Mikami, N., Hosokawa, M., 2013. Chemical and nutritional characteristics of brown seaweed lipids: A review. *J. Funct. Foods.* 5, 1507–1517.

Naseri, A., Holdt, S.L., Jacobsen, C., 2019. Biochemical and nutritional composition of industrial red seaweed used in carrageenan production. *J. Aquat. Food Product. Technol.* 28, 967–973.

Nunes, N., Valente, S., Ferraz, S., Barreto, M.C., Pinheiro de Carvalho, M.A.A., 2018. Nutraceutical potential of *Asparagopsis taxiformis* (Delile) Trevisan extracts and assessment of a downstream purification strategy. *Heliyon.* 4, e00957. 10.1016/j.heliyon.2018.e00957

Offei, F., Mensah, M., Thygesen, A., Kemausuor, F., 2018. Seaweed bioethanol production: A process selection review on hydrolysis and fermentation. *Fermentation* 4, 99. 10.33 90/fermentation4040099

Oliveira, J.V., Alves, M.M., Costa, J.C., 2014. Design of experiments to assess pre-treatment and co-digestion strategies that optimize biogas production from macroalgae *Gracilaria vermiculophylla*. *Bioresour. Technol.* 162, 323–330.

Peñuela, A., Robledo, D., Bourgougnon, N., Bedoux, G., Hernández-Núñez, E., Freile-Pelegrin, Y., 2018. Environmentally friendly valorization of *Solieria filiformis* (Gigartinales, Rhodophyta) from IMTA using a biorefinery concept. *Mar. Drugs* 16, 487. 10.3390/md16120487

Potts, T., Du, J., Paul, M., May, P., Beitle, R., Hestekin, J., 2011. The production of butanol from Jamaica bay macro algae. *Environ. Prog. Sustain. Energy.* 31, 29–36.

Purcell-Meyerink, D., Packer, M.A., Wheeler, T.T., Hayes, M., 2021. Aquaculture Production of the Brown Seaweeds *Laminaria digitata* and *Macrocystis pyrifera*: Applications in Food and Pharmaceuticals. *Molecules.* 26, 1306.

Renakabai, N., Mary Christi, R., Christy Kala, T., Shajini, R.S., 2013. Influence of seaweed liquid fertilizers on the growth and biochemical characteristics of horsegram (*Dolichos biflorus* L). *Plant Arch.* 13, 1155–1158.

del Río, P.G., Gomes-Dias, J.S., Rocha, C.M.R., Romaní, A., Garrote, G., Domingues, L., 2020. Recent trends on seaweed fractionation for liquid biofuels production. *Bioresour. Technol.* 299, 122613.

Robin, A., Sack, M., Israel, A., Frey, W., Müller, G., Goldberg, A., 2018. Deashing macroalgae biomass by pulsed electric field treatment. *Bioresour. Technol.* 255, 131–139.

Sadaka S., Boateng, A.A., 2009. Pyrolysis and bio-oil. *Agric. Nat. Resour.* FSA1052.

Sadhukhan, J., Gadkari, S., Martinez-Hernandez, E., Ng, K.S., Shemfe, M., Torres-Garcia, E., Lynch, J., 2019. Novel macroalgae (seaweed) biorefinery systems for integrated chemical, protein, salt, nutrient and mineral extractions and environmental protection by green synthesis and life cycle sustainability assessments. *Green Chem.* 21, 2635–2655.

Selvan, K.B., Piriya, S.P., Vennison, J.S., 2014. Macroalgae (*Eucheuma cottonii* and *Sargassum* sp.) are reservoirs of biodiesel and bioactive compounds. *J. Chem. Pharm. Sci.* 2, 62–70.

Shanmugam, M., Seth, A., 2018. Recovery ratio and quality of an agricultural bio-stimulant and semi-refined carrageenan co-produced from the fresh biomass of *Kappaphycus alvarezii* with respect to seasonality. *Algal Res.* 32, 362–371.

Shannon, E., Abu-Ghannam, N., 2019. Seaweeds as nutraceutricals for health and nutrition. *Phycologia* 58, 563–577.

Sharma, A., Arya, S.K., 2017. Hydrogen from algal biomass: A review of production process. *Biotechnol. Rep.* 15, 63–69.

Sharmila, S., Rebecca, L.J., Das, M.P., 2012. Production of biodiesel from *Chaetomorpha antennina* and *Gracilaria cortica*. *J. Chem. Pharm. Res.* 4, 4870–4874.

Singh, S.K., Thakur, R., Singh, M.K., Singh, C.S., Pal, S.K., 2015. Effect of fertilizer level and seaweed sap on productivity and profitability of rice (*Oryza sativa*). *Indian J. Agron.* 60, 420–425.

Škrovánková, S., 2011. Seaweed vitamins as nutraceuticals, in: Toldra, F. (Ed.), *Advances in Food and Nutrition Research*. Academic Press, pp. 357–369.

Soares, C., Švarc-Gajić, J., Oliva-Teles, M.T., Pinto, E., Nastić, N., Savić, S., Almeida, A., Delerue-Matos, C., 2020. Mineral composition of subcritical water extracts of *Saccorhiza polyschides*, a brown seaweed used as fertilizer in the North of Portugal. *J. Mar. Sci. Eng.* 8, 244. 10.3390/jmse8040244

Sudhakar, M.P., Merlyn, R., Arunkumar, K., Perumal, K., 2016. Characterization, pretreatment and saccharification of spent seaweed biomass for bioethanol production using baker's yeast. *Biomass Bioenergy* 90, 148–154.

Sudhakar, M.P., Peter, D.M., Dharani, G., 2020. Studies on the development and characterization of bioplastic film from the red seaweed (*Kappaphycus alvarezii*). *Environ. Sci. Pollut. Res.* 10.1007/s11356-020-10010-z

Sulfahri, N., Taufan, W.L., Aska, M.S., 2019. Biodiesel production from *Pichia kudriavzevii* using algae *Kappaphycus alvarezii* as a fermentation substrate. *IOP Conf. Ser.: Earth Environ. Sci.* 243, 012095.

Sunwoo, I.Y., Ra, C.H., Jeong, G.T., Kim, S.K., 2016. Evaluation of ethanol production and bioadsorption of heavy metals by various red seaweeds. *Bioprocess Biosyst. Eng.* 39, 915–923.

Tabassum, M.R., Xia, A., Murphy, J.D., 2017. Potential of seaweed as a feedstock for re-newable gaseous fuel production in Ireland. *Renew. Sustain. Energy Rev.* 68, 136–146.

Tan, I.S., Lee, K.T., 2014. Enzymatic hydrolysis and fermentation of seaweed solid wastes for bioethanol production: An optimization study. *Energy.* 78, 53–62.

Tedesco, S., Daniels, S., 2018. Optimisation of biogas generation from brown seaweed re-sidues: Compositional and geographical parameters affecting the viability of a bior-efinery concept. *Appl. Energy.* 228, 712–723. 10.1016/j.apenergy.2018.06.120

Torres, M.D., Flórez-Fernández, N., Domínguez, H., 2019a. Integral utilization of red sea-weed for bioactive production. *Mar. Drugs.* 17, 314. 10.3390/md17060313

Torres, M.D., Kraan, S., Domínguez, H., 2019b. Seaweed biorefinery. *Rev. Environ. Sci. Biotechnol.* 18, 335–388.

Trivedi, N., Baghel, R.S., Bothwell, J., Gupta, V., Reddy, C.R.K., Lali, A.M., Jha, B., 2016. An integrated process for the extraction of fuel and chemicals from marine macroalgal biomass. *Sci. Rep.* 6, 30728.

Tsubaki, S., Nishimura, H., Imai, T., Onda, A., Hiraoka, M., 2020. Probing rapid carbon fixation in fast-growing seaweed *Ulva meridionalis* using stable isotope 13 C-labelling. *Sci. Rep.* 10, 1–11.

Ubando, A.T., Felix, C.B., Chen, W.H., 2020. Biorefineries in circular bioeconomy: A comprehensive review. *Bioresour. Technol.* 299, 122585.

Venolia, C.T., Lavaud, R., Green-Gavrielidis, L.A., Thornber, C., Humphries, A.T., 2020. Modeling the growth of sugar kelp (*Saccharina latissima*) in aquaculture systems using dynamic energy budget theory. *Ecol. Modell.* 430, 109151.

Vilg, J.V., Undeland, I., 2017. pH-driven solubilization and isoelectric precipitation of proteins from the brown seaweed *Saccharina latissima*-effects of osmotic shock, water volume and temperature. *J. Appl. Phycol.* 29, 585–593.

van der Wal, H., Sperber, B.L.H.M., Houweling-Tan, B., Bakker, R.R.C., Brandenburg, W., López-Contreras, A.M., 2013. Production of acetone, butanol, and ethanol from bio-mass of the green seaweed *Ulva lactuca*. *Bioresour. Technol.* 128, 431–437.

Wijesinghe, W.A.J.P., Jeon, Y.J., 2011. Biological activities and potential cosmeceutical applications of bioactive components from brown seaweeds: A review. *Phytochem. Rev.* 10, 431–443.

Wijesinghe, W.A.J.P., Jeon, Y.J., 2012. Exploiting biological activities of brown seaweed *Ecklonia cava* for potential industrial applications: A review. *Int. J. Food Sci. Nutr.* 63, 225–235.

Wu, F.C., Wu, J.Y., Liao, Y.J., Wang, M.Y., Shih, I.L., 2014. Sequential acid and enzymatic hydrolysis *in situ* and bioethanol production from *Gracilaria* biomass. *Bioresour. Technol.* 156, 123–131.

Xing, M., Cao, Q., Wang, Y., Xiao, H., Zhao, J., Zhang, Q., Ji, A., Song, S., 2020. Advances in research on the bioactivity of alginate oligosaccharides. *Mar. Drugs* 18, 144. 10.33 90/md18030144

Yoza, B.A., Masutani, E.M., 2013. The analysis of macroalgae biomass found around Hawaii for bioethanol production. *Environ. Technol.* 34, 1859–1867.

Yuan, Y., Macquarrie, D.J., 2015. Microwave assisted step-by-step process for the produc-tion of fucoidan, alginate sodium, sugars and biochar from *Ascophyllum nodosum* through a biorefinery concept. *Bioresour. Technol.* 198, 819–827. 10.1016/j.biortech.2 015.09.090

Zhang, R., Yuen, A.K.L., Magnusson, M., Wright, J.T., de Nys, R., Maters, Maschmeyer, T., 2018. A comparative assessment of the activity and structure of phlorotannins from the brown seaweed *Carpophyllum flexuosum*. *Algal Res.* 29, 130–141. 10.1016/j.algal.201 7.11.027

Zhang, R., Yuen, A.K.L., de Nys, R., Masters, A.F., Maschmeyer, T., 2020. Step by step extraction of bio-actives from the brown seaweeds, *Carpophyllum flexuosum*, *Carpophyllum plumosum*, *Ecklonia radiata* and *Undaria pinnatifida*. *Algal Res.* 52, 102092. 10.1016/j.algal.2020.102092

Zollmann, M., Robin, A., Prabhu, M., Polikovsky, M., Gillis, A., Greiserman, S., Golberg, A., 2019. Green technology in green macroalgal biorefineries. *Phycologia.* 58, 516–534.

2 Valorization of Algal Spent Biomass into Valuable Biochemicals and Energy Resource

Saravanan Vasanthakumar and K Greeshma
Microbial Processes and Technology Division, Council of
Scientific and Industrial Research – National Institute for
Interdisciplinary Science and Technology,
Thiruvananthapuram, India

Muthu Arumugam
Microbial Processes and Technology Division, Council of
Scientific and Industrial Research – National Institute for
Interdisciplinary Science and Technology,
Thiruvananthapuram, India
and
Academy of Scientific and Innovative Research,
Ghaziabad, India

CONTENTS

2.1   Introduction..38
2.2   Algae: The Primary Producer 39
2.3   Concept of Biorefinery...40
2.4   Algal Biorefinery Approach...40
2.5   Metabolite Profiling of Algae 42
2.6   Spent Biomass...44
2.7   Green Solvent-Based Extraction...45
2.8   Valorization of Lipid-Extracted Residual Biomass for Biochemicals 46
        2.8.1   Proteins 46
        2.8.2   Polysaccharides...46
        2.8.3   Carotenoids 47
        2.8.4   Nutraceuticals 48
        2.8.5   Antimicrobial Compounds 48
        2.8.6   Antioxidant Compounds..49

DOI: 10.1201/9781003195405-2

2.9   Valorization of Lipid-Extracted Residual Biomass
         for Energy Resources 49
         2.9.1   Biodiesel 50
         2.9.2   Bioethanol..50
         2.9.3   Bio-oil 51
         2.9.4   Biochar...52
         2.9.5   Biogas 52
         2.9.6   Biomethane 53
         2.9.7   Syngas..53
         2.9.8   Biohydrogen 54
         2.9.9   Feed Supplement 54
         2.9.10 Fertilizer...55
         2.9.11 Lactic Acid 56
2.10   Valorization of Agar-Extracted Residual Biomass 56
         2.10.1 Bioactive Compounds 57
         2.10.2 Fatty Acids 57
         2.10.3 Bio-ethanol 58
         2.10.4 Paper Products..59
         2.10.5 Cellulose Nanocrystals 59
2.11   Future Research Directions...60
2.12   Conclusion 61
Acknowledgments...61
References...62

2.1   INTRODUCTION

Over the past semi-centennial period, the world population has doubled on a tremendous scale. Concurrently, the use of primary energy fuels also increased due to higher standards of living (Jones and Mayfield, 2012). The population is expected to reach 9.9 billion by 2050, requiring an estimated 70% enhanced food production. In this context, the search for renewable sources has geared up, and the best alternative is to produce energy resources and value-added compounds from photosynthetic organisms, such as algae, which have a great role as a primary producer.

Algae are an assorted group of chlorophyll-bearing eukaryotic organisms found in both marine and freshwater habitats. Based on their size, they can be micro (single-celled organism) to macro range (100 feet in length). They are known to contain interesting bioactive compounds, which have been used for centuries in several ways. They have an advantage over other terrestrial plants due to their faster growth rate, higher CO_2 fixation ability, and photosynthetic yield (Arumugam et al., 2011; Michalak and Chojnacka, 2015). Their potential in terms of energy and food production has gained a lot of attention in the last few decades. Their ability to endure the macro-panel spectrum of temperature, light intensity, pH, and salinity stress is also a remarkable feature (ElFar et al., 2020).

Nowadays, we are witnessing the use of biomass as an alternative source for fossil fuels and feed applications extensively. Advanced research by targeting a group of algal strains having maximum biomass yields under cultivation is necessary for large-scale industrial technology transfer. There are different extraction procedures for the production of each bio-molecular fraction as a major product of our interest, but the main problem arises when we talk about the efficient use of spent biomass into co-products, such as biofuels, animal feed, fertilizer, and bio-sorbent. Complete exploitation of algal biomass could be possible by biorefineries. The ultimate goal of the biorefinery is to integrate the production of value-added commodities like valuable biochemicals, energy resources, and biofuels, and to attain zero waste at the end in order to optimize the use of biomass resources while maximizing benefits and profitability (Demirbas, 2009). Biorefinery of algae was not a well-exploited field, with the relatively minimal number of publications on the algae biorefinery approach. Spent algal biomass have a huge number of applications that can add value to our economy also. However, commercialization of the process is a challenge considering current capital costs per unit of fuel production. Unlike crop biomass production systems that have large developmental history, algae biomass production systems still do not come under integrated research and developmental (R&D) strategies, which ensure the efficient production of feed, fuel, and other valuable biochemicals (Rajvanshi and Sayre, 2020). This chapter comprehensively discusses the recent advances in recycling and reuse of algal spent biomass for the production of biofuel and co-products. Integration of different fractions of compounds after each extraction process into an economically feasible zero-waste biorefinery approach is discussed.

2.2 ALGAE: THE PRIMARY PRODUCER

Macroalgae/Seaweeds are grouped into three broad classes, based on their pigmentation: brown algae (Phaeophyceae), red algae (Rhodophyceae), and green algae (Chlorophyceae). Microalgae or microphytes are microscopic organisms that may live either singly or in colonies. These minute organisms are further categorized into three specific groups of unicellular organisms: Cyanobacteria (blue-green algae), diatoms (Bacillariophyta), and dinoflagellates (Dinophyceae); they also known as phytoplankton. Apart from other land plants, microalgae lack lignin deposition in their cell wall, because they don't require support for growth in an aquatic environment (John et al., 2011). Seaweeds, on other hand, contain 50% of sugars that are suitable for fermentation purposes. In certain marine algae, such as Rhodophyceae, the carbohydrate content is due to the presence of agar, which is a polymer of galactose and galactopyranose monomers. It has been found that saccharification can be done to release monomeric sugars galactose from agar, and glucose from cellulose, respectively, so that product yields can be enhanced in terms of bioethanol (Jones and Mayfield, 2012). Algal blooms in oceans, lakes, and ponds may result in harmful effects on ecosystems, so the exclusion of those algae for biofuel production is highly appreciated. Apart from these features, microalgae possess remarkable metabolic plasticity also; they can be cultivated in both brackish and wastewaters upon providing enough nutrients at the expense of only providing sunlight and CO_2 (Michalak and Chojnacka, 2015).

2.3 CONCEPT OF BIOREFINERY

By definition, a biorefinery is a facility that assimilates the separation processes and biomass processing. It has the equipment to convert biomass into a broad spectrum of bio-based products like feed, food, and chemicals and bioenergy-related products like fuel, power, heat, etc. The biorefinery concept is analogous to our petroleum refineries. Biomass is considered one of the most significant renewable resources on earth. The biorefinery is the clearest and optimum expression of integrated bio-transformation processing, thus unfolding the aspirations of integrated technologies (Venkata Mohan et al., 2016). A biorefinery acts as a solution to enhance the techno economics of micro-algal applications through more profitable products and enhance biomass production along with low-value bio-fuels (Caporgno and Mathys, 2018; Chen et al., 2018). Furthermore, biorefinery is an integrated approach of many biological units, and the specific configuration will be obtained via process technologies and input feedstock; platforms; and output products that are required (Venkata Mohan et al., 2020).

2.4 ALGAL BIOREFINERY APPROACH

Algal biorefinery mainly targets the sequential extraction of multiple product recovery towards zero-waste processing. It valorizes its feedstock potential to exploit several bio-based products with a combined sustainable bioprocessing approach (Venkata Mohan et al., 2020). The most favorable biorefinery concept ultimately aims at utmost valorization of the algal biomass along with greater concentrations of target compounds. One of the main challenges for the biorefinery concept is to preserve the compounds (especially compounds with bioactive properties) in the remaining fractions. The algal biorefinery approach addresses waste remediation and CO_2 sequestration, and they have huge potential in resource recovery. Nowadays, algal cultivation methods are fortuitously using the single product approach; along with this, the multiple product approach can facilitate economic feasibility (Rajesh Banu et al., 2020). Substantial scientific studies have examined the use of algal biomass as a source to generate value-added products aside from wastewater remediation (Pradel et al., 2016). Thus, switching towards sustainability and a bio-based economy, along with an integrated approach, needs to be developed with effective resource recovery.

Scientists believe that the huge amount of usable carbon in algal biomass can potentially replace a greater fraction of global fossil records via sustainable biological processes (Kannengiesser et al., 2016). Algal biorefinery could effectively utilize wastewater or the intrinsic load of CO_2 and thereby extract products with high commercial interest. Methods for the valorization of the whole algae biomass include the use of disrupted whole cell content or the intact algae cells, or the fractionation of the biomass into several fractions via biorefinery (Figure 2.1). To obtain multiple marketable fractions from the single algal biomass, several cascading biorefinery approaches can be expounded (Gouveia et al., 2014).

For instance, the methodologies can be differentiated into three categories based on the value of the main target compound (Hariskos and Posten, 2014).

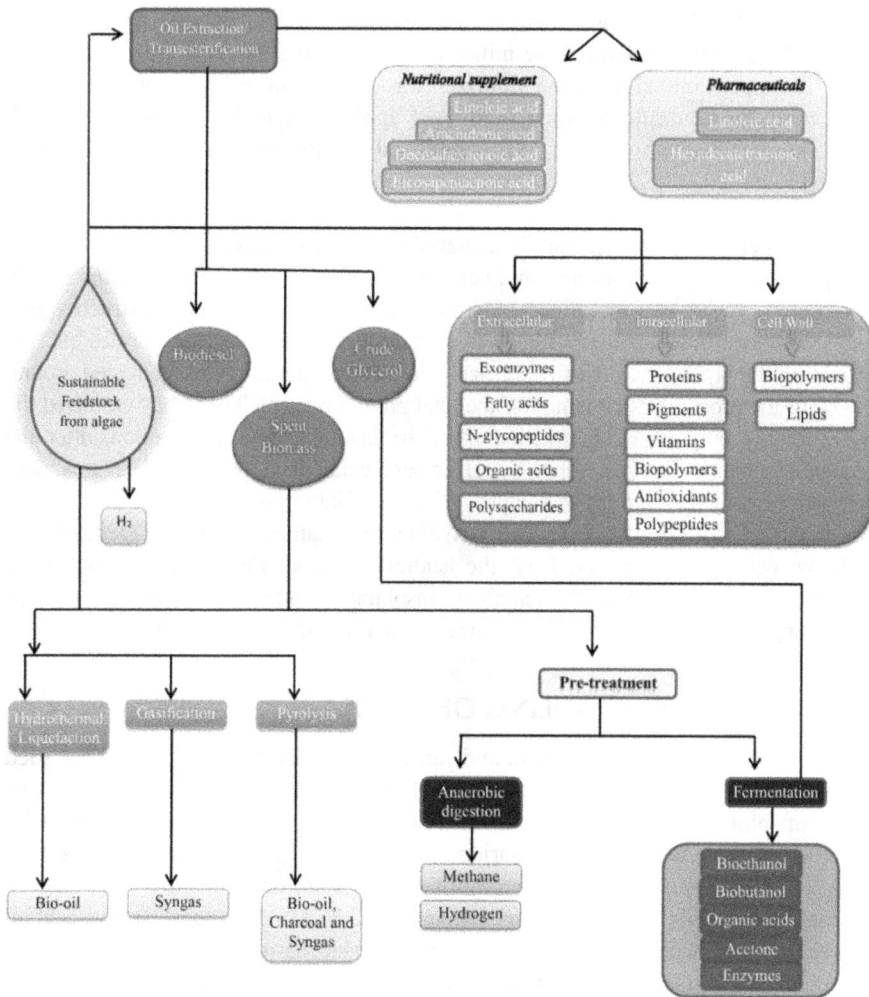

FIGURE 2.1 Schematic representaion of algal biorefinary and its possible industrial applications.

- Category 1: production of low-value compounds along with valorization of spent biomass has been reported many times (Maurya et al., 2016; Brennan and Owende, 2010).
- Category 2: production of medium-value biomolecular compounds (carbohydrates, lipids, proteins), coupled with the valorization of leftover biomass in a biorefinery approach (Figure 2.1), has received attention when biodiesel production was economically feasible only for an extended period (Wijffels et al., 2010).
- Category 3: Integrated use of high-value products in combination with spent biomass is the most economically feasible and attractive biorefinery method in terms of the minimal amount of algal biomass that is present

today. While looking for high-value products, a considerable amount of residual biomass with more minor compounds in algae will be produced. This spent biomass can be valorized through various energy applications, used for aquaculture, food, and feed application and as fertilizer, presuming that no solvent was employed in the process of extraction (Figure 2.1).

Wahlström et al. (2018) successfully achieved the sequential recovery of biomolecules like protein, carrageenan, pectin, and cellulose from the *Porphyra umbilicalis* Kuetz, toward zero waste as a biorefinery approach. They developed a non-destructive and multicomponent fractionation strategy to extract proteins and polysaccharides etc., by extracting the protein-rich fractions in ice-cold alkaline solution and recovered around 15.0% of the protein, which is half of the total protein content. Then, they extracted the water-soluble polysaccharide fraction from the insoluble residual fraction, after the sequential alkaline (Na_2CO_3 solution, pH 9.5) and acidic (0.01 M HCl, pH 2) treatment at 90°C for four hours. Further analyses of water-soluble polysaccharide fractions revealed the presence of high molecular weight carrageenan in the alkaline extraction. In addition, pectin was obtained from the acidic extraction. Finally, they isolated the cellulose-rich fractions from the remaining insoluble residues after all extraction steps by treating them in 0.5 M NaOH and heated them at 80°C for 24 hours.

2.5 METABOLITE PROFILING OF ALGAE

Algae mostly contain protein, lipid, and carbohydrate, and their proportion is varied among different algae species, as given below in Table 2.1. Algae contain a vast array of biologically active and nutritionally rich compounds, such as polyunsaturated fatty acids, polysaccharides, dietary fiber, B complex vitamins, carotenoids, polyphenols, proteins, natural plant growth-promoting substances such as auxins, gibberellins, and cytokinins (Khan et al., 2009). Marine algae are a good source of polysaccharides, especially mucopolysaccharides and structural polysaccharides. Polysaccharide proportions in some macroalgae range from 4% to 76% of their total dry weight. *Ulva, Ascophyllum, Palmaria*, and *Porphyra* are some rich sources of polysaccharides (Table 2.1).

Algal polysaccharides are known to possess pharmacological activities, such as antibacterial, antioxidant, anti-inflammatory, antitumor, and antiviral properties. The major active compounds generally present in algae, with their key pharmacological activities, are listed in Table 2.2. Apart from these, they have industrial applications such as thickeners, stabilizers, emulsifiers, beverages, food, feed, pharmaceuticals, etc.

In the nutritional aspect, seaweeds are low in energy and fat content but contain high indigestible carbohydrates, which are good for the intestine. Alginates, carrageenans, ulvans, fucoidans, and chitin have potential pharmaceutical and biomedical demand with higher usage and commercial value (Usman et al., 2017). Whole seaweeds can be supplemented to the diet of laboratory and farm animals also (Shilpi Gupta and Abu-Ghannam, 2011). They are not only beneficial to humans and animals but also plants.

TABLE 2.1
Lipid, Protein and Carbohydrate Composition (% in Dry Weight) of Various Algal Species

Algal Species	Lipids	Proteins	Carbohydrates	References
Anabaena cylindrica	4–7	43–56	25–30	(Becker and Becker, 2013)
Chaetoceros calcitrans	15	36	27	(Batista et al., 2008)
Chlamydomonas rheinhardii	21	48	17	(Spolaore et al., 2006)
Chlorella pyrenoidosa	2	57	26	(Batista et al., 2008)
Chlorella vulgaris	14–22	51–58	12–17	(Mata et al., 2010)
Dunaliela bioculata	8	49	4	(Lyons, 2009)
Dunaliela salina	6	57	32	(Batista et al., 2008)
Euglena gracilis	22–38	39–61	14–18	(Batista et al., 2008) & (Lyons, 2009)
Isochrysis galbana	12–14	50–56	10–17	(Batista et al., 2008)
Porphyridum cruentum	9–14	28–39	40–57	(Batista et al., 2008) & (Lyons, 2009)
Prymnesium parvum	22–38	28–45	25–33	(Lyons, 2009)
Scenedesmus dimorphus	16–40	8–18	21–52	(Batista et al., 2008) & (Lyons, 2009)
Scenedesmus obliquus	12–14	50–56	10–17	(Spolaore et al., 2006) & (Cai et al., 2019)
Scenedesmus quadricauda	1.9	47	21–52	(Lyons, 2009)
Spirogyra sp.	11–21	6–20	33–64	(Lyons, 2009)
Spirulina maxima	6–7	60–71	13–16	(Batista et al., 2008)
Spirulina platensis	4–9	46–63	8–14	(Batista et al., 2008)
Synechococcus sp.	11	63	15	(Spolaore et al., 2006)
Asparagopsis sp.	6.62	17.6	40.5	(Nunes, 2017)
Gelidiella sp.	1.4	14.9	24.5	(Baghel et al., 2014)
Gelidiopsis sp.	1.3	17.6	11.43	(Baghel et al., 2014)
Gelidium sp.	1.1–1.3	18.4–19.3	23.5–25.2	(Baghel et al., 2014)
Gigartina sp.	0.57	15.6	29.31	(Gómez-ordóñez et al., 2010)
Graciliaropsis sp.	0.8	10.5	77.7	(Javad et al., 2019)
Kappaphycus sp.	5.2	2.6	57.2	(Paz-cedeno et al., 2019)
Nemalion sp.	2.17	3.8	31.43	(Nunes, 2017)
Porphyra sp.	1.3	22.3–53.9	39.0–64.0	(Cian et al., 2012)

Algal lipids, especially microalgae, are valuable alternative sources of nutritionally important polyunsaturated fatty acids (PUFAs) and are, therefore, widely in use as ingredients in several food formulations (Bleakley and Hayes, 2017). Most recent biofuel research has increased the interest in algal lipid biochemistry to employ these renewable reservoirs using advanced tools of mass spectrometry and genetic engineering (Kumari et al., 2013). But still, most of the lipid research has

TABLE 2.2

Major Components Present in Algae with Their Therapeutic Activities

Antioxidative	Antimicrobial	Antiinflammatory	Antitumor
• polysaccharides	• PUFAs	• proteins	• polyphenols
• PUFAs	• proteins	• carotenoids	• carotenoids
• carotenoids	• polysaccharides	• polysaccharides	• polysaccharides
• tocopherol	• pigments:	• sterpls-fucosterols	
• ascorbate	chlorophyll and	• polyphenols-	
• glutathione	carotenoids	phlorotannins	
• proteins	• terpenes	• porphyrin derivatives –	
• mycosporine like	• polyphenols	pheophorbide a and	
amino acids		pheophytin a	
• polyphenols			

been focused on limited model organisms. The major PUFA present are eicosa-pentaenoic acid, docosahexaenoic acid, γ-linolenic acid, and arachidonic acid. PUFAs are vital for human health due to their pharmacological activities. Novel extraction techniques, such as supercritical fluid extraction (SFE), microwave-assisted extraction (MAE), ultrasound-assisted extraction (UAE), enzyme-assisted extraction (EAE), and pressurized liquid extraction (PLE), can be adopted for the easiest isolation of lipids (Michalak and Chojnacka, 2015).

Seaweed and microalgae are good sources of protein. An algal protein is known to contain many benefits over the traditional high-protein crop in terms of higher productivity and nutritional value. Some species of seaweed and microalgae are known to contain protein levels similar to meat, egg, soybean, and milk. Seaweed and microalgae have higher protein yield per unit area (2.5–7.5 tons/Ha/year and 4–15 tons/Ha/year, respectively) compared to land plants, such as soybean, pulse legumes, and wheat (0.6–1.2 tons/Ha/year, 1–2 tons/Ha/year, and 1.1 tons/Ha/year, respectively) (Krimpen et al., 2013). Land agriculture requires around 75% of the total global freshwater, particularly for animal protein, requiring 100 times more water if the same amount of protein was produced from plant sources. Unlike terrestrial plants, seaweeds don't require fresh water or arable land to grow, which makes them more suitable as an additional protein source in food industries. Apart from these, due to their unfavorable environment, algae are often exposed to higher oxidative stress, which ultimately led to the evolution of natural protective systems, such as the production of many pigments (e.g., chlorophylls, carotenes, and phy-cobiliproteins) and polyphenol compounds (e.g., flavonols, catechins, and phlor-otannins), which are known to possess good health benefits (Fabris et al., 2020).

2.6   SPENT BIOMASS

Biomass is defined as the mass of living organisms in an ecosystem or in a given area in a unit of time. Algal biomass is defined as the number of algae occupied in a water body at a particular time. Several species of algae are potent organisms for the production of biomass and could be utilized well in the production of biofuel and biogas. Biomass production depends on the cultivation methods for harvesting.

Several terms describe spent biomass: spent microalgal biomass (Rashid et al., 2013), oil-extracted algae, lipid-extracted microalgal biomass residues (Yang et al., 2011), de-fatted algae (Vardon et al., 2012), post-extracted algae residue (Bryant et al., 2012), algal biomass residues (S. Park and Li, 2012), and de-oiled algal biomass. Spent algal biomass is defined as the remaining biomass after extraction of the desirable primary targeted compound like agar, lipid, protein, etc., from the algal biomass. While defining spent biomass, the primary target metabolite from which the desired compound obtained or in which process that spent biomass are originated should be taken into account, and sometimes this process may result in functional loss of components to some extent. The compositions of spent biomass may vary from one another depending upon the processing and primary targeted compounds from the algal biomass. Often spent biomass is considered as waste or residue and is marked as environmental pollution. Spent biomass can be valorized under different categories to generate value-added products.

2.7   GREEN SOLVENT-BASED EXTRACTION

There are different procedures to extract bioactive molecules from the algal biomass. The major advantage of novel extraction techniques over conventional methods, such as SFE, MAE, UAE, EAE, and PLE, is without any functional loss we can extract our target component to a greater extent. All these novel techniques have higher yields and lesser processing time. However, the production has not been enhanced in the industrial sector due to many bottlenecks, which still hinder the commercialization of biofuel in an economically feasible way. The cost-effective and cost-efficient extraction of lipids has remained a major bottleneck for several years (Kumar et al., 2015). They are more eco-friendly than the conventional techniques using organic solvents, where large processing time and solvent volume are required (Kadam et al., 2013). When applied on an industrial scale, SFE has the highest potential as a more friendly green technique because it uses no energy expenditure from solvent removal (Halim et al., 2012). Also, high-temperature processing can lead to the degradation of thermos-labile components from biomass (Roh et al., 2008).

The use of organic solvents has so many hazards or disadvantages, especially when we talk about the food industry. Many are flammable (except chlorinated solvents), are carcinogenic (chlorinated solvents and aromatics), have higher vapor pressure (inhalation route), are narcotic (ether and chloroform), are toxic (methanol and carbon disulfide), are teratogens/mutagens (toluene), are peroxides (ethers), are smog formers, etc. The adoption of greener solvents is the only remedy to solve all issues concerned with human consumption and environmental safety. Bio-derived solvents, such as glycerol, 2-methyl tetrahydrofuran, γ-valerolactone, ethyl lactate, and cyclopentyl methyl ether, can be considered better alternatives in this context.

2.8 VALORIZATION OF LIPID-EXTRACTED RESIDUAL BIOMASS FOR BIOCHEMICALS

Lipid-extracted algae biomass is a rich source of carbohydrates and proteins with good mineral content (Maurya, Paliwal et al., 2016). Different upstream processing methods like cultivation and harvesting methods can affect the composition of this spent biomass. Various chemical and biological pretreatments can be given to enhance the bio-availability of these complex molecules. After lipid extraction, residual microalgal biomass contains mainly carbohydrates and protein, which can be determined by the carbon: nitrogen ratio. A high C/N ratio is beneficial for the production of biogases or biofuels, whereas a low C/N ratio is good to use as a high protein food source and feed supplement for animals or fish.

2.8.1 PROTEINS

Algae are rich in proteins, especially microalgae, and could be an alternative protein source because of their amino acid pattern and high protein content (Spolaore et al., 2006). The commonly known source is *Spirulina,* a single-cell protein that has many nutraceutical properties. Proteins from algae gain importance due to their potential bioactive properties. Phycobiliprotein from seaweeds is known for its antioxidant, anti-inflammatory, and hepatoprotective properties. These are water-soluble proteins that exist in certain algae (Rhodophyta) – phycoerythrobilins and (cyanobacteria) – phycocyanobilins (Michalak and Chojnacka, 2015).

Sekar and Chandramohan (2008) described antitumor, anti-inflammatory, antiviral, antioxidant, and hepatoprotective properties of phycobiliproteins. They also described that amino acids extracted from *Chlorococcum humicola* exhibit antifungal and antibacterial activity (Bhagavathy et al., 2011). In addition, phycobiliproteins have applications in food and cosmetic colorants (Eriksen, 2008). Most recently, the presence of Selenoprotein T and its structural elucidation, along with transcriptional analysis, was experimentally proved in microalgae *Scenedesmus quadricauda* (Reshma et al., 2020). Selenoproteins are a family of proteins that have selenocysteine, a non-standard amino acid residue as repeats. Selenoproteins have an important role in skeletal muscle regeneration, oxidative and calcium homeostasis, cell maintenance, thyroid hormone metabolism, and immune responses.

2.8.2 POLYSACCHARIDES

Polysaccharides from algae have potential applications in the food and pharmaceutical (cosmeceuticals, nutraceuticals) industries. Polysaccharides extracted from seaweeds have prebiotic properties. These compounds have antioxidant, antitumor, anti-inflammatory, antiviral, and antibacterial properties (Hayes, 2012). Some of the reported pharmacological activities of algal polysaccharides are given in Table 2.3 below. For instance, alginates, agar, and carrageenan are commonly used as functional ingredients, as stabilizers in the pharmaceutical industry. Algal polysaccharides are now extracted by using two common techniques, EAE and MAE. Polysaccharides extracted from algae are also useful to plants.

TABLE 2.3
Bio-Activities Present in Algal Polysaccharides

Algal Species	Active Compounds	Bioactivity	References
Chlorella stigmatophora	Crude polysaccharide	Anti-inflammatory & Immuno-modulating	(Guzmán et al., 2003)
Chlorella vulgaris	Crude polysaccharide	Anti-oxidant	(Mohamed, 2008)
Gyrodinium impudicum	p-KG03 exopolysaccharide	Anti-viral	(J. H. Yim et al., 2004)
Gyrodinium impudicum KG-03	Sulfonated polysaccharide	Anti-viral	(K. A. Lee et al., 2009)
Haematococcus lacustris	Water-soluble polysaccharide	Immuno-stimulating	(J. K. Park et al., 2011)
Navicula directa	Polysaccharide	Anti-viral	(Ee et al., 2006)
Phaeodactylum tricornutum	Crude polysaccharide	Anti-inflammatory & Immuno-modulating	(Guzmán et al., 2003)
Porphyridium sp.	Crude polysaccharide	Anti-oxidant	(Tannin-spitz et al., 2005)
Rhodella reticulata	Extracellular polysaccharide	Anti-oxidant	(B. Chen et al., 2010)
Scenedesmus quadricauda	Crude polysaccharide	Anti-oxidant	(Mohamed, 2008)

2.8.3 CAROTENOIDS

Carotenoids are pigments generally present in all algae, many photosynthetic bacteria, and higher plants. Their role is to protect organisms from light radiation in orange, yellow, or red wavelengths. Carotenoids are tetraterpene compounds by chemical nature, and they split into two classes; xanthophylls, which contain one or more oxygen molecules, and carotenes, unsaturated hydrocarbons. Green algae synthesize all the xanthophylls, which are synthesized by higher plants such as lutein, antheraxanthin, neoxanthin, violaxanthin, and zeaxanthin. Algae also synthesize canthaxanthin, astaxanthin, and loroxanthin. Fucoxanthin, diadinoxanthin, and Diatoxanthin can be produced by diatoms or brown algae (Amaro Helena et al., 2011).

β-Carotene was the first commercial product made from a microalgae *D.salina* (Borowitzka, 2013). β-carotene has large demand as pro-vitamin A (retinol); it is usually included in health and food formulation due to its therapeutic potential (Spolaore et al., 2006). Carotenoids act as a potential bioactive molecule, which can trigger the immune system, and are involved in combatting many life-threatening diseases like arthritis, coronary heart disease, cancer, and premature aging. Potential bioactivities of algal pigments are given in Table 2.4 below.

TABLE 2.4

Bioactivities Present in Algal Pigments

Algal Species	Active Compounds	Bioactivity	References
Chlorella pyrenoidosa	Lutein, violaxanthin	Anti-oxidant	(Plaza et al., 2009)
Chlorella vulgaris	Cantaxanthin, astaxanthin	Anti-oxidant	(Plaza et al., 2009)
Cryptomonads	Allophycocyanin	Anti-viral	(Shih et al., 2000)
Dunaliella salina	beta-carotene	Anti-oxidant	(Plaza et al., 2009)
Haematococcus pluvialis	Astaxanthin	Anti-tumor	(Palozza et al., 2009)
Haematococcus pluvialis	Astaxanthin, cantaxanthin, lutein	Anti-oxidant	(Plaza et al., 2009)
Haematococcus pluvialis	Astaxanthin	Anti-inflammatory	(Lee et al., 2003)

2.8.4 NUTRACEUTICALS

Algae contain a wide range of nutrients, including proteins, lipids, carbohydrates, and trace nutrients, including vitamins, antioxidants, and trace elements, which have features to serve as natural nutritional supplements in human and animal feed and have many promising health benefits. Nutrients of algae are rapidly attaining importance as a renewable source to substitute the conventional ingredients in the human diet or animal feed. Increased production of bioactive compounds from algae is considered to be an evolving area in the coming years. Different types of algae, in particular microalgae, which could be more applicable in food supplements and nutraceuticals, are *Spirulina, Chlorella, Haematococcus, Nostoc, Botryococcus, Anabaena, Chlamydomonas, Scenedesmus, Synechococcus, Parietochloris, Crypthecodinium, Porphyridium*, etc., due to the capability of producing necessary vitamins (Udayan et al., 2017). The biomass of *Chlorella* species contains Vitamin B complex, especially, α-carotene, lutein, α-tocopherol, ascorbic acid, β-carotene, and B12, and also produces polyunsaturated fatty acids and essential fatty acids. *Chlamydomonas reinhardtii* was employed in the manufacturing of pharmaceutical proteins such as immunoglobin, interferon β insulin, and erythropoietin (Scaife et al., 2015). Algae-based products are proven for combating numerous disorders and nutritional deficiencies, such as malnutrition, nutritional anemia, xerophthalmia (vitamin A deficiency), and endemic goiter (iodine deficiency). Consumption of algae promotes regular healthy functioning of major organ systems, including the humoral immune system, cardiovascular system, respiratory system, and nervous system (Udayan et al., 2017).

2.8.5 ANTIMICROBIAL COMPOUNDS

Increasing demand for pharmaceuticals in the worldwide market has increased potential use of algae. The primary and secondary metabolites extracted from algae can be utilized as raw materials for the pharmaceutical industry. Silver nanoparticles with antimicrobial properties were synthesized by the green synthesis method, exploiting the de-oiled *Saccharina japonica* powder that resulted after supercritical carbon dioxide ($Sc-CO_2$) extraction (Sivagnanam et al., 2017). The box-Behnken design was used to identify the influencing factors for the size of silver nanoparticles. The antimicrobial activity of biosynthesized silver nano-particles from de-oiled biomass was investigated against several pathogens, such as *Aspergillus brasiliensis, Candida albicans, Pseudomonas aeruginosa, Listeria monocytogenes, Staphylococcus aureus, Escherichia coli, Salmonella typhimurium,* and *Bacillus cereus,* using a well diffusion method. The antimicrobial effect from the spent biomass created a new platform in the development of novel antimicrobial drugs. The biomass obtained from *Dunaliella* remains the source of bioactive compounds like vitamins and enzymes. The extract obtained from the algae shows strong inhibitory action against the growth of harmful bacteria, which acts as good antimicrobials (Bhattacharjee, 2016).

2.8.6 Antioxidant Compounds

Antioxidant activity is defined as the prevention of oxidation of molecules by in-hibiting the formation of non-reactive stable radicals and the initiation step of oxidative chain reaction (Mittal et al., 2014). Nowadays, antioxidants play a vital role in the pharmaceutical and food industry. However, synthetic antioxidants pose danger to human health with their toxic nature. There is a need to discover a new source of natural inhibitors of the oxidation process, without any side effects. Algae are identified as a good source of antioxidant compounds. Fresh seaweed biomass contains glutathione (GSH) and ascorbate (vitamin C). Algae also produce sec-ondary metabolites exhibiting antioxidant activity, for instance, mycosporine-like amino acids (MAA), polyphenols (including catechins, lignans, tannins, flavo-noids), phlorotannins, carotenoids, and tocopherol.

Lipid-extracted spent biomass of *Acutodesmus dimorphus* was utilized for the biogenic production of silver nanoparticles with antioxidant activities (Maurya et al. 2016). The lipid-extracted spent biomass was used to synthesize microalgal water extract, which was later used for the synthesis of silver nanoparticles. The bio-synthesized Ag nanoparticles showed antioxidant potential, which was examined by using DPPH and ABTS free radicals scavenging assays. The silver nanoparticles solution showed proton-donating ability, which could efficiently serve as a free ra-dical scavenger and can act as a primary antioxidant molecule. Due to the diverse habitat of microalgae, they exhibit a significant role in the green synthesis of nano-particles that have pharmacological activities. Integration of nanotechnology and phycology leads to a new interdisciplinary approach called "phyco-nanotechnology".

2.9 VALORIZATION OF LIPID-EXTRACTED RESIDUAL BIOMASS FOR ENERGY RESOURCES

Biodiesel production from algae involves harvesting the biomass and extracting the lipids from the algal cell envelope, then converting it into biodiesel through trans-esterification (Mata et al., 2010). Some of the concerns are lack of non-standardized cultivation and harvesting methods, difficulties in cell lysis and lipid extraction protocols, and promising lipid-bearing algal strains, etc. To check this issue, the spent biomass from algae after the production of the first-generation biofuel can be utilized economically to manage the cost of the biofuels. A plenitude of strategies employed for the production of alternate microalgal biofuels and proper exploitation of spent biomass can help to solve the problems of low energy recovery and high processing costs. The significant possibilities on reuse of algal spent biomass for complimentary production of secondary biofuel will be scrutinized below.

2.9.1 BIODIESEL

Algal biomass is produced with high lipid content depending on the species. It has been reported that some species of microalgae have lipids up to 50–80% of their dry cell weight. The de-oiled or spent biomass rich in triglycerides is considered to be appropriate feedstock for biodiesel production. Algal biodiesel is third-generation fuel; it could be able to assimilate maximum CO_2 during the growth of algae. Therefore, it is a sustainable and effective solution for climate change (Patil et al., 2017). Transesterification of oils or fats resulted in biodiesel with a mixture of fatty acid alkyl esters, composed of 90–98% triglycerides, the minimal quantity of free fatty acids, monoglycerides, diglycerides, and the residual amount of carotenes, sulfur compounds, phospholipids, tocopherols, phosphatides, and water (Bozbas and Kahraman, 2008). The byproduct obtained during transesterification is glycerol, which could be effectively used as a carbon source for mixotrophic and hetero-trophic cultivation of algae for reusing this glycerol to produce oil for biodiesel (Behera et al., 2015).

Arora et al. (2016) experimented for the first time to report the use of recycled de-oiled algal biomass extract (DABE) as a growth media to enhance the lipid productivity in *Chlorella minutissima* as a cost-effective model for biomass pro-duction. Recycled DABE resulted in fast growth with a two-fold increase in bio-mass due to the presence of organic carbon (3.8 ± 0.8 g/l) compared to Bold's Basal Media (BBM). Cells that are cultivated in the recycled DABE expressed a four-fold increase in lipid productivity (126 ± 5.54 mg/l/d) compared to BBM. These cells showed large-sized lipid droplets mushrooming 54.12% of lipid content. In this regard, recycled DABE can add more value for the production of biodiesel by lipid enrichment.

2.9.2 BIOETHANOL

Bioethanol is a substitute for gasoline because of its indistinguishable chemical and physical properties (Sirajunnisa and Surendhiran, 2016). There is a food-versus-fuel dispute between the first- and second-generation biofuel resources and also the need for pretreatment to break down the recalcitrance of hemicellulose and lignin complex (Bibi et al., 2017). Due to these obstacles, industries are looking for a suitable feedstock resource for large-scale ethanol production that is available throughout the year without any competition with food. Scientists found that algae are potential organisms that could effectively act as raw material for third-generation biofuel. In the case of microalgae, starch content amounts to 60% dry weight (Dragone et al., 2011), with carbohydrate dominating up to 70–72% (Brányiková et al., 2011).

After synthesis of biodiesel from the microalgal biomass, the remaining spent biomass or oilcake is rich in carbohydrate and can be exploited for the production of ethanol via microbial fermentation after appropriate pretreatment (thermal, enzymatic, chemical, etc.). Physical pretreatments like bead-beating and sonication methods are operative in microalgal biomass. Bead-beating of *Chlorella vulgaris* cells with pectinase enzyme generates fermentable sugar release with 90% fermentation yield (O. K. Lee et al., 2013). *Dunaliella tertiolecta* de-oiled biomass can be used for bioethanol fermentation and chemo-enzymatic saccharification. By using hydrochloric acid for saccharification of the spent biomass, at 121°C for 15 min, and with enzymatic hydrolysis at 55°C, the result was 42.0% (w/w) and 29.5% (w/w) of reducing sugars, respectively. From the enzymatically saccharified spent biomass, *Saccharomyces cerevisiae* generates 0.14g ethanol per gram. Geun Goo et al. (2013) applied the defatted microalgae *Dunaliela tertiolecta* for isolation of exopolysaccharide (EPS), which could be used for the biotechnological creation of ethanol and glucose.

Mirsiaghi and Reardon (2015) used the one-step sulfuric acid process with 10% (w/w) acid concentration on the biomass of defatted *Nannochloropsis salina* at 90°C for five hours, which resulted in a maximum extraction yield of 243.2 mg of sugar released per gram of biomass, and then the generated hydrolysate was directly utilized for growing yeast without any additional nutrients. Pancha et al. (2016) achieved saccharification yields of 37.87% (w/w) by evaluating the chemo-enzymatic hydrolysis of de-oiled spent biomass of *Scenedesmus sp.* and also 43.44% for pretreatment of lipid-extracted spent biomass using Viscozyme L (20FBGU/g) and 0.513 M HCl.

2.9.3 BIO-OIL

Bio-oil can be produced by two different methods: biomass thermochemical liquefaction and biomass pyrolysis. The pyrolysis procedure is an anaerobic process under very high temperatures (200°C and 750°C). Pyrolysis of spent algal biomass might lead to the production of oil with quality between bio-oil and petroleum tar from lingo-cellulosic biomass along with mass yields of gas, oil, and bio-char of 281%, 282%, 441%, respectively (Miao et al., 2004). Pyrolysis may take place slowly or quickly. Quick pyrolysis produces bio-char and 19–58% of bio-oil (Miao et al., 2004). Slow pyrolysis results in bio-char and biogas with CO_2 and methane accounting for much gaseous product. During slow pyrolysis, 10–30% of biomass is renewed into gas, whereas 10–25% becomes char (Biller and Ross, 2011). Bio-oil extracted from microalgal spent biomass is more stable than those obtained from traditional crops; still, it is not as stable as fossil fuel (Mohan et al., 2006). This kind of bio-oil is composed of aromatic and aliphatic hydrocarbons, nitrogenous compounds, long-chain fatty acids, and phenols.

The composition of biomass and the processing conditions like pressure, temperature, catalyst, and time residence can adversely affect the profile of products. Depending on the composition of carbohydrates, the bio-oil yield can be up to 25% greater than the lipid content of original microalgae. For instance, *Spirulina sp.* produces bio-oil yield ranging up to 54% (Balasubramaniam et al., 2021).

Thermochemical liquefaction of biomass is an additional method for bio-oil production. This method requires heating the biomass under high pressure above 20 bar, at a temperature ranging from 200°C and 500°C in the presence of a suitable catalyst. This process resulted in 9–72% of bio-oil along with the gaseous mixture ranging from 6–20% (Biller and Ross, 2011). The remaining ash content ranges from 0.2–0.5%. The final product obtained from biomass liquefaction, when compared to the crude fuel, has a characterized ratio of water to solid of 10:1 leads to 37% yield of bio-oil (Zou et al., 2010). The aqueous co-product obtained from biomass liquefaction can be utilized in the microalgal culture, as it is rich in N, P, and K elements. This culture resulted in one-half of the growth rate in a medium containing 0.1% of aqueous co-product compared to the established media, BG11 (Jena et al., 2011).

2.9.4 BIOCHAR

Biochar is a solid material produced as the result of the carbonization of biomass. The biochar obtained through the combustion of spent algal biomass at 700°C has an ash content of 455%. Without the ash, the mass yields of gas, oil, and bio-char were 361%, 389%, and 245%, respectively (Torri et al., 2011). Bio-char can adsorb fatty acids, thus having an effective application in air and water purification systems as hydrophobic adsorbents. Though bio-char has a hydrophobic core, it gets wet by water due to the presence of hydrophilic functional groups, such as hydroxyl, aldehydes, and carboxylic acids on the surface. Due to these properties, it could be utilized as an additive in organic polymers and cement. It can be potentially exploited as a carbon source for the production of syngas or to replace coke in the steel manufacturing process due to its low ash content (Heilmann et al., 2011). Bio-char retained major portions of carbon (45%), nitrogen (80%) and phosphorous (100%), thus utilized as fertilizer to nourish the soil, which diminishes greenhouse gases while making it ready to transfer carbon-neutral energy into carbon-negative bioenergy. Since it is easily attacked by soil microbes, this material can be used in soil amendment. It can be easily hoarded in subterranean areas, thus necessitating a form of carbon sequestration (Heilmann et al., 2011).

2.9.5 BIOGAS

Biogas is used as an alternative energy source for supporting consumer and industrial activities, even being exploited as a renewable fuel for transportation. As fossil fuels are deteriorating at an alarming rate, the discovery of biogas remains an essential one in the current scenario, where it has been predicted the fossil fuel reserve will remain for only 50 years. To scale up biogas production to the industrial level, prominent sources have to be examined to optimize the extraction potential and biogas production. In addition, fermentation of spent biomass resulted in the production of biogas and also opened wide the doors for recycling the original nutrients in large proportions. The lipid-free residue in the spent biomass after biodiesel production could be anaerobically digested into biogas (Sialve et al., 2009).

2.9.6 BIOMETHANE

Spent algal biomass after lipid extraction and even biomass alone could be a potential raw material for biogas production through anaerobic digestion. Anaerobic digestion is a chemical process where anaerobic microbes react on complex organic substrate molecules to generate CO_2 (25–45%) and methane (55–75%) (Jankowska et al., 2017). Biogas is a mixture made up of CO_2 (20–40%), CH_4 (60–70%) and trace amounts of N_2, H_2S, and NH_3. There are various steps in biogas production, such as methanogenesis, acetogenesis, fermentation, and hydrolysis (Parsaee et al., 2019). Microalgae such as *Scenedesmus sp.* and *Isochrysis sp.* resulted in methane production with the yield of 400 ml CH_4/g VS (González-Delgado and Kafarov, 2011). The non-fermentable spent biomass during the production of bioethanol, encompasses the lipids, proteins, and organic acids, which could be effectively used as feedstock for biomethane synthesis up to 10% additionally; the cells may be ruptured to produce their enzymes or proteins as good byproducts (Suali and Sarbatly, 2012).

2.9.7 SYNGAS

Syngas, a mixture of hydrogen as well as carbon monoxide (CO), could be an alternative to fuel gas. It can be extracted by gasification of algal biomass by reacting carbonaceous compounds with steam, oxygen, or atmospheric air at high temperature (200°C to 700°C) in a gasifier (Suali and Sarbatly, 2012). This resulted in CO with yield from 9–52% and clean H_2 with yield ranging from 5–56%, (Abuadala et al., 2010). Methane is also obtained as a co-product with a yield from 2–25%. The hydrocarbon products obtained can be further processed to extract methanol; 64%(w/w) of methanol can be extracted at 1000°C on a biomass weight basis.

Gasification is a methodology for extracting biofuel from microalgae, where biomass is subjected to high temperature (>700°C) without combustion and steam for the production of traces of methane gas, hydrogen, CO_2, and a controlled amount of oxygen; this mixture is called syngas. Due to nitrogen content in microalgae, ammonia could be easily recovered in the aqueous phase during the process of gasification and can be exploited as a nitrogen source for microalgal cultivation (Dote et al., 1994). Catalytic gasification of de-oiled *Botryococcus braunii* was experimented with using Ni–Fe/Mg/Al catalyst at 750°C, in which 91% of carbon was transformed into gaseous products like CO_2, CH_4, and CO; 74.7 mmol/g biomass of H_2 yield was also obtained (Watanabe et al., 2015).

2.9.8 BIOHYDROGEN

Biohydrogen is considered to be the cleanest fuel because there is no CO_2 emission to the atmosphere during combustion. Hydrogen can be biologically produced from biomass via anaerobic fermentation, photo-fermentation, bio photolysis, or dark fermentation. It has more specific energy (142 MJ/kg) compared to other fuels such as gasoline (Phwan et al., 2018). These qualities made hydrogen a good option for next-generation fuel.

There has been some research done to valorize the spent algal biomass for the production of biohydrogen. Yang et al. (2011) conducted a thermo-alkaline pretreatment of *Scenedesmus* spent biomass after lipid extraction, and they optimized the production of biohydrogen, which resulted in higher yield, and the production rate of hydrogen increased from 160 and 500%, with 6.7% (w/v) solid content and 8 g/L dosages of NaOH in 2.5 hours as pretreatment time. Pan et al. (2010) reported various pretreatment methods, such as thermo-alkaline, alkaline, and thermal methods applied to de-oiled spent algal biomass in which the thermo-alkaline method was found to be most appropriate to increase the hydrogen production three-fold to 45.54 mL/g VS, in comparison with untreated DMB.

Venkata Subhash and Venkata Mohan (2014) examined the production of biohydrogen from the spent biomass of mixed microalgae by utilizing the acidogenic consortia of bacteria. They exploited acid-treated and untreated DMB to obtain specific hydrogen yields of 3.0, 3.3, 4.9, and 2.4 mol/kg COD of DMB and untreated DMB, respectively.

Microalgae are rich in proteins, while macroalgae are rich in carbohydrates. Thus, microalgae are not most suitable for fermentation due to their low C/N ratio. To alleviate this, N_2 rich micro-algae (*Nannochloropsis oceanica and Chlorella pyrenoidosa*) mixed with carbon-rich macroalgae (*Laminaria digitata*) with a C/N ratio of 20 was utilized in a two-stage batch co-fermentation process. This combination potentially generated about 97 mL of biohydrogen/g of volatile solids. The algal biomass could be transferred into biohydrogen by thermochemical methods such as gasification. *Fucus sp.* was subjected to supercritical water gasification and generated 1036 ml of biohydrogen/g of algae biomass, whereas biomass from *Saccharina latissima* could generate 15.1 mol of biohydrogen/kg of algae, which is comparatively higher than *Nannochloropsis oceanica* and *Chlorella pyrenoidosa* (Duman et al., 2014).

2.9.9 FEED SUPPLEMENT

Algae are widely recognized as a food supplement due to their significant nutritional content for the food source. It has been investigated that 75% of algal biomass is exploited for producing algal tablets, capsules, and powder for human and animal food. Algal biomass can serve as an alternative protein source for fish and animals (Chisti, 2008). The addition of a minimum amount of algal biomass to animal and fish diet can enhance their growth (Singh et al., 2011). The production of algal biomass for this purpose is not economically feasible; thus, spent algal biomass produced as waste would be a promising option.

Patterson and Gatlin (2013) attempted to replace the soy and fish meal protein isolates in the diet of *Sciaenops ocellatus* with de-oiled spent biomass of *Chlorella;* spent algal biomass could replace crude protein (5–25%) in the reference diet. When the soy and fishmeal protein was substituted at 20–25% with spent algal biomass, it affected survival, feed efficiency, and weight gain. The substitution level of 10% was effective, causing regular fish growth (Patterson and Gatlin, 2013).

Comparative studies on traditional food and poultry feed of algae resulted in low cholesterol, improved color of the yolk, and enhanced resistance to disease from the algal feed. Madeira et al. (2017) reported that upstream processes, such as cultivation, harvesting, and extraction, can significantly affect the feasibility of spent algal biomass to be used as feed for animals and fish (Bryant et al., 2012). After adopting several methods of harvesting (spray drying, belt drying, flocculation, and coagulation) and lipid extraction (hexane or ethanol), they quantified the mineral content of the lipid-extracted spent algal biomass of *Nannochloropsis oculata*. Na, S, and Cu are the minerals in spent biomass that limited the use of spent biomass as cattle feed. Spent biomass obtained after coagulation and flocculation contained Mn, P, Ca, Na, Al, and S. The reduction of S and Na in the spent algal biomass (SAB) can make it suitable for animal feed.

2.9.10 FERTILIZER

Intensive use of chemical fertilizer can cause a great impact on the environment and lead to the destruction of the local ecosystem and a drop in crop yield. Chemical fertilizers could jeopardize the whole agricultural land. Exploring an alternative source for biofertilizer is an essential shift toward viable production of crops in a pollution-free safe environment.

The de-oiled microalgal biomass can be applied as a soil fertilizer and conditioner. The major needs of the crop as macronutrients are nitrogen, phosphorus, and potassium, for growth and development. The spent biomass contains a high amount of protein codes for nitrogen. It is also rich in phosphorous and other minerals that could enhance the performance of spent biomass as a fertilizer. The spent biomass itself is extracted from the photosynthetic source, so it is obvious that it has the requisite amount of nutrient source for crops, such as magnesium, an essential moiety of chlorophyll.

Gupta et al. (2012) documented the utilization of spent biomass to reclaim barren soil. Maurya et al. (2016) divulged the utilization of spent algal biomass fertilizer as a substitution of chemical fertilizer and to promote growth. Bohutskyi et al. (2015) reported that 30–60% of S, Ca, Mg, N, and P, and 15–25% of Fe and Mn remained for recycling after biogas synthesis from *Auxenochlorella protothecoides* spent biomass. Recently, it was documented that the spent biomass of *L. majuscule* and *C.vulgaris* could be employed as a fertilizer substitute for *Zea mays L.* Besides, they have estimated that the fertilizer cost might be reduced up to 45% by nutrient recycling with biogas slurry.

Nayak et al. (2019) published a report on improved and sustainable agriculture of rice crops (cv. IR 36) by valorizing the spent biomass after lipid extraction of *Scenedesmus sp.,* as a biofertilizer. This biofertilizer exhibits superiority over the vermicompost and commercial chemical fertilizer concerning the yield and growth of rice plants. In their experiment, five nutrient management treatments (CF100, VC100, MA100, MA50 + CF50, and MA50 + VC50) met the 100% nitrogen demand in the soil. In addition, the combination of microalgal fertilizer with chemical fertilizer (MA50+CF50) resulted in greater performance in terms of tiller number, plant height, grain yield, and biomass due to the increased amount of nutrients, such as N, P, and K. During the harvest stage, this combination also produced up to 1000 grain weight, panicle weight, and plant dry weight compared to other treatments. The nutritional quality of rice was also enhanced due to the presence of protein and amylose content. They concluded that the use of spent algal biomass after lipid extraction is effective and potentially sustainable in increasing the yield of rice crop with minimum dependence on chemical fertilizers and also improved the agronomic competence as well as developed the economy of algal-based biofuel production.

2.9.11 LACTIC ACID

During algal biodiesel production, the spent biomass/algal cake produced as a waste byproduct is utilized as a fermentable substrate for the production of value-added chemicals in the bio-refinery loop. Overbeck et al. (2016) investigated the ability of *Lactobacillus casei* 12 A to ferment the spent biomass for the co-production of lactic acid. The nutritional requirements for amino acid and carbohydrate were attained through enzymatic hydrolysis of spent biomass using pepsin, cellulase (endo-1,4-b-D-glucanase), and α-amylase enabled rapid growth. Alternatively, heterologous expression of cellulose and α-amylase by *Lactobacilli* have been reported earlier (Narita et al., 2006) and can provide a potential means for optimizing lactic acid production from the spent biomass (algal cake) without any enzymatic pretreatments or costly chemicals. Production of value-added chemicals as a byproduct of algal cake fermentation effectively manages the cost competitiveness of biodiesel production from algae.

In another approach towards de-oiled spent algal biomass utilization, defatted spent biomass from *Nannochloropsis salina* were pretreated with acid-catalyzed hot water extraction and exploited as a growth substrate for oleaginous yeast *Cryptococcus sp.* obtained final lipid content 23 ± 2.2% (Seo et al., 2015). Similarly, Talukder et al. (2012) examined the synthesis of lactic acid from lipid-free *Nannochloropsis salina* biomass using *Lactobacillus pentosus*.

2.10 VALORIZATION OF AGAR-EXTRACTED RESIDUAL BIOMASS

Agar is one of the important phycocolloids utilized as gelling agents in cosmetics, biotechnology, and the food industry. It is generally extracted from agarophytes. It is also called red seaweed. Agar is commonly extracted from two algae with the genera *Gracilaria* and *Gelidium*. Agar is composed of agaropectin and agarose and is exploited as a stabilizer for suspension and emulsion. Agar is a polysaccharide unit composed of agarobiose units, i.e., 3,6-anhydro-4-O-β-D-galactopyranosyl-L-galactose. Around 90% of agar produced is exploited for food applications, and the remaining is utilized as a medium for cell culture and for medical applications.

After the agar is successfully extracted from these two commercial sources, *Gracilaria* and *Gelidium,* through various methods – for instance, Agar Traditional Water Extraction (TWE) and MAE – the remaining spent biomass is valorized for producing multiple value-added products.

2.10.1 Bioactive Compounds

Algae act as a good source for the production of various bioactive compounds. Several methods can be employed to extract bioactive molecules from algal biomass. The novel extraction techniques are PLE, SFE, EAE, UAE, and MAE. The main advantage of these methods is they extract the bioactive compounds without much loss of their biological activity and are eco-friendlier than the traditional methods of extraction.

Agar extraction in industries consists of several steps. Initially, alkaline treatment was carried out for desulfation of polymers for improving the gelling properties. Before solubilization of agar, several rinses with hot water under pressure are necessary. This process generates a high amount of spent biomass, which could generate a high number of byproducts. Algae contain a wide array of bioactive compounds, such as polysaccharides, carotenoids, vitamins [B group], polyunsaturated fatty acids [PUFAs], polyphenols, proteins, dietary fiber, and plant growth regulating compounds like cytokinin, gibberellins, and auxin (Hayes, 2012). Several food processing by-products from algae are an important and potential source of valuable compounds. Among them, commonly found low molecular weight carbohydrates (LMWC) in red algae are floridoside (2-O-α-Dgalactopyranosylglycerol) and isofloridoside (α-D-galactopyranosyl-(1-1)-glycerol.

Lebbar et al. (2018) designed a fractionation process to obtain the oligo-saccharidic fraction in order to isolate glycerol-galactosides. They reported that the spent biomass of agar extraction from *G. sesquipedale* could represent a new source of value-added molecules. As the smallest of glycerol-galactoside family, florido-side was already described. They reported a series of Gal2–4-glycerol harboring β1→3 and/or β1→4 linkages between galactose residues were found and constituted original derivatives of galactosyl-glycerol in algae or plant kingdoms. This compound floridoside is involved in osmoregulation and also acts as a precursor of cell wall polysaccharides.

2.10.2 Fatty Acids

Lipids and valuable materials can be effectively extracted from algae. Microalgae are a prominent source of commercial high-value chemicals, including PUFAs. It has been investigated that PUFAs play a vital role in the functioning of the human body due to their antioxidant, antifungal, and antibacterial properties (Ibañez et al., 2012). The lipid fraction primarily consists of PUFAs, for instance, ω-6 fatty acid: arachidonic acid and γ-linolenic acid, and ω-3 fatty acids: docosahexaenoic acid (DHA) and eicosapentaenoic acid (EPA). Marine algae are a well-defined source of lipids with greater nutritional value and with significant bioactive properties, such as phospholipids, ω-3 and ω-6 polyunsaturated fatty acids, and glycolipids. The

TABLE 2.5
Bioactivities Present in Algal Lipids

Algal Species	Active Compounds	Bioactivity	References
Chlorella stigmatophora	Sterols	Anti-inflammatory	(Gato et al., 2001)
Chlorococcum HS-101	α-Linolenic acid	Antimicrobial	(Dote et al., 1994)
Haematococcus pluvialis	Short chain fatty acids	Antimicrobial	(Cifuentes et al., 2010)
Phaeodactylum tricornutum	Eicosapentaenoic acid	Antibacterial	(Desbois et al., 2009)
Skeletonema costatum	Saturated long chain fatty acid	Antibacterial	(Naviner et al., 1999)

nitrogen stress-mediated targeted biochemical and metabolomic changes can lead to lipid accumulation in *Scenesesmus quadricauda* (Sulochana and Arumugam 2020). It has been investigated that inorganic carbon source addition can improve the biomass yield, lipid accumulation, and fatty acid composition in *S. quadricauda* species under nitrogen starved condition (Anusree et al., 2017). Some of the algal strains with their reported pharmacological activities are given in Table 2.5.

Nonpolar or neutral lipids other than triacylglycerides, such as fatty acids and sterols have a great interest in the valorization of microalgae in biofuel production. The spent biomass after agar extraction could yield a good amount of lipids and oils. Yuan et al. 2020 reported that During the synthesis of agar from the red seaweed *Gracilaria lemaneiformis*, the lipid content demolished and left out thing as spent biomass, can be valorized as a suitable alternative feedstock for lipid production. Ethanol can be used as an extraction solvent in the green UAE method for the enrichment of lipids from the spent biomass. It resulted in a maximum yield of lipids (0.71 wt%) than the conventional Soxhlet extraction (CSE) (0.44 wt%). The major fatty acid compositions were arachidonic acid (9.44–15.12%), oleic acid (1.85–9.43%), palmitic acid (59.20–63.25%), while the free fatty acid content of the ultrasound-assisted extraction (94.67%) was comparatively larger than that of the conventional Soxhlet extraction (73.12%). Guo et al. (2017) reported that lipids of *Gracilaria lemaneiformis* exhibit tyrosine phosphatase 1B inhibitory activity, these lipids could be convenient in cosmetics, pharmaceuticals, and food industries

2.10.3 BIO-ETHANOL

Macroalgae are more applicable as ethanol feedstock due to their rich source of polysaccharide content such as glucan and mannitol, which could be effectively converted into ethanol by industrial yeast (Roesijadi et al., 2008). Marine macroalgae (seaweeds) are a potential source of hydrocolloids and are used to extract biofuels. Recently, the production of bioethanol has been encouraged from algal sources rather than crops. This can be achieved through the fermentation of its sugar, starch, and cellulose. Sudhakar et al. (2016) found that macroalgae spent biomass collected from alginate manufacture industry were found to have a considerable amount of phycocolloids. They have reported that the spent biomass of red seaweed of *Gracilaria corticata var corticata* exhibits greater polysaccharide content (190.71 ± 30.67 mg/g dry weight). They found that ethanol can be effectively produced through a fermentation process using the spent biomass with baker's yeast at pH 5.3. the ethanol yield from the red seaweed spent biomass was 0.02 ± 0.003 g/g compared with standard YPD (0.42 ± 0.03 g/g) and D-galactose (0.37 ± 0.04 g/g) respectively. The spent biomass after ethanol production could be recycled, and the CO_2 obtained can be fed to the culture of microalgae, while nonfermentable cellulose can be further utilized as a feed supplement for anim als (Suali and Sarbatly, 2012). Thus, the agar extracted spent biomass from the seaweed industry can be effectively utilized in bioethanol production systems.

2.10.4 PAPER PRODUCTS

China is one of the leading producers of seaweed aquaculture. *Gracilariopsis lemaneiformis* farming in the fifth largest algal cultivation industry in China resulted in the production of millions of tons annually. They were primarily utilized for agar extraction in agar mills (Miao et al., 2004). The extraction of agar from *Gracilariopsis lemaneiformis* generates a huge amount of spent biomass, which is not valorized, and they are often cleared into the pubic environment (Seo et al., 2015). It has been suggested that certain species of algae *Gracilariopsis* could be used as a good raw material for paper manufacture. Evaluation of the paper handsheets obtained from the spent biomass resulted in lower strength and permeability due to large quantities of algal material, but it has higher greaseproof and waterproof characteristics, as well as stronger microbial effects. This spent biomass could be utilized as functional fillers to generate paper products, which have powerful applications in the food packaging industry.

2.10.5 CELLULOSE NANOCRYSTALS

Cellulose is one of the most plentiful compounds on the earth. It is widely used for several industrial applications due to its unique characteristics, such as biodegradability, cost-effectiveness, light weight, stiffness and high tensile strength, renewability, and environmental benefits. Cellulose is naturally present in marine animals, invertebrates, fungi, marine biomass, bacteria, and plants, among others. Cellulose can be produced from cellulose-rich materials in different forms, such as fibers, microfibers, nanofibers, and nanocrystals, by exploiting a top-down process.

CNC can be extracted by several methods, such as TEMPO-mediated oxidation, enzyme-assisted hydrolysis, mechanical disintegration, and acid hydrolysis (Trache et al., 2016). Among these methods, the acid hydrolysis procedure remains the most effective method. Recently, CNC was potentially extracted from the marine biomass of *Gelidium elegans* and *Posidonia oceanica* (Bettaieb et al., 2015). Algal biomass usually contains a minimum amount of natural physicochemical barriers, which led to cellulose available without severe chemical pretreatment. It has a higher yield of carbohydrates and grows faster than classic terrestrial lignocellulosic biomass (Y. W. Chen et al., 2016). Hence, marine algal biomass is considered to be a valuable resource for the production of cellulose fibers and their derivatives, such as CNC.

The utilization of red algae, *Gelidium sesquipedale,* in the agar industry for the synthesis of agar products generates a greater amount of spent biomass with fibrous content, which could cause environmental problems (Yarnpakdee et al., 2015). The agar-extracted biomass can be used as fertilizer or a soil conditioner in many coastal regions (Cocozza et al., 2011).

El Miri et al. (2016) explored the re-valorization of the agar extracted spent biomass for isolation of high-quality CNC using alkali, bleaching, and acid hydrolysis. The length and diameter of the isolated needle-like shaped CNC ranged from 285.4 ± 36.5 to 315.7 ± 30.3 nm, and from 5.2 ± 2.9 to 9.1 ± 3.1 nm and based on the hydrolysis time (30, 40, and 80 minutes); the crystallinity index ranged between 81–87%. After extraction of CNC, it was used to reinforce nanofillers for polymer-based nanocomposites manufacture using polyvinyl alcohol (PVA) as a polymeric matrix. The solvent casting method was employed for the production of PVA-CNC nanocomposites. It was reported that young's module was increased by 215%, the toughness by 45%, and tensile strength by 150% when 8wt % CNC was added into PVA matrix. Additionally, PVA-CNC nanocomposite films maintained the same level of optical transparency, indicating that the CNC was disseminated at the nanoscale. The nanocomposite films obtained with superior properties could be potentially used in food packaging.

2.11 FUTURE RESEARCH DIRECTIONS

- For competitiveness of the algal-based scenario, algal-based biorefinery should be implemented as an integrated approach by the industries, thus optimizing the economic return of all the compounds present in the algal biomass, and resulting in zero residues.
- Recently, several researchers reported on various algal biorefinery approaches that focus on integrated valorization of algal spent biomass utilization and bioenergy production.
- Microalgal bioenergy production is one of the clean and sustainable processes; however, there are many factors that hinder its commercialization, such as selection of the potential algal strain, the extraction methodology for bulk production, the requirement of water, harvesting of biomass, availability of feedstock in winter and rainy seasons, and selection of a suitable biofuel conversion process.
- Extensive cultivation of algae and potential application of spent biomass could play a crucial role in the process of multilateral alternative energy developments to address the CO_2 emission objectives of the Kyoto Protocol and the Copenhagen Agreement.
- Focused research is needed for optimizing the algal-based biodiesel production on a large scale.

- Freshwater-based algal biorefinery is not a wise option due to the foreseen demand for freshwater in the near future. In the future algal biomass, cultivation must depend on wastewater for biomass production of low value-added products like biogas, biodiesel, and bioethanol. Seawater could be efficiently used for algal cultivation to extract high-value products like pigments, animal feed, and PUFAs from the algal biomass.
- Algal bio-refineries should be attached with seawater and wastewater utilization besides CO_2 from industries with complete utilization of the biomass to balance the sustainability and viability of the technology. The challenges about the algal-based biofuel sector entail several associated benefits, which may contribute to economic competitiveness.
- The exploitation of SCFA synthesis facilitates effective carbon fixation for biopolymer, and this addresses the plastic management issues.
- An intensive analysis of the present scenario pointed out that it is challenging to produce algal biofuel to the stage where it can completely replace conventional fossils fuels, either in developed or developing economies. Authorities must indeed follow an affirmative action to fund R&D efforts, subsidize investment, and encourage consumption of renewable energies as well as enforce carbon taxes to limit the consumption of fossil fuels.

2.12   CONCLUSION

There is great demand for energy sources due to the increasing population size and several environmental issues triggered by fossil fuels. These factors urged the development of algal-based biofuel. The integrated approach of algal biofuel production, along with the valorization of spent biomass, can increase the economy of the process. Utilization of spent algal biomass after lipid extraction is a cost-effective way to make algal-based biofuels sustainable. The spent algal biomass with high carbohydrate content can be potentially utilized as renewable feedstock. Spent algal biomass could be utilized as fermentation substrate to produce value-added products, and it has huge applications from feed formulations, through secondary biofuels, bio-fertilizer, and fine chemicals, to the bioremediation process; it has great economic interest. The powerful, lucrative biorefineries should be developed to optimize the production of value-added co-products from the biomass.

ACKNOWLEDGMENTS

The work was supported by the grant from the Council of Scientific and Industrial Research (CSIR), Government of India through the theme Agriculture, Nutrition and Biotechnology (ANB) "MLP-0051: Investigation on the separation, composition and utilization of deoiled microalgal biomass as value added nutraceuticals" to Muthu Arumugam.

Brányiková, I., Maršálková, B., Doucha, J., Brányik, T., Bišová, K., Zachleder, V., Vítová, M. 2011. Microalgae-novel highly efficient starch producers. *Biotechnol. Bioeng.* 108 (4), 766–776. 10.1002/bit.23016.

Brennan, L., Owende, P. 2010. Biofuels from microalgae – A review of technologies for production, processing, and extractions of biofuels and co-products. *Renew. Sustain. Energy Rev.* 14(2), 557–577. 10.1016/j.rser.2009.10.009. Pergamon.

Bryant, H.L., Gogichaishvili, I., Anderson, D., Richardson, J.W., Sawyer, J., Wickersham, T., Drewery, M.L. 2012. The value of post-extracted algae residue. *Algal Res.* 1(2), 185–193. 10.1016/j.algal.2012.06.001.

Cai, T., Park, S.Y., Li, Y. 2019. Nutrient recovery from wastewater streams by microalgae: Status and prospects nutrient recovery from wastewater streams by microalgae: Status and prospects. *Renew. Sustain. Energy Rev.* 19 (November), 360–369. 10.1016/j.rser.2012.11.030.

Caporgno, M.P., Mathys A. 2018. Trends in microalgae incorporation into innovative food products with potential health benefits. *Front. Nutr.* 5, 58. 10.3389/fnut.2018.00058

Chen, B., You, W., Huang, J. 2010. Isolation and antioxidant property of the extracellular polysaccharide from Rhodella Reticulata. *World J. Microbiol. Biotechnol.* 26(5), 833–840. 10.1007/s11274-009-0240-y

Chen, Y., Ho, S.H., Nagarajan, D., Ren, N., Chang, J.S. 2018. Waste biorefineries – Integrating anaerobic digestion and microalgae cultivation for bioenergy production. *Curr. Opin. Biotechnol.* 50, 101–110. 10.1016/j.copbio.2017.11.017

Chen, Y. Wei, H.V. Lee, J.C.J., Phang, S.M. 2016. Production of new cellulose nanomaterial from red algae marine biomass Gelidium elegans. *Carbohydr. Polym.* 151 (October), 1210–1219. 10.1016/j.carbpol.2016.06.083.

Chisti, Y. 2008. Biodiesel from microalgae beats bioethanol. *Trends Biotechnol.* 26(3), 126–131. 10.1016/j.tibtech.2007.12.002.

Cian, R.E., Martínez-Augustin, O., Drago, S.R. 2012. Bioactive properties of peptides obtained by enzymatic hydrolysis from protein byproducts of Porphyra columbina. *FRIN.* 49(1), 364–372. 10.1016/j.foodres.2012.07.003.

Cifuentes, A., Santoyo S., Suarez, S., Jaime, L., Rodrı, I., Sen, F.J., Iba, E. 2010. LWT – Food science and technology pressurized liquids as an alternative process to anti-oxidant carotenoids' extraction from Haematococcus pluvialis microalgae 43, 105–112. 10.1016/j.lwt.2009.06.023.

Cocozza, C., Parente, A., Zaccone, C., Mininni, C., Santamaria, P., Miano, T. 2011. Comparative management of offshore Posidonia residues: Composting vs. energy recovery. *Waste Manag.* 31 (1), 78–84. 10.1016/j.wasman.2010.08.016.

Demirbas, A. 2009. Biorefineries: Current activities and future developments. *Energy Convers. Manag.* 50(11), 2782–2801. 10.1016/j.enconman.2009.06.035.

Desbois, A.P., Mearns-Spragg, A., Smith, V.J. 2009. A fatty acid from the diatom Phaeodactylum tricornutum is antibacterial against diverse bacteria including multi-resistant Staphylococcus aureus (MRSA). *Mar. Biotechnol.* 11(1), 45–52. 10.1007/s10126-008-9118-5

Dote, Y., Sawayama, S., Inoue, S., Minowa, T., Yokoyama, S. 1994. Recovery of liquid fuel from hydrocarbon-rich microalgae by thermochemical liquefaction. *Fuel.* 73(12), 1855–1857. 10.1016/0016-2361(94)90211-9.

Dragone, G., Fernandes, B.D., Abreu, A.P., Vicente, A.A., Teixeira, J.A. 2011. Nutrient limitation as a strategy for increasing starch accumulation in microalgae. *Appl. Energy.* 88 (10), 3331–3335. 10.1016/j.apenergy.2011.03.012.

Duman, G., Uddin, M.A., Yanik, J. 2014. Hydrogen production from algal biomass via steam gasification. *Bioresour. Technol.* 166, 24–30. 10.1016/j.biortech.2014.04.096.

ElFar, O.A., Chang, C.K., Leong, H.Y., Peter, A.P., Chew, K.W., Show, P.L. 2020. Prospects of Industry 5.0 in algae: Customization of production and new advance technology for clean bioenergy generation. *Energy Convers. Manag. X.* 10 (June), 100048. 10.1016/j.ecmx.2020.100048.

Eriksen, N.T., 2008. Production of phycocyanin – A pigment with applications in biology, biotechnology, foods and medicine. *Appl. Microbiol. Biotechnol.* 80(1), 1–14. 10.1 007/s00253-008-1542-y

Fabris, M., Abbriano, R.M., Pernice, M., Sutherland, D.L., Commault, A.S., Hall, C.C., Labeeuw, L., et al. 2020. Emerging technologies in algal biotechnology: Toward the establishment of a sustainable, algae-based bioeconomy. *Front. Plant Sci.* 11, 279. 10.3389/fpls.2020.00279

Gato, A, Calleja, J.M., Guzma, S. 2001. Antiinflammatory, analgesic and free radical scavenging activities of the marine microalgae Chlorella stigmatophora and Phaeodactylum tricornutum. *Phytother Res.* 15(3), 224–230.

Gómez-Ordóñez, E., Jiménez-Escrig, A., Rupérez, P. 2010. Dietary fibre and physico-chemical properties of several edible seaweeds from the Northwestern Spanish Coast. *FRIN.* 43(9), 2289–2294. 10.1016/j.foodres.2010.08.005.

González-Delgado, Á.D., Kafarov, V. 2011. Microalgae based biorefinery: Issues to con-sider. *CTyF – Ciencia, Tecnologia y Futuro.* 4(4), 5–22. 10.29047/01225383.225

Goo G., Bon, G.B., Choi, D.J., Park, Y.I., Synytsya, A., Bleha, R., Seong, D.H., Lee, C.G., Park, J.K. 2013. Characterization of a renewable extracellular polysaccharide from defatted microalgae Dunaliella tertiolecta. *Bioresour. Technol.* 129, 343–350. 10.1016/j.biortech.2012.11.077.

Gouveia, L., Neves, C., Sebastião, D., Nobre, B.P., Matos, C.T. 2014. Effect of light on the production of bioelectricity and added-value microalgae biomass in a photosynthetic alga microbial fuel cell. *Bioresour. Technol.* 154, 171–177. 10.1016/j.biortech.2013.12.049.

Guo, X., Gu, D., Wang, M., Huang, Y., Li, H., Dong, Y., Tian, J., Wang, Y., Yang, Y. 2017. Characterization of active compounds from: Gracilaria lemaneiformis inhibiting the protein tyrosine phosphatase 1B activity. *Food Funct.* 8(9), 3271–3275. 10.1039/c7fo00376e.

Gupta, S., Abu-Ghannam, N. 2011. Bioactive potential and possible health effects of edible brown seaweeds. *Trends Food Sci. Technol.* 22(6), 315–326. 10.1016/j.tifs.2011.03.011

Gupta, S., Gupta, S., Sharma, S. 2012. Management of barren land soil using waste algal residue and agricultural residue. *J. Algal Biomass Utln.* 9(6), 739–752

Guzmán, S., Gato, A., Lamela, M., Freire-Garabal, M., Calleja, J. M. 2003. Anti-inflammatory and immunomodulatory activities of polysaccharide from Chlorella stigmatophora and Phaeodactylum tricornutum. *Phytother. Res.* 17(6), 665–670. 10.1 002/ptr.1227.

Halim, R., Danquah, M.K., Webley, P.A. 2012. Extraction of oil from microalgae for biodiesel production: A review. *Biotechnol. Adv.* 30(3), 709–732. 10.1016/j.biotechadv.2012.01.001

Hariskos, I., Posten C. 2014. Biorefinery of microalgae – Opportunities and constraints for different production scenarios. *Biotechnol. J.* 9(6), 739–752. 10.1002/biot.201300142

Hayes, M. 2012. *Marine Bioactive Compounds: Sources, Characterization and Applications.* Springer Science and Business Media. 9781461412. 10.1007/978-1-4614-1247-2

Heilmann, S.M., Jader, L.R., Harned, L.A., Sadowsky, M.J., Schendel, F.J., Lefebvre, P.A., von Keitz, M.G., Valentas, K.J., 2011. Hydrothermal carbonization of microalgae II. Fatty acid, char, and algal nutrient products. *Appl. Energy.* 88(10), 3286–3290. 10.101 6/j.apenergy.2010.12.041.

Ibanez, E., Herrero, M., Mendiola, J.A., Castro-Puyana, M. 2012. Extraction and char-acterization of bioactive compounds with health benefits from marine resources: Macro and micro algae, cyanobacteria, and invertebrates, in: *Marine Bioactive Compounds: Sources, Characterization and Applications.* Springer, Boston, MA, pp. 55–98, 9781461412472, 10.1007/978-1-4614-1247-2_2

Jankowska, E., Sahu, A.K., Oleskowicz-Popiel, P. 2017. Biogas from microalgae: Review on microalgae's cultivation, harvesting and pretreatment for anaerobic digestion. *Renew. Sustain. Energy Rev.* 75, 692–709. 10.1016/j.rser.2016.11.045. Springier, Boston, MA.

Javad, M., Moi, S., Sohrabipoor, J., Fazeli, F. 2019. The bio-methane potential of whole plant and solid residues of two species of red seaweeds: Gracilaria manilaensis and Gracilariopsis persica. *Algal Res.* 42(November), 101581. 10.1016/j.algal.2019.101581.

Jena, U., Vaidyanathan, N., Chinnasamy, S., Das, K.C., 2011. Evaluation of microalgae cultivation using recovered aqueous co-product from thermochemical liquefaction of algal biomass. *Bioresour. Technol.* 102(3), 3380–3387. 10.1016/j.biortech.2010.09.111.

John, R.P., Anisha, G.S., Nampoothiri K.M., Pandey A. 2011. Micro and macroalgal biomass: A renewable source for bioethanol. *Bioresour. Technol.* 102(1), 186–193. 10.1016/j.biortech.2010.06.139.

Jones, C.S., Mayfield S.P., 2012. Algae biofuels: Versatility for the future of bioenergy. *Curr. Opin. Biotechnol.* 10.1016/j.copbio.2011.10.013.

Kadam, S.U., Tiwari, B.K., O'Donnell, C.P., 2013. Application of novel extraction technologies for bioactives from marine algae. *J. Agric. Food Chem.* 61(20), 4667–4675. 10.1021/jf400819p

Kannengiesser, J., Sakaguchi-Söder, K., Mrukwia, T., Jager, J., Schebek, L., 2016. Extraction of medium chain fatty acids from organic municipal waste and subsequent production of bio-based fuels. *Waste Manag.* 47(Pt A), 78–83. 10.1016/j.wasman.2015.05.030.

Khan, W., Rayirath U.P., Subramanian S., Jithesh M.N., Rayorath P., Mark Hodges D., Critchley A.T., Craigie J.S., Norrie J., Prithiviraj B.. 2009. Seaweed extracts as biostimulants of plant growth and development. *J. Plant Growth Regul.* 28(4), 386–399. 10.1007/s00344-009-9103-x

Krimpen, M.M. van, Bikker P., van der Meer I.M., van der Peet-Schwering C.M.C., Vereijken J.M. 2013. Cultivation, processing and nutritional aspects for pigs and poultry of European protein sources as alternatives for imported soybean products. Wageningen UR Livestock Research. https://research.wur.nl/en/publications/cultivation-processing-and-nutritional-aspects-for-pigs-and-poult.

Kumar, R.R., Rao, P.H., Arumugam, M., 2015. Lipid extraction methods from microalgae: A comprehensive review. *Front. Energy Res.* 10.3389/fenrg.2014.00061. Frontiers Media S.A., Lausanne, Switzerland.

Kumari, P., Kumar, M., Reddy, C.R.K., Jha, B. 2013. Algal lipids, fatty acids and sterols, in: *Functional Ingredients from Algae for Foods and Nutraceuticals*, Elsevier Ltd, Amsterdam, Netherlands. pp. 87–134 10.1533/9780857098689.1.87.

Lee, J.L., Ayashi, K.H., Irata, M.H., Uroda, E.K., Uzuki, E.S. 2006. Antiviral sulfated polysaccharide from Navicula directa, a diatom collected from deep-sea water in Toyama bay. *Biol. Pharm. Bull.* 29 (10), 2135–2139.

Lee, S.J., Bai, S.K., Lee, K.S., Namkoong, S., Na, H.J., Ha, K.S., Han, J.A., Yim, S.V., Chang, K., Kwon, Y.G., Lee, S.K. 2003. Anti-Inflammatory. *Mol. Cells* 16(1), 97–105.

Lebbar, S., Fanuel, M., Gall, S.L., Falourd, X., Ropartz, D., Bressollier, P., Gloaguen, V., Faugeron-Girard, C. 2018. Agar extraction by-products from Gelidium Sesquipedale as a source of glycerol-galactosides. *Molecules.* 23(12). 10.3390/molecules23123364.

Lee, K.A., Zhang, H., Qian, D.Z., Rey, S., Liu, J.O., Semenza, G.L. 2009. Acriflavine inhibits HIF-1 dimerization, tumor growth, and vascularization. *Proc. Natl. Acad. Sci. U.S.A.* 106(42), 17910–17915. 10.1073/pnas.0909353106.

Lee, O.K., Leum Kim, A., Seong, D.H., Lee, C.G., Jung, Y.T., Lee, J.W., Lee, E.Y. 2013. Chemo-enzymatic saccharification and bioethanol fermentation of lipid-extracted residual biomass of the microalga, Dunaliella Tertiolecta. *Bioresour. Technol.* 132, 197–201. 10.1016/j.biortech.2013.01.007.

Lyons, H. 2009. A review of the potential of marine algae as a source of biofuel in Ireland. Sustainable Energy Ireland, 1–88.

Madeira, M.S., Cardoso, C., Lopes, P.A., Coelho, D., Afonso, C., Bandarra, N.M., Prates, J.A.M. 2017. Microalgae as feed ingredients for livestock production and meat quality: A review. *Livest. Sci.* 205, 111–121. 10.1016/j.livsci.2017.09.020. Elsevier B.V.

Mata, T.M., Martins, A.A., Caetano, N.S. 2010. Microalgae for biodiesel production and other applications: A review. *Renew. Sustain. Energy Rev.* 14(1), 217–232. 10.1016/j.rser.2009.07.020. Pergamon.

Maurya, R., Chokshi, K., Ghosh, T., Trivedi, K., Pancha, I., Kubavat, D., Mishra, S., Ghosh, A. 2016. Lipid extracted microalgal biomass residue as a fertilizer substitute for Zea Mays L. *Front. Plant Sci.* 6 (January), 1266. 10.3389/fpls.2015.01266.

Maurya, R., Paliwal, C., Ghosh, T., Pancha, I., Chokshi, K., Mitra, M., Ghosh, A., Mishra, S. 2016. Applications of de-oiled microalgal biomass towards development of sustainable biorefinery. *Bioresour. Technol.* 214, 787–796. 10.1016/j.biortech.2016.04.115.

Miao, X., Wu Q., Yang C. 2004. Fast pyrolysis of microalgae to produce renewable fuels. *J. Anal. Appl. Pyrolysis.* 71(2), 855–863. 10.1016/j.jaap.2003.11.004.

Michalak, I., Chojnacka K. 2015. Algae as production systems of bioactive compounds. *Eng. Life Sci.* 15(2), 160–176. 10.1002/elsc.201400191.

Miri, N.E., Achaby M.E., Fihri A., Larzek M., Zahouily M., Abdelouahdi K., Barakat A., Solhy A. 2016. Synergistic effect of cellulose nanocrystals/graphene oxide nanosheets as functional hybrid nanofiller for enhancing properties of PVA nanocomposites. *Carbohydr. Polym.* 137 (February), 239–248. 10.1016/j.carbpol.2015.10.072.

Mirsiaghi, M., Reardon K.F. 2015. Conversion of lipid-extracted nannochloropsis salina biomass into fermentable sugars. *Algal Res.* 8(8), 145–152. 10.1016/j.algal.2015.01.013.

Mittal, A.K., Bhaumik J., Kumar S., Banerjee U.C. 2014. Biosynthesis of silver nanoparticles: Elucidation of prospective mechanism and therapeutic potential. *J. Colloid Interface Sci.* 415 (February), 39–47. 10.1016/j.jcis.2013.10.018.

Mohamed, Z.A. 2008. Polysaccharides as a protective response against microcystin-induced oxidative stress in Chlorella vulgaris and Scenedesmus quadricauda and their possible significance in the aquatic ecosystem. *Ecotoxicology* 17(6), 504–516. 10.1007/s10646-008-0204-2

Mohan, D., Pittman, C.U., Steele, P.H. 2006. Pyrolysis of wood/biomass for bio-oil: A critical review. *Energy Fuels* 20(3), 848–889. 10.1021/ef0502397. American Chemical Society.

Narita, J., Okano, K., Kitao, T., Ishida, S., Sewaki, T., Sung, M.H., Fukuda, H., Kondo, A. 2006. Display of α-Amylase on the surface of Lactobacillus casei cells by use of the PgsA anchor protein, and production of lactic acid from starch. *Appl. Environ. Microbiol.* 72(1), 269–275. 10.1128/AEM.72.1.269-275.2006.

Naviner, M, Durand, P, Le Bris, H. 1999. Antibacterial activity of the marine diatom Skeletonema costatum against aquacultural pathogens. *Aquaculture* 174(1–2), 15–24.

Nayak, Manoranjan, Dillip Kumar Swain, Ramkrishna Sen. 2019. Strategic valorization of de-oiled microalgal biomass waste as biofertilizer for sustainable and improved agriculture of rice (Oryza sativa L.) crop. *Sci. Total Environ.* 682(September), 475–484. 10.1016/j.scitotenv.2019.05.123.

Nunes, N. 2017. Biochemical composition, nutritional value, and antioxidant properties of seven seaweed species from the Madeira Archipelago. *J. Appl. Phycol.* 29(5), 2427–2437. 10.1007/s10811-017-1074-x

Overbeck, T., Steele, J.L., Broadbent, J.R., 2016. Fermentation of de-oiled algal biomass by Lactobacillus casei for production of lactic acid. *Bioprocess Biosyst. Eng.* 39(12), 1817–1823. 10.1007/s00449-016-1656-z.

Palozza, P., Torelli, C., Boninsegna, A., Simone, R., Catalano, A., Cristina, M., Picci, N. 2009. Growth-inhibitory effects of the astaxanthin-rich alga Haematococcus pluvialis in human colon cancer cells. *Cancer Lett.* 283(1), 108–117. 10.1016/j.canlet.2009.03.031.

Pan, P., Hu, C., Yang, W., Li, Y., Dong, L., Zhu, L., Tong, D., Qing, R., Fan, Y. 2010. The direct pyrolysis and catalytic pyrolysis of Nannochloropsis Sp. residue for renewable bio-oils. *Bioresour. Technol.* 101(12), 4593–4599. 10.1016/j.biortech.2010.01.070.

Pancha, I., Chokshi, K., Maurya, R., Bhattacharya, S., Bachani, P., Mishra, S. 2016. Comparative evaluation of chemical and enzymatic saccharification of mixotrophically grown de-oiled microalgal biomass for reducing sugar production. *Bioresour. Technol.* 204(March), 9–16. 10.1016/j.biortech.2015.12.078

Park, J.K., Kim, Z., Lee, C.G., Synytsya, A., Jo, H.S., Kim, S.O., Park, J.W. 2011. Characterization and immunostimulating activity of a water-soluble polysaccharide isolated from *Haematococcus lacustris. Biotechnol. Bioprocess Eng.* 16(6), 1090–1098. 10.1007/s12257-011-0173-9

Park, S., Li Y. 2012. Evaluation of methane production and macronutrient degradation in the anaerobic co-digestion of algae biomass residue and lipid waste. *Bioresour. Technol.* 111(May), 42–48. 10.1016/j.biortech.2012.01.160.

Parsaee, M., Kiani, M.K.D., Karimi, K. 2019. A review of biogas production from sugarcane vinasse. *Biomass Bioenerg.* 122, 117–125. 10.1016/j.biombioe.2019.01.034. Elsevier Ltd.

Patil, P.D., Reddy, H., Muppaneni, T., Deng, S. 2017. Biodiesel fuel production from algal lipids using supercritical methyl acetate (glycerin-free) technology. *Fuel.* 195, 201–207. 10.1016/j.fuel.2016.12.060.

Patterson, D., Gatlin D.M. 2013. Evaluation of whole and lipid-extracted algae meals in the diets of juvenile red drum (Sciaenops ocellatus). *Aquaculture.* 416–417 (December), 92–98. 10.1016/j.aquaculture.2013.08.033.

Paz-Cedeno, F.R., Solórzano-Chávez, E.G., Oliveira, L.E.De, Gelli, V.C., Monti, R., Oliveira, S.C.De, Masarin, F. 2019. Sequential enzymatic and mild-acid hydrolysis of by-product of Carrageenan process from *Kappaphycus alvarezii*. Bioenerg. Res. 12 (2), 419–432, 14800.

Phwan, C.K., Ong, H.C., Chen, W.H., Ling, T.C., Ng, E.P., Show, P.L. 2018. Overview: Comparison of pretreatment technologies and fermentation processes of bioethanol from microalgae. *Energy Convers. Manag.* 173, 81–94. 10.1016/j.enconman.201 8.07.054. Elsevier Ltd.

Plaza, M., Herrero, M., Cifuentes, A., Ibáñez, E. 2009. Innovative natural functional ingredients from microalgae. *J. Agric. Food Chem.* 57, 7159–7170. 10.1021/jf901070g

Pradel, M., Aissani, L., Villot, J., Baudez, J.C., Laforest, V. 2016. From waste to added value product: Towards a paradigm shift in life cycle assessment applied to wastewater sludge – A review. *J. Clean. Prod.* 131, 60–75. 10.1016/j.jclepro.2016.05.076Elsevier Ltd.

Rajesh Banu, J., Preethi, S. Kavitha, M.G., Kumar, G. 2020. Microalgae based biorefinery promoting circular bioeconomy-techno economic and life-cycle analysis. *Bioresour. Technol.* 302, April 2020, 122822. 10.1016/j.biortech.2020.122822. Elsevier Ltd.

Rajvanshi, M., Sayre, R. 2020. *Recent Advances in Algal Biomass Production*, 0–36. London, United Kingdom: IntechOpen. 10.5772/intechopen.94218

Rashid, N., Rehman, M.S.U., Han, J.I. 2013. Recycling and reuse of spent microalgal biomass for sustainable biofuels. *Biochem. Eng. J.* 75, 101–107. 10.1016/j.bej.2013.04.001

Reshma, R., Kumari, S., Arumugam, M. 2020. Structural elucidation of selenocysteine insertion machinery of microalgal Selenoprotein T and its transcriptional analysis. *Biotechnol. Appl. Biochem.* 68(3), 636–647. 10.1002/bab.1974

Roesijadi, G., Copping, A.E., Huesemann, M.H., Forster, J, Benemann, J.R. 2008. Techno-economic feasibility analysis of offshore seaweed farming for bioenergy and biobased products, Battelle Pacific Northwest Division Report Number PNWD-3931, 1e115.

Roh, M.K., Uddin, Md S., Chun, B.S. 2008. Extraction of fucoxanthin and polyphenol from undaria pinnatifida using supercritical carbon dioxide with co-solvent. *Biotechnol. Bioprocess Eng.* 13(6), 724–729. 10.1007/s12257-008-0104-6.

Scaife, M.A., Nguyen, G.T.D.T., Rico, J., Lambert, D., Helliwell, K.E., Smith, A.G. 2015. Establishing *Chlamydomonas reinhardtii* as an industrial biotechnology host. *Plant J.* 82(3), 532–546. 10.1111/tpj.12781.

Sekar, S., Chandramohan, M. 2008. Phycobiliproteins as a commodity: Trends in applied research, patents and commercialization. *J. Appl. Phycol.* 20(2), 113–136. 10.1007/s10811-007-9188-1. Springer.

Seo, Y.H., Sung, M., Han, J.I. 2015. Recycle of algal residue suspension from acid-catalyzed hot-water extraction (AHE) as substrate of oleaginous yeast Cryptococcus Sp. *Fuel.* 141(February), 222–225. 10.1016/j.fuel.2014.10.043.

Shih, S., Ho, M., Lin, K., Wu, S. 2000. Genetic analysis of Enterovirus 71 isolated from fatal and non-fatal cases of hand, foot and mouth disease during an epidemic in Taiwan, 1998. *Virus Res.* 68(3), 127–136.

Sialve, B., Bernet, N., Bernard, O. 2009. Anaerobic digestion of microalgae as a necessary step to make microalgal biodiesel sustainable. *Biotechnol. Adv.* 21(4), 409–416. 10.1016/j.biotechadv.2009.03.001

Singh, A., Nigam, P.S., Murphy, J.D. 2011. Mechanism and challenges in commercialisation of algal biofuels. *Bioresour. Technol.* 102(1), 26–34. 10.1016/j.biortech.2010.06.057.

Sirajunnisa, A.R., Surendhiran, D. 2016. Algae – A quintessential and positive resource of bioethanol production: A comprehensive review. *Renew. Sustain. Energy Rev.* 66, 248–267. 10.1016/j.rser.2016.07.024. Elsevier Ltd.

Sivagnanam, S.P., Getachew, A.T., Choi, J.H., Park, Y.B., Woo, H.C., Chun, B.S.. 2017. Green synthesis of silver nanoparticles from deoiled brown algal extract via Box-Behnken based design and their antimicrobial and sensing properties. *Green Process. Synth.* 6(2), 147–160. 10.1515/gps-2016-0052.

Sousa, I., Gouveia, L., Batista, A.P., Raymundo, A., Bandarra, N. 2008. Microalgae in novel food products. *Food Chem. Res. Dev.* 75–112.

Spolaore, P., Joannis-Cassan, C., Duran, E., Isambert, A. 2006. Commercial applications of microalgae. *J. Biosci. Bioeng.* 101(2), 87–96. 10.1263/jbb.101.87.

Suali, E., Sarbatly, R. 2012. Conversion of microalgae to biofuel. *Renew. Sustain. Energy Rev.* 16(6), 4316–4342. 10.1016/j.rser.2012.03.047. Pergamon.

Sudhakar, M.P., Ravel Merlyn, K.A., Perumal, K. 2016. Characterization, pretreatment and saccharification of spent seaweed biomass for bioethanol production using baker's yeast. *Biomass Bioenerg.* 90(July), 148–154. 10.1016/j.biombioe.2016.03.031.

Talukder, Md M.R., Das P., Wu, J.C. 2012. Microalgae (Nannochloropsis salina) biomass to lactic acid and lipid. *Biochem. Eng. J.* 68(October), 109–113. 10.1016/j.bej.2012.07.001.

Tannin-Spitz, T., Bergman, M., Grossman, S., Arad, S.M. 2005. Antioxidant activity of the polysaccharide of the red microalga Porphyridium Sp. .*J Applied Phycol.* 17(3), 215–222. 10.1007/s10811-005-0679-7

Torri, C., Samorì, C., Adamiano, A., Fabbri, D., Faraloni, C., Torzillo, G. 2011. Preliminary investigation on the production of fuels and bio-char from chlamydomonas reinhardtii biomass residue after bio-hydrogen production. *Bioresour. Technol.* 102(18), 8707–8713. 10.1016/j.biortech.2011.01.064.

Trache, D., Hazwan Hussin, M., Chuin, C.T.H., Sabar, S., Nurul Fazita, M.R., Taiwo, O.F.A., Hassan, T.M., Mohamad Haafiz, M.K. 2016. Microcrystalline cellulose: Isolation, characterization and bio-composites application – A review. *Int. J. Biol. Macromol.* 93, 789–804. Elsevier B.V. 10.1016/j.ijbiomac.2016.09.056

Udayan, A., Arumugam, M., Pandey, A. 2017. Nutraceuticals from algae and cyanobacteria. *Algal Green Chemistry: Recent Progress in Biotechnology*, pp. 65–89. 10.1016/B978-0-444-63784-0.00004-7

Usman, A., Khalid, S., Usman, A., Hussain, Z., Wang, Y. 2017. Algal polysaccharides, novel application, and outlook, in: *Algae Based Polymers, Blends, and Composites: Chemistry, Biotechnology and Materials Science.* Elsevier, pp. 115–153. 10.1016/B978-0-12-812360-7.00005-7. Amsterdam, Netherlands.

Vardon, D.R., Sharma, B.K., Blazina, G.V., Rajagopalan, K., Strathmann, T.J. 2012. Thermochemical conversion of raw and defatted algal biomass via hydrothermal liquefaction and slow pyrolysis. *Bioresour. Technol.* 109(April), 178–187. 10.1016/j.biortech.2012.01.008.

Venkata Mohan, S., Hemalatha, M., Chakraborty, D., Chatterjee, S., Ranadheer, P., Kona, R. 2020. Algal biorefinery models with self-sustainable closed loop approach: Trends and prospective for blue-bioeconomy. *Bioresour. Technol.* 295, 122128. 10.1016/j.biortech.2019.122128. Elsevier Ltd.

Venkata Mohan, S., Nikhil, G.N., Chiranjeevi, P., Nagendranatha Reddy, C., Rohit, M.V., Naresh Kumar, A., Sarkar, O. 2016. Waste biorefinery models towards sustainable circular bioeconomy: Critical review and future perspectives. *Bioresour. Technol.* 215, 2–12. 10.1016/j.biortech.2016.03.130. Elsevier Ltd.

Venkata Subhash, G., Venkata Mohan, S. 2014. Deoiled algal cake as feedstock for dark fermentative biohydrogen production: An integrated biorefinery approach. *Int. J. Hydrog. Energy.* 39(18), 9573–9579. 10.1016/j.ijhydene.2014.04.003.

Wahlström, N., Harrysson H., Undeland I., Edlund U. 2018. A strategy for the sequential recovery of biomacromolecules from red macroalgae *Porphyra umbilicalis* Kützing. *Ind. Eng. Chem. Res.* 57(1), 42–53. 10.1021/acs.iecr.7b03768.

Watanabe, H., Li, D., Nakagawa, Y., Tomishige, K., Watanabe, M.M. 2015. Catalytic gasification of oil-extracted residue biomass of *Botryococcus braunii*. *Bioresour. Technol.* 191(July), 452–459. 10.1016/j.biortech.2015.03.034.

Wijffels, R.H., Barbosa, M.J., Eppink, M.H.M. 2010. Microalgae for the production of bulk chemicals and biofuels. *Biofuel. Bioprod. Biorefin.* 4(3), 287–295. 10.1002/bbb.215. John Wiley & Sons, Ltd.

Yang, Z., Guo, R., Xu, X., Fan, X., Li, X. 2011. Thermo-alkaline pretreatment of lipid-extracted microalgal biomass residues enhances hydrogen production. *J. Chem. Technol. Biotechnol.* 86(3), 454–460. 10.1002/jctb.2537.

Yang, Z., Guo, R., Xu, X., Fan, X., Luo, S. 2011. Fermentative hydrogen production from lipid-extracted microalgal biomass residues. *Appl. Energy.* 88(10), 3468–3472. 10.1016/j.apenergy.2010.09.009.

Yarnpakdee, S., Benjakul, S., Kingwascharapong, P. 2015. Physico-chemical and gel properties of agar from *Gracilaria tenuistipitata* from the Lake of Songkhla, Thailand. *Food Hydrocoll.* 51(October), 217–226. 10.1016/j.foodhyd.2015.05.004.

Yim, J.H., Kim, S.J., Ahn, S.H., Lee, C.K., Rhie, K.T., Lee, H.K. 2004. Antiviral effects of sulfated exopolysaccharide from the marine microalga *Gyrodinium impudicum* strain. *Mar. Biotechnol.* 6(1), 17–25. 10.1007/s10126-003-0002-z.

Yuan, S., Wang, P., Xiao, L., Liang, Y., Huang, Y., Ye, H., Wu, K., Lu, Y.. 2020. Enrichment of lipids from agar production wastes of Gracilaria lemaneiformis by ultrasonication: A green sustainable process. *Biomass Convers. Biorefin.* 1-10, 10.1007/s13399-020-00674-5.

Zou, S., Wu, Y., Yang, M., Li, C., Tong, J. 2010. Bio-oil production from sub-and super-critical water liquefaction of microalgae Dunaliella tertiolecta and related properties. *Energy Environ. Sci.* 3(8), 1073–1078.

3 Algal Role in Microbial Fuel Cells

Ghadir Aly El-Chaghaby
Bioanalysis Laboratory, Regional Center for Food and
Feed (RCFF), Agricultural Research Center (ARC),
Giza, Egypt

Sayed Rashad
El-Fostat Laboratory, Cairo Water Company, Cairo, Egypt

Meenakshi Singh
Department of Ecology and Biodiversity, Terracon Ecotech
Pvt. Ltd., Mumbai, Maharashtra, India

K. Chandrasekhar
School of Civil and Environmental Engineering, Yonsei
University, Seoul, Republic of Korea

Murthy Chavali
Office of the Dean (Research) and Department of
Chemistry, Alliance College of Engineering and Design,
Faculty of Science and Technology, Alliance University
(Central Campus), Bengaluru, Karnataka, India

CONTENTS

3.1 Introduction 72
3.2 Algae and their Biomass 73
3.3 Microbial Fuel Cells (MFCs) 74
    3.3.1 Types of Electrode Materials 74
    3.3.2 Operational Conditions 74
    3.3.3 Factors Affecting MFCs' Performance 75
3.4 Algae Used as Culture and Substrate in Microbial Fuel Cell 76
    3.4.1 Different Algal Substrates as Autotrophic Mode MFC 77
    3.4.2 Different Algal Substrates as Heterotrophic Mode MFC 77
    3.4.3 Algal Role in Microbial Fuel Cell 78
3.5 Applications of Algal MFC 80

DOI: 10.1201/9781003195405-3

3.5.1   Bioelectricity Generation 80
3.5.2   Wastewater Treatment...80
3.5.3   Biohydrogen Generation 81
3.5.4   Robots...82
3.5.5   Biosensors...82
3.6   Challenges and Prospects of Algal MFCs......................................83
3.7   Conclusion and Outlook..83
References...83

3.1   INTRODUCTION

Nearly 80% of world energy consumption is from the combustion of fossil fuels. Fossil fuels pollute the environment by releasing a huge amount of CO_2. The risks of over-dependence on fossil fuels can be avoided by using renewable and carbon-neutral energy sources in a large amount. Life on Earth was threatened by two major issues: environmental pollution and global warming. These extreme variations are either directly or indirectly caused by CO_2 emission and have become a very severe problem worldwide (Chandrasekhar et al., 2020c; Park et al., 2021). Global energy needs are increasing day by day, and the quest for using green energy sources is a major sustainability goal (Chandrasekhar et al., 2020a). Most of the world's energy consumption (85%) comes from the combustion of fossil fuels; with the disturbing reduction of non-renewable resources (natural gas, coal and petroleum), an essential understanding of renewable energy sources towards synchronization is necessary (Mohan et al., 2011b; Singh et al., 2020). For decades, fossil fuels were the main supplier of the energy needed for all human activities (Chaturvedi and Verma, 2016).

The CO_2 fixation technologies are necessary to alleviate the environmental concerns generated by CO_2 emissions (Bhatia et al., 2021; Mehariya et al., 2020, 2019; Molino et al., 2019a, 2019b; Siciliano et al., 2019). Chemical absorption, membrane separation, physical adsorption, and cryogenic sequestration techniques are regarded as more costly and problematic because of secondary contamination. CO_2 is biologically fixed by photosynthesis in terrestrial plants and an abundance of photosynthetic microbes (e.g., microalgae and cyanobacteria). Microalgae and cyanobacteria grow considerably quicker than plants on land. Their CO_2 fixation efficiency is 10–50 times greater (Enamala et al., 2018). Thus, phototrophic microalgae and cyanobacteria can be used to sequester CO_2 (Obileke et al., 2021; Patel et al., 2017). Fossil fuels are a major contributor to environmental degradation because of CO_2 released (Arun et al., 2020). However, because of the finite and non-sustainable nature of fossil fuels, other renewable energy sources are now being explored.

Microbial fuel cells (MFCs) have been a major topic of research interest since the early 2000s. MFCs are devices capable of producing bioelectricity from different sources of substrates (Rezaei et al., 2007). The novel idea of generating electric energy from the biological route, using microorganisms to generate electrical power, was studied (Rahimnejad et al., 2014). This process occurs at ambient pressure and temperature through the metabolic activity of microorganisms (Daniel et al., 2009). The microbial fuel cells (MFCs) seem a very eye-catching option for bioelectricity generation that converts chemical energy to electric energy through a biochemical reaction catalyzed by microorganisms (Rasep et al., 2016; Zhang et al., 2018). Currently, microalgae-fixed CO_2 and a bioelectrochemical system are considered viable and ecologically acceptable solutions for CO_2 sequestration and energy generation (Saratale et al., 2017).

Algal microbial fuel cells (AMFCs) oxidize glucose present in synthetic wastewater, feeding this to the anode chamber to produce electrons and protons. Cathode chamber algal photosynthesis oxidizes CO_2 to biomass, releasing oxygen, which is an electron acceptor for the O_2 reduction process. Using *Chlorella vulgaris* as a biodiesel feedstock, microalgae culture in AMFC offers a sustainable means of biodiesel generation (Saratale et al., 2017). This has led to AMFC getting a lot of interest in the scientific disciplines of wastewater treatment, bioenergy generation, and CO_2 sequestration. The solar radiation trapped by algae when incorporated with cathodic and anodic chambers can produce bioelectricity in an efficient manner (Enamala et al., 2020b). Recently, different autotrophic and heterotrophic algae integrated into MFCs have represented the potential to convert biomass into bioelectricity using organic contaminants and also efficiently remove various by-products (Kakarla and Min, 2014a; Pandit et al., 2017). This chapter provides a brief description of MFC designs and thoroughly describes the simple setup of types of MFCs and their electrode materials used in preparing cathodes, anodes, and membranes. Also, it allows the reader to recognize the essential information concerning MFCs in general and the role of algae in MFC, in particular; emphasis was also laid on their essential applications (Deval et al., 2017). The importance, design, and uses of algal MFCs are discussed, along with future challenges and perspectives.

3.2 ALGAE AND THEIR BIOMASS

Algae represent a large diversity of photosynthetic oxygenic microorganisms that produce many important metabolites (Mohamed et al., 2020) as those produced by higher plants (Koushalya et al., 2021). Algae comprise two groups: microalgae and macroalgae; both groups of algae can be cultivated using freshwater, seawater, and even wastewater. Microalgae are the oldest living organisms in aquatic and freshwater habitats (Kakarla et al., 2017). They are an array of microorganisms (prokaryotic and eukaryotic) with a single cell or multi-cell structure, which supports them to grow steadily and survive under harmful circumstances (Baicha et al., 2016). On the other hand, macroalgae, also referred to as seaweed, are plant-like organisms that range in size from a few centimeters to several meters in length. Marine macroalgae are generally categorized based on their pigments into brown macroalgae, green macroalgae, and red macroalgae (Sayed and Ghadir, 2020). Algal biomass contains protein, carbohydrates, and lipids as major metabolites, which make it an important feedstock for bioenergy production (Enamala et al., 2019; Karthik et al., 2020b; Sudhakar et al., 2019; Yadavalli et al., 2021, 2020; Yukesh Kannah et al., 2021).

Although various forms of feedstock are being used to generate renewable energy, algae-based bioenergy is highly attractive due to its strong photosynthesis capacity, higher rate of CO_2 fixation, higher biomass productivity, high lipid content, and faster growth rate (Sun et al., 2018). Both microalgae and macroalgae have been explored as sources of bioenergy in several studies. Different algal species were used to produce biofuel (Serrà et al., 2020), bio-ethanol (Kim et al., 2017), biogas (Bruhn et al., 2011; Ndayisenga et al., 2018), biomethane (Hessami et al., 2019; Klassen et al., 2017), bio-oil (Bae et al., 2011), and biodiesel (Mureed et al., 2018).

3.3 MICROBIAL FUEL CELLS (MFCS)

Finding economical and sustainable alternative energy sources has risen to the top of the priority list (Chandrasekhar et al., 2021; Gambino et al., 2021). MFC is a promising methodology among emerging renewable energy production technology because of its capacity to directly transform biomaterial via certain reactions into electrical energy (Saba et al., 2017). MFC technology, which captures energy from microbe metabolism, appears to be appealing as a means of generating energy (Gambino et al., 2021). Using MFC for power production is advantageous as it is a safe, clean, and efficient process. It is made with renewable resources and produces no hazardous waste (Chaturvedi and Verma, 2016).

3.3.1 TYPES OF ELECTRODE MATERIALS

MFC is a bioelectrochemical device that encourages microorganisms to change organic or inorganic substrates into energy through the microorganism's metabolic activities. A conventional MFC has two compartments – anode and cathode – with an ion-exchange membrane that divides the anodic and the cathodic compartments (Chandrasekhar et al., 2020b). The microorganisms in the anodic compartment oxidize the carbon-based material and release protons and electrons that react in the cathodic chamber, producing water. The electrons are subsequently transmitted from the surface of the anode to the cathode via the electrical circuit, whereas the protons (H^+) travel through the electrolyte and then the membrane. Thus, electricity is generated when the electrons are mobile between the chambers through an outside circuit (Alatraktchi et al., 2014; Chaturvedi and Verma, 2016; Obileke et al., 2021; Rahimnejad et al., 2014).

3.3.2 OPERATIONAL CONDITIONS

The key factors for optimizing efficient energy output from MFCs include micro-organism's consortium, substrate(s) existing in the medium, electrode type, and extenuation of methanogenic species from the microbial population (Mekuto et al., 2020). In MFCs, microorganisms break down various organic compounds to yield electrons and protons, allowing them to power their respiration system, and they also demonstrate an ability to produce a flow of electrons across the electrodes. Microorganisms, on the other hand, can transmit electrons to electron acceptors that are insoluble (Gustave et al., 2021). Several strains of microorganisms may be capable of transporting electrons and protons across electrodes in MFCs chambers. In the MFCs chamber, it was discovered that five major microorganism species: firmicutes, proteobacteria, acid bacteria, fungi, and algae display distinctive, electricity generation (Yaqoob et al., 2021). Some bacteria were previously described as electron exchangers with electrodes. Furthermore, some bacterial species without the need for an exogenous acceptor are capable of directly transferring electrons to the anode. Exoelectrogens are the bacterial species that accomplish this activity (Srivastava, 2019).

The substrate is one of the most essential biological components influencing MFCs in generating electricity. MFCs generating power can be made out of a wide range of substrates. Inoculums, electrode material, ionic strength of the solution, use of diverse sorts of real wastewater or synthetic solution (substrate material), ion-exchange membrane, and temperature have all been investigated to improve the performance of MFCs (Ullah and Zeshan, 2020). In an MFC, the biodegradable substrate (or electron donor) is critical for the generation of power. Various organic and inorganic substrates have been added to MFCs, most typically in the form of liquid or particulate solid, and the results have been published. These substrates range from simple chemicals like acetate and glucose to more complicated chemical molecules like ethanol. MFCs' output power varies based on the substrate's availability and the microorganisms' capacity to consume it (Liu et al., 2013).

3.3.3 FACTORS AFFECTING MFCs' PERFORMANCE

Membrane, temperature, pH, nature of the substrate, and loading are the significant aspects to consider while optimizing MFC performance (Kusmayadi et al., 2020). Membrane material is particularly important in MFC since it affects the degree of chemical species transfer in the cell (Chandrasekhar et al., 2018). The voltage gradient across the membrane allows cations to travel from the anode to the cathode compartment. MFC has been used with a variety of membranes, including membranes for anion exchange, bipolar membrane, cation exchange, ceramic type, salt bridge, and others (Rahimnejad et al., 2015).

Temperature is an important factor in microorganism growth, and temperature changes have an impact on the microbial growth rate, activity, and dispersal. The operating temperature is another factor that affects electricity generation. Higher operating temperatures can advance reaction kinetics, mass transfer, Coulombic efficiency, power density, and resistance while also lowering resistance. The pace of biochemical reactions increases when the temperature rises, which aids microorganism growth and substrate assimilation (Zhang et al., 2016).

pH is crucial for achieving high power output in an MFC. For optimal metabolic activity, microorganisms require a pH close to neutral, as well as low ion concentrations (Jannelli et al., 2017). The cathode compartment, like the anode compartment, needs to be kept at the proper pH level (Puig et al., 2010). The kind of substrate and its characteristics have a significant impact on MFC performance. Advantages of maintaining a pH balance in the context of MFCs include maximum harvest, good resistance to toxic substances, less influence of environmental fluctuations, good and stable long-term performance, increased efficiency, development of sustainable wastewater treatment technologies, increase in the pH of the cathode, and decrease in anodic pH, which ultimately results in reducing the working efficiency of the MFC (Gil et al., 2003).

Microorganisms may easily use simple molecules as a substrate to make reducing equivalents, resulting in improved system performance. Some wastewaters, on the other hand, include high amounts of organic chemicals that are difficult to decompose into simple molecules. As a consequence, selecting a substrate based on its kind and nature is critical for MFC system performance. Because these factors influence biofilm morphology and bacteria growth rates, both retention duration and loading rates have an impact on MFC power generation and Coulombic efficiency (Munoz-Cupa et al., 2021).

To put it another way, MFCs use microorganisms as biocatalysts to oxidize the organic matter and transmit electrons to the anodic surface for bioelectricity production via substrate oxidation. To power MFCs, carbon-based oxidizable substrates have included starch, cellulose, simple carbohydrates, organic acids, proteins/amino acids, and chitin, as well as toxic waste chemicals like phenol, PAHs, indole, ethanolamine, p-nitrophenol, nitrobenzene, and sulfide. Wastewater contaminated with heavy metals, swine wastewater, brewery/distillery waste, and marine sediments have all been effectively used in laboratory-based MFC systems to produce bioelectricity and the creation of novel biomaterials (Zhang et al., 2018).

3.4   ALGAE USED AS CULTURE AND SUBSTRATE IN MICROBIAL FUEL CELL

Use of algae for bioenergy production offers several advantages; hence, algal biomass is considered a new bio-material, which holds the promise to fulfil the rising demand for energy. Microbial fuel cells offer new opportunities for the sustainable production of energy from biodegradable, reduced compounds, a novel biotechnology for energy production. Algae are microscopic ubiquitous living entities, which can be unicellular as well as multicellular. Some autotrophic algae can produce their food from the sunlight that they get from their environment, whereas heterotrophic algae use CO_2 as a carbon source from substrates, to prepare food in the absence of light (Jareonsin and Pumas, 2021; Kondaveeti et al., 2020). Therefore, algae play an important role in the food chain and the provision of O_2 on Earth. Furthermore, they have rapid growth rates as well as rapid CO_2 fixation rates (Cheng et al., 2006; Cheng and Logan, 2007). Algae, through their photosynthetic output, play a critical role in converting solar energy into various types of biological energy, which is essential for human survival (Mohan et al., 2011a).

Figure 3.1 illustrates photosynthetic redox reaction (autotrophic) mode in algae, which is a complicated biological redox process that allows algae to exploit solar energy to generate oxygen, carbohydrates, and a variety of other chemicals. *Canna indica* is a lignocellulosic aquatic plant rich in cellulose, hemicellulose, and lignin used as an air-cathode. *Chlorella pyrenoidosa*, a unicellular green alga, is used as anode material. *Microcystis aeruginosa* serves as an electron donor. *Chlorella regularis* are some of the algae used for producing bioelectricity in a microbial fuel cell. *Nannochloropis salina* is used as a cathode in a three-chambered MFC to transport electrons via ion-exchange

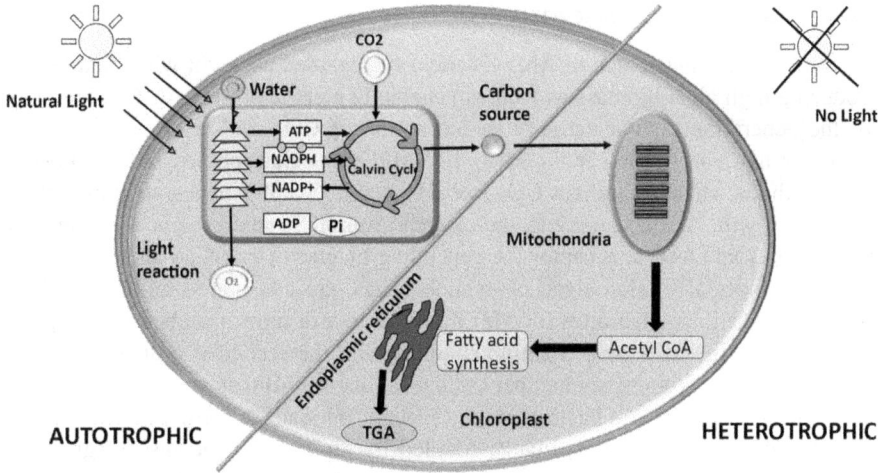

FIGURE 3.1 A schematic representation of Autotrophic and Heterotrophic mode of algae.

membrane and purify saline water while generating minute power output in the bio-catalysis reaction (Girme et al., 2014).

3.4.1 Different Algal Substrates as Autotrophic Mode MFC

Combining the autotrophic and heterotrophic growth modes can result in the formation of a mixed culture (mixotrophic) growth system, in which photosynthetic and respiratory metabolisms work in tandem to absorb organic carbon and CO_2 from the environment. For instance, a novel study conducted by (Fleury, 2017) on photosynthetic MFC used *Nannochloropsis, Spirulina,* and *Chlorella* to evaluate energy efficiency (electrical output) and the magnitude of absorption of MFCs. The result suggested higher metabolic activity of *Nannochloropsis,* which generated 35 mW of bioelectricity, followed by *Spirulina* (31 mW) and *Chlorella* (30.2 mW), before reaching the stationary phase.

3.4.2 Different Algal Substrates as Heterotrophic Mode MFC

The heterotrophic growth system has the additional benefit of permitting the use of any type of bioreactor without the need for any specialized design, which is a significant advantage. The growth rate of algal biomass, as well as the generation of ATP, is extremely rapid when the algae are in heterotrophic mode. In addition, the nitrogen output and fat content are significantly higher than they are in the autotrophic phase. Because the microalgal species that may be grown in heterotrophic algal cultures are restricted, there are numerous disadvantages to using them in research. Figure 3.1 shows when carbon-based substrates are added to a heterotrophic system, the energy expenditure is considerable, and the system is also susceptible to contamination by other bacteria (Yang et al., 2000).

3.4.3 ALGAL ROLE IN MICROBIAL FUEL CELL

Algae cultivation and usage in AMFCs could be regarded as a win-win situation for producing high algal biomass as well as generating electric power. AMFCs are used for the generation of bioelectricity, sequestration of CO_2, and removal of organic matters or nutrients (Zhang et al., 2018). In AMFCs, the algae photosynthesis process in the cathodic chamber utilizes CO_2 from the anode chamber as a source of carbon for their growth. At the same time, the oxygen resulting from algae photosynthesis is consumed as an electron acceptor for generating bioelectricity (Xiao and He, 2014).

Algae, especially microalgae or cyanobacteria, are considered among the best biologically active metabolites for MFCs. Two major designs have been recognized for AMFCs. The first one is a single chamber, and the second is a dual or double chamber design. In a single-chamber configuration, the MFC is a one-compartment where the microalgae and bacteria grow together, microalgae form a biofilm on the anode, and the chamber is typically coupled by air cathode. In this design, CO_2 created by autotrophic and heterotrophic microorganisms is utilized concurrently by algae in the same chamber, and the protons are transported to the cathode from the anode solution. Single-chamber algal fuel cells (Figure 3.2) are straightforward to operate, profitable in scaling up, and may be utilized commercially. The main drawback of such MFCs is connecting with the back diffusion of oxygen from cathode to anode (Kannan and Donnellan, 2021; Karthik et al., 2020a; Obileke et al., 2021; Yaqoob et al., 2021).

The double chamber configuration of microbial fuel cells (Figure 3.3) comprises an anode chamber and a cathode chamber for microalgae and bacteria that are divided by a proton exchange membrane. In this cell design, the algae culture in the cathode use illumination to perform photosynthesis, and also the anode is illuminated to make the algae covering the bacterial chamber. The main advantage of the two-chambered cell is that its working conditions (e.g., pH, oxygen purge, flow rate, etc.) can be optimized to obtain maximum fuel cell performance. On the contrary,

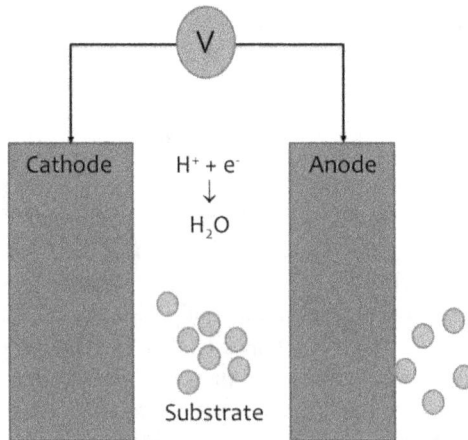

FIGURE 3.2 Single-chambered MFC.

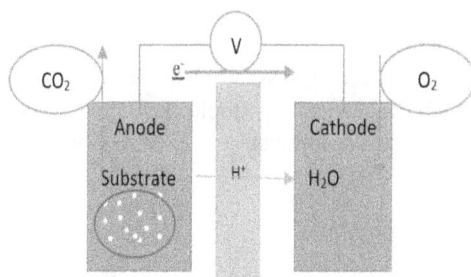

FIGURE 3.3 Double chambered MFC.

this type of MFC has some limitations associated with high relative internal resistance and membrane crossover (Kannan and Donnellan, 2021; Liu et al., 2015; Ullah and Zeshan, 2020).

Table 3.1 describes different algae and their consortium used as a substrate, or assist at the anodic chamber and cathodic chamber of single- and double-chambered MFCs. In these systems, power is produced at the anode as well as at the cathode via algal photosynthesis, liberating oxygen and renewable biomass that can be used for various applications (Shukla and Kumar, 2018).

Integrating algae in MFCs proved several advantages, e.g., carbon dioxide mitigation and global warming reduction, treatment of wastewater, production of several valuable products such as biofuels, biofertilizer, pigments, biofilters, food additives, and others. Another advantage is that in two-chamber cell algae can be used at the biocathode to block energy-intensive reflexive ventilation in the bioreactor. Also, at the bioanode, algal feedstock, which is rich in carbohydrates, lipids, and vitamins, could be further used to generate biofuel (Goswami et al., 2021b; Mehariya et al., 2021a; Mekuto et al., 2020).

TABLE 3.1
Power Generation Capacity of Algae MFCs

Algae	Function	Type of MFC	Power Density (W/m^2)	References
Cyanobacteria	Substrate	Single chamber	0.114	(Yuan et al., 2011)
Chlorella vulgaris	Substrate	Single chamber	2.77	(Velasquez-Orta et al., 2009)
Chlamydomonas reinhardtii	Assisting anode	Single chamber	0.075	(Nishio et al., 2013)
Chlorella vulgaris	Assisting cathode	Dual chamber	0.188	(Liu et al., 2015)
Chlorella pyrenoidosa	Assisting anode	Dual chamber	0.031	(Xu et al., 2015)
Scenedesmus	Substrate	Dual chamber	1.78	(Rashid et al., 2013)
Mixed algae	Assisting anode	Single chamber	3.56×10^{-6}	(Venkata Subhash et al., 2013)

3.5   APPLICATIONS OF ALGAL MFC

Several algal species have been successfully used in MFC, and the literature cites many pieces of research concerning this topic. By controlling the cell conditions, AMFCs have many outstanding applications listed below:

3.5.1   Bioelectricity Generation

Gadhamshetty and co-authors were capable of obtaining a extreme power density per unit anode volume that was equal to 6030 mW/m^2 using living microalgae (green) *Chlorella pyrenoidosa* to act as an electron donor in a double chamber MFC (Gadhamshetty et al., 2013). Bioelectricity was produced using the powder of *Chlorella vulgaris* and *Ulva lactuca* in single-chamber MFCs; maximum power was achieved as 0.98 W/m^2 using *Chlorella vulgaris*, and 0.76 W/m^2 using *Ulva lactuca* (Velasquez-Orta et al., 2009).

Hou and co-authors have successfully constructed a single chamber MFC for bioelectricity generation using *Golenkinia* SDEC-16 algal strain and showed maximum power of 6255 mW/m^2 (Hou et al., 2016). Photoautotrophic algae *Scenedesmus obliquus* attached to the cathode surface of MFC was skilled at producing a full power density of 153 mW/m^2 (Kakarla and Min, 2014b). The biomass of *Scenedesmus obliquus* was used as a substrate for generating electricity with a full power density of 102 mW/m^2 in a two-chambered MFC (Kondaveeti et al., 2020, 2014). Other algae used in MFCs are summarized in Table 3.2.

3.5.2   Wastewater Treatment

Wastewater treatment has become a major problem in most human settlements. Primarily wastewater contains organics; nature has been taking care of organics for billions of years by converting them into energy. Microbial fuel cells convert organic matter in wastewater directly into electrical energy using the power of

TABLE 3.2
Algae Used in Microbial Fuel Cells to Generate Bioelectricity

Algae	Maximum Power Density (mW/m^2)	Reference
Laminaria saccharina	2.50	(Gadhamshetty et al., 2013)
Chlamydomonas reinhardtii	7.5	(Walter et al., 2015)
Scenedesmus obliquus	1.02	(Rashid et al., 2013)
Dunaliella tertiolecta	1.5	(Subadri et al., 2020)
Chlorella pyrenoidosa	3.02	(Subadri et al., 2020)
Arthrospira maxima	1.0	(Subadri et al., 2020)
Chlorella vulgaris	8.60	(Ndayisenga et al., 2018)
Chlorella vulgaris	1.35	(González del Campo et al., 2013)

respiring microbes. Several types of wastewater containing organic matter can be treated, including domestic wastewater, brewery effluent, anaerobic digestor effluent, abattoir wastewater, brewery wastewater, and sewerage sludge. Several operational plants have shown good results. The use of MFCs for wastewater requires a design that allows the wastewater to flow through the cell over the anode surface. Out of the several applications of algal-MFCs, their applications in wastewater treatment stand as a "state of the art" technique that offers the dual benefit of treating wastewater along with generating electricity (Arora et al., 2021), producing biofuel (Mehariya et al., 2021b, 2021c, 2021d) as well as for sustainable resource recovery from wastewater (Goswami et al., 2021b). Many wastewaters were treated, and different pollutants have been successfully remediated by applying MFCs, as shown in Table 3.3.

3.5.3 Biohydrogen Generation

MFCs may be quickly changed to the collection of biohydrogen, instead of producing energy for subsequent application. Algal-based MFCs provide sources of renewable hydrogen, which may be given in a hydrogen economy to the total need for hydrogen (Singh et al., 2021). To convert it to hydrogen production, it is just necessary to turn off the oxygen supply and apply a little amount of voltage (V) to decrease the number of protons in the cathodic chamber. Bio-electrochemically

TABLE 3.3
Pollutant Remediation Using MFCs

Wastewater Type	Pollutant / (Removal)	Reference
Olive mill	COD / (65%)	(Obileke et al., 2021)
Domestic	COD / (88%)	(Obileke et al., 2021)
Chemical	COD / (63%)	(Obileke et al., 2021)
Real Urban	COD/ (85%)	(Obileke et al., 2021)
Food	COD/ (90%)	(Obileke et al., 2021)
Landfill Leachate	TN/ (80%); COD/ (52.8%)	(Karthik et al., 2020b)
Domestic	COD/ (71%)	(Kusmayadi et al., 2020)
Swine	COD / (86%); NH_3 / (83%)	(Kusmayadi et al., 2020)
Heavy Metals Industry	Cd, Pb, Cr, Cu, Hg and Zn (8.81%, 5.59%, 1.95%, 12.6%, 14.0%, and 8.81%, respectively)	(Kusmayadi et al., 2020)
Cane Sugar Mill Effluent	COD / (56%)	(Kusmayadi et al., 2020)
Starch Processing	COD / (98%)	(Kusmayadi et al., 2020)

aided microbial reactors (BEAMRs) are regarded to be unique reactors and referred to as the bio-catalyzed electrolysis process (Chandrasekhar and Ahn, 2017). Practically speaking, ≥ 0.25 V is essential for the production of H_2 gas. Additionally, utilizing glucose as a substrate in the BEAMR method can result in the production of around 8–9 mol of hydrogen. The use of carbohydrates as a substrate for hydrogen production is limited, but remaining decomposable dissolved organic materials may be utilized as a substrate, making it a practically sustainable and ecologically acceptable technique (Goswami et al., 2021a).

3.5.4 ROBOTS

To maximize power usage, the majority of robots nowadays rely on alkaline cells or storage battery packs. Despite the past development of numerous robots via MFC, only EcoBot-I (bioinspired ecological robots) was the world's first successful robot that only used sugar as a substrate. Recently, a series of EcoBot-II, III, IV was created to transform refined biomass (insects or plant materials) into useable energy, resulting in increased power production and improved stability (Sarma et al., 2021). A variety of tasks were performed by these robots, including actuation, sensing, and communicating data to the computer (Melhuish et al., 2006). Another biologically inspired robot belongs to the Gastrobot family, which can be called a "robot with a stomach." Gastrobot is the name used to describe a new generation of intelligent food-powered robots that have been developed. In addition to the capability of self-locating the food, ingestion system, digesting (energy assimilation), and faeces (waste disposal), these robots are also self-sustaining (Wilkinson, 2000).

3.5.5 BIOSENSORS

Another use of biofuel cells is the use of MFC technology as a transmission sensor for pollution measurement and process monitoring. MFCs are therefore appropriate for electrical-chemical sensors and are tiny telemetry systems to send signals received to remote receivers, which are of limited-service life and need to be updated or recharged. It has also been shown that the BOD kind of sensor has great operational durability and repeatability and can remain operative for five years, and MFC's may be employed as a sensor for biological oxygen requirement (Rahimnejad et al., 2014).

With the fast development of biosensors as an alternate ecological assessment tool for the rapid and receptive detection of various analytes, they have gained great traction in this quickly evolving time. The timely detection and monitoring of toxicants solutes dissolved in water may be accomplished with relative ease using MFC, which is extremely beneficial in the treatment of wastewater and the processing of portable water. The high level of toxicants will result in high energy generation by the MFC, which will allow for the monitoring of the water's overall quality (Sarma et al., 2021). It is not only restricted to wastewater, but MFC sensors may also be used to detect organic acids, primarily succinic, acetic and muconic acids, surfactants, formaldehyde, and other chemicals in real time (Kaur et al., 2013). Also possible with MFC is the determination of the biological oxygen demand levels of the effluent. The output power production of the substrate is affected by changes in the BOD content of the substrate. It has a fast reaction time and good operating stability, making it ideal for monitoring.

3.6 CHALLENGES AND PROSPECTS OF ALGAL MFCS

Though there are numerous benefits of spending algae in MFCs as discussed in this chapter, there are still some limitations and obstacles that pose challenges to scientists in this field. After examining several studies, it can be postulated that the complicated electricity generation mechanism using algae-based MFCs is not yet completely understood. A limitation of employing algae in MFCs in some studies was the lower growth of algae, which led to lower harvested biomass. Since several factors contribute to the performance of algae-MFCs, extensive process optimization studies should be addressed. Finally, the use of algal biomass represents a clean, green and sustainable bioenergy study area that is attracting increasing attention (Do et al., 2018; Xiao and He, 2014). Synchronization of the utilization of a single type of algae-based MFC or a mixture of algal MFC designs for a single type of energy generation and request will necessitate a simplified R&D effort focused on more field size experimentation and validation (Khandelwal et al., 2020).

3.7 CONCLUSION AND OUTLOOK

Integrating algae in MFCs could offer an excellent solution for obtaining sustainable bio-energy. The usage of algae in MFCs is an outstanding approach for extracting energy relying on a process that employs sunlight economically and sustainably. Besides the cost-effectiveness of using algae in MFCs, the recovery of valuable products from algae biomass after its harvest is another important economic benefit that can be achieved. There are various advantages over other biomass sources when it comes to the use of algae for energy generation as they can grow in a variety of environments, including wastewater, they have a high biomass production rate, and most importantly, they require somewhat of a small cultivation area. For future achievement of sustainability goals, more researches and applications should be focused on setting new and insightful perspectives for the uses of algae in MFCs to replace non-renewable energy sources with greener, economic, and sustainable sources such as MFCs. The existing algae-assisted MFCs' practical approach is still in the nascent stage of bioengineering technology, leading to low electricity generation, which needs to be addressed to increase the output. This can be possible by building robust genetic engineered algae by designing integrated MFC that scale up power generation (Kannan and Donnellan, 2021).

REFERENCES

Alatraktchi, F.A.Z.A., Zhang, Y., Angelidaki, I., 2014. Nanomodification of the electrodes in microbial fuel cell: Impact of nanoparticle density on electricity production and microbial community. *Appl. Energy* 116, 216–222. 10.1016/j.apenergy.2013.11.058

Arora, K., Kumar, P., Bose, D., Li, X., Kulshrestha, S., 2021. Potential applications of algae in biochemical and bioenergy sector. *3 Biotech* 11, 296. 10.1007/s13205-021-02825-5

Arun, S., Sinharoy, A., Pakshirajan, K., Lens, P.N.L., 2020. Algae based microbial fuel cells for wastewater treatment and recovery of value-added products. *Renew. Sustain. Energy Rev.* 132, 110041. 10.1016/j.rser.2020.110041

Bae, Y.J., Ryu, C., Jeon, J.-K., Park, J., Suh, D.J., Suh, Y.-W., Chang, D., Park, Y.-K., 2011. The characteristics of bio-oil produced from the pyrolysis of three marine macroalgae. *Bioresour. Technol.* 102, 3512–3520. 10.1016/j.biortech.2010.11.023

Baicha Z.; Salar-García, M.J.; Ortiz-Martínez, V.M.; Hernández-Fernández, F.J.; de los Ríos, A.P.; Labjar, N.; Lotfi, E.; Elmahi, M., 2016. A critical review on microalgae as an alternative source for bioenergy production: A promising low cost substrate for microbial fuel cells. *Fuel Process. Technol.* 154, 104–116. 10.1016/j.fuproc.2016.08.017

Bhatia, S.K., Mehariya, S., Bhatia, R.K., Kumar, M., Pugazhendhi, A., Awasthi, M.K., Atabani, A.E., Kumar, G., Kim, W., Seo, S.-O., Yang, Y.-H., 2021. Wastewater based microalgal biorefinery for bioenergy production: Progress and challenges. *Sci. Total Environ.* 751, 141599. 10.1016/j.scitotenv.2020.141599

Bruhn, A., Dahl, J., Nielsen, H.B., Nikolaisen, L., Rasmussen, M.B., Markager, S., Olesen, B., Arias, C., Jensen, P.D., 2011. Bioenergy potential of Ulva lactuca: biomass yield, methane production and combustion. *Bioresour. Technol.* 102, 2595–2604. 10.1016/j.biortech.2010.10.010

Chandrasekhar, K., Ahn, Y.H., 2017. Effectiveness of piggery waste treatment using microbial fuel cells coupled with elutriated-phased acid fermentation. *Bioresour. Technol.* 244, 650–657. 10.1016/j.biortech.2017.08.021

Chandrasekhar, K., Cayetano, R.D.A., Mehrez, I., Kumar, G., Kim, S.-H., 2020a. Evaluation of the biochemical methane potential of different sorts of Algerian date biomass. *Environ. Technol. Innov.* 20, 101180. 10.1016/j.eti.2020.101180

Chandrasekhar, K., Kadier, A., Kumar, G., Nastro, R.A., Jeevitha, V., 2018. Challenges in microbial fuel cell and future scope, in: Das, D. (Ed.), *Microbial Fuel Cell: A Bioelectrochemical System that Converts Waste to Watts*. Springer International Publishing, Cham, pp. 483–499. 10.1007/978-3-319-66793-5_25

Chandrasekhar, K., Kumar, G., Venkata Mohan, S., Pandey, A., Jeon, B.H., Jang, M., Kim, S.H., 2020b. Microbial Electro-Remediation (MER) of hazardous waste in aid of sustainable energy generation and resource recovery. *Environ. Technol. Innov.* 19. Article no. 100997. 10.1016/j.eti.2020.100997

Chandrasekhar, K., Kumar, S., Lee, B.-D., Kim, S.-H., 2020c. Waste based hydrogen production for circular bioeconomy: Current status and future directions. *Bioresour. Technol.* 302, 122920. 10.1016/j.biortech.2020.122920

Chandrasekhar, K., Velvizhi, G., Venkata Mohan, S., 2021. Bio-electrocatalytic remediation of hydrocarbons contaminated soil with integrated natural attenuation and chemical oxidant. *Chemosphere* 280, 130649. 10.1016/j.chemosphere.2021.130649

Chaturvedi, V., Verma, P., 2016. Microbial fuel cell: a green approach for the utilization of waste for the generation of bioelectricity. *Bioresour. Bioprocess.* 3. Article number 38. 10.1186/s40643-016-0116-6

Cheng, S., Liu, H., Logan, B.E., 2006. Increased performance of single-chamber microbial fuel cells using an improved cathode structure. *Electrochem. Commun.* 8, 489–494. 10.1016/j.elecom.2006.01.010

Cheng, S., Logan, B.E., 2007. Ammonia treatment of carbon cloth anodes to enhance power generation of microbial fuel cells. *Electrochem. Commun.* 9, 492–496. 10.1016/j.elecom.2006.10.023

Daniel, D.K., Mankidy, B.D., Ambarish, K., Manogari, R. 2009. Construction and operation of a microbial fuel cell for electricity generation from wastewater. *Int. J. Hydrog. Energy* 34(17), 7555–7560.

Deval, A.S., Parikh, H.A., Kadier, A., Chandrasekhar, K., Bhagwat, A.M., Dikshit, A.K., 2017. Sequential microbial activities mediated bioelectricity production from distillery wastewater using bio-electrochemical system with simultaneous waste remediation. *Int. J. Hydrog. Energy* 42, 1130–1141. 10.1016/j.ijhydene.2016.11.114

Do, M.H., Ngo, H.H., Guo, W.S., Liu, Y., Chang, S.W., Nguyen, D.D., Nghiem, L.D., Ni, B.J., 2018. Challenges in the application of microbial fuel cells to wastewater treatment and energy production: A mini review. *Sci. Total Environ.* 639, 910–920. 10.1016/j.scitotenv.2018.05.136

Enamala, M.K., Dixit, R., Tangellapally, A., Singh, M., Dinakarrao, S.M.P., Chavali, M., Pamanji, S.R., Ashokkumar, V., Kadier, A., Chandrasekhar, K., 2020. Photosynthetic microorganisms (Algae) mediated bioelectricity generation in microbial fuel cell: Concise review. *Environ. Technol. Innov.* 19, 100959. 10.1016/j.eti.2020.100959

Enamala, M.K., Enamala, S., Chavali, M., Donepudi, J., Yadavalli, R., Kolapalli, B., Aradhyula, T.V., Velpuri, J., Kuppam, C., 2018. Production of biofuels from microalgae – A review on cultivation, harvesting, lipid extraction, and numerous applications of microalgae. *Renew. Sustain. Energy Rev.* 94, 49–68. 10.1016/j.rser.2018.05.012

Enamala, M.K., Pasumarthy, D.S., Gandrapu, P.K., Chavali, M., Mudumbai, H., Kuppam, C., 2019. Production of a variety of industrially significant products by biological sources through fermentation, in: Arora, P.K. (Ed.), *Microbial Technology for the Welfare of Society*. Springer Singapore, Singapore, pp. 201–221. 10.1007/978-981-13-8844-6_9

Fleury, D., 2017. A modular photosynthetic microbial fuel cell with interchangeable algae solar compartments. *bioRxiv* 166793, 1–16. 10.1101/166793

Gadhamshetty, V., Belanger, D., Gardiner, C.J., Cummings, A., Hynes, A., 2013. Evaluation of Laminaria-based microbial fuel cells (LbMs) for electricity production. *Bioresour. Technol.* 127, 378–385. 10.1016/j.biortech.2012.09.079

Gambino, E., Chandrasekhar, K., Nastro, R.A., 2021. SMFC as a tool for the removal of hydrocarbons and metals in the marine environment: A concise research update. *Environ. Sci. Pollut. Res.* 1–16. 10.1007/s11356-021-13593-3

Gil, G.C., Chang, I.S., Kim, B.H., Kim, M., Jang, J.K., Park, H.S., Kim, H.J., 2003. Operational parameters affecting the performannce of a mediator-less microbial fuel cell. *Biosens. Bioelectron.* 18(4), 327–334. 10.1016/s0956-5663(02)00110-0

Girme, G.M., Faze, N.R., Bower, T.A., Christy, A. 2014. Algae powered Microbial Desalination Cells. American Society of Agricultural and Biological Engineers Annual International Meeting 2014, ASABE 2014. 7, 4672–4685.

González del Campo, A., Cañizares, P., Rodrigo, M.A., Fernández, F.J., Lobato, J., 2013. Microbial fuel cell with an algae-assisted cathode: A preliminary assessment. *J. Power Sources* 242, 638–645. 10.1016/j.jpowsour.2013.05.110

Goswami, R.K., Mehariya, S., Obulisamy, P.K., Verma, P., 2021a. Advanced microalgae-based renewable biohydrogen production systems: A review. *Bioresour. Technol.* 320 (Pt A), 124301. 10.1016/j.biortech.2020.124301

Goswami, R.K., Mehariya, S., Verma, P., Lavecchia, R., Zuorro, A., 2021b. Microalgae-based biorefineries for sustainable resource recovery from wastewater. *J. Water Process Eng.* 40(3), 101747. 10.1016/j.jwpe.2020.101747

Gustave, W., Yuan, Z., Liu, F., Chen, Z., 2021. Mechanisms and challenges of microbial fuel cells for soil heavy metal(loid)s remediation. *Sci. Total Environ.* 756, 143865. 10.1016/j.scitotenv.2020.143865

Hessami, M.J., Phang, S.M., Sohrabipoor, J., Zafar, F.F., Aslanzadeh, S., 2019. The bio-methane potential of whole plant and solid residues of two species of red seaweeds: Gracilaria manilaensis and Gracilariopsis persica. *Algal Res.* 42, 101581. 10.1016/j.algal.2019.101581

Hou, Q., Nie, C., Pei, H., Hu, W., Jiang, L., Yang, Z., 2016. The effect of algae species on the bioelectricity and biodiesel generation through open-air cathode microbial fuel cell with kitchen waste anaerobically digested effluent as substrate. *Bioresour. Technol.* 218, 902–908. 10.1016/j.biortech.2016.07.035

Jannelli, N., Anna Nastro, R., Cigolotti, V., Minutillo, M., Falcucci, G., 2017. Low pH, high salinity: Too much for microbial fuel cells? *Appl. Energy* 192, 543–550. 10.1016/j.apenergy.2016.07.079

Jareonsin, S., Pumas, C., 2021. Advantages of Heterotrophic Microalgae as a Host for Phytochemicals Production. *Front. Bioeng. Biotechnol.* 9, 628597.

Kakarla, R., Kuppam, C., Pandit, S., Kadier, A., Velpuri, J., 2017. Algae-the potential future fuel: Challenges and prospects, in: Kalia, V.C., Kumar, P. (Eds.), *Microbial Applications Vol. 1: Bioremediation and Bioenergy*. Springer International Publishing, Cham, pp. 239–251. 10.1007/978-3-319-52666-9_11

Kakarla, R., Min, B., 2014a. Evaluation of microbial fuel cell operation using algae as an oxygen supplier: Carbon paper cathode vs. carbon brush cathode. 37(12), 2453–2461. 10.1007/s00449-014-1223-4

Kakarla, R., Min, B., 2014b. Photoautotrophic microalgae Scenedesmus obliquus attached on a cathode as oxygen producers for microbial fuel cell (MFC) operation. *Int. J. Hydrog. Energy* 39, 10275–10283. 10.1016/j.ijhydene.2014.04.158

Kannan, N., Donnellan, P., 2021. Algae-assisted microbial fuel cells: A practical overview. *Bioresour. Technol. Rep.* 15. 100747. 10.1016/j.biteb.2021.100747

Karthik, O., Nagendranatha Reddy, C., Mehariya, S., Banu, R., Kannah R., Y., Kavitha, Jayaprakash, K., Rajashri, Y., 2020a. Electro-fermentation of biomass for high-value organic acids, In: *Biorefineries: A Step Towards Renewable and Clean Energy*, Springer Nature,Singapore 10.1007/978-981-15-9593-6_16

Karthik, O., Reddy, C.N., Mehariya, S., Banu, R.J., 2020b. Electro-fermentation of biomass for high-value organic acids metadata of the chapter that will be visualized online, in: *Biorefineries: A Step Towards Renewable and Clean Energy*, Springer Nature, Singapore. 10.1007/978-981-15-9593-6

Kaur, A., Kim, J.R., Michie, I., Dinsdale, R.M., Guwy, A.J., Premier, G.C.(SERC), =Sustainable Environment Research Centre, 2013. Microbial fuel cell type biosensor for specific volatile fatty acids using acclimated bacterial communities. *Biosens. Bioelectron.* 47, 50–55. 10.1016/j.bios.2013.02.033

Khandelwal, A., Chhabra, M., Yadav, P., 2020. Performance evaluation of algae assisted microbial fuel cell under outdoor conditions. *Bioresour. Technol.* 310, 123418. 10.1016/j.biortech.2020.123418

Kim, H.M., Oh, C.H., Bae, H.J., 2017. Comparison of red microalgae (Porphyridium cruentum) culture conditions for bioethanol production. *Bioresour. Technol.* 233, 44–50. 10.1016/j.biortech.2017.02.040

Klassen, V., Blifernez-Klassen, O., Wibberg, D., Winkler, A., Kalinowski, J., Posten, C., Kruse, O., 2017. Highly efficient methane generation from untreated microalgae biomass. *Biotechnol. Biofuels* 10, 186. 10.1186/s13068-017-0871-4

Kondaveeti, S., Abu-Reesh, I.M., Mohanakrishna, G., Bulut, M., Pant, D., 2020. Advanced routes of biological and bio-electrocatalytic carbon dioxide (CO2) mitigation toward carbon neutrality. *Front. Energy Res.* 8, 00094.

Kondaveeti, S., Choi, K.S., Kakarla, R., Min, B., 2014. Microalgae Scenedesmus obliquus as renewable biomass feedstock for electricity generation in microbial fuel cells (MFCs). *Front. Environ. Sci. Eng.* 8, 784–791. 10.1007/s11783-013-0590-4

Koushalya, S., Vishwakarma, R., Malik, A., 2021. Unraveling the diversity of algae and its biomacromolecules, In: *Microbial and Natural Macromolecules*. Elsevier, pp. 179–204. 10.1016/b978-0-12-820084-1.00008-9

Kusmayadi, A., Leong, Y.K., Yen, H.W., Huang, C.Y., Dong, C. Di, Chang, J.S., 2020. Microalgae-microbial fuel cell (mMFC): an integrated process for electricity generation, wastewater treatment, CO2 sequestration and biomass production. *Int. J. Energy Res.* 44, 9254–9265. 10.1002/er.5531

Liu, H., Hu, T.J., Zeng, G.M., Yuan, X.Z., Wu, J.J., Shen, Y., Yin, L., 2013. Electricity generation using p-nitrophenol as substrate in microbial fuel cell. *Int. Biodeterior. Biodegradation* 76, 108–111. 10.1016/j.ibiod.2012.06.015

Liu, T., Rao, L., Yuan, Y., Zhuang, L., 2015. Bioelectricity generation in a microbial fuel cell with a self-sustainable photocathode. *Sci. World J.* 2015, 864568. 10.1155/2015/864568

Mehariya, S., Fratini, F., Lavecchia, R., Zuorro, A., 2021c. Green extraction of value-added compounds form microalgae: A short review on natural deep eutectic solvents (NaDES) and related pre-treatments. *J. Environ. Chem. Eng.* 9, 105989. 10.1016/j.jece.2021.105989

Mehariya, S., Goswami, R.K., Karthikeysan, O.P., Verma, P., 2021a. Microalgae for high-value products: A way towards green nutraceutical and pharmaceutical compounds. *Chemosphere* 280, 130553. 10.1016/j.chemosphere.2021.130553

Mehariya, S., Goswami, R.K., Verma, P., Lavecchia, R., Zuorro, A., 2021b. Integrated approach for wastewater treatment and biofuel production in microalgae biorefineries. *Energies* 14, 2282. 10.3390/en14082282

Mehariya, S., Iovine, A., Casella, P., Musmarra, D., Chianese, S., Marino, T., Figoli, A., Sharma, N., Molino, A., 2020. Chapter 12 – Bio-based and agriculture resources for production of bioproducts, in: Figoli, A., Li, Y., Basile, A. (Eds.), *Current Trends and Future Developments on (Bio-) Membranes.* Elsevier, pp. 263–282. 10.1016/B978-0-12-816778-6.00012-6

Mehariya, S., Iovine, A., Di Sanzo, G., Larocca, V., Martino, M., Leone, G., Casella, P., Karatza, D., Marino, T., Musmarra, D., Molino, A., 2019. Supercritical fluid extraction of lutein from Scenedesmus almeriensis. *Molecules* 24, 1324. 10.3390/molecules24071324

Mehariya, S., Kumar, P., Marino, T., Casella, P., Iovine, A., Verma, P., Musmarra, D., Molino, A., 2021d. Aquatic weeds: A potential pollutant removing agent from wastewater and polluted soil and valuable biofuel feedstock, in: Pant, D., Bhatia, S.K., Patel, A.K., Giri, A. (Eds.), *Bioremediation Using Weeds.* Springer Singapore, Singapore, pp. 59–77. 10.1007/978-981-33-6552-0_3

Mekuto, L., Olowolafe, A.V.A., Pandit, S., Dyantyi, N., Nomngongo, P., Huberts, R., 2020. Microalgae as a biocathode and feedstock in anode chamber for a self-sustainable microbial fuel cell technology: A review. *S. Afr. J. Chem. Eng.* 31, 7–16. 10.1016/j.sajce.2019.10.002

Melhuish, C., Ieropoulos, I., Greenman, J., Horsfield, I., 2006. Energetically autonomous robots: Food for thought. *Auton. Robot.* 21, 187–198. 10.1007/s10514-006-6574-5

Mohamed, S.N., Hiraman, P.A., Muthukumar, K., Jayabalan, T. 2020. Bioelectricity production from kitchen wastewater using microbial fuel cell with photosynthetic algal cathode. *Bioresour. Technol.* 295, 122226. 10.1016/j.biortech.2019.122226

Mohan, S.V., Devi, M.P., Mohanakrishna, G., Amarnath, N., Babu, M.L., Sarma, P.N., 2011a. Potential of mixed microalgae to harness biodiesel from ecological water-bodies with simultaneous treatment. *Bioresour. Technol.* 102, 1109–1117. 10.1016/j.biortech.2010.08.103

Mohan, S.V., Devi, M.P., Reddy, M.V., Chandrasekhar, K., 2011b. Bioremediation of petroleum sludge under anaerobic microenvironment: Influence of biostimulation and bioaugmentation. *Environmental Engineering and Management Journal* 10, 1609–1616.

Molino, A., Larocca, V., Di Sanzo, G., Martino, M., Casella, P., Marino, T., Karatza, D., Musmarra, D., 2019a. Extraction of Bioactive Compounds Using Supercritical Carbon Dioxide. *Molecules* 24(4), 782. 10.3390/molecules24040782

Molino, A., Mehariya, S., Karatza, D., Chianese, S., Iovine, A., Casella, P., Marino, T., Musmarra, D., 2019b. Bench-Scale Cultivation of Microalgae Scenedesmus almeriensis for CO_2 Capture and Lutein Production. *Energies* 12(14), 2806. 10.3390/en12142806

Munoz-Cupa, C., Hu, Y., Xu, C., Bassi, A., 2021. An overview of microbial fuel cell usage in wastewater treatment, resource recovery and energy production. *Sci. Total Environ.* 10.1016/j.scitotenv.2020.142429

Mureed, K., Kanwal, S., Hussain, A., Noureen, S., Hussain, S., Ahmad, S., Ahmad, M., Waqas, R., 2018. Biodiesel production from algae grown on food industry wastewater. *Environ. Monit. Assess.* 190, 3–13.10.1007/s10661-018-6641-3

Ndayisenga, F., Yu, Z., Yu, Y., Lay, C.-H., Zhou, D., 2018. Bioelectricity generation using microalgal biomass as electron donor in a bio-anode microbial fuel cell. *Bioresource Technol.* 270, 286–293. 10.1016/j.biortech.2018.09.052

Nishio, K., Hashimoto, K., Watanabe, K., 2013. Light/electricity conversion by defined cocultures of Chlamydomonas and Geobacter. *Journal of bioscience and bioengineering* 115, 412–417. 10.1016/j.jbiosc.2012.10.015

Obileke, K.C., Onyeaka, H., Meyer, E.L., Nwokolo, N., 2021. Microbial fuel cells, a renewable energy technology for bio-electricity generation: A mini-review. *Electrochemistry Communications* 125, 107003. 10.1016/j.elecom.2021.107003

Pandit, S., Chandrasekhar, K., Kakarla, R., Kadier, A., Jeevitha, V., 2017. Basic principles of microbial fuel cell: technical challenges and economic feasibility, in: *Microbial Applications Vol.1*. Springer International Publishing, Cham, pp. 165–188. 10.1007/978-3-319-52666-9_8

Park, J.-H., Chandrasekhar, K., Jeon, B.-H., Jang, M., Liu, Y., Kim, S.-H., 2021. State-of-the-art technologies for continuous high-rate biohydrogen production. *Bioresour. Technol.* 320, *124304*. 10.1016/j.biortech.2020.124304

Patel, V., Pandit, S., Chandrasekhar, K., 2017. Basics of methanogenesis in anaerobic digester, in: *Microbial Applications*. Springer International Publishing, Cham, pp. 291–314. 10.1007/978-3-319-52669-0_16

Puig, S., Serra, M., Coma, M., Cabré, M., Balaguer, M.D., Colprim, J., 2010. Effect of pH on nutrient dynamics and electricity production using microbial fuel cells. *Bioresour. Technol.* 101, 9594–9599. 10.1016/j.biortech.2010.07.082

Rahimnejad, M., Adhami, A., Darvari, S., Zirepour, A., Oh, S.-E., 2015. Microbial fuel cell as new technology for bioelectricity generation: A review. *Alex. Eng. J.* 54, 745–756. 10.1016/j.aej.2015.03.031

Rahimnejad, M., Bakeri, G., Ghasemi, M., Zirepour, A., 2014. A review on the role of proton exchange membrane on the performance of microbial fuel cell. *Polym. Adv. Technol.* 25, 1426–1432. 10.1002/pat.3383

Rasep, Z., Aripen, N.S.M., Ghazali, M.S.M., Yahya, N., Arida, A.S., Som, A.M., Mustaza, M.F., 2016. Microbial fuel cell for conversion of chemical energy to electrical energy from food industry wastewater. *J. Environ. Sci. Technol.* 9, 481–485. 10.3923/jest.201 6.481.485

Rashid, N., Cui, Y.-F., Saif Ur Rehman, M., Han, J.-I., 2013. Enhanced electricity generation by using algae biomass and activated sludge in microbial fuel cell. *Sci. Total Environ.* 456–457, 91–94. 10.1016/j.scitotenv.2013.03.067

Rezaei, F., Richard, T.L., Brennan, R.A., Logan, B.E., 2007. Substrate-enhanced microbial fuel cells for improved remote power generation from sediment-based systems. *Environ. Sci. Technol.* 41, 4053–4058.

Saba, B., Christy, A.D., Yu, Z., Co, A.C., 2017. Sustainable power generation from bacterio-algal microbial fuel cells (MFCs): An overview. *Renew. Sustain. Energy Rev.* 73, 75–84. 10.1016/j.rser.2017.01.115

Saratale, R.G., Kuppam, C., Mudhoo, A., Saratale, G.D., Periyasamy, S., Zhen, G., Koók, L., Bakonyi, P., Nemestóthy, N., Kumar, G., 2017. Bioelectrochemical systems using microalgae – A concise research update. *Chemosphere* 177, 35–43. 10.1016/j.chemosphere.2017.02.132

Sarma, R., Tamuly, A., Kakati, B.K., 2021. Recent developments in electricity generation by Microbial Fuel Cell using different substrates. *Mater. Today Proc.* 10.1016/j.matpr.2021.02.522

Sayed, R., Ghadir, A.E.-C., 2020. Marine Algae in Egypt: distribution, phytochemical composition and biological uses as bioactive resources (a review). *Egypt. J. Aquat. Biol. Fish.* 24, 147–160.

Serrà, A., Artal, R., García-Amorós, J., Gómez, E., Philippe, L., 2020. Circular zero-residue process using microalgae for efficient water decontamination, biofuel production, and carbon dioxide fixation. *Chem. Eng. J.* 388, 124278. 10.1016/j.cej.2020.124278

Shukla, M., Kumar, S., 2018. Algal growth in photosynthetic algal microbial fuel cell and its subsequent utilization for biofuels. *Renew. Sustain. Energy Rev.* 82, 402–414. 10.1016/j.rser.2017.09.067

Siciliano, A., Limonti, C., Mehariya, S., Molino, A., Calabrò, V., 2019. Biofuel production and phosphorus recovery through an integrated treatment of agro-industrial waste. *Sustainability* 11(1), 52. 10.3390/su11010052

Singh, M., Chavali, M., Enamala, M., Karthik, O., Dixit, R., Kuppam, C., 2021. Algal Bioeconomy: A Platform for Clean Energy and Fuel, in: Verma, P. (Ed.), *Biorefineries: A Step Towards Renewable and Clean Energy, Clean Energy Production Technologies.* Springer Nature, Singapore, pp. 335–370. 10.1007/978-981-15-9593-6_13

Singh, M., Chavali, M., Enamala, M.K., Obulisamy, P.K., Dixit, R., Kuppam, C., 2020. *Algal Bioeconomy: A Platform for Clean Energy and Fuel.* Springer, Singapore, pp. 335–370. 10.1007/978-981-15-9593-6_13

Srivastava, R.K., 2019. Bio-energy production by contribution of effective and suitable microbial system. *Mater. Sci. Energy Technol.* 2, 308–318. 10.1016/j.mset.2018.12.007

Subadri, I., Adhi, S., Ignatius, D.M.B., 2020. Facing Indonesia's Future Energy with Bacterio-Algal Fuel Cells. *Indones. J. Energy* 3, 68–82. doi:10.33116/ije.v3i2.87

Sudhakar, M.P., Kumar, B.R., Mathimani, T., Arunkumar, K., 2019. A review on bioenergy and bioactive compounds from microalgae and macroalgae-sustainable energy perspective. *J. Clean. Prod.* 228, 1320–1333. 10.1016/j.jclepro.2019.04.287

Sun, H., Zhao, W., Mao, X., Li, Y., Wu, T., Chen, F., 2018. High-value biomass from microalgae production platforms: strategies and progress based on carbon metabolism and energy conversion. *Biotechnol. Biofuels* 11, 227. 10.1186/s13068-018-1225-6

Ullah, Z., Zeshan, S., 2020. Effect of substrate type and concentration on the performance of a double chamber microbial fuel cell. *Water Sci. Technol.: a Journal of the International Association on Water Pollution Research* 81, 1336–1344. 10.2166/wst.2019.387

Velasquez-Orta, S.B., Curtis, T.P., Logan, B.E., 2009. Energy from algae using microbial fuel cells. *Biotechnol. Bioeng.* 103, 1068–1076. 10.1002/bit.22346

Venkata Subhash, G., Chandra, R., Venkata Mohan, S., 2013. Microalgae mediated bioelectrocatalytic fuel cell facilitates bioelectricity generation through oxygenic photomixotrophic mechanism. *Bioresour. Technol.* 136: 644–653. 10.1016/j.biortech.2013.02.035

Walter, X.A., Greenman, J., Taylor, B., Ieropoulos, I.A., 2015. Microbial fuel cells continuously fuelled by untreated fresh algal biomass. *Algal Res.* 11, 103–107. 10.1016/j.algal.2015.06.003

Wilkinson, S., 2000. "Gastrobots"—Benefits and Challenges of Microbial Fuel Cells in FoodPowered Robot Applications. *Auton. Robot.* 9, 99–111. 10.1023/A:1008984516499

Xiao, L., He, Z., 2014. Applications and perspectives of phototrophic microorganisms for electricity generation from organic compounds in microbial fuel cells. *Renew. Sustain. Energy Rev.* 37, 550–559. 10.1016/j.rser.2014.05.066

Xu, C., Poon, K., Choi, M.M.F., Wang, R., 2015. Using live algae at the anode of a microbial fuel cell to generate electricity. *Environ. Sci. Pollut. Res.* 22, 15621–15635. 10.1007/s11356-015-4744-8

Yadavalli, R., Ratnapuram, H., Motamarry, S., Reddy, C.N., Ashokkumar, V., Kuppam, C., 2020. Simultaneous production of flavonoids and lipids from Chlorella vulgaris and Chlorella pyrenoidosa. *Biomass Convers. Biorefin.* 10.1007/s13399-020-01044-x

Yadavalli, R., Ratnapuram, H., Peasari, J.R., Reddy, C.N., Ashokkumar, V., Kuppam, C., 2021. Simultaneous production of astaxanthin and lipids from Chlorella sorokiniana in the presence of reactive oxygen species: A biorefinery approach. *Biomass Convers. Biorefin.* 1–9. 10.1007/s13399-021-01276-5

Yang, C., Hua, Q., Shimizu, K., 2000. Energetics and carbon metabolism during growth of microalgal cells under photoautotrophic, mixotrophic and cyclic light-autotrophic/dark-heterotrophic conditions. *Biochem. Eng. J.* 6, 87–102. 10.1016/s1369-703x(00)00080-2

Yaqoob, A.A., Ibrahim, M.N.M., Guerrero-Barajas, C., 2021. Modern trend of anodes in microbial fuel cells (MFCs): An overview. *Environ. Technol. Innov.* 23, 101579. 10.1016/j.eti.2021.101579

Yuan, Y., Chen, Q., Zhou, S., Zhuang, L., Hu, P., 2011. Bioelectricity generation and microcystins removal in a blue-green algae powered microbial fuel cell. *J. Hazard. Mater.* 187, 591–595. 10.1016/j.jhazmat.2011.01.042

Yukesh Kannah, R., Kavitha, S., Parthiba Karthikeyan, O., Rene, E.R., Kumar, G., Rajesh Banu, J., 2021. A review on anaerobic digestion of energy and cost effective microalgae pretreatment for biogas production. *Bioresour. Technol.* 332, 125055. 10.1016/j.biortech.2021.125055

Zhang, Q., Hu, J., Lee, D.J., 2016. Microbial fuel cells as pollutant treatment units: Research updates. *Bioresour. Technol.* 217, 121–128. 10.1016/j.biortech.2016.02.006

Zhang, Y., He, Q., Xia, L., Li, Y., Song, S., 2018. Algae cathode microbial fuel cells for cadmium removal with simultaneous electricity production using nickel foam/graphene electrode. *Biochem. Eng. J.* 138, 179–187. 10.1016/j.bej.2018.07.021

4 Potential of Microalgae for Protein Production

Elena M. Rojo and Silvia Bolado
Department of Chemical Engineering and Environmental
Technology, School of Industrial Engineering, University of
Valladolid, Valladolid, Spain
and
Institute of Sustainable Processes, University of Valladolid,
Valladolid, Spain

Alejandro Filipigh
Department of Chemical Engineering and Environmental
Technology, School of Industrial Engineering, University of
Valladolid, Valladolid, Spain

David Moldes
Department of Analytical Chemistry, Faculty of Sciences,
University of Valladolid, Valladolid, Spain

Marisol Vega
Institute of Sustainable Processes, University of Valladolid,
Valladolid, Spain
and
Department of Analytical Chemistry, Faculty of Sciences,
University of Valladolid, Valladolid, Spain

CONTENTS

4.1 Introduction..92
4.2 Effect of Microalgae Culture in Protein Production....................96
    4.2.1 Light...98
    4.2.2 Temperature..100
    4.2.3 pH..101
    4.2.4 Salinity...101
    4.2.5 Nutrients..102
          4.2.5.1 Carbon...103
          4.2.5.2 Nitrogen...104

DOI: 10.1201/9781003195405-4

4.3　Protein Extraction Methods: Effect of Different Methods and
　　　Operational Parameters on Extraction Yields and Protein Recovery.........105
　　　4.3.1　Physical Methods 107
　　　　　　　4.3.1.1　Bead Milling..107
　　　　　　　4.3.1.2　High Pressure Homogenization.......................................107
　　　　　　　4.3.1.3　Ultrasonication..108
　　　　　　　4.3.1.4　Electric Field...109
　　　　　　　4.3.1.5　Microwave 109
　　　4.3.2　Chemical Methods..110
　　　　　　　4.3.2.1　Acid and Alkaline Treatment..110
　　　　　　　4.3.2.2　Oxidative Treatment...111
　　　4.3.3　Biological Methods 111
　　　　　　　4.3.3.1　Enzymatic Hydrolysis..111
　　　4.3.4　Novel Methods 112
　　　　　　　4.3.4.1　Ionic Liquids...112
　　　　　　　4.3.4.2　Osmotic Shock..112
4.4　Protein Separation Methods...113
　　　4.4.1　Three-Phase Partitioning 113
　　　4.4.2　Membrane Technology...114
　　　4.4.3　Protein Dispersion – pH Shift and Salting Out 115
　　　4.4.4　Electrophoresis 116
　　　4.4.5　Column Chromatography 116
　　　　　　　4.4.5.1　Size Exclusion (SEC)...117
　　　　　　　4.4.5.2　Ion Exchange (IEC)..118
　　　　　　　4.4.5.3　Affinity Chromatography (AC).......................................118
　　　　　　　4.4.5.4　Hydrophobic Interactions (HIC) 118
　　　　　　　4.4.5.5　Thin Layer Chromatography (TLC) 118
4.5　Potential Application of Proteins from Microalgae..................................119
　　　4.5.1　Microalgae Proteins in Feeding and Nutrition.............................119
　　　　　　　4.5.1.1　Microalgae Proteins for Humans 119
　　　　　　　4.5.1.2　Microalgae Proteins as Animal Feed...............................120
　　　4.5.2　Microalgae Proteins as Nutraceuticals.......................................121
　　　4.5.3　Other Industrial Applications of Microalgae Proteins 123
4.6　Challenges in the Production of Microalgae Proteins 124
4.7　Conclusions..125
Acknowledgments...125
Bibliography 125

4.1　INTRODUCTION

Microalgae are microorganisms known for their high productivity. They can produce and accumulate significant amounts of biomolecules such as lipids, carbohydrates, antioxidants, pigments, and proteins (Alavijeh et al., 2020). From an industrial point of view, microalgae culture has received significant interest due to the high growth rate of microalgae. However, despite microalgae's remarkable capacity to produce bioactive molecules, the extraction and recovery of microalgae components is not straightforward, and extensive research is currently being developed in this field (Costa et al., 2020). These drawbacks are related to microalgae cytology, especially to the rigid cell wall of these microorganisms. Microalgae cells have a cell wall and a plasma membrane encapsulating the cytosol, which contains a defined nucleus. While the main organelles show a similar chemical composition compared to other organisms, the cell wall's chemistry varies significantly among microalgae species (Alhattab et al., 2019). However, some general trends can be stated: the plasma membrane is made of phospholipids and transmembrane proteins, while the cell wall is formed by cellulose fibers, hemicellulose, β-glucan, and proteins, which confers its harshness. Sometimes a mucilage layer can also be found, composed of alginate and extracellular matrix, acting as a defense layer (Borowitzka, 2018).

The cell contains bioactive compounds, which are usually classified into primary and secondary metabolites, depending on their biosynthetic origin, chemical composition, or function. Primary metabolites are produced as a result of cell growth, cell development, and microalgae reproduction and mainly include protein, carbohydrate, lipids, and photosynthetic pigments. Secondary metabolites are uniquely accumulated to relieve cellular injuries under stress condition and consist mainly of carotenoid, phytosterols, and phenolic compounds. Some carotenoids, such as lutein and fucoxanthin, are components of the light-harvesting complex for photosynthesis and photo-protection and can thus be considered primary metabolites (de Morais et al., 2015). On the other hand, microalgal lipids can be mainly divided into membrane lipids (consisting of polar lipids) and storage lipids (consisting of neutral lipids mainly in the form of triglyceride). Similarly, carbohydrates can be divided into structural carbohydrates and storage carbohydrate (such as starch, glycogen, and glucan). Generally, the membrane lipids and structural carbohydrates are related to cell growth, while storage lipids and carbohydrate accumulation are enhanced under stress conditions (Ma et al., 2020).

The microalgae composition depends mainly on the strain and cultivation conditions. The concentration of the three main macronutrients in microalgae can vary widely: the concentration of crude proteins can lay between 17 and 71%, lipids can represent up to 47% (although values around 5–15% are the most common), and the concentration of carbohydrates can be anywhere from 10 to 57% (Grossmann et al., 2020a).

Pigments are sometimes considered the fourth principal component of microalgae as they harvest and transfer energy from the sun, but their concentration is always lower than the other three (around 0–5%) (Ma et al., 2020). The three primary pigments are chlorophylls, phycobilins, and carotenoids, which differ in their absorption range in the visible spectrum (Pagels et al. 2020). Chlorophylls are the most critical pigment in eukaryotic cells, and phycobilins are more relevant in procaryotic cells and eukaryotic red algae. Phycobilins are noncyclic tetrapyrroles linked by alpha, beta, and gamma single carbon bridges as shown in Figure 4.1. The phycobilins are chromophores bound to a protein, forming phycobiliproteins, which are water-soluble proteins with a distinct color given by the phycobilins that they possess (Ma et al., 2020).

The last type of pigments are carotenoids. Carotenoids are lipophilic compounds that have light absorbance capacities and are essential in some cell defense

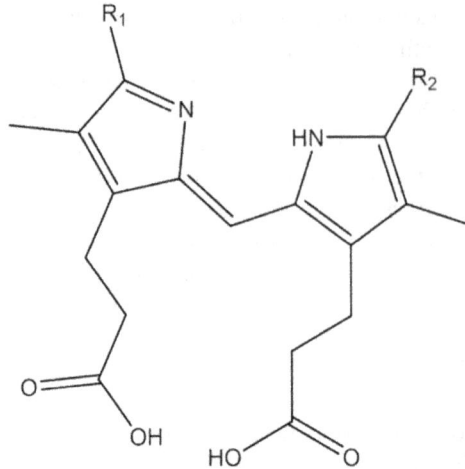

FIGURE 4.1   General skeleton of phycobilin.

mechanisms due to their antioxidant properties. Based on their biosynthetic origin, carotenoids can be classified as primary (essential in photosynthesis) and secondary (possessing a defensive role in the cell) (Ma et al., 2020).

Among the common macronutrients found in microalgae, proteins are almost half of the dry weight, and water-soluble proteins that can be liberated from algal cells varies between 21–90% (Grossmann et al., 2020a), and phycobiliproteins may account for 50% of the total crude protein (Pagels et al., 2019). It is worth noting that the "crude protein" is usually overestimated, as it includes other non-protein nitrogen compounds, such as nucleic acids (which for some species account for 3–6%), pigments, glucosamides, or inorganic components (Becker, 2007). Proteins are crucial in microalgae metabolism and are involved in growth and maintenance processes. They also act as chemical messengers, regulators, and provide defense against other microorganisms. Nowadays, microalgae's protein quality is beyond doubt due it is known for its excellent digestibility, and it provides in high amounts the nine essential amino acids that humans do not synthesize (Acquah et al., 2020). This, coupled with their high biomass productivity rates, makes microalgae a potential answer to worldwide proteins demands.

Proteins may be found within the plasma membrane and in the cell wall (as transmembrane proteins), or bound to the membrane's lipids (as periphery proteins) (Figure 4.2). They are also found in the cytoplasm or as part of many organelles such as chloroplast, mitochondria, the endoplasmic reticulum, or inside the cell's nucleus (Safi et al., 2014a). Transmembrane proteins have a hydrophobic region in contact with the bilayer membrane that is tightly bound (Safi et al., 2014a).

The protein concentration of the cell depends on their place in the cell and the microalgae species studied. For instance, *Chlorella vulgaris* accounts for 42–58% of proteins in dry biomass weight, and it is estimated that 20% of its total proteins are found in the cell wall, around 30% actively migrate through the cell, and 50% are located in the cytoplasm (Berliner, 1986). Hence, as roughly 50% of the proteins

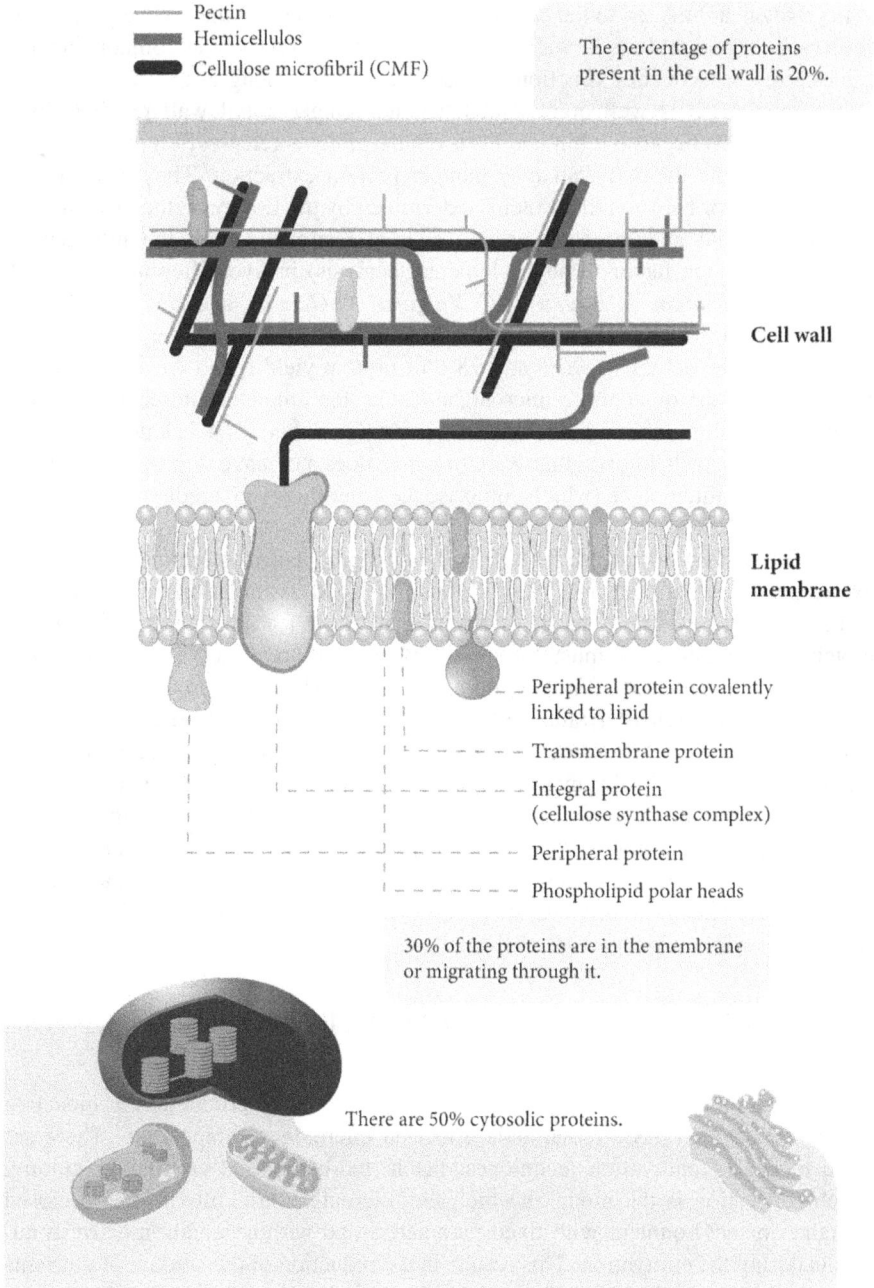

FIGURE 4.2   Protein localization on microalgal structure.

in microalgae are related to the cell membrane and to increase the recovery rate, it is necessary to disrupt the cell wall's multiples layers to release the components and enhance the subsequent extraction and isolation steps (Phong et al., 2018a).

In literature, it is well documented that microalgae's cell wall prevents high extraction yields because it hinders intracellular proteins release. Safi et al. (2014b) investigated how the cell wall may hamper protein extraction. They calculated a proportion factor between the proteins determined by the Lowry method (the hydro-soluble proteins) and the total protein content (estimated with the nitrogen-to-protein conversion factor through elemental analysis) in five different microalgae species: *P. Cruentum, A. Platensis, C. Vulgaris, N. Oculate,* and *H. Pluvialis.* The results showed that more hydro-soluble proteins were obtained when the cell wall was more labile, achieving 90% and 78% of protein yield for *P. Cruentum* and *A. Platensis.* For the other three microalgae tested, the amounts diminish, obtaining lower ratios of 52.8%, 52.3%, and 41% for *C. Vulgaris, N. Oculate,* and *H. Pluvialis.* It is well known that *P. Cruentum* does not have a proper cell wall, whereas green microalgae (which comprise the other three) are credited for having a more rigid cell wall.

Hence, the results indicate that the cell wall significantly determines the process extractability output, and it is vital in obtaining good assimilation, bioavailability, and solubilization of proteins. It is crucial that to get the advantage of the whole proteins' potential, they must be smoothly released to preserve their structural identity and functionality (Grossmann et al., 2018). However, proteins are not the only components released after cell wall breakthrough. The protein recovery requires separating proteins from other cell components like lipids, carbohydrates, and less concentrated substances such as pigments or nucleic acids, which are not commonly considered when dealing with microalgae composition. Additionally, an optimal extraction method should be easy to perform, consume small amounts of energy, and provide high disruption yields in short times (Soto-Sierra et al., 2018). From an industrial point of view, the process should ideally produce as little reagent waste as possible (Phong et al., 2018a).

4.2 EFFECT OF MICROALGAE CULTURE IN PROTEIN PRODUCTION

The cultivation of microalgae is the preliminary step in the production of bioactive compounds, which has a remarkable affect on microalgae composition. There are three modes of cultivation techniques: batch, fed-batch, and continuous culture. Batch cultivation is the mode at which the microalgae are cultivated in a closed container or environment with fixed parameters and with no addition of fresh nutrients during the cultivation. This results in the reduction of the amount of nutrients over time (Tang et al., 2020). This mode of culture has the advantage of being easy to handle and is suitable for studying growth kinetics and parameters affecting cell growth. However, this closed system will cause a self-shading effect of microalgal cells and non-homogeneous irradiance due to the continuous consumption of nutrients and high biomass production (Zhu, 2015). On the other hand, in fed-batch or semi-continuous culture, part of the volume is collected for use, typically at the end of the exponential growth phase, and the amount that is removed is replaced with fresh culture medium, which makes this process more efficient and reliable than batch cultivation. However, the drawbacks of this mode include the difficulty in controlling the microalgal growth pattern in the recirculation stream. In continuous cultivation (as in batch mode) the population of microalgae is kept in the exponential phase for long periods of time. This culture mode provides high product yields or production rates and easy manipulation of the nutrient concentration and pH. Hence, continuous cultivation would be more advantageous when compared to batch and semi-continuous cultivation conditions. However, this culture mode has not been favored at the industrial scale due to the high risk of contamination as well as the difficulty and complexity of the process (Tan et al., 2020).

There are two main types of photobioreactors (PBRs) used for microalgae cultivation: open systems and closed systems. In open cultivation, such as open ponds, tanks, and raceway ponds, the crop is exposed to the atmosphere and the physicochemical parameters are difficult to control. Some of the major advantages of an open cultivation system are minimal capital and operating cost, and a lower energy requirement for culture mixing. However, open systems require large areas to scale up, are susceptible to contamination, and are affected by adverse weather conditions. On the other hand, closed systems are reactors typically used to produce large quantities of microalgal biomass under controlled conditions and is considered the most appropriate system to cultivate pure microalgae (Tang et al., 2020).

The growth media and cultivation conditions (nutrients, light intensity, temperature, pH, etc.) can produce a remarkable effect on the biochemical composition of microalgae (Figure 4.3). The conditions for microalgal cultivation are important factors that influence the metabolism of these microorganisms, thus directing the synthesis of specific compounds of interest. In this context, the study of the

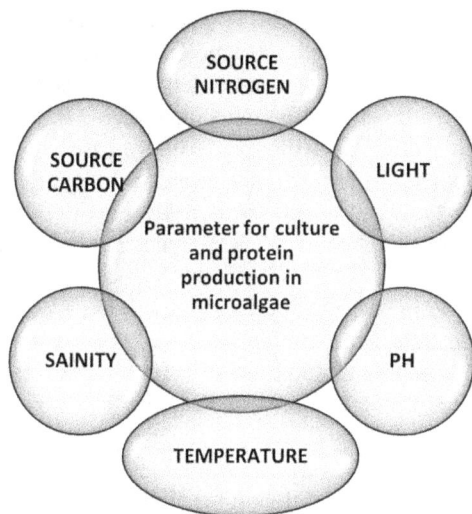

FIGURE 4.3   Growth medium parameters that influence biochemical composition of microalgae.

cultivation stage is crucial in obtaining a suitable microalgal biomass. Culture conditions promoting microalgae growth result in the accumulation of primary metabolites such as protein and structural carbohydrate. Culture conditions inhibiting cell growth, such as a decline in the nutrient concentration during the cultivation process, can limit the co-production of multiple compounds. Hence, changes in operational parameters can lead to higher productivity of the targeted products (de Morais et al., 2015). Unfortunately, very scarce information is available about the effect of these parameters on both biomass productivity and protein content of microalgae. A systematic study of how operational parameters affect protein productivity cannot be developed yet (Borella et al., 2021). So, this section focuses on the effect of the main operational and environmental parameters on the protein content of microalgal biomass. Table 4.1 shows some of the best conditions for the cultivation of microalgae with the objective of obtaining high concentrations of proteins.

4.2.1 LIGHT

Light is one of the three components essential for photosynthesis in microalgal cells, along with CO_2 and water. The chosen light intensity to culture microalgal cells should be high enough to penetrate through a dense culture but low enough not to reach the light threshold of the cells. Above the light threshold, excessive light damages the light-absorbing systems, causing photoinhibition and restricting the photosynthetic capability of the microalgal cells (Tan et al., 2020). There is a direct relationship between microalgae growth, light intensity, and cultivation duration because the variation of the latter two is able to directly affect photosynthesis and the related biochemical composition and biomass yield. For example, the effects of different light regimes on the biomass and protein productions of *Chlorella vulgaris* were investigated by Seyfabadi et al. (2011) in a batch photobioreactor. Operating at different irradiances (37.5, 62.5, and 100 μmol/m²/s), the increase in light duration (8:16, 12:12, and 16:8 h light/dark) was associated with an increase in specific growth rates. At 37.5 μmol/m²/s, the minimum growth rate (0.60 ± 0.02 day⁻¹) was observed under the 8:16 h photoperiod, and the exponential phase continued to day 10 of cultivation. At this point, the growth rate decreased slightly, and the maximum growth rate (0.81 ± 0.04 day⁻¹) was observed on day 8 in the 16:8 h photoperiod. The growth rate increased with the light irradiance, achieving the maximum value (1.13 ± 0.04 day⁻¹) at 100 μmol/m²/s with a 16:8 h photoperiod, and the exponential phase continued to day 7. The increase in irradiance and duration was associated with an increase in the growth rate and therefore in protein concentration. The maximum protein concentration was 46% at 100 μmol/m²/s during the 16:8 h photoperiod, and the minimum was 33% at 37.5 μmol/m²/s during the 8:16 h photoperiod.

Similar results were found by Tan et al. (2020), who studied the biomass and protein productivity of *Chlorella sorokinina* CY1 and *Chlorella vulgaris* ESP-31 in a batch photobioreactor at different light intensities (from 100 to 1000 μmol/m²/s). The biomass and protein productivity of both microalgae achieved maximum values under 750 μmol/m²/s of light intensity, which reduced the culture time of both

TABLE 4.1

Best Results on Different Culture Experiments for Protein Production in Microalgae (Source nitrogen in g/L; LI = light intensity (µmolphotonscm-1s-1; PP = Photoperiod (light/dark) in h; Cultivation time (days), Salinity (g/L) of NaCl; Temperature (°C); Biomass productivity (g/L/d); Protein, Carbohydrate and Lipid content (%)

Microalgae	Cultivation System and Operation Mode	Source of Carbon	Source of Nitrogen	LI	PP (h)	Cultivation Time (day)	pH	Salinity	Temperature	Biomass Productivity	Protein Content	Carbohydrate Content	Lipid Content	Reference
C. sorokiniana CY1 C. vulgaris ESP-31	Closed photobioreactor: batch	5% CO$_2$	0.75 g/L NaNO$_3$	750	N/A	2	7.5	N/A	35	4.35 4.64	25.9 26.8	– –	– –	(Tan et al., 2020)
Chlorella F&M-M49 Chlorella IAM-C212	Closed photobioreactor, fed-batch	Air/CO$_2$ mixture (98/2, v/v)	1.5 g/L NaNO$_3$	400	8:16	5	7.5	30	25	0.64 0.59	45.4 44.1	24.2 26.6	22.8 22.0	(Guccione et al., 2014)
Chlorella PROD1 Chlorella CH2 Chlorella PAW2PS										0.73 0.75 0.71	39.9 39.3 45.3	28.3 28.8 27.1	28.1 23.1 20.0	
D. salina SAG 184.80	Closed photobioreactor: batch	Air enriched with 2% CO$_2$	1 g/L KNO$_3$; 0.15 g/L NaNO$_3$	55	12:12	7	7.5	20	20	–	64.6	20	N/A	(Sui and Vlaeminck, 2019)
C. pyrenoidosa	Closed photobioreactor: batch	Air enriched with 4% CO$_2$	250 mg/L NaNO$_3$	127	12:12	N/A	6.8 6.8	25	25 35	1.92 1.92	57.9 63.7	12.5 14.3	19.3 12.7	(Zhao et al., 2019)
C. vulgaris	Closed photobioreactor: batch	N/A	N/A	100	16:8	10	N/A	N/A	25	1.13	46	N/A	33.38	(Seyfabadi et al., 2011)
Spirulina sp.	Closed photobioreactor: batch	N/A	N/A	150	N/A	N/A	9	N/A	30	–	48.23	16.26	4.38	(Ogbonda et al., 2007)
I. galbana	Closed photobioreactor: batch	Air/ 0.04% CO$_2$	0.072 g/L NaNO$_3$	80	12:12	13	N/A	2.5	25	0.33	36.3	16	29.4	(Zarrinmehr et al., 2020)

species by two days. The increase in light intensity enhanced light penetration into the culture, which in turn raised the biomass growth rate of both species. No effects of photoinhibition were found as the biomass concentrations of both species were high and the protein contents were stable at high light intensities of 750 and 1000 $\mu mol/m^2/s$. The one exception (decrease in protein content at 750 $\mu mol/m^2/s$ for *Chlorella sorokiniana* CY1) could be due to the accumulation of biomass at the expense of protein. The light intensity of 750 $\mu mol/m^2/s$ was chosen as the optimum parameter since the biomass and protein productivities at light intensity of 750 $\mu mol/m^2/s$ were similar to those obtained with 1000 $\mu mol/m^2/s$.

In summary, the light intensity and the photoperiod, which set the mechanism for many biological cycles, affect microalgal growth and metabolism. When cells are growing in light-dark cycles, protein and other macromolecule synthesis continue in the dark phase at the expense of the carbon and energy stored in carbohydrates (Zachleder et al., 2016).

4.2.2 Temperature

One of the most important factors for the growth of all living organisms is temperature. The specific growth rate of microalgae is directly correlated with the rate of CO_2 fixation/O_2 production (photosynthesis) and the respiration rate. Photosynthesis and respiration are temperature-dependent, with the respiration rate increasing exponentially as the temperature increases (de Morais et al., 2015). The optimum growth temperature for the most commonly used microalgae, such as *Chlorella, Chalamydomonas, Botryococcus, Scenedesmus, Neochloris, Haematococcus,* and *Nannochloropsis,* is in the range of 15–35°C depending on the strain (Dolganyuk et al., 2020). The recommended cultivation temperature for the maximum biomass yield of various types of microalgae ranges from 27°C to 30°C. Varshney et al. (2018) evaluated the effect of temperature on algal growth for *A. quadricellulare* and *C. sorokinina* by culturing the strain under optimal light (200 $\mu mol/m^2/s$) and atmospheric CO_2 condition at five temperatures (23.5, 30, 37, 40, and 43°C) in batch photobioreactors. It was observed that *A. quadricellulare* did not grow at 23.5°C but could survive and grow at a temperature of up to 43°C (μ=0.013 h^{-1}). *C. sorokinina* showed a low but measurable growth rate at 23.5°C (μ = 0.013 h^{-1}). However, it could not sustain growth beyond 40°C (μ= 0.030 h^{-1}). The optimum temperature for both strains was 37°C, where the maximum growth rates of 0.052 h^{-1} and 0.057 h^{-1} were obtained for *A. quadricellulare* and *C. sorokiniana,* respectively.

Temperature greatly influences the production of microalgal biomass. The protein content of microalgae has been associated with cell growth. The optimum temperature for cultivation of microalgae is 35–37°C (de Morais et al., 2015). Under sub-optimal temperatures, changes in the cytoplasmic viscosity occur, which results in low efficiency in the utilization of carbon and nitrogen followed by a reduction in the carboxylase activity. On the other hand, high temperatures reduce the enzymatic activity and adversely affect the protein structure and the efficiency of protein synthesis. This in turn promotes lipids storage and carbohydrate accumulation (Varshney et al., 2018). Ogbonda et al. (2007) studied the influence of temperature on biomass and protein biosynthesis for *Spirulina* sp. in a batch photobioreactor and measured the protein content and amino acid composition of *Spirulina* at different temperatures (25, 30, 35, and 40°C), obtaining few variations near the optimal temperature (30°C). Concretely, the highest biomass concentration (4.4 mg/ml), protein content (46.39 g/100 g), and amino acid content (76.06 g/16 gN) were obtained at 30°C. By decreasing the temperature to 25°C, a 40% decrease in the *Spirulina* sp. biomass concentration was observed and consequently a 10% and 34% decrease in protein and amino acid content. The same effect was observed at 40°C, reducing the biomass concentration in approximately the same proportion at 25°C, with a 17% and 30.5% decrease in protein and amino acid content. This showed that for this microalga, 30°C is the optimum temperature for biomass production and protein biosynthesis.

4.2.3 pH

The pH of the culture medium can be influenced by factors such as the composition and buffer capacity, the amount of dissolved carbon dioxide, the metabolic activity of the cells, and temperature. Different types of microalgae have different levels of tolerance to the pH of the culture medium, which may affect their growth rate (Dolganyuk et al., 2020). For some microalgae, such as *C. sorokiniana, C. reinhardtii, B. braunii, S. obliquss, N. coherens,* and *H. pluvialis,* the optimal pH range is from 6 to 8. However, in several cases, significant growth rates have been reported at extremely low or high pH values, for example, for *Chlorella vulgaris* (pH = 4), *Chlorella protothecoides* (pH = 2.5), or *Spirulina* sp (pH = 9) (Dolganyuk et al., 2020; Ogbonda et al., 2007). The pH is also considered to be a critical parameter in microalgal growth and metabolism. In general, protein contents are highest at the optimal pH for cell growth, while the accumulation of storage carbohydrates and lipids show the opposite trend with the biomass (Ma et al., 2020).

Ogbonda et al. (2007) reported a protein content of 48.23% for *Spirulina* sp. at optimum pH and temperature conditions (9 and 30°C) in a 250 mL closed batch photobioreactor. The protein content of *Spirulina* sp. decreased by 25% when the pH increased from 9 to 10. On the other hand, the protein content decreased by 29% as the pH decreased from 9 to 8.5. Zhao et al. (2019) reported a protein content of 57.9% for *C. pyrenoidosa* grown in anaerobic wastewater under optimal growth conditions (pH = 6.8 and 25°C). The protein content was significantly reduced by 19.8–23.3% under pH 7.8-8.5. Contrary to the decrease in protein content, the lipid and carbohydrate contents increased significantly under high pH conditions. In addition, the low pH values of the medium caused a decrease in the internal pH. This decrease increases the energy requirement for metabolic maintenance. However, high pH can cause denaturation of proteins (by changing the charge of the protein) due to the loss of the proteins' bounds (Keithellakpam et al., 2015). Moreover, high pH stress can inhibit the cell cycle and triggers lipid accumulation (Zuccaro et al., 2020).

4.2.4 SALINITY

The salinity of the solution is another factor that affects the growth and development of microalgae. In general, microalgae can be classified according to its salinity resistance into oligogaline (when they can grow only in media with low salinity (maximum salinity 0.5–5.0 g/L), mesogaline (when they can grow in media with moderately saltwater of 5–18 mg/L) and polyhaline (when they can grow in salt-water with a salinity of 18–30 g/L) (Dolganyuk et al., 2020). An optimal salinity condition is essential for microalgal growth and the accumulation of primary me-tabolites, whereas extreme salinity causes osmotic stress, ion stress, and alteration of cellular ionic ratios, thus affecting cellular metabolism. For example, Sui and Vlaeminck (2019), working with *Dunaliella salina*, reported a protein productivity of 3.5 mg protein/L/d under optimal salinity (2 M NaCl) for a cell growth of 60.7 mg/L/d in a batch photobioreactor. However, the protein content and cell growth decreased by 97% and 44%, respectively, when increasing the salinity of the medium to 3 M NaCl. Additionally, the accumulation of storage lipids and car-bohydrates was enhanced under low/high salinity stress for *Dunaliella tertiolecta* (Rizwan et al., 2018). On the other hand, the accumulation of carotenoids, such as β-carotene, was enhanced under salinity stress, which can be attributed to the generation of oxidative stress (Hashemi et al., 2020).

Hence, optimal salinity is a crucial culture parameter for the biosynthesis of primary metabolites and protein. The salinity stress results in the accumulation of storage lipids, carbohydrates, and secondary metabolites such as carotenoids (Ma et al., 2020).

4.2.5 NUTRIENTS

The metabolism of microalgae can be autotrophic or heterotrophic. The former re-quires only inorganic compounds, such as carbon dioxide, salts, and solar energy (*Spirulina platensis, Chlamydomonas reinhardtii*); the latter is not photosynthetic and requires an external source of organic compounds to use as a nutrient and energy source (*Gladieria sulphuraria, Nitzschia alba*). Some photosynthetic species are mixotrophic, such as *C. sorokiniana* or *D. salina*, which has the ability to perform photosynthesis and use exogenic organic sources simultaneously. In any case, nutrient limitation has a direct impact on the synthesis of biologically active substances, biomass growth, and photosynthesis processes in microalgae (Kadkhodaei et al., 2015). Stress caused by nutrient deficiency also leads to the formation of free radicals in the cell and changes in antioxidant content (de Morais et al., 2015).

Besides trace metals and vitamins, C, N, and P are the three most important macronutrients to maintain microalgal growth in the culture medium. Nitrogen is directly associated with the primary metabolism of microalgae because it is ne-cessary for the construction of nucleic acids and proteins. In synthetic culture media for microalgae cultivation, nitrogen is usually provided as nitrate (NO_3^-), nitrite (NO_2^{2-}), and ammonia salts, such as urea (NH_4^{4+}), while phosphorus is added as hydrogen phosphate (HPO_4^{2-}) and dihydric phosphate ($H_2PO_4^{4-}$). The effect of N and P on the microalga growth, the biosynthesis of compounds, and carbon fixation is related to the compounds used as the nitrogen source, in addition to the N and P concentrations (Zhao and Su, 2014). In the formulation of a culture medium, the nutrient ratio is also important; an extremely low N:P concentration would inhibit microalgal growth, while a high N:P concentration would produce toxic effects on the microalgae, decreasing the growth rate (de Morais et al., 2015). Rasdi and Qin (2015) studied the impact of the N:P ratio in the range from 5:1 to 120:1 on the growth and protein content of *Tisochrysis lutea* and *N. oculate*. The results revealed that the N:P ratio of 20:1 favored algal growth and a higher protein content while the N:P ratio of 120:1 reduced the growth of these microalgae. The critical N:P ratio that marks the transition between N and P limitation appears to be in the range of 10–30 as described by Choi and Lee (2015) for *C. vulgaris* in wastewater and 5–20 for *Scenedesmus* (Park et al., 2020).

4.2.5.1 Carbon

Carbon is the basic element of most metabolites and constitutes approximately 50% of the element fraction in microalgae biomass. In autotrophic cultures, CO_2, and bicarbonate (HCO^{3-}) are the carbon sources absorbed by microalgae, and HCO^{3-} is converted into CO_2 through the action of carbonic anhydrase (Ma et al., 2020). During autotrophic growth, microalgae carry out oxygen photosynthesis and fix carbon dioxide. One part of fixed carbon is used to maintain cells and growth, while the other part is stored in several forms, depending on the type of microalgae. Generally, adequate carbon concentration is essential for cell growth and metabolite accumulation (Ma et al., 2020). CO_2 provides the carbon necessary for the synthesis of glucose by photosynthesis and the energy obtained from glucose enables synthesis of protein for cell growth and cell division. For example, Tan et al. (2020) selected four different concentrations of CO_2 (1.25%, 2.5%, 5%, and 7.5% CO_2 in air) to find out their effects on the biomass and protein production for *Chlorella sorokinina* CY1 and *Chlorella vulgaris* ESP-31 in closed batch photobioreactors. Biomass productivities increased up to 5% with the CO_2 concentration, achieving values of 1.95 and 2.36 g biomass/L/d for *C. sorokinina* and *C. vulgaris*, respectively. At 7.5% CO_2, both microalgal species required one more day to reach the nitrogen starvation phase when compared to 5% CO_2, with a 28 and 30% reduction in biomass productivity for *C. sorokinina* and *C. vulgaris*, respectively. This negative impact of excessive CO_2 concentration could be related to a decrease in the pH of the culture medium (from 7 for 5% CO_2 to 6.5 for 7.5% CO_2) which would decrease both the activity of extracellular carbonic anhydrase and the uptake of CO_2 by the microalgae. Depending on the species, different effects of the CO_2 concentration on the protein content of the microalgae were found in this study. The protein content of *C. sorokiniana* increased by 33% when the CO_2 concentration was increased from 1.5 to 5%, while the CO_2 concentration did not influence the protein content of *C. vulgaris* (approximately 25% for all CO_2 concentrations studied). After comparing the biomass and protein productivity, it was clear that 5% CO_2 was the optimum for both species. Finally, these authors concluded that at the 5% CO_2 concentration, *C. sorokiniana* accumulated proteins preferentially, whereas *C. vulgaris* accumulated biomass while keeping proteins at stable levels.

In other studies, Kareya et al. (2020) and Sabia et al. (2018) reported increases in biomass productivity and the contents of protein, chlorophyll, and carotenoids when the CO_2 concentration was increased from 0.03% to 3% in *Microchloropsis gaditana* and *Thalassiosira pseudonana*. Nevertheless, under the high CO_2 concentration (5%), the lipid content increased by decreasing the protein and structural carbohydrate contents. This trend can be attributed to the promotion of carbon fixation by the high CO_2 concentration, leading to more precursors, energy, and reductants for the biosynthesis of these metabolites. Moreover, the accumulation of secondary carotenoids, as well as storage carbohydrates and lipids, increased under excessive CO_2 and/or HCO^{3-} concentrations in order to handle the low pH condition. Chu et al. (1996) showed that the protein content of *Nitzschia incospicua* decreased in favor of both lipid and storage carbohydrate contents when the CO_2 concentration in the air increased from 1% to 5%.

The supply of CO_2 to microalgae cultures can increase biomass productivity; however, the excess CO_2 results in acidification of the medium, inhibiting the growth and the biosynthesis of protein for certain types of microalgae. Therefore, in microalgae, the contents of chlorophylls, primary carotenoids, protein, structural carbohydrates, and membrane lipids increase under the optimal carbon concentration, whereas the accumulation of secondary carotenoids, as well as storage carbohydrates and lipids, are enhanced under excessive carbon source (Ma et al., 2020).

4.2.5.2 Nitrogen

Nitrogen is the second most abundant element in the microalgae biomass (with a concentration of 1% to 14% of the dry mass (Dolganyuk et al., 2020)). Nitrogen is an essential macronutrient for microalgal growth and plays and important role in protein, lipid, and carbohydrate synthesis. Depletion of nitrogen in the cultivation medium causes a decrease in growth and an increase in lipid productivities (Yaakob et al., 2021). Yang et al. (2018) observed that biomass accumulation in *Chlamydomonas reinhardtii* was inhibited up to 31.7% under nitrogen deficiency, with simultaneous increases in the total fatty acid yield (up to 93%), coupled with an enhancement in lipid production of up to 113.46 mg/L.

The growth of four *Chlorella* strains (F&M-M49, IAM C-212, PROD1, and CH2) was investigated by Guccione et al. (2014) in a fed-batch photobioreactor. Under a nutrient sufficient medium (1.5 g/L $NaNO_3$), all strains presented high biomass productivities in the range of 0.60 to 0.75 g/L/d and high protein contents from 39 to 45%. On the other hand, in all the strains, nitrogen starvation strongly reduced productivity. It decreased the protein contents and induced the accumulation of carbohydrates (about 50%) in strains F&FM-M49 and IAMC-212, and lipids (40–45%) in strains PROD1 and CH2.

Nitrogen starvation induces the accumulation of storage carbohydrates and lipids, probably due to the transformation of protein or peptides into these energy-rich metabolites (Ho et al., 2012). The effect of different nitrogen concentrations (0, 36, 72, 144, and 288 mg/L) on biomass and protein productivity for *I. galbana* in a batch photobioreactor was studied by Zarrinmehr et al. (2020). The results revealed that the percentage of accumulated protein at different nitrogen concentrations ranged from 17.1 to 36.3%. The protein percentage was significantly higher in the groups where the medium contained 36 and 72 mg/L compared to those containing 0 and 288 mg/L, but did not statistically differ from those containing 144 mg/L. The maximum and minimum protein concentration obtained, 326.1 and 56.9 mg/L, were observed under 72 and 0 mg/L nitrogen concentration, respectively. On the other hand, cell grown in 0 mg/L nitrogen showed the highest amount of carbohydrate and lipid content, 47% and 17.2%, respectively. The results obtained by Zarrinmehr et al. (2020) were similar to those of Zhu et al. (2014) who reported that *Chlorella zonfingiensis* showed rapid growth in nitrogen sufficient culture medium, whereas growth inhibition was observed under nitrogen starvation conditions. In reference to nitrogen starvation conditions, lipid accumulation increased greatly in *C. zongingiensis*. Finally, they concluded that, in general, microalgae preferentially degrade nitrogen-containing macromolecules, such as proteins, in response to a shortage of nitrogen in the culture medium, resulting in a decrease in protein content and an accumulation of reserve carbon compounds, such as lipids and/or carbohydrates.

Microalgae can assimilate nitrogen in the form of nitrate, nitrite, urea, and ammonium (Ma et al., 2020). Nevertheless, nitrate is more widely used for microalgae culture compared to ammonium salts, as nitrate is more stable and has a lesser likelihood of a pH shift (Tan et al., 2020). In nitrate compounds, nitrogen is in the oxidized +5 state and must be reduced to the oxidation state of the amino group (-NH$_2$) present in proteins (−3). These redox transformations consume a large amount of reducing power (NADPH) and raise the pH by consuming H$^+$.

On the other hand, ammonium (NH^{4+}) can be toxic at concentrations above 25 µM by acidification of the medium (Yaakob et al., 2021). pH changes the NH^{4+}/NH$_3$ buffer system and induces free ammonium content in the culture, which is particularly toxic for the photosynthetic apparatus, affecting the accumulation of primary metabolites, such as chlorophylls and protein (Ma et al., 2020).

The use of organic nitrogen sources such as urea has also been studied. For example, Piña et al. (2007) found higher biomass and protein concentrations in *Chaetoceros muelleri, Thalassiosira weisflogii,* and *Isochrysis* sp. in culture medium with urea as the nitrogen source. Serpa Ibáñez and Calderón Rodríguez (2006) reported that the use of urea resulted in higher concentrations of carotenoids in *Dunaliella salina* than in a culture medium with NaNO$_3$. Probably the urease activity decreased the energy requirements to produce biomass, proteins, and carotenoids.

Based on these observations, the accumulation of primary metabolites – such as proteins, chlorophylls, lutein, and fucoxanthin – increase under nitrogen sufficient conditions, while storage carbohydrates and lipids enhance under nitrogen depleted conditions (Ma et al., 2020).

4.3 PROTEIN EXTRACTION METHODS: EFFECT OF DIFFERENT METHODS AND OPERATIONAL PARAMETERS ON EXTRACTION YIELDS AND PROTEIN RECOVERY

The choice of the extraction treatment depends on the resistance of the raw biomass, among other economic, technical, and environmental aspects. Mild disruption methods are usually used to avoid damage to the proteins and to preserve their techno functional properties (Callejo-López et al., 2020). Among them are several different treatments, classified into physical, chemical, and biological methods (Figure 4.4).

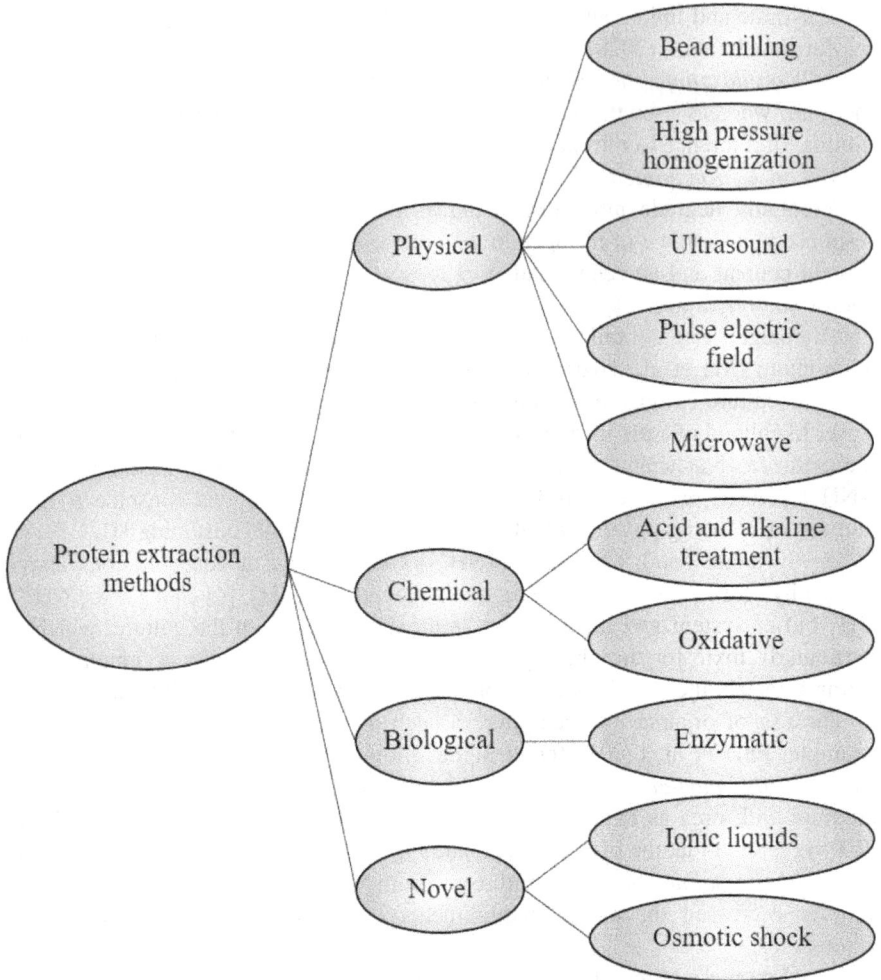

FIGURE 4.4 Classification of methods used for protein extraction.

Selecting one method or a combination of methods will be conditioned by the microalgae cell wall and the proteins application. For example, if proteins must maintain their structural integrity and functionality, physical methods are recommended as they are milder and usually avoid damage to the proteins and preserve their techno-functional properties (Callejo-López et al., 2020). In this case, special attention must be paid to the possible overheating and, as consequence, the protein degradation. The main disadvantage is that these techniques are not usually severe enough and provide low extraction yields compared to other methods. For example, comparing ultrasonication and chemical treatments for *H. Pluvialis*, the protein yield obtained was 13.5% and 31.1%, respectively (Safi et al., 2014b).

On the contrary, if looking for a higher protein output, the use chemical agents, enzymes, or physicochemical techniques are preferred to break covalent bonds of the cell wall more efficiently. However, these methods can degrade proteins through denaturation, cross-linking, racemization, or hydrolysis reactions (Amorim et al., 2020). Therefore, a compromise solution must be found between a good recovery yield, energy consumption, and product degradation (Alhattab et al., 2019).

4.3.1 PHYSICAL METHODS

4.3.1.1 Bead Milling

Bead milling is one of the best methods for cell disruption of microalgae to extract different internal compounds, inducing direct mechanical damage on the cell wall. It uses beads inside the milling chamber and the cell wall disruption is caused by several processes: collision of cells with these beads due to differences in velocities, shear stress due to the acceleration of beads towards the milling chamber wall, and centrifugal acceleration of the mill wall (Nitsos et al., 2020). Advantages such as high cell wall disruption efficacy, high biomass loading, temperature control, commercially available equipment, quick and easy scale-up, and low labor intensity make bead milling an efficient technique for protein extraction (Timira et al., 2021).

The most important parameters are milling chamber geometry, microalgal biomass concentration, agitator speed, suspension flow rate, bead filling ratio, bead type, and bead diameter (Postma et al., 2017). These parameters influence the efficiency of cell disintegration and, therefore, of protein extraction. Several articles have investigated this method for the extraction and recovery of proteins from microalgae and have confirmed that it is one of the most effective techniques. Alavijeh et al. (2020) obtained a protein recovery yield of 40% from *Chlorella vulgaris* after only 10 min in a horizontal 75 mL bead mill chamber with a 65% filling percentage by 0.4 mm Y_2O_3 stabilized ZrO_2 beads, a constant agitation speed of 2039 rpm. and a biomass concentration of 25 g/L. Also, subunits of RuBisCo, a protein enzyme with sizes between 14 and 56 kDa, was recovered after the process. They concluded that the bead milling process consumed less time and required minimal effort in extracting proteins under mild conditions (room temperature and no addition of chemicals).

4.3.1.2 High Pressure Homogenization

High pressure homogenization (HPH) is a mechanical process, during which a microalgae biomass suspension is forced by high pressure (50–300 MPa) through a micrometric disruption chamber, where the velocity increases rapidly (Carullo et al., 2020). As a consequence, cell wall rupture occurs due pressure drop, shear stress, cavitation, turbulence, and impingement of the cells to the surface of the valve at high velocities (Nitsos et al., 2020). The most important parameters that affect the process are the operation pressure, number of cycles, and fluid dynamics such as flow rate (Timira et al., 2021). HPH is one of the most effective rupture techniques for compound extraction from microalgae, including proteins. It can be scaled, is easily applicable to highly concentrated algal pastes, is relatively energy-efficient, and the cell disruption rates are high compared to PEF, acid, and alkaline treatment (Timira et al., 2021).

Using HPH has several disadvantages and drawbacks (low dry cell concentrations, difficulties in breaking rigid cell walls, and nonselective intracellular compound release) (Timira et al., 2021). This method always requires an efficient heat depletion at the homogenization valve because of the high temperature increase that occurs (Nitsos et al., 2020), which can degrade the extracted proteins and can cause reversible or irreversible alteration of the tertiary and quaternary structure of proteins (Carullo et al., 2020). Also, the shear stress and high pressure used can damage protein properties.

Carullo et al. (2018) extracted proteins from *Chlorella vulgaris* with a yield of 54.1% after 5 HPH passes at 150 MPa and 155 mL/min of flow rate, and the protein release had already peaked at a pressure of 100 MPa, with a yield of 50%, indicating that partial cell breakage is enough for the sufficient extraction. Elain et al. (2020) achieved a protein yield recovery of 62.7% using *Arthrospira platensis* and a two-stage homogenizer with an inlet pressure of 500 bar and 9 L/h of flow rate, solid to liquid ratio of 1:6 w/v, and 7 passes.

4.3.1.3 Ultrasonication

Ultrasonication is considered a green extraction technique that presents several advantages in terms of shortening the extraction time, decreasing solvent volumes, and increasing the yield of targeted compounds (like proteins) in comparison with conventional methods (Vernès et al., 2019a). In ultrasound assisted extraction (UAE), ultrasound waves of 20–100 MHz are used to create localized high-pressure bubbles in the liquid that collapse and generate shock waves, which causes high shear forces and thus lead to cell wall disruption (Shahid et al., 2020). The main factors affecting the process are the solvent's physical properties, such as viscosity, saturation, vapor pressure, surface tension (Vernès et al., 2019a), and process temperature (which can be significantly increased if not controlled, affecting the quality and properties of the extracted proteins).

The most important parameters are ultrasound power and frequency, process time, microalgal biomass concentration, and type of solvent. For protein extraction, the most used frequencies are between 20 and 40 kHz (Vernès et al., 2019a), with treatment times ranging from 10 min to 2 h. For example, Hildebrand et al. (2020) achieved a maximum protein recovery of 76.6% from a *Chlorella vulgaris* suspension (0.2 g/mL) using an ultrasonic probe at maximum power (1000 W) for 10 min and NaOH 0.4 M as solvent. Using water or HCL 0.4 M as solvents, the yield was only 35%. The amino acid recovery was between 40% and 50% with water, HCL 0.4 M, and NaOH 0.4 M, with the highest being the alkaline solvent. Vernès et al. (2019b) obtained a protein recovery yield of 26.7% after 20 min of process from 1:20 (g/g) of *Arthrospira platensis* suspension in the phosphate buffer using an ultrasonic device at low frequency (20 kHz). The value increased by 6% with the application of 2 bar pressure.

Due to the highly resistant cell wall of most microalgal species, ultrasonication alone is not very effective for the complete extraction of proteins and must be accompanied by other methods, like enzymatic hydrolysis or bead milling (Nitsos et al., 2020; Soto-Sierra et al., 2018). The combination of various treatments can significantly improve the extraction yield, as shown by Hildebrand et al. (2020) who obtained a protein recovery of 82.1% when a protease treatment of 1 h was combined with 10 minutes of ultrasound pretreatment at 1000 W of power and water as solvent.

4.3.1.4 Electric Field

Pulse electric field (PEF) is a non-thermal method that disrupts the lipid bilayer of cell membranes allowing molecules of certain sizes, such as small proteins, to enter and/or diffuse out of the cells (Soto-Sierra et al., 2018). PEF involves applying an external high electric current to increase the transmembrane voltage to perforate and permeabilize the microalgal cell wall (Matos, 2019).

The most important operational parameters are the electric field strength, wave shape, number and duration of pulses, temperature, and the product and media characteristics (Rocha et al., 2018). Gateau et al. (2021) extracted 10.2 µg/mL of proteins from *H. pluvialis* using an electric filed strength of 1 kV/cm and five pulses from a microalgae suspension of 10^5 cells/mL at 10°C. Buchmann et al. (2019a) achieved an extracted protein concentration of 0.5 g/L using a filed strength of 20 kV/cm from a microalgae concentration of 6 g/L, a lower value than the concentration reached after HPH treatment (2.75 g/L) of the same microalgae.

As can be concluded from the results described above, PEF could be used as a supplementary treatment, and it has been successfully applied for protein extraction at low energy intensities (Timira et al., 2021). However, it is not an efficient disruption method for complete protein extraction in comparison with other methods, like HPH or enzymatic hydrolysis. When complete solubilization and extraction of microalgal proteins is required, energy-intensive cell disruption methods, or a combination of more than one method, is recommended (Soto-Sierra et al., 2018).

4.3.1.5 Microwave

In microwave-assisted extraction (MAE), the process acceleration and high extraction yields are the result of a synergistic combination of two transport phenomena: heat and mass gradients (Vernès et al., 2019a). Microwaves offer fast heating in comparison to conventional heating and selective energy dissipation. It cuts down working times and increases the yield and quality of the protein extracts (Shahid et al., 2020). MAE consists of applying a microwave irradiance at a frequency near 2.45 GHz (Kapoore et al., 2018), causing dielectric heating by absorption of the energy in water and other polar compounds (Timira et al., 2021). It induces the vibration of water and polar molecules within wet microalgae biomass, resulting in temperature increases in the intracellular liquids, which causes the solvents to evaporate and exert pressure on the cell walls, leading to disruption (Costa et al., 2020; Kapoore et al., 2018). MAE consumes less solvents; presents higher extraction yields and enhanced efficiency; is nontoxic; can be used for larger volumes with high uniformity, selectivity, and low energy consumption; uses short reaction time; and has low operation costs (Costa et al., 2020).

The most important operational parameters are microwave irradiation time, duty cycle, microwave power, and solvents (Chew et al., 2019). Passos et al. (2015) obtained a protein concentration of 193 mg/L when they applied a microwave power of 900 W during 3 min from a mixed culture of microalgae (*Stigeoclonium* sp., *Monoraphidium* sp., *Nitzschia* sp. *and Navicula* sp.) with water as solvent. It is commonly used in combination with other treatments, increasing their efficacy. For example, the protein recovery yield from microalgae *Chlorella vulgaris* obtained by Chew et al. (2019) was 63.2%, which represented a 2.54-fold increase over the three phase partitioning (TPP) extraction process. The operational conditions were the microwave time of 120 s, duty cycle of 80%, and 100 W of power. This method has been investigated to extract protein, although it is necessary to explore MAE's full potential in extracting valuable products from microalgae (Timira et al., 2021).

4.3.2 Chemical Methods

4.3.2.1 Acid and Alkaline Treatment

Acid and alkaline treatments involve exposing the microalgal biomass to an acid or basic aqueous medium. In some cases, this pretreatment is combined with elevated temperature (120-160°C), in which case it can also be considered a hydrothermal pretreatment variation with the addition of an acid and basic solvent that acts as catalyst (Nitsos et al., 2020). The main acid solvents used are HCl and H_2SO_4 with a concentration between 1% and 5%. These solvents cause the biomass to swell and degrade the cell wall polymers (Salakkam et al., 2021). In contrast, alkaline treatment consists of treating the algal biomass with an alkaline solvent (NaOH), which is used as a catalyst and is more used for microalgae biomass. This method is used to disintegrate and disrupt the microalgal cell wall and solubilize organic molecules, particularly protein (Salakkam et al., 2021). It has been demonstrated that acid and alkaline treatments disrupt the cell walls of microalgae and facilitate protein extraction, although when combined with heat, these methods can degrade and modify protein properties due to denaturation and racemization (Callejo-López et al., 2020). They can also lead to the formation of amino acid complexes through Maillard reactions and limiting the availability of amino acids in the extracts (Timira et al., 2021).

Advantages of these chemical methods include high efficiency, low energy input, and easy scalability. Nevertheless, they are not considered to be mild and can have serious effects on protein, as described above. Therefore, due to the aggressive nature of these methods, careful process conditions are required to avoid the degradation of the extracted protein (Nitsos et al., 2020). In addition, they show low selectivity, causing the release of multiple components, which results in difficult posterior separation. These are the principal drawbacks that need to be investigated before applying them.

Callejo-López et al. (2020) designed an alkaline process for the extraction of proteins from fresh microalgae biomass at 60 mg/mL and 100 mg/mL (consortium formed with *Chlorella vulgaris, Nannochloropsis gaditana, and Scenedesmus obliquus*) with a final protein recovery yield of 87.5%, after an alkaline process of 2 hours at 50°C and pH 13.4 (50 mM Na_2HPO_4 and NaOH titration). For the acid method, Martin Juárez et al. (2021) obtained protein solubilizations of 47.3% and 75.5% after and acid treatment of 1 hour and 121°C with 0.5 M HCl and 2 M HCl using fresh algal-bacterial biomass composed mainly of *Scenedesmaceae* grown in pig manure.

4.3.2.2 Oxidative Treatment

This method consists of exposing the microalgae to oxidative agents, such as ozone or hydrogen peroxide (Nitsos et al., 2020), which can be very aggressive to the biological biomass. Ozone is a strong oxidant that can destroy several types of microorganisms and attack the cell wall (especially the double bonds in membranes), resulting in damage to the cell structure (Keris-Sen and Gurol, 2017). It has proven to be attractive alternative over other extraction methods due to its low production of inhibitor compounds, low chemical consumption, mild operational conditions, and generation of easily degradable subproducts (González-Balderas et al., 2020). Among the disadvantages are the highly reactive, flammable, corrosive, and toxic characteristics of ozone, the exothermic process, the necessity of special construction materials, and the high costs generation (Travaini et al., 2016). One recent investigation achieved a protein release of 58% at an ozone concentration of 45 mg/L, a contact time of 35 min, and alkaline conditions (pH 11) from *Desmodesmus* sp. (González-Balderas et al., 2020). On the other hand, hydrogen peroxide (H_2O_2) is an alternative oxidizing agent for the pretreatment of microalgae with a similar mechanism to ozonation. Duan et al. (2017) obtained a protein extract content of around 25 μg/cm^3 from *Chlorella pyrenoidosa* at a cell concentration of 12.5×10^7 cells/mL by applying a combined treatment of ultrasounds (20 min and 35 Hz) and H_2O_2 (0.1 mM).

4.3.3 BIOLOGICAL METHODS

4.3.3.1 Enzymatic Hydrolysis

Enzymatic hydrolysis is a green alternative to traditional physical and chemical methods that permits a selective and a more efficient extraction of all the biomass components, including proteins (Timira et al., 2021). This method is based on the use of different types of enzymes under mild conditions (proteases, cellulases, and lipases) to degrade the complex cell wall, which is composed mainly of carbohydrates, proteins, and organic polymers (Nitsos et al., 2020).

Since this method can be operated under mild and gentle operating conditions, serious damage to the intracellular compounds can be avoided with high efficiency yield extraction and high selectivity while operating at low temperatures with low energy demand (Nitsos et al., 2020). There are many variables that could affect the enzymatic activity. These variables include the characteristics and concentrations of the enzymes, the intracellular composition, the cell wall composition, the type of microalgae, and temperature, although the most important variable is the type of enzyme used (Phong et al., 2018b). Enzymatic methods that involve the use of proteases have been proven to be highly efficient in amino acid production from microalgae (Callejo-López et al., 2020). The potential limitations of this method include the cost of commercial enzymes, the lack of knowledge about optimal or compatible enzyme formulations for cell disruption, and the requirement for holding tanks to accommodate long incubation periods (Dixon and Wilken, 2018; Salakkam et al., 2021).

Various studies have probed the efficiency of this method. Corrêa et al. (2021) obtained a protein solubilization yield of 11.6% with fresh microalgae biomass mainly formed by Scenedesmaceae growing with pig manure. In this case, enzymatic hydrolysis was performed with Celluclast 1.5 L and Novozyme 188 and an enzymatic hydrolysis time of 12 h. Callejo-López et al. (2019) achieved an optimum protein recovery yield of 80.3% from a pure consortium with *Nannochloropsis* sp. by applying enzymatic hydrolysis with Alcalase 2.5 L, a temperature of 60 °C, and a hydrolysis time of 2 hours. Finally, compared to other methods, such as microwave or PEF, an enzymatic treatment can result in a better protein extraction yield (Corrêa et al., 2021).

4.3.4 NOVEL METHODS

4.3.4.1 Ionic Liquids

Ionic liquids are organic salts in the liquid state with low melting points below 100°C (Nitsos et al., 2020). They consist of a large asymmetric organic cation and an organic or inorganic anion. They have excellent properties for cellulosic biomass treatment due to their high hydrogen bond accepting ability that could disrupt the extensive hydrogen bonding network of polymers, leading to the breakdown of complex networks of lignin, cellulose, and hemicelluloses in the cell wall (Matos, 2019). Protein extraction enhancement is expected due to the modification of the cell wall. This is because the hydrogen bonds of the microalgae's cell walls are affected by IL's ions. (Timira et al., 2021). Ionic liquids present a few attractive characteristics such as a low-melting point, extremely low volatility under atmospheric conditions, the capability of dissolving a wide range of polar to non-polar compounds, low flammability, and high thermal and chemical stability (Phong et al., 2018b). Despite these advantages, few articles have used this method due it is a complexity, the fact that some ILs are not environmentally friendly, and the laborious purification process (Timira et al., 2021). Pojić et al. (2018) reported that the extraction of proteins by an IL aqueous two-phase system based on the guanidine IL and hydrogen phosphate is possible. The results showed that the extraction efficiency could reach up to 99.6% under the optimum conditions and that the protein properties did not change after the extraction.

4.3.4.2 Osmotic Shock

Osmotic shock is a disruption technique based on the rapid increase or decrease of salt concentrations in the solution (Dixon and Wilken, 2018). The cell wall ruptures as a result of permeation in order to attain equilibrium with the medium that enables protein extraction (Krishna Koyande et al., 2020). The type and concentration of salts, incubation time, and biomass concentration are all important factors that affect the efficiency of the method (Nitsos et al., 2020). For the extraction of bioproducts, hypotonic conditions (low osmotic pressure) are ideal but have disadvantages, including inefficiency and the high salinity of the resulting wastewater (Dixon and Wilken, 2018). This method also takes longer than other processes (such as autoclaving and microwave irradiation) and is economically unfeasible on a large scale (Corrêa et al., 2021). Krishna Koyande et al. (2020) studied the recovery of whole proteins from *Chlorella vulgaris* using osmotic shock through a liquid biphasic flotation (LBF) system. The study concluded that a protein recovery yield of 93% could be achieved using osmotic shock (as opposed to only 84.8% without osmotic shock). In this work, 0.1 g of microalgae biomass was dissolved in 40 mL of salt solution with a 500 g/L concentration. The salt solution was then diluted to 200 g/L, causing the osmotic shock.

4.4 PROTEIN SEPARATION METHODS

Depending on the disruption methods applied to the microalgae biomass, the extracts can present fine particles, intact cells, disrupted or damaged cells, or other undesired compounds and by-products that must be removed from the proteins with different techniques (Amorim et al., 2020). Centrifugation is mainly used in the separation of the protein-rich liquid phase from the solid phase after the cell wall disruption (Grossmann et al., 2020a) and then proteins can be further separated and purified from other small impurities with several methods explained below. Separation methods (three-phase partitioning, membranes, and protein dispersion) allow for the separation of proteins from other molecules, while purification methods (electrophoresis and column chromatography) permit the isolation of a single type of protein from a complex mixture.

4.4.1 THREE-PHASE PARTITIONING

Three-phase partitioning (TPP) is a recent method useful for protein extraction and separation with high recovery yields. This method has generated a lot of interest due to its simplicity, scalability, and fast extraction times, which can be used to concentrate or isolate proteins from different raw materials (Waghmare et al., 2016). TPP fractionates the main components of microalgae in the upper phase (with nonpolar molecules such as lipids) and the lower phase (with carbohydrates), while proteins are kept in the middle phase through a combination of ammonium sulphate for protein precipitation, and t-butanol to increase the protein buoyancy (Amorim et al., 2020). Some advantages of this method include its simple operation with short processing times, mild conditions, which avoid protein denaturation, competitive costs and large-scale feasibility, and final high purity levels (Corrêa et al., 2021). The partitioning behavior of the targeted product depends on the mass transfer, which can be enhanced with other technologies, such as ultrasound, microwave, or enzymes (Chew et al., 2019; Chia et al., 2019).

Waghmare et al. (2016) studied the effect of various parameters of TPP to optimize the concentration of proteins at 28°C from *Chlorella pyrenoidosa*, obtaining a protein concentration of 78.1% w/w in the middle concentrate phase after 20 min. Recently, articles have been published combining this process with other techniques mentioned above. Chia et al. (2019) compared ultrasonic-assisted three phase partitioning (UATPP) with TPP for extracting proteins from *Chlorella vulgaris* and found that ultrasounds improved the efficiency of the process up to 74.6% with the optimized conditions (salt saturation of 50%, a microalgae solution and t-butanol ratio of 1:2, sonication frequency of 35 kHz, irradiation time of 10 min, duty cycle of 80%, frequency of 35 kHz, and biomass loading of 0.75%). Also, Chew et al. (2019)

developed an effective method of microwave-assisted three phase partitioning (MWTPP) and achieved a protein separation efficiency of 67.2% of proteins from *Chlorella vulgaris* with a sulphate ammonium concentration of 30%, a microalgae and t-butanol ratio of 1:1, microwave time of 2 min, duty cycle of 80%, microwave power of 100 W, and microalgae biomass concentration of 0.5%.

4.4.2 MEMBRANE TECHNOLOGY

Membrane filtration, including ultrafiltration or dialysis, are green technologies widely applied in various industrial fields, such as the purification and desalinization of water. This method can be considered one of the best for preserving the functional properties of proteins (Gifuni et al., 2020). Advantages include the lack of additional chemicals, the mild operating conditions, its usefulness in large volumes, and the ease of automation and scalation at the industrial level (Balti et al., 2018). Nevertheless, there are various drawbacks, which include the prolonged processing time, susceptibility of rapid membrane clogging, and the use of expensive membrane filters (Waghmare et al., 2016). The principle is quite simple: the membrane acts as a very selective filter, which lets some molecules pass through while retaining other larger substances (Figure 4.5).

For protein separation, membranes concentrate extracted microalgal proteins based on their sizes and remove other soluble non-protein molecules (Soto-Sierra et al., 2018). The filtration process depends on several parameters: the membrane characteristics (material and molecular weight cut-off), the feed protein concentration, the protein size, the physicochemical attributes of solutions to be filtered (pH, ionic strength), and the operating conditions (pressure, cross flow velocity, temperature, etc.) (Balti et al., 2021).

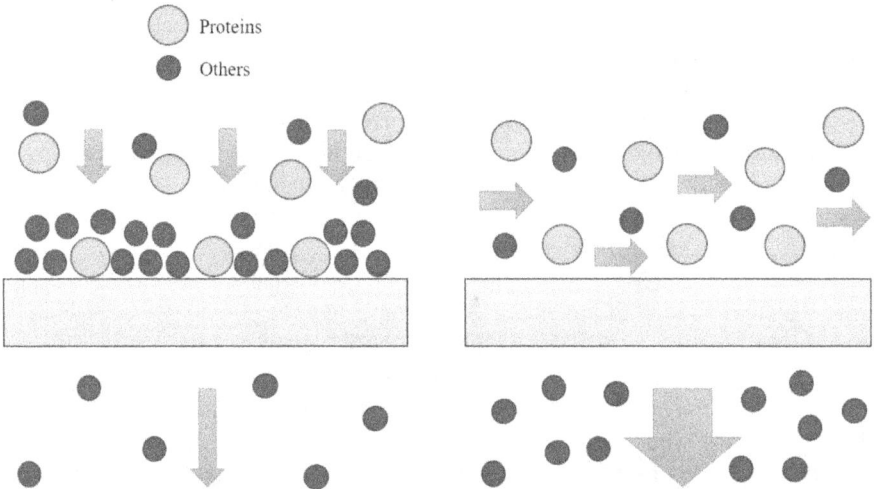

FIGURE 4.5 Schematic process of membrane separation.

Diafiltration is a purification method that is based on the concentration gradient as the driving force because the larger molecules, such as proteins, cannot pass through the semi-permeable membrane while ultrafiltration is based on the pressure-driven to force the selective filtration by the membrane (Liu et al., 2020). By using reduced molecular weight cut-off membranes, high protein retention can be achieved. However, it also can be susceptible to fouling and absorbing of target compounds, decreasing the efficiency of the process (Corrêa et al., 2021). Within ultrafiltration technologies, cross flow filtration, where the flow passes tangentially along the surface of the membrane, is commonly employed (Grossmann et al., 2020a).

Gifuni et al. (2020) designed a three-step cross-flow filtration (prefiltration with 0.22 µm, diafiltration with 0.22 µm, ultrafiltration with 3 kDa) that allowed for the recovery of 12% of the initial proteins extracted by a bead milling process (1 g/L of biomass concentration, 20°C, a flow rate of 200 mL/min, and a filling ratio of 80% with glass beads of 0.6 mm) from *C. sorokiniana.* Balti et al. (2021) fractionated soluble proteins by diafiltration with 50 kDa, 150 kDa, 300 kDa, and 0.2 µm membranes after bead milling disruption (20 g/L of biomass concentration with 0.5-0.75 mm glass beads at an 80% filling rate) from *Arthospira platensis,* producing various protein fractions with different sizes, purity, and high protein concentration levels (more than 80%).

4.4.3 PROTEIN DISPERSION – pH SHIFT AND SALTING OUT

The most used method for protein separation is based on the protein precipitation by shifting the pH (adding HCl or NaOH) to the isoelectric point. Most proteins present a high dispersion rate (which results in greater interactions between these proteins and renders them insoluble) around their isoelectric point and can thereby be separated from other compounds whose solubility is not pH dependent (Grossmann et al., 2020a). The opposite trend is observed for pH values that are too different from the isoelectric points of the proteins, resulting in a lower dispersion rate (Amorim et al., 2020). Microalgae proteins have different isoelectric points depending on the species. These points vary from four for *Chlorella vulgaris* (Ursu et al., 2014) and *Arthrospira platensis* proteins (Sánchez-Zurano et al., 2020) to seven for *Haematococcus pluvialis* (Ba et al., 2016).

pH-shift for protein separation is a simple and low-cost process with great scalability. However, harsh operating conditions due to extreme pH can affect the structure of proteins and lead to protein denaturation and functionality losses (Matos, 2019; Waghmare et al., 2016).

Several attempts have been made to precipitate protein using this technique. Sánchez-Zurano et al. (2020) concluded that protein precipitation was affected by two parameters: pH and extraction time. They obtained a protein recovery of 75.2% from *Arthrospira platensis* at the optimized conditions (pH of 3.89 over 45 min and after a protein solubilization with sonication at 400 W, 24 kHz, and 2 min).

On the other hand, the salting out effect is also used for protein separation. This method is based on the addition of a salt (usually ammonium sulphate) into an aqueous solution containing the proteins. The salt dissociates and the water that once offered a great solvating power to proteins, form aggregates and precipitating (Corrêa et al., 2021). Sankaran et al. (2018) used this method in combination with ultrasounds (20 kHz) to extract proteins from *Chlorella vulgaris,* obtaining a recovery yield of 93.3% of the total initial protein content. The operational parameters were as follows: a biomass concentration of 0.6%, a salt concentration of 200 g/L, an acetonitrile concentration of 100% with 5 min of 5 s ON/10 s OFF pulse mode, and at a flow rate of 100 cc/min.

4.4.4　ELECTROPHORESIS

The electrophoresis process uses the charge of the molecules to make them migrate through an electric field and is commonly used for protein and other macromolecule separation (Corrêa et al., 2021). The advantage of this approach is that the molecules are first separated based on their charges and then according to their molar masses. This can be achieved by sodium dodecyl sulphate polyacrylamide gel electrophoresis (SDS-PAGE). It can isolate bands of protein in a simple and relatively rapid manner when used with a staining method (Shen, 2019). Vizcaíno et al. (2019) carried out an electrophoresis with SDS-PAGE to separate different proteins from 4 types of microalgae (*Tisochrysis lutea, Nannochloropsis gaditana, Tetraselmis suecica,* and *Scenedesmus almeriensis*) after an enzymatic hydrolysis time of 90 min with digestive enzymes from fish extracts at 25°C, obtaining protein bands of 11.7 to 44.5 kDa for *Tisochrysis lutea,* 41 to 57.3 kDa for *Nannochloropsis gaditana,* 24.1 to 55 kDa for *Tetraselmis suecica,* and 26.8 to 149.2 for *Scenedesmus almeriensis.*

4.4.5　COLUMN CHROMATOGRAPHY

Chromatography separates proteins based on differences in their characteristics in a mixture sample. Basically, it uses a glass or plastic tube filled with a resin or solid (stationary phase) that can separate proteins based on their physical characteristics (Figure 4.6). The sample is first added into the top of the column. While the mobile

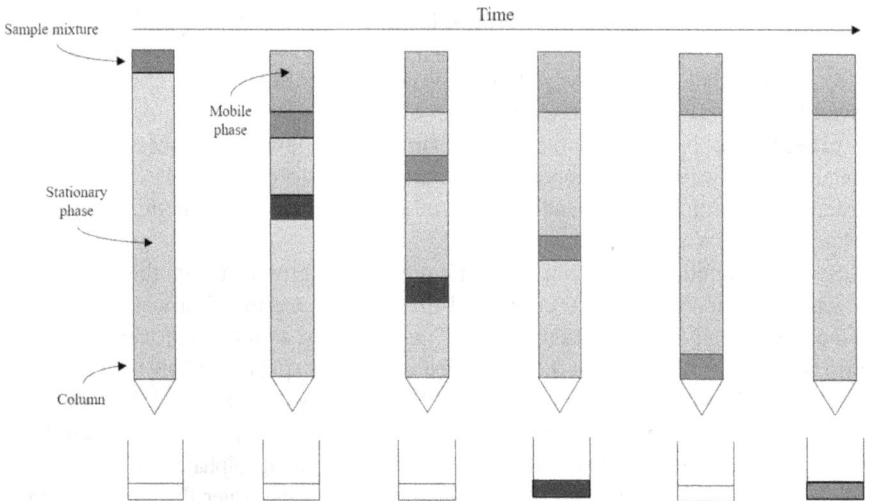

FIGURE 4.6　Protein chromatography separation process.

phase flows continuously through the column, proteins in the sample migrate through the column at different rates depending on the nature of the sample and the properties of the proteins. Therefore, the targeted proteins can be separated from others (Shen, 2019). Different principles can be used for protein separation in column chromatography, such as size exclusion (according to their size), ion-exchange (according to the magnitude of their electric charge), affinity (according to the interaction of the protein with a binder of the stationary phase), and hydrophobic interaction (Amorim et al., 2020).

However, the purification of proteins based on chromatographic columns is a costly and complex process, which requires the careful development of a specific purification method according to the characteristics of the target protein and the protein extract or isolate (Amorim et al., 2020). Several investigations that have used these techniques for microalgal protein separation are shown in Table 4.2.

4.4.5.1 Size Exclusion (SEC)

This type of chromatography can separate proteins based on their size and shape (Shen, 2019). Size exclusion column packing is porous, has no adsorption

TABLE 4.2
Purification Techniques Applied to Microalgae Proteins

Technique	Microalgae	Column Conditions	Target Protein Molecule	Reference
SEC	Nannochloropsis gaditana Tetraselmis impellucida Arthorpira	Superdex 200 Increase 10/300 GLFlow rate of 0.75 mL/min. Absorbance was measured at 214 nm and 280 nm.	Algae soluble protein isolates	(Teuling et al., 2019)
IEC	Spirulina platensis	A GE XK 16/20 column (i.d. 1.6 cm) equilibrated with 20 mM Tris-HCl buffer pH 7 Flow rate of 1 cm/h	C-Phycocyanin	(Corrêa et al., 2021)
AC	Chlamydomonas reinhardtii	StrepII-Tag affinity chromatography	Glycoproteins	(Baier et al., 2018)
HIC	Phormidium fragile	Methyl macro-prep HIC column (1.5 cm × 5 cm) equilibrated with 20 mM Tris–Cl buffer (pH 8.1) supplemented with 1.5 M $(NH4)_2SO_4$	Phycocyanin	(Pagels et al., 2020)
TLC	Chlorella vulgaris Spirulina platensis	HPTLC silica gel 60 F254 Mobile phase: petroleum ether:acetone (75:25, v/v)	Carotenoids Astaxanthin	(Hynstova et al., 2018)

properties, and can separate particles and molecules from others of approximately the same size to achieve purification (Liu et al., 2020). In this type of column, molecules that are smaller than the cut-off threshold size, are more likely to go through the pore of the matrix, and thus are trapped in the resin and travel through the column more slowly (Shen, 2019).

4.4.5.2   Ion Exchange (IEC)

Ion-exchange chromatography (IEC) uses the difference between the charges of molecules to promote separation and can separate biomolecules with ionizable groups, such as proteins or peptides, from others (Corrêa et al., 2021). In this case, the column contains a resin bearing either positively or negatively charged chemical groups (Shen, 2019), but the electric charge of the ion exchangers should be made according to the pH so that the protein is more stable and soluble (Amorim et al., 2020).

4.4.5.3   Affinity Chromatography (AC)

Affinity chromatography (AC) is a method of purification based on the specific biological properties of molecules, such as proteins, which can be specifically and reversibly bound to the ligand, which is immobilized in the stationary phase (Liu et al., 2020). When the sample goes through the column, the targeted protein becomes bound and immobilized and, as a result, it is removed from the mixture of sample. Finally, the protein is dissociated or eluted from the resin by the addition of high concentrations of free ligand in the solution (Shen, 2019). To select the most appropriate stationary phase, besides the specific interaction of the target molecule, other characteristics should be considered (high stability, low adsorption of undesirable substances, and good properties to maintain the mobile phase flow throughout) (Corrêa et al., 2021). It is an alternative to separate complex mixtures and produce proteins with a high degree of purity, conserving their structure and functions (Amorim et al., 2020).

4.4.5.4   Hydrophobic Interactions (HIC)

In this purification technique, proteins are passed through a chromatographic column packed with a support resin to which the hydrophobic groups are covalently linked (Shen, 2019). Hydrophobic interaction chromatography (HIC) exploits hydrophobic regions present in macromolecules, which bind to hydrophobic ligands on chromatography adsorbents. The separation of proteins occurs by hydrophobicity derived from hydrophobic amino acids on the surface of each type of protein. These interactions occur in an environment that favors hydrophobic interactions, such as an aqueous solution with a high salt concentration (Amorim et al., 2020).

4.4.5.5   Thin Layer Chromatography (TLC)

This method is a type of liquid chromatography in which the stationary phase is supported by a flat surface (e.g., a glass plate, a plastic sheet, or paper). The mobile phase and sample pass through the stationary phase by capillary action and/or gravity. The speed at which the solute moves through the stationary phase depends on the force of the mobile phase as it dissolves the solute and moves it up the plate, and the resistance of the sorbent as it pulls the solute out of solution and back into the sorbent.

4.5 POTENTIAL APPLICATION OF PROTEINS FROM MICROALGAE

Once the proteins from microalgae have been extracted and purified, there is a wide range of different potential applications for microalgae proteins in the food science field (as an alternative to meet increasing protein demand) as functional ingredients, supplements, or products with techno functional properties. Proteins not only play a crucial role in our metabolism, but they also have attractive techno-functional features that transcend their nutritional properties so that they can be exploited by industry (Grossmann et al., 2020b). Traditionally, proteins have been an underestimated resource compared to lipids or pigments, but now new applications are emerging. And thanks to the high protein content of microalgae, many investigations are shifting their attention to these biomolecules. This section will look at a range of these potential applications.

4.5.1 MICROALGAE PROTEINS IN FEEDING AND NUTRITION

Microalgae stand as an excellent source of nutrients, proteins in particular: they usually present from 30% to 70% of protein in dry weight (Dolganyuk et al., 2020). For thousands of years, seaweed has been consumed worldwide, especially in Eastern societies, although microalgae cultivation and the evaluation of their nutritional value began only a few decades ago. New protein sources started to be searched for during the last century instead of terrestrial plants because of the notable increase in the world population (Acquah et al., 2020). As a result, microalgae emerged as a good alternative because they require less arable land, fewer pesticides and resources, and are ten times more efficient at using solar energy (Chew et al., 2019). Moreover, by 2030, it is estimated that the world will have a 40% water deficit; thus, replacing animal and terrestrial plant protein sources with microalgae may reduce water consumption (Li et al., 2019). Besides its environmental advantages, microalgae can also be interesting option for those who follow a vegetarian or vegan diet, as it stands as another protein source along with eggs or pulses (Koyande et al., 2019).

Notwithstanding, microalgae proteins are interesting not only for human consumption but also for animal livestock. In the following sections, these two markets will be analyzed.

4.5.1.1 Microalgae Proteins for Humans

Many microalgae species have proved to be nutritionally superior compared to soybean or corn since they display both a high protein content and a high diversity of amino acids. Some examples are *Arthorspira maxima, Synechococcus* sp., *Chlorella Vulgaris, Dunaliella salina, H. Pluvialis,* and *Scenedesmus obliquus.* The tests conducted in observational studies have proved that microalgae are a safe and sound food ingredient with no evidence of producing toxic or acute side effects (Becker, 2007).

However, there is a concern about regular microalgae consumption among nutritionists, which is related to their high levels of iodine and heavy metals. Regarding heavy metals (As, Cu, Zn, Hg, Mn, Cd), it is well known that microalgae are outstanding biosorbents, as they can tolerate and accumulate heavy metals efficiently and thus remove them from wastewater streams (Leong and Chang, 2020). Iodine is an element essential for certain hormones and plays an important role in the proper growth and development of mammals. Even though iodine deficiency entails several health risks, its excess is not desirable, and microalgae cultivated in the sea has high iodine content. Although Eastern societies are accustomed to microalgae consumption (and can eliminate excess iodine), in Europe and America, nutritionists suggest limiting their intake, as it can lead to hyperthyroidism or other diseases (Bouga and Combet, 2015).

Additionally, the scores of qualities of microalgae as protein sources, that is calculated by tools such as the Protein Digestibility Corrected Amino Acid Score (PDCAAS) (Consultation, 2011) are usually low due to the rigid cell wall, which prevents proper digestion. Finally, microalgae are sometimes not suitable from a sensorial point of view, because even small amounts of microalgae may lead to unattractive colors in food or incorporate some unusual taste (Becker, 2007).

In order to overcome these drawbacks, proteins can be extracted, isolated, and purified, transforming them into edible supplements or foods, avoiding the iodine or the prospective contaminants because they are removed, and at the same time exploiting their unique benefits (Amorim et al., 2020). Thus, using extracts of proteins entails several advantages, such as improving the PDCAAS index (protein hydrolysate boosts up to 90–97% in-vitro digestibility, while the whole cell accounts for 70%) or bettering consumer acceptance (Amorim et al., 2020). For instance, *Chlorella Prototheroides* have shown promising results in this sense and might be added to foods shortly (Grossmann et al., 2020b).

Microalgae *Tetraselmis Chuii* has recently been investigated as a protein fortification supplement in wheat bread or flour, giving those foods new rheological properties without significantly affecting their sensory characteristics (Qazi et al., 2021). Another striking application of protein concentrates, isolates, or hydrolysates may be in the field of plant-based meat analogs. However, despite studies concerning their use, only a few deals with the microalgae protein as the main ingredient. The central problem is the disparity of the cell tissue between terrestrial plants and microalgae. Thus, more research must be performed in this field.

4.5.1.2   Microalgae Proteins as Animal Feed

Feeding is the most expensive cost derived from farming activities. Making "meat" from animals is not an ideal process, since in the first place the feed must be cultivated (consuming water and using pesticides) and then delivered to the animals, which will require an additional and higher resource investment. Microalgae can be an excellent sustainable feed because they generally require fewer resources (no freshwater nor arable land) than terrestrial plants and crops. Currently, 50% of the total algae global production is oriented to the animal feed industry (Kusmayadi et al., 2021). Although the use of the whole microalgae biomass as feed displays several advantages, improved PUFAs and ω-3 FA profiles or higher mineral intakes, has been thoroughly studied.It shows improvements in the nutritional quality of some products.

The use of protein extracts can be attractive when applying a biorefinery approach. Lipids from microalgae can be employed in biofuel production, and the protein fraction can be exploited as feed because of its high nutritional quality. Moreover, protein extracts may have several health benefits (Hayes et al., 2017). However, the isolation and purification process increase the final cost of the product. Therefore, it is currently preferred to use the whole microalgae as a supplement to replace traditional plant-based feed.

4.5.2 Microalgae Proteins as Nutraceuticals

Microalgae biomass has proved to have beneficial health properties by itself. For example, *C. Vulgaris* accumulates carotenoids, and after being used as feed for fish and poultry, the life expectancy of those animals increased. Moreover, *C. Vulgaris* showed a protective effect against heavy metals and other harmful compounds by reducing their oxidative stress and increasing their antioxidant activity (Safi et al., 2014a).

Nevertheless, we will focus here on specific compounds, generally known as nutraceuticals. Nutraceutical components are defined as products that have benefits on a physiological level, lower the risk of illness, or improve an individuals' overall health, going beyond human nutrition (Costa et al., 2020). According to the World Health Organization (WHO), microalgae stand as a fascinating source of nutraceuticals, such as phycobiliproteins, bioactive peptides, enzymes, vitamins, γ-linolenic acid (GLA), fibers, and antioxidants, which have proven antitumor, antiviral, antifungal, antidiabetic, and antiseptic effects, among others.

Among the bioactive compounds that can be extracted from microalgae, proteins are one of the most important, and in particular, the family of phycobiliproteins has shown promising results in pharmaceutical industry. One of the most relevant phycobiliproteins is c-phycocyanin, but multiple types exist that differ in size and the chromophores which they attach to (Bertsch et al., 2021). For instance, *S. platensis* contains a high amount of this pigment and its protein fraction contains about 200 g/kg phycocyanin (Lupatini et al., 2017).

Other interesting nutraceuticals are bioactive peptides, which are molecules between 2 and 30 amino acids in length. They are inactive when they are in their native protein, but when they are freed, they influence the physiology of cells, showing promising results for treating and preventing cancer, cardiovascular diseases, diabetes, inflammation, etc. (Hayes et al., 2018). Nowadays, it is accepted that bioactive peptides from microalgae might be even more interesting than their animal counterparts, as these microorganisms are accustomed to living in extreme environments, which have conditioned their amino acid sequences, resulting in significant interest in the treatment of certain diseases (Larsen et al., 2011).

Regarding antibacterial activities, the Kawaguchipeptin B peptide (extracted from *Microcystis Aeruginosa*) showed promising antibacterial activity results against Staphylococcus aureus (Ishida et al., 1997). The peptide Pahayokolides A (isolated from *Lyngbya* sp.) proved to target *Bacillus Megaterium* and *Bacillus Subtilis* effectively (Berry et al., 2004). Guzmán et al. (2019) identified antimicrobial peptides from *Tetraselmis suecica,* which acted against gram-positive and gram-negative bacteria. A bioactive peptide hydrolysate from *Nannochloropsis oculate* protein has shown an inhibitory effect on angiotensin-converting enzyme-I (ACE-I), a metalloprotease that catalyzes a dipeptide's cleavage and could be an effective antihypertensive treatment or used to manage cardiovascular diseases.

An example of a bioactive peptide with anti-inflammatory activity is Leu-Asp-Ala-Val-Asn-Arg, which is derived from *Spirulina maxima* (Vo et al., 2013). Another application of bioactive peptides deals with reducing oxidative stress. A recent study investigated an antioxidant biopeptide from *Spirulina* and compared its antioxidant capacities when the biomolecule was introduced in snacks. Although the antioxidant activity was higher for the isolated biopeptide, the addition of a 2% fraction to snacks significantly increased radical inhibition, which causes a reduction power compared to the control sample (with no bioactive peptide). Some sensory properties were altered, such as hardness or fracturing, but overall there were no significant differences (Costa da Silva et al., 2021).

Many studies have proven the effectiveness of bioactive peptides regarding anticancer properties, although most of them have been trialed exclusively in vitro and further investigation is needed (Saadaoui et al., 2020). An extract from *Synechococcus* sp. VDW demonstrated high antioxidant activity, was able to prevent cell proliferation, and was a helpful tool to favor apoptotic phenomena. Diabetes and obesity are diseases for which bioactive peptides have also shown promising results. Some studies demonstrate that bioactive peptides can mimic cholecystokinin (CCK), an anorexigenic gut hormone that positively affects satiety. Hence, adding particular biopeptides to foods could prevent and treat obesity (Hayes et al., 2015). Concerning diabetes, peptides from *A. Platensis* demonstrated encouraging results as they inhibit a serine exopeptidase DPP-IV, which is related to glucose metabolism and enhances insulin secretion. Furthermore, this peptide also proved to influence starch digestion, diminishing the glycemia peak after ingestion (Hu et al., 2019).

Further uses of protein extracts from microalgae are related to anti-aging and UV protection related to the cosmetic industry. For instance, multifunctional mycosporine-like amino acids, which are molecules based on an amino acid bound to a chromophore, have proven high UV radiation absorption comparable to commercial UV filters. Experiments performed on mouse skin demonstrate significant protection against UV radiation (Sun et al., 2020). A biopeptide derived from *Chlorella* sp. reduced UV damage, and Met-Pro-Asp-Trp (extracted from *Nannochloropsis oculate*) has also demonstrated promising results (Nguyen et al., 2013). Notwithstanding, more than 25 different peptides can be found in skincare products; some even mimic Botox (Hayes et al., 2015).

The structural diversity of bioactive peptides (the amino acid sequence, the charge, or the conformation) rules their functionality and, consequently, their beneficial properties. To identify these protein fragments, proteomic tools combined with bioinformatics perform high-throughput screening, looking at the structure-function relationship of interest.

4.5.3 Other Industrial Applications of Microalgae Proteins

Other potential applications of microalgae proteins are related to their techno-functional properties which go beyond their nutritional value. Due to their unique physicochemical features, proteins from microalgae can produce foam, emulsification, or gelation, among others, as a multitude of studies show (Amorim et al., 2020). Although proteins extracted from animals or other plant sources may also present these techno-functional properties, the global aim is to reduce animal protein sources as much as possible and avoid the use of other synthetic products. Besides, terrestrial plant proteins are relatively poor in techno-functional capabilities (Nasrabadia et al., 2021); therefore, finding an alternative plant source like microalgae for these techno-functional applications is of great interest, and proteins derived from microalgae have arisen as a green option.

Foams are dispersions of gases in a liquid or solid phase. They are found in many common household items: alcoholic beverages (beers, champagne), mousses, ice creams, meringue bread, and shampoos. Foams are also the precursors of solid foams, which have attracted scientists' attention due to their exceptional mechanical properties (polyurethane, glass, and metal foams). Many studies have proven the capability of microalgae proteins to form and stabilize foams. For instance, proteins extracted from *Arthrospira platensis* demonstrated superior foaming properties when compared to whey protein isolate (Buchmann et al., 2019b). Waghmare et al. (2016) measured the foaming capacity of proteins from *Chlorella pyrenoidosa*, obtaining outstanding values of 95% and 97% of foam stability (much higher than winged bean (36%) or mucuna bean concentrates) and achieved stability after 180 minutes. Foaming properties of proteins from *Arthrospira platensis* were also investigated, achieving a foaming capacity of 56.5% at pH 10 (Benelhadj et al., 2016).

Emulsions are liquid-liquid dispersions. These colloids are very important as they are applied in many food systems which consist of fat and aqueous phases. Ursu et al. (2014) obtained comparable or higher emulsion capacity of protein extracts from *Chlorella vulgaris* compared to commercial proteins, such as whey proteins. Regarding emulsion stability, they reach a maximum of 82–89%, and after the first 24 h, they reach a constant value of 72–79%, which is 10% higher than their commercial counterparts. Additional studies proved the superior emulsion activities of proteins from *Tetraselmis suecica* compared to whey protein isolate (Suarez-Garcia et al., 2018).

Marti-Quijal et al. (2019) replaced soy proteins with *Chlorella* sp. and *Spirulina* sp. protein's isolates 1% (w/w) in turkey meat and studied the changes in the physicochemical properties. The results showed that the inclusion of microalgae proteins improved the amino acid profile, including more essential amino acids. However, textural parameters (given by the emulsification capacities) and color differed significantly from soy proteins. Benelhadj et al. (2016) studied *A. platensis* protein extracts' emulsification properties, reaching a constant value of 62.5% when pH was greater than 7. A recent review made by Bertsch et al. (2021) compiles many studies where microalgae/cyanobacteria have been used for emulsion, foam stabilization, or in adsorption trials. It also points out that the crude protein behaves correctly in foam and emulsion formation, and maybe there is no need to purify further and isolate protein extracts. Furthermore, as a general trend, it can be stated that foams and emulsions based on microalgae proteins are usually less pH-dependent than those derived from surfactants (Hayes et al., 2018).

In "haute cuisine", they have also been used as thickeners, stabilizers, and more recently as gelling agents, applied to yogurts and bakery products (Deniz et al., 2017). For instance, Parniakov et al. (2018) added several protein extracts to chicken roti, including soy, bean, lentil, broad bean, *Spirulina*, and *Chlorella*, and then characterized their physicochemical composition and textural properties. A Principal Components Analysis (PCA) found significant differences, and microalgae chicken roti presented a better ratio of essential/non-essential amino acids. However, the soy protein demonstrated the best sensory acceptability.

Focusing on the gelation properties, Suarez-Garcia et al. (2018) compared the protein isolate of *Arthrospira platensis* to RuBisCo extracted from spinach, lupine, soy, and two commercial protein isolates derived from whey and egg. The results demonstrate a higher similar storage modulus of microalgae proteins as opposed to RuBisCo, lupine, and soy, although the egg and whey extracts provided the most rigid and brittle gels. Other studies proved fruitful gel formation of proteins extracted from *Tetraselmis suecica* and *Chlorella sorokiniana* (Suarez-Garcia et al., 2018; Grossmann et al., 2019).

Furthermore, enzymes have been obtained from microalgae. Microalgae have been described as a great synthesizer of enzymes for industrial utilization, and there are many examples of enzymes such as proteases, laccases, lipases, phytases, etc. (Tang et al., 2020).

4.6 CHALLENGES IN THE PRODUCTION OF MICROALGAE PROTEINS

Despite the applications and benefits of microalgae usage in the food and feed industry, the production costs exceed their profits. For instance, fish oil and meal production is around 7 million tonnes/year with a price of 2 €/kg, while microalgae production is 0.025 tonnes/year, and the market price is between 20-50 €/kg. It is hoped that by improving the cultivation processes, a higher amount of biomass can be produced so that the prices will be diminished (Fernández et al., 2021).

There is an ongoing investment based on the long-term potential applications of microalgae (beyond the ones derived from proteins), using them as a new "crop". However, most of the studies performed are at a laboratory scale, so there is a need to investigate microalgae production at an industrial level. The main drawbacks of the process are the elevated costs and the challenging extraction of the bio-compounds. The use of wastewater would be an ideal solution as growth medium for microalgae cultivation by taking advantage of the proteins formed in the microalgae biomass. However, the pollutant's concentration (heavy metals and emerging contaminants) must be monitored and carefully analyzed since microalgae are excellent adsorbents. Also, microalgae grown in wastewater form a consortium with bacteria, which can influence the entire process of protein extraction, but few articles have studied its impact. In this sense, microalgae cultivation conditions and downstream processes will determine the final application of the extracted proteins, depending on the pollutants or compounds detected (Koyande et al., 2019). Nonetheless, there is a lack of a more exhaustive systematic study of their application as a food and feed supplement (Acquah et al., 2020), and some issues like the pharmacokinetic and sensory properties require further research.

Nevertheless, there are two issues preventing broader use of the bioactive components of microalgae: (i) the upstream and downstream process must improve, and (ii) the extraction and separation process of proteins must be optimized because of the rigidity of the cell walls.

4.7 CONCLUSIONS

Microalgae have become a promising biomass to produce proteins and peptides which can be used in many applications (industrial, feed, nutraceuticals). However, the costs are still very high, and their production is not economically feasible. To increase the benefits, they can be cultivated in wastewater streams, which have a high amount of nutrients (carbon and nitrogen) necessary for growth. This has two benefits: wastewater treatment and utilization of nutrients in the production of algal biomass with higher value proteins. Nevertheless, it is necessary to optimize the process, choosing the most appropriate extraction and separation method to obtain quality proteins according to their future use.

ACKNOWLEDGMENTS

This work was supported by the "Ministerio de Ciencia, Innovación y Universidades" of Spain (CTQ2017-84006-C3-1-R). The authors also thank the regional government of Castilla y León (UIC 071, CLU 2017-09 and VA080G18) and the EU-FEDER (CLU 2017-09 and CTQ2017-84006-C3-1-R) for the financial support of this work. Elena M. Rojo would like to thank the "Ministerio de Ciencia, Innovación y Universidades" for her doctorate scholarship. David Moldes would like to thank the University of Valladolid for his doctorate scholarship.

REFERENCES

Acquah, C., Tibbetts, S.M., Pan, S., Udenigwe, C., 2020. Chapter 19: Nutritional quality and bioactive properties of proteins and peptides from microalgae, in: Jacob-Lopes, E., Maroneze, M.M., Queiroz, M.I., Zepka, L.Q. (Eds.), *Handbook of Microalgae-Based Processes and Products*. Academic Press, United Kingdom, pp. 493–531.
Alavijeh, R.S., Karimi, K., Wijffels, R.H., van den Berg, C., Eppink, M., 2020. Combined bead milling and enzymatic hydrolysis for efficient fractionation of lipids, proteins, and carbohydrates of Chlorella vulgaris microalgae. *Bioresour. Technol.* 309, 123321.
Alhattab, M., Kermanshahi-Pour, A., Brooks, M.S.L., 2019. Microalgae disruption techniques for product recovery: Influence of cell wall composition. *J. Appl. Phycol.* 31, 61–88.

Amorim, M.L., Soares, J., Coimbra, J.S. dos R., Leite, M. de O., Albino, L.F.T., Martins, M.A., 2020. Microalgae proteins: production, separation, isolation, quantification, and application in food and feed. *Crit. Rev. Food Sci. Nutr.* 61(12), 1976–2002.

Ba, F., Ursu, A., Laroche, C., Djelveh, G. 2016. Haematococcus pluvialis soluble proteins: extraction, characterization, concentration/fractionation and emulsifying properties. *Bioresour. Technol* 200, 147–152.

Baier, T., Kros, D., Feiner, R.C., Lauersen, K.J., Müller, K.M., Kruse, O. 2018. Engineered fusion proteins for efficient protein secretion and purification of a human growth factor from the green microalga *Chlamydomonas reinhardtii. ACS Synth. Biol.* 7, 2547–2557.

Balti, R., Le Balc'h, R., Brodu, N., Gilbert, M., Le Gouic, B., Le Gall, S., Sinquin, C., Massé, A. 2018. Concentration and purification of Porphyridium cruentum exopolysaccharides by membrane filtration at various cross-flow velocities. *Process Biochem.* 74, 175–184.

Balti, R., Zayoud, N., Hubert, F., Beaulieu, L., Massé, A. 2021. Fractionation of Arthrospira platensis (Spirulina) water soluble proteins by membrane diafiltration. *Sep. Purif. Technol.* 256, 117756.

Becker, E.W., 2007. Micro-algae as a source of protein. *Biotechnol. Adv.* 25, 207–210.

Benelhadj S., Gharsallaoui A., Degraeve P., Attia H., Ghorbel D. 2016. Effect of pH on the functional properties of Arthrospira (Spirulina) platensis protein isolate. *Food Chem.* 194, 1056–1063.

Berliner, M., 1986. Proteins in Chlorella vulgaris. *Microbios.* 46, 199–203.

Berry J.P., Gantar M., Gawley R.E., Wang M., Rein K.S. 2004. Pharmacology and toxicology of pahayokolide A, a bioactive metabolite from a freshwater species of Lyngbya isolated from the Florida Everglades. *Comp. Biochem. Physiol. C. Toxicol. Pharmacol.* 139, 4, 231–238.

Bertsch, P., Böcker, L., Mathys, A., Fischer, P. 2021. Proteins from microalgae for the stabilization of fluid interfaces, emulsions, and foams. *Trends Food Sci. Technol.* 108, 326–342.

Borella, L., Sforza, E., Bertucco, A., 2021. Effect of residence time in continuous photobioreactor on mass and energy balance of microalgal protein production. *N. Biotechnol.* 64, 46–53.

Borowitzka, M.A., 2018. Chapter 3: Biology of microalgae, in: Levine, I.R. and Fleurence, J. (Eds.), *Microalgae in Health and Disease Prevention.* Academic Press, United Kingdom, pp. 23–72.

Bouga M., Combet E. 2015. Emergence of seaweed and seaweed-containing foods in the UK: Focus on labeling, iodine content, toxicity and nutrition. *Foods.* 4, 2, 240–253.

Buchmann, L., Bertsch, P., Böcker, L., Krähenmann, U., Fischer, P., Mathys, A. 2019b. Adsorption kinetics and foaming properties of soluble microalgae fractions at the air/water interface. *Food Hydrocoll.* 97, 105182.

Buchmann, L., Brändle, I., Haberkorn, I., Hiestand, M., Mathys, A., 2019a. Pulsed electric field based cyclic protein extraction of microalgae towards closed-loop biorefinery concepts. *Bioresour. Technol.* 291, 121870.

Callejo-López, J.A., Ramírez, M., Bolívar, J., Cantero, D. 2019. Main variables affecting a chemical-enzymatic method to obtain protein and amino acids from resistant microalgae. *J. Chem.* 2019, 1390463.

Callejo-López, J.A., Ramírez, M., Cantero, D., Bolívar, J., 2020. Versatile method to obtain protein- and/or amino acid-enriched extracts from fresh biomass of recalcitrant microalgae without mechanical pretreatment. *Algal Res.* 50, 102010.

Carullo, D., Abera, B.D., Casazza, A.A., Donsì, F., Perego, P., Ferrari, G., Pataro, G., 2018. Effect of pulsed electric fields and high pressure homogenization on the aqueous extraction of intracellular compounds from the microalgae*Chlorella vulgaris. Algal Res.* 31, 60–69.

Carullo, D., Donsì, F., Ferrari, G., 2020. Influence of high-pressure homogenization on structural properties and enzymatic hydrolysis of milk proteins. *LWT*. 130, 109657.

Chew, K.W., Shir, R.C., Sze, Y.L., Liandong, Z., Pau, L.S. 2019. Enhanced microalgal protein extraction and purification using sustainable microwave-assisted multiphase partitioning technique. *Chem*. 367, 1–8.

Chia, S. R., Kit, W.C., Hayyiratul, F.M.Z., Dinh, T. C., Yang T., Pau L. S. 2019. Microalgal protein extraction from Chlorella vulgaris FSP-E using triphasic partitioning technique with sonication. *Front. Bioeng. Biotechnol*. 7, 396.

Choi, H.J., Lee, S.M., 2015. Effect of the N/P ratio on biomass productivity and nutrient removal from municipal wastewater. *Bioprocess Biosyst. Eng*. 38, 761–766.

Chu, W.-L., Phang, S.-M., Goh, S.-H., 1996. Environmental effects on growth and bio-chemical composition of Nitzschia inconspicua Grunow. *J. Appl. Phycol*. 8, 389–396.

Consultation, R.E. 2011. Dietary protein quality evaluation in human nutrition. *FAO Food Nutr. Pap*. 92, 1–66.

Corrêa, P.S., Júnior, M., Wilson, G., Martins, A.A., Caetano, N.S., Mata, T.M. 2021. Microalgae biomolecules: Extraction separation and purification methods. *Processes*. 9, 1, 10.

Costa, J., Freitas, B., Moraes, L., Zaparoli, M., Morais, M. G. 2020. Progress in the physi-cochemical treatment of microalgae biomass for value-added product recovery. *Bioresour. Technol*. 301, 122727.

Costa da Silva, P., Toledo, T., Brião, V., Bertolin, T.E., and Vieira Costa, J.A. 2021. Development of extruded snacks enriched by bioactive peptides from microalga Spirulina Sp. LEB 18. *Food Biosci*. 42, 101031.

de Morais, M.G., Vaz, B. da S., de Morais, E.G., Costa, J.A.V., 2015. Biologically active metabolites synthesized by microalgae. *Biomed. Res. Int*. 2015, 835761.

Deniz, I., García-Vaquero, M., Imamoglu, E. 2017. Trends in red biotechnology: Microalgae for pharmaceutical applications. In: Gonzalez-Fernandez, C. and Muñoz, R. (Eds.), *Microalgae-Based Biofuels and Bioproducts: From Feedstock Cultivation to End-Products*. Woodhead Publishing, United Kingdom, pp. 429–460. 10.1016/b978-0-08-101023-5.00018-2

Dixon, C., Wilken, L.R. 2018. Green microalgae biomolecule separations and recovery. *Bioresour. Bioprocess*. 5, 14.

Dolganyuk, V., Belova, D., Babich, O., Prosekov, A., Ivanova, S., Katserov, D., Patyukov, N., Sukhikh, S., 2020. Microalgae: A promising source of valuable bioproducts. *Biomolecules*. 10, 8, 1153.

Duan, Z., Tan, X., Dai, K., Gu, H., Yang, H. 2017 Evaluation on H2O2-aided ultrasonic pretreatment for cell disruption of Chlorella pyrenoidosa. *Asia-Pac. J. Chem. Eng*., 12, 502–510.

Elain, A., Nkounkou, C., Le Fellic, M., Donnart, K., 2020. Green extraction of poly-saccharides from Arthrospira platensis using high pressure homogenization. *J. Appl. Phycol*. 32, 1719–1727.

Fernández F.G.A., Reis A., Wijffels R.H., Barbosa M., Verdelho V., Llamas B. 2021. The role of microalgae in the bioeconomy. *N Biotechnol*. 61, 99–107

Gateau, H., Blanckaert, V., Veidl, B., Burlet-Schiltz, O., Pichereaux, C., Gargaros, A., Marchand, J., Schoefs, B., 2021. Application of pulsed electric fields for the bio-compatible extraction of proteins from the microalga Haematococcus pluvialis. *Bioelectrochemistry*. 137, 107588.

Gifuni, I., Lavenant L., Pruvost J., Masse A. 2020. Recovery of microalgal protein by three-steps membrane filtration: Advancements and feasibility. *Algal Res*. 51, 102082.

González-Balderas, R.M., Velásquez-Orta, S.B., Valdez-Vazquez, I., Orta Ledesma, M.T. 2020. Sequential pretreatment to recover carbohydrates and phosphorus from Desmodesmus Sp. cultivated in municipal wastewater. *Water Sci. Technol*. 82(6), 1237–1246.

Grossmann, L., Ebert, S., Hinrichs, J., Weiss, J. 2018. Production of protein-rich extracts from disrupted microalgae cells: Impact of solvent treatment and lyophilization. *Algal Res.* 36, 67–76.

Grossmann, L., Hinrichs, J., Goff, H.D., Weiss, J. 2019. Heat-induced gel formation of a protein-rich extract from the microalga Chlorella sorokiniana. *Innov. Food Sci. Emerg. Technol.* 56, 1–9.

Grossmann, L., Hinrichs, J., Weiss, J., 2020a. Cultivation and downstream processing of microalgae and cyanobacteria to generate protein-based techno functional food ingredients, *Crit. Rev. Food Sci. Nutr.* 60(17), 2961–2989.

Grossmann, L., Wörner, V., Hinrichs, J., Weiss, J., 2020b. Sensory properties of aqueous dispersions of protein-rich extracts from Chlorella prototothecoides at neutral and acidic pH. *J. Sci. Food Agric.*, 100, 1344–1349.

Guccione, A., Biondi, N., Sampietro, G., Rodolfi, L., Bassi, N., Tredici, M.R., 2014. Chlorella for protein and biofuels: From strain selection to outdoor cultivation in a Green Wall Panel photobioreactor. *Biotechnol. Biofuels* 7, 84.

Guzmán F., Wong G., Román T, et al. 2019. Identification of antimicrobial peptides from the microalgae Tetraselmis suecica (Kylin) Butcher and Bactericidal activity improvement. *Mar Drugs.* 17, 8, 453.

Hashemi, A., Moslemi, M., Pajoum Shariati, F., Delavari Amrei, H., 2020. Beta-carotene production within Dunaliella salina cells under salt stress condition in an indoor hybrid helical-tubular photobioreactor. *Can. J. Chem. Eng.* 98, 69–74.

Hayes, M., Bastiaens, L., Gouveia, L., Gkelis, S., Skomedal, H., Skjanes, K., Murray, P., García-Vaquero, M., Hosoglu, M.I., Dodd, J., Konstantinou, D., Safarik, I., Zittelli, G.C., Rimkus, V., del Pino, V., Muylaert, K., Edwards, C., Laake, M., da Silva, J.G.L., Pereira, H. and Abelho, J. 2018. Microalgal Bioactive Compounds Including Protein, Peptides, and Pigments: Applications, Opportunities, and Challenges During Biorefinery Processes, in: Hayes M. (Ed.), *Novel Proteins for Food, Pharmaceuticals and Agriculture.* John Wiley & Sons, USA, pp. 239–255.

Hayes, M., García-García, M., Fitzgerald, C., Lafarga, T. 2015. Seaweed and milk derived bioactive peptides and small molecules in functional foods and cosmeceuticals, in: Gupta, V.K., Tuohy, M.G. (Eds.), *Biotechnology of Bioactive Compounds.* John Wiley & Sons, United Kingdom.

Hayes, M., Skomedal, H., Skjånes, K., Mazur-Marzec, H., Toruńska-Sitarz, A., Catala, M., Isleten Hosoglu, M., García-Vaquero, M. 2017. Microalgal proteins for feed, food and health, in: Gonzalez-Fernandez, C., Muñoz, R. (Eds.), *Microalgae-Based Biofuels and Bioproducts: From Feedstock Cultivation to End-Products.* Woodhead Publishing, United Kingdom, pp. 347–368.

Hildebrand, G., Poojary, M.M., O'Donnell, C., Lund, M.N., Garcia-Vaquero, M., Tiwari, B.K., 2020. Ultrasound-assisted processing of Chlorella vulgaris for enhanced protein extraction. *J. Appl. Phycol.* 32, 1709–1718.

Ho, S.-H., Chen, C.-Y., Chang, J.-S., 2012. Effect of light intensity and nitrogen starvation on CO_2 fixation and lipid/carbohydrate production of an indigenous microalga Scenedesmus obliquus CNW-N. *Bioresour. Technol.* 113, 244–252.

Hu, S., Fan, X., Qi, P., Zhang, X. 2019. Identification of anti-diabetes peptides from Spirulina platensis. *J. Funct. Foods.* 56: 333–341.

Hynstova, V., Sterbova, D., Klejdus, B., Hedbavny, J., Huska, D., Adam, V. 2018. Separation, identification and quantification of carotenoids and chlorophylls in dietary supplements containing Chlorella vulgaris and Spirulina platensis using high performance thin layer chromatography. *J. Pharm. Biomed. Anal.* 148, 108–118.

Ishida K, Matsuda H, Murakami M, Yamaguchi K. 1997. Kawaguchipeptin B, an antibacterial cyclic undecapeptide from the cyanobacterium Microcystis aeruginosa. *J. Nat. Prod.* 60, 7, 724–726.

Kadkhodaei, S., Abbasiliasi, S., Shun, T.~J., Fard Masoumi, H.~R., Mohamed, M.~S., Movahedi, A., Rahim, R., Ariff, A.~B., 2015. Enhancement of protein production by microalgae Dunaliella salina under mixotrophic conditions using response surface methodology. *RSC Adv.* 5, 38141–38151.

Kapoore, R. V., Butler, T. O., Pandhal, J., V., Seetharaman. 2018. Microwave-assisted extraction for microalgae: From biofuels to biorefinery. *Biology.* 7, 1: 18.

Kareya, M.S., Mariam, I., Shaikh, K.M., Nesamma, A.A., Jutur, P.P., 2020. Photosynthetic carbon partitioning and metabolic regulation in response to very-low and high CO2 in Microchloropsis gaditana NIES 2587. *Front. Plant Sci.* 11, 981.

Keithellakpam, O.S., Nath, T.O., Oinam, A.S., Thingujam, I., Oinam, G., Dutt, S.G., 2015. Effect of external pH on cyanobacterial phycobiliproteins production and ammonium excretion. *J. Appl. Biol. Biotechnol.* 3, 38–42.

Keris-Sen, U. D., Gurol, M.D. 2017. Using ozone for microalgal cell disruption to improve enzymatic saccharification of cellular carbohydrates. *Biomass Bioenergy* 105, 59–65.

Koyande, A.K., Chew, K.W., Rambabu, K., Tao, Y., Chu, D., Show, P.L. 2019. Microalgae: A potential alternative to health supplementation for humans. *Food Sci. Hum. Wellness.* 8, 1, 16–24.

Krishna Koyande, A., Vera, T., Haridharan, M.D., Manivarman, S., Ryann, N.R., Phei, L.L., Ianatul, K., Pau, L.S. 2020. Integration of osmotic shock assisted liquid biphasic system for protein extraction from microalgae Chlorella vulgaris. *BioChem.* 157, 107532.

Kusmayadi A., Leong Y.K., Yen H.W., Huang C.Y., Chang J.S. 2021. Microalgae as sustainable food and feed sources for animals and humans - Biotechnological and environmental aspects. *Chemosphere.* 271, 129800.

Larsen R., Eilertsen K.E., Elvevoll E.O. 2011. Health benefits of marine foods and ingredients. *Biotechnol. Adv.* 29, 5, 508–518.

Leong Y.K., Chang J.S. 2020. Bioremediation of heavy metals using microalgae: Recent advances and mechanisms. *Bioresour Technol.* 303, 122886.

Li, K., Liu, Q., Fang, F., Luo, R., Lu, Q., Zhou, W., Huo, S., et al. 2019. Microalgae-based wastewater treatment for nutrients recovery: A review. *Bioresour. Technol.* 291, 121934.

Liu S., Li Z., Yu B., Wang S., Shen Y., Cong H. 2020. Recent advances on protein separation and purification methods. *Adv. Colloid Interface Sci.* 284, 102254.

Lupatini, A.L., Colla, L.M., Canan, C., Colla, E. 2017. Potential application of microalga Spirulina platensis as a protein source. *J. Sci. Food Agric.* 97, 3, 724–732.

Ma, R., Wang, B., Chua, E.T., Zhao, X., Lu, K., Ho, S.-H., Shi, X., Liu, L., Xie, Y., Lu, Y., Chen, J., 2020. Comprehensive utilization of marine microalgae for enhanced co-production of multiple compounds. *Mar. Drugs.* 18, 9, 467.

Marti-Quijal, F.J., Zamuz, S., Tomašević, I., Rocchetti, G., Lucini, L., Marszałek, K., Barba, F.J., Lorenzo, J.M. 2019, A chemometric approach to evaluate the impact of pulses, Chlorella and Spirulina on proximate composition, amino acid, and physicochemical properties of turkey burgers. *J. Sci. Food Agric.,* 99, 3672–3680.

Matos, Â.P., 2019. Chapter 3 Microalgae as a potential source of proteins, in: Galanakis, C.M. (Ed.), *Proteins: Sustainable Source Processing and Applications.* Academic Press, United Kingdom, pp. 63–96.

Nasrabadia, M.N., Doost, A.S., Mezzenga, R. 2021. Modification approaches of plant-based proteins to improve their techno-functionality and use in food products. *Food Hydrocoll.* 118, 106789.

Nguyen, M.H.T., Qian Z.J., Van Tinh Nguyen, I.l Choi W., Heo S.J., Oh C.H., Kang D.H., Kim G.H., Jung W.K.. 2013. Tetrameric peptide purified from hydrolysates of biodiesel byproducts of nannochloropsis oculata induces osteoblastic differentiation through MAPK and Smad Pathway on MG-63 and D1 cells. *Process Biochem.* 48, 9, 1387–1394.

Nitsos, C., Filali, R., Taidi, B., Lemaire, J., 2020. Current and novel approaches to downstream processing of microalgae: A review. *Biotechnol. Adv.* 45, 107650.

Ogbonda, K.H., Aminigo, R.E., Abu, G.O., 2007. Influence of temperature and pH on biomass production and protein biosynthesis in a putative Spirulina sp. *Bioresour. Technol.* 98, 2207–2211.

Pagels, F., Guedes, A.C., Amaro, H.M., Kijjoa, A., Vasconcelos,V. 2019. Phycobiliproteins from Cyanobacteria: Chemistry and biotechnological applications. *Biotechnol. Adv.* 37, 3, 422–443.

Park, M.-H., Park, C.-H., Sim, Y.B., Hwang, S.-J., 2020. Response of Scenedesmus quadricauda (Chlorophyceae) to salt stress considering nutrient enrichment and intracellular proline accumulation. *Int. J. Environ. Res. Public Health.* 17.

Parniakov O., Toepfl S., Barba F.J., et al. 2018. Impact of the soy protein replacement by legumes and algae-based proteins on the quality of chicken rotti. *J Food Sci Technol.* 55(7), 2552–2559.

Passos F., Carretero J., Ferrer I., 2015. Comparing pretreatment methods for improving microalgae anaerobic digestion: Thermal, hydrothermal, microwave and ultrasound, *Chem.* 279, 1, 667–672.

Phong, W.N., Pau, L.S., Tau, C.L., Joon, C.J., Eng, P.N., Jo, S.C., 2018b. Mild cell disruption methods for bio-functional proteins recovery from microalgae—Recent developments and future perspectives. *Algal Res.* 31, 506–516.

Phong, W.N., Show, P.L., Le, C.F., Tao, Y., Chang, J.S., Ling, T.C., 2018a. Improving cell disruption efficiency to facilitate protein release from microalgae using chemical and mechanical integrated method. *BioChem.* 135, 83–90.

Piña, P., Medina, M.A., Nieves, M., Leal, S., López-Elías, J.A., Guerrero, M.A., 2007. Cultivo De Cuatro Especies De Microalgas Con Diferentes Fertilizantes Utilizados En Acuicultura. *Rev. Invest.* 28(3), 225–236.

Pojić, M., Aleksandra M., Brijesh T., 2018. Eco-innovative technologies for extraction of proteins for human consumption from renewable protein sources of plant origin. *Trends Food Sci. Technol.* 75, 93–104.

Postma, P.R., Suarez-Garcia, E., Safi, C., Olivieri, G., Olivieri, G., Wijffels, R.H., Wijffels, R.H., 2017. Energy efficient bead milling of microalgae: Effect of bead size on disintegration and release of proteins and carbohydrates. *Bioresour. Technol.* 224, 670–679.

Qazi, W.M., Balance, S., Uhlen, A.K., Kousoulaki, K., Haugen, J.E., Rieder, A. 2021. Protein enrichment of wheat bread with the marine green microalgae Tetraselmis chuii – Impact on dough rheology and bread quality. *LWT.* 143, 111115.

Rasdi, N.W., Qin, J.G., 2015. Effect of N:P ratio on growth and chemical composition of Nannochloropsis oculata and Tisochrysis lutea. *J. Appl. Phycol.* 27(6), 2221–2230.

Rizwan, M., Mujtaba, G., Memon, S.A., Lee, K., Rashid, N., 2018. Exploring the potential of microalgae for new biotechnology applications and beyond: A review. *Renew. Sustain. Energy Rev.* 92, 394–404.

Rocha, C.M.R., Genisheva, Z., Ferreira-Santos, P., Rodrigues, R., Vicente, A.A., Teixeira, J.A., Pereira, R.N. 2018. Electric field-based technologies for valorization of bioresources. *Bioresour. Technol.* 254, 325–339. 10.1016/j.biortech.2018.01.068

Saadaoui I., Rasheed R., Abdulrahman N., et al. 2020. Algae-derived bioactive compounds with anti-lung cancer potential. *Mar. Drugs.* 18(4), 97.

Sabia, A., Clavero, E., Pancaldi, S., Salvadó Rovira, J., 2018. Effect of different CO2 concentrations on biomass, pigment content, and lipid production of the marine diatom Thalassiosira pseudonana. *Appl. Microbiol. Biotechnol.* 102, 1945–1954.

Safi, C., Ursu, A.V., Laroche,C., Zebib, B., Merah, O., Pontalier, P.Y., Vaca-Garcia,C., 2014b. Aqueous extraction of proteins from microalgae: Effect of different cell disruption methods. *Algal Res.* 3(1), 61–65.

Safi, C., Zebib, B., Merah, O., Pontalier, P.Y., Vaca-Garcia,C. 2014a. Morphology, composition, production, processing and applications of Chlorella vulgaris: A review. *Renew. Sustain. Energy Rev.* 35, 265–278.

Salakkam, A., Sureewan S., Chonticha M., Orawan P., Alissara R., 2021. Valorization of microalgal biomass for biohydrogen generation: A review. *Bioresour. Technol.* 322, 124533.

Sánchez-Zurano, A., Morillas-España, A., González-López, C. V., Lafarga, T., 2020. Optimisation of protein recovery from *Arthrospira platensis* by ultrasound-assisted isoelectric solubilisation/precipitation. *Processes*, 8, 12, 1586.

Sankaran, R., Manickam, S., Yap, Y.J., Ling, T.C., Chang, J.S., Show, P.L., 2018. Extraction of proteins from microalgae using integrated method of sugaring-out assisted liquid biphasic flotation (LBF) and ultrasound. *Ultrason. Sonochem.* 48, 231–239.

Serpa Ibáñez, R.F., Calderón Rodríguez, A., 2006. Efecto de diferentes fuentes de nitrógeno en el contenido de carotenoides y clorofila de cuatro cepas peruanas de Dunaliella salina Teod. *Ecol. Apl.* 5, 93–99.

Seyfabadi, J., Ramezanpour, Z., Amini Khoeyi, Z., 2011. Protein, fatty acid, and pigment content of Chlorella vulgaris under different light regimes. *J. Appl. Phycol.* 23, 721–726.

Shahid, A., Khan, F., Ahmad, N., Farooq, M., Mehmood, M.A., 2020. Microalgal carbohydrates and proteins: Synthesis, extraction, applications, and challenges, in: Alam, Md.A , Xu, J.-L., Wang, Z. (Eds.), *Microalgae Biotechnology for Food Health and High Value Products*. Springer, Singapore, pp. 433–468.

Shen, C. 2019. Chapter 8 – Quantification and analysis of proteins, in: Shen, C.-H. (Ed.), *Diagnostic Molecular Biology*. Academic Press, United Kingdom, pp. 187–214.

Soto-Sierra, L., Stoykova, P., Nikolov, Z.L. 2018. Extraction and fractionation of microalgae-based protein products. *Algal Res.* 36, 175–192.

Suarez-Garcia, E., van Leeuwen, J. J. A., Safi, C., Sijtsma, L., van den Broek, L. A. M., Eppink, M. H. M., Wijffels, R. H., van den Berg, C. 2018. Techno-functional properties of crude extracts from the green microalga Tetraselmis suecica. *J. Agric. Food Chem.* 66, 29, 7831–7838.

Sui, Y., Vlaeminck, S.E., 2019. Effects of salinity, pH and growth phase on the protein productivity by Dunaliella salina. *J. Chem. Technol. Biotechnol.* 94, 1032–1040.

Sun, Y., Zhang, N., Zhou, J., Dong,S., Zhang, X., Guo, L., Guo, G., 2020. Distribution, contents, and types of mycosporine-like amino acids (MAAs) in marine macroalgae and a database for MAAs based on these characteristics. *Mar. Drugs.* 18, 1, 43.

Tan, C.H., Show, P.L., Lam, M.K., Fu, X., Ling, T.C., Chen, C.-Y., Chang, J.-S., 2020. Examination of indigenous microalgal species for maximal protein synthesis. *Biochem. Eng. J.* 154, 107425.

Tang, D.Y.Y., Khoo, K.S., Chew, K.W., Tao, Y., Ho, S.H., Show, P.L., 2020. Potential utilization of bioproducts from microalgae for the quality enhancement of natural products. *Bioresour. Technol.* 304, 122997.

Teuling, E., Schrama, J.W., Gruppen, H., Wierenga, P.A., 2019. Characterizing emulsion properties of microalgal and cyanobacterial protein isolates. *Algal Res.* 39, 101471.

Timira, V., Meki, K., Li, Z., Lin, H., Xu, M., Pramod, S.N., 2021. A comprehensive review on the application of novel disruption techniques for proteins release from microalgae. *Crit. Rev. Food Sci. Nutr.* 1–17.

Travaini R, Martín-Juárez J, Lorenzo-Hernando A, Bolado-Rodríguez S. 2016. Ozonolysis: An advantageous pretreatment for lignocellulosic biomass revisited. *Bioresour. Technol.* 199, 2–12.

Ursu, A. V., Marcati, A., Sayd, T., Sante-Lhoutellier, V., Djelveh, G., Michaud, P. 2014. Extraction, fractionation and functional properties of proteins from the microalgae Chlorella vulgaris. *Bioresour. Technol.* 157, 134–139.

Varshney, P., Beardall, J., Bhattacharya, S., Wangikar, P.P., 2018. Isolation and biochemical characterisation of two thermophilic green algal species-Asterarcys quadricellulare and Chlorella sorokiniana, which are tolerant to high levels of carbon dioxide and nitric oxide. *Algal Res.* 30, 28–37.

Vernès, L., Abert-Vian, M., El Maâtaoui, M., Tao, Y., Bornard, I., Chemat, F., 2019b. Application of ultrasound for green extraction of proteins from spirulina. Mechanism, optimization, modeling, and industrial prospects. *Ultrason. Sonochem.* 54, 48–60.

Vernès, L., Vian, M., Chemat, F., 2019a. Ultrasound and microwave as green tools for solid-liquid extraction, in: Poole, C.F. (Ed.), *Liquid-Phase Extraction*. Elsevier, Netherlands, pp. 355–374.

Vizcaíno, A. J., Sáez, M. I., Martínez, T. F., Acién, F. G., Alarcón, F. J. 2019. Differential hydrolysis of proteins of four microalgae by the digestive enzymes of gilthead sea bream and Senegalese sole. *Algal Res.* 37, 145–153.

Vo, T.S., Ryu, B.M., Kim, S.K. 2013. Purification of novel anti-inflammatory peptides from enzymatic hydrolysate of the edible microalgal Spirulina Maxima. *J. Funct. Foods.* 5, 3, 1336–1346.

Waghmare, A.G., Salve, M.K., LeBlanc, J.G. et al. 2016. Concentration and characterization of microalgae proteins from Chlorella pyrenoidosa. *Bioresour. Bioprocess.* 3, 16.

Yaakob, M.A., Mohamed, R.M.S.R., Al-Gheethi, A., Aswathnarayana Gokare, R., Ambati, R.R., 2021. Influence of nitrogen and phosphorus on microalgal growth, biomass, lipid, and fatty acid production: An overview. *Cells* 10(2), 393.

Yang, L., Chen, J., Qin, S., Zeng, M., Jiang, Y., Hu, L., Xiao, P., Hao, W., Hu, Z., Lei, A., Wang, J., 2018. Growth and lipid accumulation by different nutrients in the microalga Chlamydomonas reinhardtii. *Biotechnol. Biofuels* 11, 40.

Zachleder, V., Bišová, K., Vítová, M., 2016. The cell cycle of microalgae, in: Borowitzka, M.A., Beardall, J., Raven, J.A. (Eds.), *The Physiology of Microalgae*. Springer International Publishing, Cham, pp. 3–46.

Zarrinmehr, M.J., Farhadian, O., Heyrati, F.P., Keramat, J., Koutra, E., Kornaros, M., Daneshvar, E., 2020. Effect of nitrogen concentration on the growth rate and bio-chemical composition of the microalga, Isochrysis galbana. *Egypt. J. Aquat. Res.* 46, 153–158.

Zhao, B., Su, Y., 2014. Process effect of microalgal-carbon dioxide fixation and biomass production: A review. *Renew. Sustain. Energy Rev.* 31, 121–132.

Zhao, X.C., Tan, X.B., Yang, L. Bin, Liao, J.Y., Li, X.Y., 2019. Cultivation of Chlorella pyrenoidosa in anaerobic wastewater: The coupled effects of ammonium, temperature and pH conditions on lipids compositions. *Bioresour. Technol.* 284, 90–97.

Zhu, L., 2015. Microalgal culture strategies for biofuel production: A review. Biofuels, *Bioprod. Biorefining.* 9, 801–814.

Zhu, S., Huang, W., Xu, Jin, Wang, Z., Xu, Jingliang, Yuan, Z., 2014. Metabolic changes of starch and lipid triggered by nitrogen starvation in the microalga Chlorella zofingiensis. *Bioresour. Technol.* 152, 292–298.

Zuccaro, G., Yousuf, A., Pollio, A., Steyer, J.-P., 2020. Chapter 2 - Microalgae cultivation systems, in: Yousuf, A. (Ed.), *Microalgae Cultivation for Biofuels Production*. Academic Press, United States, pp. 11–29.

5 Microalgae for Pigments and Cosmetics

Nídia S. Caetano
LEPABE-Laboratory for Process Engineering, Environment, Biotechnology and Energy, Faculty of Engineering, University of Porto (FEUP), Porto, Porto, Portugal
and
CIETI – Centre for Innovation in Engineering and Industrial Technology, ISEP/P.Porto – School of Engineering/Polytechnic of Porto, Porto, Portugal

Priscila S. Corrêa and Wilson G. de Morais Júnior
CIETI – Centre for Innovation in Engineering and Industrial Technology, ISEP/P.Porto – School of Engineering/Polytechnic of Porto, Porto, Portugal

Gisela M. Oliveira
UFP Energy, Environment and Health Research Unit, University Fernando Pessoa, Porto, Portugal

António A.A. Martins
LEPABE-Laboratory for Process Engineering, Environment, Biotechnology and Energy, Faculty of Engineering, University of Porto (FEUP), Porto, Portugal

Teresa M. Mata
INEGI-Institute of Science and Innovation in Mechanical and Industrial Engineering, Porto, Portugal

Monique Branco-Vieira
LEPABE-Laboratory for Process Engineering, Environment, Biotechnology and Energy, Faculty of Engineering, University of Porto (FEUP), Porto, Portugal
and

DOI: 10.1201/9781003195405-5

PPE – Energy Planning Program, Alberto Luiz Coimbra
Institute for Graduate Studies and Research in
Engineering (COPPE), Federal University of Rio de Janeiro
(UFRJ), Rio de Janeiro, Brazil
and
IPK – Leibniz Institute of Plant Genetics and Crop Plant
Research, Gatersleben, Germany

CONTENTS

5.1   Introduction..134
5.2   Pigments in Microalgae 136
        5.2.1   Chlorophylls 137
        5.2.2   Carotenoids..138
        5.2.3   Phycobiliproteins 140
5.3   Biosynthesis..141
5.4   Strategies to Enhance the Production of Pigments.....................................144
        5.4.1   Nutrimental, Physical, and Culture Regime....................................144
        5.4.2   Genetic and Metabolic Engineering 147
5.5   Extraction, Separation, and Purification 150
5.6   Industrial Production 152
        5.6.1   Microalgae Cultivation Systems...154
        5.6.2   Harvesting and Biomass Pre-Processing 156
        5.6.3   Extraction/Separation and Purification Processes 156
5.7   Importance and Commercial Applications of Microalgae Pigments..........157
        5.7.1   Nutraceutical and Pharmaceutical Uses..157
                    5.7.1.1   Cosmetics and Cosmeceutical Use 158
5.8   Uses of Microalgal Pigments Boosting Bioeconomy and Circular
        Economy 161
5.9   Conclusions...163
Acknowledgments..163
References..164

5.1   INTRODUCTION

The variety of known microalgae (eukaryotic) and cyanobacteria (prokaryotic) are estimated to be over 70,000 species (Guiry, 2012), and they can be found in every kind of environment as long as there is water available. However harsh the conditions, these microorganisms seem to be adaptable to every habitat on Earth. Because of this variety of species and adaptation facility, there is a considerable diversity of compounds produced by microalgae, often with corresponding diverse metabolic pathways that originate those substances (Borowitzka, 2019).

Microalgae can also be explored to host the production and delivery of vaccines. An example is the research by Márquez-Escobar et al. (2018), who transformed *Schizochytrium* sp. to produce an oral vaccine candidate against Zika virus (ZIKV) infection. This virus had several outbreaks and dissemination in recent years, thus finding vaccines against it and/or treatment emerged as urgent. In their research, Márquez-Escobar et al. (2018) applied the Algevir technology to express the ZK antigenic protein in *Schizochytrium* sp. with up to $365\,\mu g{\cdot}g^{-1}$ microalgae fresh weight yield. The *in vivo* assays of the microalgae-made ZK protein in mice demonstrated that the oral administration caused significantly higher humoral responses than those induced by subcutaneous immunization.

Microalgae and cyanobacteria are the major photosynthesizers on Earth, being responsible for the important balance of oxygen and carbon dioxide flows between water bodies and the atmosphere. In every photoautotrophic cell, the most important reaction is the carbon dioxide (CO_2) biofixation (some species may use hydrogen sulfide instead of CO_2): the conversion of CO_2 and water by light into oxygen and organic substances obtaining energy for cell metabolism.

Microalgae and cyanobacteria are mainly composed of carbohydrates, proteins, and lipids, but all cells produce pigments, and several species produce vitamins, polyphenols, and other antioxidants (do Nascimento et al., 2020). Generally, three distinct groups of pigments may be considered: chlorophylls – greenish, photosynthetic, and lipid-soluble; carotenoids – hydrocarbons (lipids) in a range of yellow to red colours; and phycobiliproteins – blue-green pigments that are only known to exist in cyanobacteria and the phylum Rhodophyta (Cuellar-Bermudez et al., 2015). Still, the diversity of chemicals produced by so many species and their potential as valuable compounds, especially as bioactive compounds is yet to be unveiled (Del Mondo et al., 2020; Tudor et al., 2021).

Microalgae ubiquity, diversity, completeness in nutrition value, and availability makes them attractive as a food source, especially as functional foods and nutraceuticals (Galasso et al., 2019; García et al., 2017; Torres-Tiji et al., 2020).

The interest of microalgae biomass as a food source is quite ancient: the Aztecs, from Lake Texcoco (Mexico), and the indigenous Kanembu tribe, from Lake Kossorom (Chad), harvested *Spirulina* (cyanobacteria) and included it in their diet in the form of a dried cake (Habib and Ahsan, 2008; Vonshak and Richmond, 1988). Other microalgae, such as *Chlorella,* were also commonly consumed as food in other areas of the globe, especially in Asia (Borowitzka, 2013a; García et al., 2017), because algae-dried biomass is very well digested and tolerated by humans (Gurev et al., 2020; Hemantkumar and Rahimbhai, 2019).

Notwithstanding, industrial mass cultivation of microalgae in photobioreactors has started only in the 1950s of the 20th century, propelled by the first Algal Mass-Culture Symposium held at the Stanford University and by a dedicated project of the Carnegie Institute (Burlew, 1953) that led to the establishment, in California, of one of the world's first research institutions and microalgae laboratories. The 1950s were the starting point of the recognition of the microalgae value as a mean to capture and store solar energy, but also as an edible source of important nutrients for human health. *Spirulina* species are widely recognized as important sources of protein and have gained popularity among food supplements and nutraceuticals (Habib and Ahsan, 2008; Hemantkumar and Rahimbhai, 2019; Molino et al., 2018). In the 1960s, *Chlorella* was the first microalga to be industrially produced in Asia (China, Taiwan, and Thailand) for food purposes (Benemann, 2016). Soon after, *Dunaliella* and *Spirulina* (Ben-Amotz et al., 1991; Raja et al., 2007; Vonshak and Richmond, 1988) also became industrially produced in large open ponds – raceways in Australia (Borowitzka and Borowitzka,

1990); in Israel – Eilat, the Red Sea at Nature Beta Technologies (Ben-Amotz, 1995), and then, in the 1980s, at Cyanotech – Kailua-Kona, Hawaii (Cysewski, 2011), *Spirulina* and *Haematococcus* also started commercial production.

5.2 PIGMENTS IN MICROALGAE

Light-harvesting and CO_2 biofixation efficiency determines microalgae biomass growth and the number of valuable products accumulated in cells (Eloka-Eboka and Inambao, 2017; Yarnold, 2016); however, these outputs are mainly limited by the microalgae strain being cultivated. Industrial cultivation of microalgae has demonstrated that CO_2 fixation efficiency depends on many external factors like the photobioreactor design and operation, which influences mixing, light, and mass transfer conditions in the culture (Dolganyuk et al., 2020; Fernandes et al., 2015; Li et al., 2013; Nwoba et al., 2019). The knowledge on the biochemical processes involved in photosynthesis is becoming enriched by new findings on carbon metabolism and energy conversion by microalgae (Borowitzka, 2019). It is recognized that by modulating abiotic factors, such as the regulation of light intensity and spectra or nutrients starvation, some microalgae biochemical processes may be either inhibited or triggered to synthesize or convert carbohydrates into other valuable substances, such as antioxidants (fatty acids, pigments, vitamins), that can be accumulated in the cells at considerable concentrations. Due to the possibility of biochemical and metabolic engineering manipulations to enhance valuable substances production in cells, sometimes, microalgae are called as cell factories (Mitchell and Goacher, 2017). At the heart of these factories is the photochemical apparatus of photosynthesis where photons are captured by an assemblage of pigments in organelles, followed by an orchestrated electrons transfer and consequential biochemical chain reactions with proton transport and energy synthesis that lead to CO_2 fixation and metabolic reactions (Esland et al., 2018; Rubio et al., 2003; Stirbet et al., 2020).

The production of pigments and other valuable substances is directly related to the harvesting of light and to the specific cells photosynthetic machinery, which is characteristic of each microorganism species (Kalaji et al., 2016; Stirbet et al., 2020). The photosynthetic system, with its organelles and processes, is so relevant that, decades ago, this was one of the major criteria to systematize and organize microalgae and cyanobacteria into taxonomical classes, among other cellular characteristics. Recently, the classification of microalgae and cyanobacteria has undergone major developments due to the easiness of genome sequence; thus, microalgae taxonomy is constantly being revised.

The harvesting of light in microalgae and cyanobacteria is classified as oxygenic photosynthesis, due to the involvement of water to produce oxygen, and is considered to be processed by two different but, complementary, photosynthetic systems – Photosystem I (PSI) and Photosystem II (PSII) – that operate in series in the electron and proton transfer process (Larkum, 2016; Müh and Zouni, 2020). In different species and because of their divergent evolutionary pathways, it is considered that the roles of PSI and PSII also may have different contributions in the process and in the efficiency of light harvesting (Liu et al., 2019; Müh and Zouni, 2020; Stirbet et al., 2020). The variety of pigments involved in the light-harvesting processes is high as well: chlorophylls act as the major primary pigments, but others, like phycobilins and carotenoids, are associated with pigment-binding proteins that are also determinant in the harvesting of light. The roles of PSI and PSII and the intervention of pigments are also influenced by the environmental conditions that microalgae are exposed to, namely, and as previously referred: the light intensity and spectra, deprivation of some nutrients, and the daily light-dark cycle (Dolganyuk et al., 2020; Xu et al., 2016).

As referred, the manipulation of the environmental growth conditions, specifically by nutrient deprivation, such as nitrogen and phosphorous (Cuellar-Bermudez et al., 2015), is often used as a form to stimulate microalgae metabolism to enhance photosynthesis rate and to influence the type of substances being produced and accumulated in cells, such as lipids, mostly polyunsaturated fatty acids (PUFA), or pigments like carotenoids (Dickinson et al., 2016; Fu et al., 2014; Sarkar and Shimizu, 2015).

5.2.1 CHLOROPHYLLS

Chlorophylls are an extraordinary group of green pigments that exist in every microalgae, cyanobacteria, and plants as well (Masuda and Fujita, 2008; Rubio et al., 2003). They are the most abundant and valuable natural pigment, essential for the microalgae life cycle that absorbs and converts solar/light energy into chemical energy through photosynthesis for the functioning of its metabolism and reproduction (Li et al., 2015). Chlorophylls are complex molecules based on tetrapyrroles organized in cycles around a central magnesium atom that is bonded to the four nitrogen atoms of the pyrroles (Fujita et al., 2015). According to variations of the tetrapyrrole structure of the molecule, chlorophylls are identified by letters *a*, *b*, *c*, *d*, *e*, and *f* (Barkia et al., 2019; Li et al., 2012). Other types of chlorophyll that differ in the molecular structure, altering the form of light absorption, have been identified in other microorganisms: Protochlorophyll and Bacteriochlorophyll *a*, *b*, *c*, *d*, *e*, and *g* (Silva and Lombardi, 2020).

Concerning chlorophyll *c*, two main structures may occur, recognized as c_1 and c_2 (Larkum, 2016). Very little is known about chlorophyll *e*; however, the other five chlorophylls have different but complementary absorption properties as they can harvest light at different wavelengths from the ultraviolet end of the spectrum (300 to 380 nm), the visible (380 to 700 nm), and NIR – Near Red Infrared (700 to 750 nm) part of the spectra (Li et al., 2012). Chlorophylls *a* and *b* are the most common primary photosynthetic pigments in microalgae and cyanobacteria (Barkia et al., 2019; Dolganyuk et al., 2020). Their structure has a linear chain of saturated hydrocarbons named phytol (Masojídek et al., 2013). Chlorophyll a is known to be the most abundant form of chlorophyll in cyanobacteria, green and red algae, and plays a crucial role in both light-harvesting systems PSI and PSII (Larkum, 2016). Its green color makes it absorb mainly red light from the solar spectrum, showing absorption peaks at 420 and 660 nm (Pareek et al., 2017).

In some species of microalgae that predominantly absorb light in the NIR, chlorophylls *d* and *f*, probably dominate the photosynthetic process (Larkum, 2016; Li et al., 2012). Chlorophylls *b*, *c*, *d*, and *f* are considered accessory pigments that expand the light spectrum used in photosynthesis, transferring all absorbed energy to chlorophyll *a* (Chen and Blankenship, 2011).

Chlorophyll *b* is found in Euglenophyta, Chlorophyta, and Charophyta, and, despite being yellow in color, absorbs all blue light from the solar spectrum, showing absorption peaks at 453 and 625 nm (Larkum, 2016). Chlorophyll *c* is golden brown, has a molecular structure with different C8 group bonds, and is therefore classified as *C1* (with absorption peaks at 444, 457, and 626 nm), *C2* (with absorption peaks at 447, 580 and 627 nm), and *C3* (with absorption peaks at 452, 585, and 625 nm), and is present in Chrysophyceae, Xanthophyceae, Raphidophyceae, Phaeophyceae, Bacillarophyta, Haptophyta, Cryptophyta, and Dinophyta (Chen, 2014). Chlorophyll *d* is a green pigment with a smaller structure than chlorophyll *a*; it is found in the red algae Rhodophyta and absorbs red light at the end of the sunlight spectrum, presenting absorption peaks between 700 and 750 nm (Schliep et al., 2013). Chlorophyll *f* was recently discovered in cyanobacteria from deep aquatic regions of Australia, is a red pigment, and allows light absorption to reach the infrared range with absorption peaks between 700 and 800 nm (Shen et al., 2019).

Thus, being photoreceptors, chlorophylls' concentration in microalgal biomass is directly influenced by the spectrum color and the light intensity incident in the culture medium (Begum et al., 2015). Chlorophyll molecules are formed by complexes derived from porphyrin, stored in the microalgae chloroplasts, and can be easily extracted from microalgal biomass (Schoefs, 2002), to be used as valuable compounds.

The cosmetics industry is one of the sectors that has been widely using this natural pigment. In addition to its high pigmentation, chlorophyll has been used as an additive agent due to its antioxidant and anti-inflammatory properties, as well as its ability to mask unwanted odors in personal care products, such as toothpaste and deodorants (Levasseur et al., 2020).

5.2.2 CAROTENOIDS

Carotenoids are yellow to red pigments, formed by isoprenoid molecules that occur as isomers with 5-, 9-, 13-, and 15-*cis*, and all *trans* forms, commonly found in chloroplasts, endoplasmic reticulum, or in the mitochondria of microalgae (Kaur et al., 2009). Carotenoids are a group of lipophilic compounds characterized by molecules with a tetraterpene carbon structure: a long chain of forty C atoms having several conjugated double C=C bonds, which creates chromophores where visible light may excite the double bond electrons to high energy levels. The antioxidant effect of carotenoids is due to reactive electrons suitable for scavenging reactive oxygen species (ROS) and free radicals that are made relatively available due to the long chains of conjugated double carbon bonds. The antioxidant properties of carotenoids play an important role in the mitigation of oxidative reactions as they induce the release of reactive oxygen species in the metabolic activity (Santos-Sánchez et al., 2019) and photoprotection against light damage of different biological systems (Galasso et al., 2019).

It is estimated that over 500 different carotenoids exist, so usually they are distinguished in two main families – carotenes and xanthophylls (Galasso et al., 2019; Santos-Sánchez et al., 2019). While carotenes – like α-carotene, β-carotene, lycopene, and phytoene, among others – are hydrocarbon molecules without oxygen in their composition, xanthophylls have oxygen in their molecular structure (positioned in the form of methoxy, hydroxy, keto, carboxy, and epoxy groups), as is the case of lutein, astaxanthin, zeaxanthin, neoxanthin, violaxanthin, flavoxanthin, among other substances (Pereira et al., 2021). With exception of lutein, the presence of oxygen in xanthophylls make these molecules more polar than carotenes and thus slightly more soluble.

When microalgae are exposed to intensive light, β-carotene also acts as an accessory pigment in the light-harvesting processes, modulating defense mechanisms that protect cells from photooxidation caused by high light stress (Morocho-Jácome et al., 2020); therefore, the production of β-carotene may be stimulated by submitting microalgae cultures to extreme light conditions.

At the current state-of-the-art, carotenoids known for having the strongest biological activity are considered to be the ketocarotenoid astaxanthin (a red pigment) that can be accumulated in the aquatic *Haematococcus pluvialis* at 4–7% (w/w), followed by β-carotene (an orange pigment) produced by hypersaline species *Dunaliella salina /bardawil,* reaching an intracellular concentration as high as 10–13% (w/w) (Pick et al., 2018; Rammuni et al., 2019). These two carotenoids have been approved as natural food colorants, animal feed additives for aquaculture production, functional foods, and nutraceutical supplements and are in the market for several years as they have been produced in large industrial production facilities (Camacho et al., 2019; Cuellar-Bermudez et al., 2015). Both astaxanthin and β-carotene are precursors of vitamin A, being converted in animal livers, as is the case of humans. Although β-carotene is the most commonly found in microalgae and cyanobacteria, there are species such as *Dunaliella salina* and *Chlorella pyrenoidosa* that are capable of producing α-carotene (Novoveská et al., 2019), and *Dunaliella bardawill* that is capable of producing lycopene (Liang et al., 2016).

Among the many known xanthophylls, the main ones like zeaxanthin are present in all classes of oxygenic phototrophs microalgae and cyanobacteria and, astaxanthin and lutein are present in Chlorophyta and some Rhodophyta (Takaichi, 2020). In addition to these with higher concentration, some species of microalgae can produce alloxanthin, antheraxanthin, caloxanthin, canthaxanthin, crocoxanthin, diadinoxanthin, diathoxanthin, dinoxanthin, echinenone, fucoxanthin, heteroxanthin, loroxanthin, perinan, nin pyrrhoxanthin, siphonaxanthin, vaucheriaxanthin, and violaxanthin, all showing absorption peaks in the visible region of the sunlight spectrum, except phytene and phytofluene, which present absorption peaks over the UVA and UVB regions (Stahl and Sies, 2012).

Other carotenoids, known to have high human bioavailability are lutein and lycopene; however, the absorption and bioaccessiblity of carotenoids is highly dependent on their formulation (oil capsules, pills) and also affected by human lifestyles (for example, smoking) or previous health conditions (Pereira et al., 2021).

For decades, carotenoids have been widely used in food, pharmaceutical, and animal feed industries, initially as natural sources of colorants and then as antioxidants (Camacho et al., 2019; Goiris et al., 2012), playing a role as photoprotectors, directly absorbing ultraviolet light (Yeager and Lim, 2019).

Recently, a growing number of studies demonstrated other important biological effects of carotenoids, such as anti-inflammatory properties (Pereira et al., 2021; Ramos-Romero et al., 2021); growth inhibition of tumor cells (Barkia et al., 2019;

Ferrazzano et al., 2020; Galasso et al., 2019; Pereira et al., 2021), and repair and protection from chemotherapy treatments (Ferdous and Yusof, 2021). An exploratory study advanced the possibility that carotenoids may as well have anti-viral properties (Rosales-Mendoza et al., 2020).

The biosynthesis of microalgae carotenoids is directly influenced by the concentration of nutrients in the culture medium (mainly the nitrogen sources and salinity) and by the lighting conditions, such as intensity, photoperiod, and light spectra (Takaichi, 2020).

Many types of carotenoids are known, have different molecular structures varying in chemical properties. and their light-absorbing characteristics, classified as carotenes (compounds with chains terminated by cyclic compounds complemented by carotenoid hydrocarbons, e.g., β-carotene) and xanthophylls (compounds with chains terminated by cyclic compounds complemented by functional groups containing oxygen, e.g., astaxanthin and lutein) (Sathasivam and Ki, 2018).

5.2.3 PHYCOBILIPROTEINS

The other important group of pigments, organized in supramolecular complexes called phycobilisomes, only found in cyanobacteria and in the stroma of chloroplast organelles of some microalgae phyla (such as Rhodophyta or Glaucophyta), are phycobiliproteins, often called only as phycobilins, because their molecular structure is also open-chain tetrapyrrole chromophores (Cuellar-Bermudez et al., 2015). Like carotenoids, phycobiliproteins (PBPs) are accessory light-capturing molecules (Dolganyuk et al., 2020), capable of light-harvesting at distinct wavelengths of chlorophylls, increasing photosynthesis efficiency, and also extending species-adaptive skills by expanding its spectrum of sunlight absorption (Zilinskas and Greenwald, 1986). However, Luimstra et al. (2019) suggest that blue light, which is not absorbed by phycobilisomes, creates an imbalance between photosystem I and II, reducing the photosynthetic efficiency of cyanobacteria when in presence of this range of the light spectrum. Phycobilins have an active role in the photosynthetic processes, in syntony with chlorophylls, especially chlorophyll *a*, complementing the absorption of light in several wavelengths of the visible spectrum (450–650nm), depending on the microalgae characteristics and the light conditions of its natural habitat, so they are considered antennae-protein pigments (Cuellar-Bermudez et al., 2015; Khanra et al., 2018; Rammuni et al., 2019).

Phycobiliproteins are water-soluble and are attached by polypeptide linkers, forming the phycobilisome, which in turn is associated with the cytoplasmic surface of the thylakoid membrane (Glazer, 1994). Phycobiliproteins have brilliant fluorescent colours depending on the target light absorption wavelength: phycocyanin (PC, λ_{max} = 610–620nm; dark blue) is present in cyanobacteria and is responsible for their blue-green colour, while phycoerythrin (PE, λ_{max} = 540–570nm; bright pink or red) exists in Rhodophyta and gives these microalgae a characteristic red color (Amarante et al., 2020). Besides these more common phycobiliproteins, there are allophycocyanin (APC, λ_{max} = 650–665nm; brighter blue) and phycoerythrocyanin (PEC, λ_{max} = 560–600nm; magenta) as well (Amarante et al., 2020; Pagels et al., 2019).

The composition and properties of PBPs can be modulated according to growth conditions, such as temperature, light intensity and composition of the growth medium (Kannaujiya and Sinha, 2016). The genus *Arthrospira* (*Spirulina*) is well-known as the main source of C-phycocyanin in cyanobacteria, although this pigment can also be found in other species, such as *Galdieria sulphuraria*, *Limnotrhix* sp., *Nostoc* sp., and *Synechococcus bacillaris* (Kannaujiya and Sinha, 2016; Halim, 2020).

Due to high water solubility and its bright colors, phycobiliproteins are quite attractive as colorants and are widely used in the food and cosmetic industries and molecular biology as fluorescent markers and pharmaceuticals (Dolganyuk et al., 2020; Kannaujiya et al., 2019).

The knowledge on the potential biological activity of phycobiliproteins is not yet so developed as it is for carotenoids, but they are recognized as promising molecules due to their antioxidant, anti-inflammatory, and anticancer effects (Ferdous and Yusof, 2021; Pagels et al., 2019), immunomodulatory and fluorescent properties, among others (Braune et al., 2021; Galasso et al., 2019; Galetovic et al., 2020; Grover et al., 2021; Kaur et al., 2019; Pourhajibagher and Bahador, 2021; Sun et al., 2003; Wang et al., 2021). In particular, phycocyanin is well-known for its antioxidant effect, protective role against oxidative damage, and selective anticancer activity (Ferdous and Yusof, 2021). However, the low stability of phycobiliproteins is one limiting factor for its application. C-phycocyanin, for example, requires mild storage conditions to avoid degradation, such as low humidity, low temperature, pH 5.5–6.0, and absence of light (Gustiningtyas et al., 2020). In addition to molecule stabilization, the industrial establishment of selective and high-yield techniques for PBPs extraction is another aspect to be improved for its complete insertion in the market.

Overall, there is no doubt that certain microalgae biomass or species extracts may have a beneficial effect on human health with probable advantages on the prevention, control, or even remission of several diseases of the metabolic, inflammatory, chronic, and oncological type (Barkia et al., 2019; Ferdous and Yusof, 2021; Ferrazzano et al., 2020; Galasso et al., 2019; Pereira et al., 2021; Ramos-Romero et al., 2021).

5.3 BIOSYNTHESIS

The biosynthesis of main microalgae pigments is briefly described in this section, focusing on the biosynthesis of chlorophylls and carotenoids, the main primary pigments in plants and microalgae.

Chlorophylls take part in the photochemical reactions of photosystem I (PSII) and photosystem II (PSII). Chlorophyll is the primary photosynthetic pigment of microalgae. While Chlorophyll *a* (Chl *a*) is important for light harvesting and electron transfer, Chlorophyll *b* (Chl *b*) mostly exist associated with PSII antenna and peripheral antenna complexes (Wang and Grimm, 2021).

The biosynthesis of chlorophyll occurs at the plastid stroma and is catalyzed by soluble enzymes. The biosynthesis of Chl *a* in microalgae can take place in dark or light conditions. This characteristic is due to the existence of two different nonhomologous enzymes for Chl *a* production: one light-dependent and one light-independent.

Besides the complexity of chlorophyll biosynthesis in microalgae, the production of this pigment is a result of the reduction of protochlorophyllide into chlorophyllide that is then converted to Chl *a* or *b*. However, the first intermediate of this pathway is the amino acid glutamate that is converted to the precursor of chlorophyll, glutamate 1-semialdehyde (GSA), and is rapidly converted to 5-aminolevulinic acid (ALA).

The chlorophyll biosynthesis can be divided into two different pathways. The first step is the biosynthesis of chlorophyll and heme precursor, called proto-porphyrin IX. This compound is synthesized from the ALA at the stroma of the chloroplast. The second step is the conversion of protoporphyrin IX into chlorophyllide, the precursor of Chl *a* and Chl *b* (Figure 5.1). Chl *a* and Chl *b* share the same biosynthetic pathway, differing on the transformation of the methyl group in Chl *a* and an aldehyde group in Chl *b* (Willows, 2007).

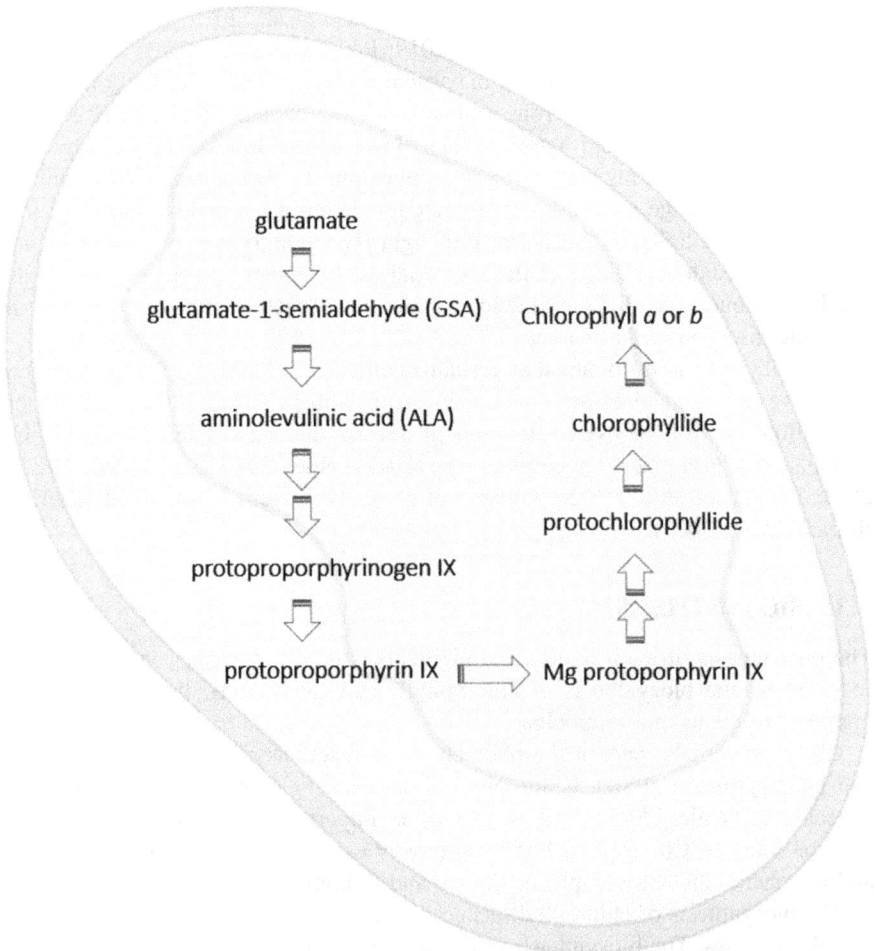

FIGURE 5.1   Scheme of chlorophyll biosynthesis pathway in microalgae.

On the other hand, carotenoids are natural lipid-soluble pigments and compose a large family of isoprenoid molecules, involved in the photosynthetic process and serving as precursors of other compounds (Liu et al., 2021). In the photosynthetic process, they play an important role as light-harvesting complexes and as photo-protective compounds.

The carotenoids biosynthesis has the same precursor as all isoprenoids and takes place in two different cell compartments: cytosol and plastid. The cytosolic via encompasses the mevalonate pathway and the plastid one involves the methylery-thriol phosphate (MEP) pathway. Besides the biosynthesis of carotenoids that can occur by these two pathways, there is no evidence that green microalgae show the cytosolic mevalonate pathway (Kato and Shinomura, 2020).

The precursors of the carotenoids biosynthesis in microalgae are the metabolites isopentyl pyrophosphate (IPP) and dimethylallyl pyrophosphate (DMAPP) that are condensed to form the geranyl pyrophosphate (GPP), the first step of the biosynthetic pathway (Kato and Shinomura, 2020). Further reactions culminate in the production of lycopene, which will drive the carotenoid biosynthesis into two branches: the formation of the pigment lutein, and the pigment β-carotene (Figure 5.2). Consequently, these two pigments can be precursors of other carotenoids, depending on the microalgae strain considered and the culture conditions submitted.

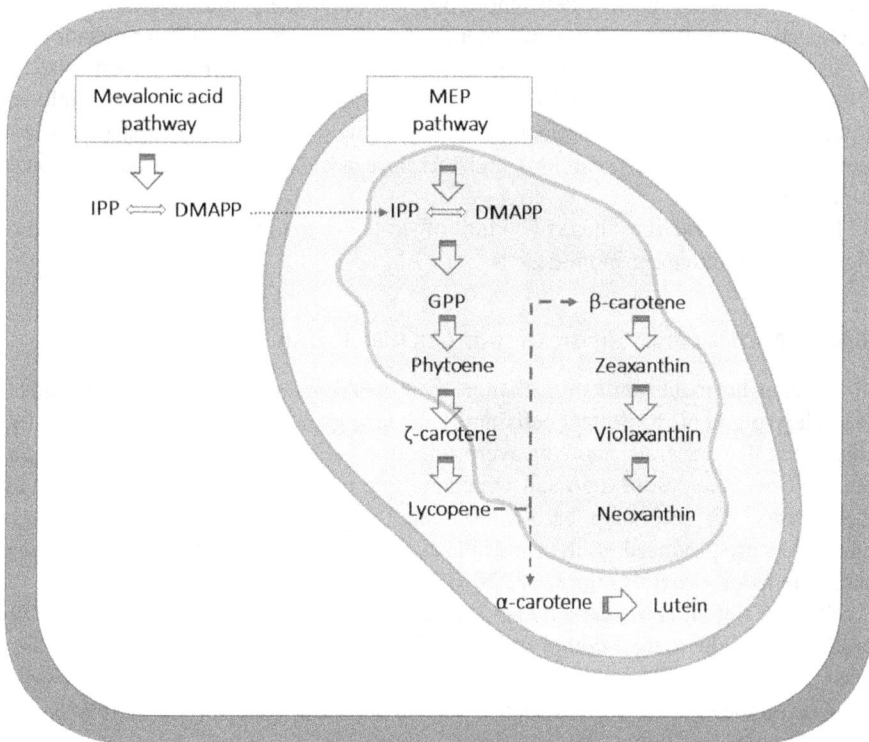

FIGURE 5.2 Scheme of carotenoid biosynthesis pathway in microalgae. The mevalonic acid pathway occurs on the cellular cytosol and the MEP pathway occurs on the chloroplast. IPP Isopentenyl diphosphate, DMAPP dimethylallyl diphosphate, GPP geranylgeranyl di-phosphate.

5.4 STRATEGIES TO ENHANCE THE PRODUCTION OF PIGMENTS

The microalgae biochemical composition varies hugely according to different growth conditions. While certain adverse conditions stimulate the production of some pigments, the stress environments usually have an inhibitory effect on growth. The balance of achieving high biomass concentration and the production of target compounds needs to be balanced to obtain the correct trade-off to increase the total product yields (Pereira et al., 2021).

Besides light, water, and carbon dioxide, inorganic nutrients are required for the photoautotrophic growth of microalgae. Most of the photosynthetic pigments are present in low concentrations in microalgae biomass, considering their growth under optimal conditions. Chlorophylls are usually found ranging from 1–2% w/w content of dry weight (Mulders et al., 2014b), which are considered a high concentration related to other pigments content. On the other hand, the concentration of the accessory pigments is below 0.5% w/w content of dry weight (Mulders et al., 2014b) under optimal growth conditions. In this regard, the production of microalgae pigments is economically unprofitable compared with chemical production. To overcome this issue, it is necessary to increase the pigment content of the microalgae cells. Figure 5.3 shows the strategies that can be used to increase the microalgae pigments productivity, which involve two different approaches: (i) the adjustments of culture conditions by using nutrient deficiency, physical parameters modifications, and implementation of different growth regimen, and (ii) modifying the cellular metabolism by using genetic and metabolic engineering toolbox. Some primary pigments, as certain classes of carotenoids, degrade under harmful environmental conditions; however, several primary carotenoids, such as β-carotene, can act as a secondary metabolite under stress conditions and therefore accumulate on the cell (Rabbani et al., 1998).

This section describes these two kinds of strategies to enhance the photosynthetic pigments in microalgae biomass.

5.4.1 NUTRIMENTAL, PHYSICAL, AND CULTURE REGIME

Microalgae have the ability to change their metabolic machinery to accumulate secondary pigments to protect cells under adverse growth conditions. The stressful conditions that generate pigment overproduction can be a nutrient limitation, light intensity oversaturation, high salt concentration, and different culture regimes. In these cases, the secondary carotenoids can be transported from the thylakoids, where they are produced, to the chloroplast stroma or cytosol of the cell, where they assemble as oil droplets (Jin et al., 2006).

Nutrient limitation or depletion is one of the most used methods to increase the production of microalgal compounds. Nutrient limitation and depletion affect biomass production and biochemical composition of the microalgal cell. The stress induced on the cells by nutrient starvation can drive the production of diverse types of compounds, including photosynthetic pigments.

Nitrogen (ammonia or nitrate) is an essential macronutrient to microalgal cell growth and the metabolic pathways are modulated according to the different

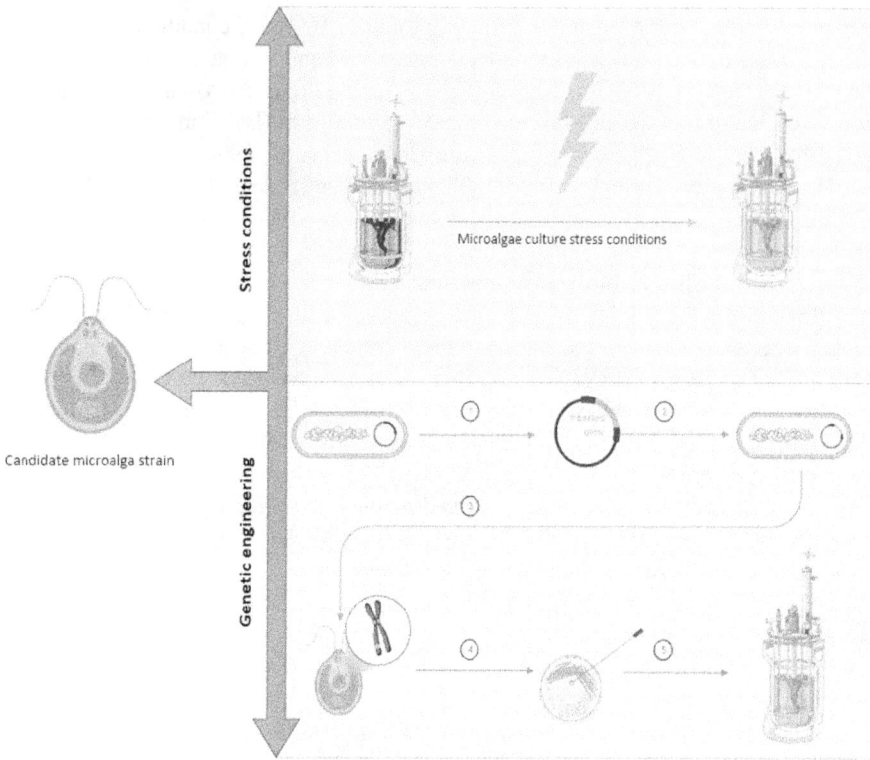

FIGURE 5.3 Strategies for increasing microalgae pigment production. The upper box represents the strategy of applying stress conditions in microalgae culture to overproduce the pigment. The lower box demonstrates a genetic engineering strategy for delivering heterologous DNA into microalgae via *Agrobacterium tumefaciens* transfection mediated: (1) *Agrobacterium tumefaciens* containing a plasmid; (2) the plasmid was cut and the target gene was inserted; (3) the recombinant plasmid carrying the inserted gene is incorporated into microalgae genome; (4) the transfected microalgae culturing in Petri plates; and (5) cultivation of genetically engineered microalgae.

concentrations of nitrogen in the culture medium (Yaakob et al., 2021). Nitrogen plays an important role in the synthesis of protein, lipid, and carbohydrates. Plants and microalgae can assimilate nitrogen in the form of nitrate, nitrite, urea, and ammonium. The form of assimilation and regulation processes varies widely in algae (Kumar and Bera, 2020). The nitrogen stress can be divided into nitrogen limitation, in which the culture medium is supported with an insufficient amount of nitrogen, and nitrogen depletion, characterized by the absence of nitrogen in the culture medium (Benavente-Valdés et al., 2016). Generically, nitrogen limitation leads to a decrease in biomass production, the photosynthetic process continues; however, a smaller content of nitrogen-rich components is produced, and the cells show overexpression of lipids. On the other hand, nitrogen depletion leads to a decrease in photosynthetic mechanisms that results in a smaller content of chlorophylls and an increase in non-photochemically active carotenoid pigments. The enhancement of carotenoid production is a cell response to alleviate the oxidative damage prompted by stress conditions (Shi et al., 2020). Nitrogen stress has been reported as an enhancement strategy to increase carotenoids contents in *Neochloris oleoabundans* (Urreta et al., 2014) and *Chlorella zofingiensis* (Mulders et al., 2014a), lutein in *Muriellopsis sp* (Del Campo et al., 2000), and astaxanthin in *Haematococcus pluvialis* (Fábregas et al., 1998).

Microalgae growth is highly driven by light conditions and photoperiod. Light is one of the main parameters affecting microalgal physiology and photosynthesis. Photoacclimation is another situation that leads to the accumulation of photosynthetic pigments in the cells. The prolonged subsaturating light conditions promote the increase of cell photosystems amount to adapt the organism to that light regime. The cellular content of primary pigment increases in minor content in consequence of acclimation. These minor increases usually are so-called "accumulation" in contrast to the expression "overproduction" used to describe major increases of the cellular pigment content as a consequence of the stressful growth conditions. Besides the accumulation of primary pigments is widely observed in microalgae, the accumulation of secondary carotenoids is observed exclusively in a limited number of Chlorophytes.

On the other hand, photoinhibition must be considered when high light stress is used. Usually, it is used a combination of nitrogen starvation, salinity, temperature stress, and light stress to enhance pigment production. To this respect, the combination of nitrogen limitation and high light intensities was analyzed by Zhao et al. (2018), who found an increase in the accumulation of lipids and astaxanthin in *Haematococcus pluvialis*.

A study conducted by Seyfabadi et al. (2011) showed the effects of combining different photoperiod and light intensities on the growth rate, chlorophyll *a*, β-carotene, total protein, and fatty acid content of *C. vulgaris*, identifying optimal conditions for the production of β-carotene, growth rate, total protein, and saturated fatty acids when cells were submitted to 100 μmol photons/m^2·s and 16:8 h light/dark photoperiod, while the content of chlorophyll *a*, monounsaturated and polyunsaturated fatty acids decreased with increasing light intensities and photoperiod. Studies have reported that carotenoids protect the cell from photoinhibition and damage to photosynthesis units (Vaquero et al., 2014), acting as antioxidant agents by accumulating outside of the thylakoid membrane in some microalgae strains after high UV exposure (Rastogi et al., 2020).

Temperature is another factor that plays an important role in carotenoids accumulation. Several studies have reported the enhancement of carotenoid production in microalgae grown under higher temperatures due to increased photo-oxidative stress, as observed in *H. pluvialis* and *C. zofingiensis* (Chekanov et al., 2014; Del Campo et al., 2004). Temperature has been described as the only stress factor that enhances the production of lutein, despite the molecular mechanism involved in the process still is not well known. The overall lutein production in *Muriellopsis* and *S. almeriensis* was affected when the temperature was increased (Del Campo et al., 2007), and it was observed an accumulation of lutein in *D. salina* under high-temperature culture conditions (Gómez and González, 2005).

Salinity stress can stimulate the production of total carotenoids in some microalgae species; however, they show a different range of tolerance to salinity levels. The mechanism that underlines the salinity tolerance involves the increase of nitrogen and carbon content in cells under higher salinity stress, which leads to the production of glycerol and amino acids (Jiménez and Niell, 1991). An example is the increased content of astaxanthin in *H. pluvialis* under salinity stress; however, the tolerance of this strain to salt concentration is above 10% w/v (Borowitzka et al., 1991).

The microalgae cultivation regime is another factor that influences the pigment content. Two-stage culture techniques have been used for improving microalgae pigment production on a commercial scale.

Culture regimen is also used as a pigment enhancement strategy in microalgae. The biochemical composition of microalgae is greatly influenced by culture conditions. However, since the culture conditions promote the rapid growth of microalgae biomass, the concentration of metabolites is usually low under optimal conditions. Conversely, the stress environments lead to an accumulation or overexpression of target compounds in microalgal cells besides the inhibitory effect on growth. To increase the overall microalgae product yield, it is required to find a trade-off to better suit both conditions.

The two-stage microalgae cultivation regime has been identified as a good method for enhancing the target compounds' productivity (Liyanaarachchi et al., 2021). The two-stage approach refers to microalgae cultivation under two different growth conditions. Usually, the first stage is characterized by microalgae growth under optimal conditions to provide high biomass productivities. In the second stage, the cells are submitted to stress conditions to accumulate target compounds (Benavente-Valdés et al., 2016). A two-stage strategy with light stress has been described as a viable approach to enhance pigment production. The enhancement of lutein productivity in *Scenedesmus incrassatulus* (Flórez-Miranda et al., 2017) and maximum β-carotene productivity in *D. salina* (Lamers et al., 2010) were obtained by increasing the light intensity.

It has been investigated the use of sequential cultivation regimes, by applying heterotrophic and autotrophic cultivation methods to analyze the production and accumulation of microalgal biomass and pigments. A study conducted by Ogbonna et al. (1997) for achieving a high concentration of biomass, protein, and chlorophyll contents applied the heterotrophic and autotrophic cultivation stages. The authors obtained a high biomass concentration (14 g/L), with 60.1% of protein and 3.6% of chlorophyll. Moreover, a method called "Sequential Heterotrophy–Dilution–Photoinduction" has been utilized for biomass and astaxanthin production in *H. pluvialis* (Wan et al., 2015). Microalgal cells were first cultivated heterotrophically to achieve high biomass density, and then transferred to a high light intensity environment for astaxanthin accumulation. Hence, the two-stage cultivation strategy is a viable method to enhance the production of target compounds and is feasible from the economic point of view when applying in an industrial context.

5.4.2 Genetic and Metabolic Engineering

The overexpression of microalgae target compounds through stress conditions is quite reasonable, but it is not entirely effective in acquiring huge amounts of products, considering the economic feasibility of the production. To overcome this hurdle, genetic and metabolic engineering tools have emerged as an alternative to enhance algae-based bioproducts, achieve high yields of microalgae biomass, and reduce costs as well.

The improvement of synthetic biology and molecular techniques associated with the availability of a huge amount of "omics" (genomics, transcriptomics, metabolomics, lipidomics) information have driven the development of customized bioproducts, with specific functions.

The process design of microalgae pigment production requires a holistic and multidisciplinary integration. The screening of a potential strain to produce a product of interest usually involves the knowledge of biological characteristics, the content of biochemical compounds, and the metabolic pathways involved in their synthesis. Despite the available knowledge regarding the pigment composition and function of a specific strain of microalgae, the information is dispersed in the literature and an effort to integrate them is essential (Mulders et al., 2014a).

One of the first microalgae sequenced was a strain of *Chlamydomonas reinhardtii* that has been considered as the model study of photosynthesis and chloroplast biogenesis (Rochaix, 2002). Some studies characterized the carotenoid production in *H. pluvialis* (Ota et al., 2018), *D. salina* (Elleuch et al., 2019), and *Chlorella zofingiensis* (Y. Zhang et al., 2019), revealing the genes involving in the biosynthesis.

The microalgal nuclear, plastid, or mitochondrial genomes can be engineered for a desired metabolic trait. Most genes associated with secondary metabolism are coded in the nuclear genome, addressed to the chloroplast to post-translational modifications and acquiring their function (Gimpel et al., 2015).

The first step for engineering the microalgae genome is delivering heterologous DNA into the cells. Highly diverse methods are available, and they differ in efficacy, expression patterns, and applicability to different purposes. The heterologous DNA delivering into the microalgae cell results in a stable chromosomal integration or a plasmid extrachromosomal replication, the latter resulting in a fractional cloning or complete foreign transgene synthesis (Figure 5.4).

Besides the importance of genetic transformation of microalgae, there is no extensive information available about the approach, strategies, and techniques applied to these microorganisms. The main methods used for delivering DNA to the microalgae genome are electroporation, glass bead agitation, microparticle bombardment, natural transformation, bacterial DNA transfer (conjugation or *Agrobacterium*-mediated), mechanical agitation, and surfactant permeabilization (Gutiérrez and Lauersen, 2021). However, other promised tools that have already been applied in plant transfections also have shown the potential to be applied in microalgae transformation. It is the case of cell-penetrating peptides, polymers, metal-organic frameworks, and liposome-based transfection technology (Gutiérrez and Lauersen, 2021) (Figure 5.4).

The transformation method applied to microalgae engineering directly impacts the overall effectiveness of the process and the levels of expression of the heterologous DNA. On the other hand, it is essential to consider the number of desired transgene copies and the target genome (nuclear, plastidial, or mitochondrial) that would like to address the heterologous DNA before deciding which transformation technique better suits the project aims. For example, Galarza et al. (2018) analyzed

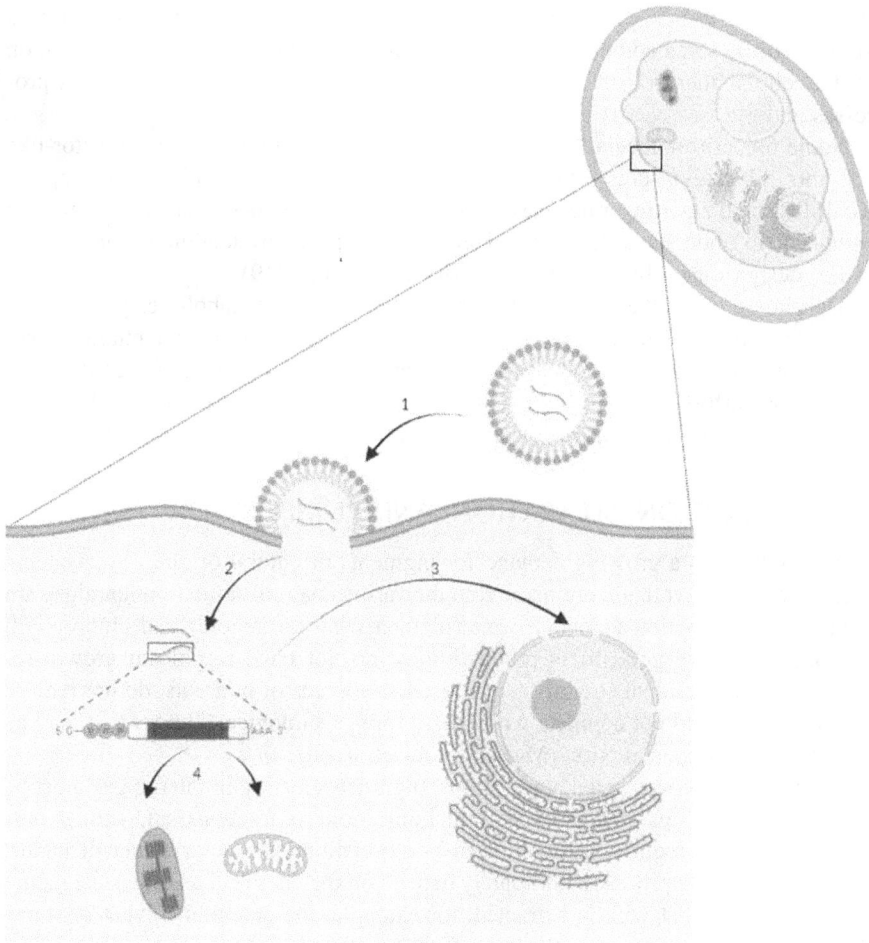

FIGURE 5.4   Heterologous DNA delivering into microalgae cells. The scheme represents a liposome-based transfection technology for delivering foreign DNA into microalgae. (1) The liposome carrying the foreign DNA integrates with the cell membrane, (2) the foreign DNA is exposed in the cell cytoplasm and is directed to the (3) cellular nucleus to be integrated into the chromosomal DNA or (4) expressed into the chloroplast or mitochondria.

the expression of nuclear gene engineered into chloroplast of *H. pluvialis* and found an over-accumulation of astaxanthin up to 90% per culture volume.

Besides the transformation techniques to microalgae genetic engineering, the "omics" era brings a wide range of possibilities to explore the potentialities of microalgae strains. The differential expression patterns of genes are analyzed by diverse kinds of "omics" approaches, such as genomics, transcriptomics, proteomics, metabolomics, and lipidomics. These tools allow one to know the regulation and the expression mechanisms of genes under certain conditions of cultivation and evaluating the key factors involved in the overproduction of a target compound. This mechanism can be used to identify target genes for genetic engineering or to manipulate the metabolic pathways to enforce the cellular fluxes to overexpress some compounds of interest. However, interfering and manipulation of the cellular metabolism requires a deep understanding of the biosynthesis processes and the involved regulation mechanism.

Some modern molecular biology techniques, such as transcription activator-like effectors (TALEs), clustered regularly interspaced short palindromic repeats (CRISPR), and zinc-finger nucleases (ZFN) can be used to decipher some metabolic pathways, to understand the mechanism involved in the production of compounds, and to design a new biological system (Kumar et al., 2019).

Despite some of the successful cases of genetic and metabolic engineering of microalgae that have been described in the literature, transformation methods do not guarantee a long-term expression pattern and stability of the engineered strain. Thus, more efforts in research and development are still required for addressing high productivities and expression stability of strain and of the desired compounds.

5.5   EXTRACTION, SEPARATION, AND PURIFICATION

Currently there is a growing demand for pigments of natural origin, as consumers prefer them over synthetic options. From the various natural sources, microalgae are seen as a good option to obtain some pigments, as carotenoids and chlorophyll, because they have good mass productivities, do not have significant growth requirements and residual streams can be used as sources of nutrients, do not require arable land, and do not compete with other crops or biological sources that may be used for human consumption (Morais Junior et al., 2020).

To be used in commercial applications, pigments existing in microalgae must be extracted and/or separated from the remaining biomass fractions, and further purified to meet the requirements imposed by regulations and/or customer demands. The most adequate processes ultimately depend on the chemical and physical nature of the pigments that will be extracted, for example, the potential solvent mixtures that may be used will have to show an affinity for the compounds or interest (Corrêa et al., 2020).

An extensive review of the extraction, separation and purification methods, protocols, and processes that can be used to obtain pigments from microalgae is outside the scope of this chapter, and good reviews on the subject can be found in the literature (Corrêa et al., 2020; Halim et al., 2010). In this work, a brief overview of the available methods is given, focusing on the key aspects that control the extraction, separation, and purification processes.

From a purely practical point of view, excretion of the pigments of interest would be the best possible situation, as it involves only the separation of the pigments from the growth media and potential purification. However, to the best of the authors knowledge there are no known species of microalgae that do this. Therefore, to obtain the pigments it is necessary to harvest the microalgae, remove the excess water, and disrupt the cell wall to allow an easier and complete contact of the extracting agents with the biomass. After obtaining the extract, it usually needs to be further processed to remove impurities or other compounds that are also extracted, to ensure that the quality requirements are met (Corrêa et al., 2020; Cuellar-Bermudez et al., 2015).

In this section, attention will be paid to the extraction processes themselves. As noted before, the various pigments that may be present in microalgae can be divided into three groups (Cuellar-Bermudez et al., 2015; Khanra et al., 2018): Chlorophylls, Carotenoids, and Phycobiliproteins. Each group of pigments has particular characteristics, as described below:

- Chlorophylls are non-polar compounds insoluble in water, but with good solvability in organic solvents, such as ethanol, methanol, acetone; ionic liquids, or supercritical carbon dioxide (Arashiro et al., 2020; Halim et al., 2010).
- Carotenoids absorb light in the range 400 to 550 nm, for which chlorophylls have a low absorption efficiency, and protect the cell from damage from solar radiation (Cuellar-Bermudez et al., 2015; Khanra et al., 2018). Similar to chlorophylls, carotenoids are also soluble in organic solvents and supercritical carbon dioxide, although some care must be considered in their selection, as carotenoids can be non-polar (carotenes) or polar (xanthophylls). Thus, a mixture of non-polar and polar solvents should be used, depending on the relative amounts of the carotenoids composition of the specific microalga (Aberoumand, 2011).
- Phycobiliproteins include two very interesting compounds: phycoerythrin (red color), and phycocyanin (blue color) (Furuki et al., 2003). They are polar compounds that are soluble in water or polar solvents, thus avoiding the need to use organic solvents that need to be recovered and treated afterwards. Aqueous solutions can be used as solvents, in which the value of the pH is controlled to avoid the potential degradation of the pigments or other compounds of interest.

The extraction process can be seen as a sequential process. After the harvesting of microalgae biomass, the initial step common to all extraction/separation processes is the biomass pre-treatment, whose objective is to deal with the presence of the cell wall, which limits the access to the microalgal pigments that are intracellular compounds. Only when the cell wall is destroyed or sufficiently weakened will the intracellular products be released, improving the contact between phases and ultimately increasing the extraction yield (Corrêa et al., 2020).

The cell wall is a complex structure that protects the cells and gives them their shape and structural strength. Various methods/processes have been proposed and are applied in practice to break the cell walls, based on various physical and chemical phenomena, and can be classified in two types (Corrêa et al., 2020; Furuki et al., 2003):

- Mechanical, such as high-pressure homogenization, bead milling, ultrasonication, pulsed-electric field treatment, microwaving, freezing-thawing cycles, and osmotic shock treatment;
- Chemical, such as enzymes or acid or alkali solutions.

The most adequate method/process depends on the characteristics of the cell walls, in particular its rigidity and thickness. For species that have a thick and rigid wall, such as *Haematococcus pluvialis* (a source of astaxanthin), mechanical treatments (Shah et al., 2016), or enzymes (Machado et al., 2016), are effective forms of disrupting the wall. Nevertheless, several other methods can be used (Cheng and Shah, 2021). For other species without robust or even with no cell wall (e.g., *Dunaliella salina*) simple physical methods, such as freezing-thawing cycles, is enough to rupture the wall.

Additionally, the pre-treatment should avoid as much as possible the formation of emulsions, which would limit the mass transfer between phases. They should be efficient, both in terms of energy and materials consumption, and avoid the degradation, not only of the pigments but also of the other biomass components. A full description of all the potential pre-treatment methods proposed and/or applied in practice is outside the scope of this work, and excellent reviews on the matter can be found in the literature (Corrêa et al., 2020; Kim et al., 2016).

After the rupture of the cell walls, the extraction process itself is performed. For extraction, a solvent, or mixtures of solvents, are mixed with the biomass. The choice of solvent(s) depends on the nature of the pigments to be extracted (Gorgich et al., 2020). To promote the extraction process, microwave or ultrasonication systems may be coupled to promote the mixing and enhance mass transfer, to increase the pigment recovery.

The pigments extracted from the microalgae biomass usually contain other compounds that are also solubilized by the solvents. These contaminants reduce the purity of the pigments extracted and, therefore, limit the range of applications in which they can be used. An example is the presence of chlorophylls and/or lipids in carotenoids, due to the affinity of those compounds to the solvents used for their extraction. Hence, it is necessary to further purify the pigments, to improve their quality and consequently their market value.

Depending on the pigments extracted and the potential impurities present, various purification chemical or chromatographic techniques can be used. The most adequate method, or combination of methods, depends on the pigments and impurities involved and in the degree of purity intended, which depends on the applications, and this is currently an intense area of research. For example, to remove chlorophylls from carotenoid extracts (either astaxanthin or β-carotene) chemical precipitation using alkali or acid can be used. This method precipitates the chlorophylls in the form of water-soluble salts that can be separated from the pigments by centrifugation or other suitable solid-liquid separation technique (Fujii, 2012; Rammuni et al., 2019). To purify phycobiliproteins various techniques can be applied, combined, or isolated, involving the impurities precipitation and removal, or ionic chromatography (Benavides and Rito-Palomares, 2006; Gantar et al., 2012; Román et al., 2002; Sørensen 2013; Rodrigues et al., 2020).

5.6 INDUSTRIAL PRODUCTION

In the previous section, extraction/separation and purification methods to obtain pigments from microalgae were discussed in detail. Most of them were designed, developed, and optimized at experimental or small scales, in which the operating conditions can be better controlled. This allows the identification of the key parameters that control the extraction/separation and purification steps, and to get a better insight into the physicochemical and thermodynamic phenomena that occur in the processes. This information is fundamental when moving from experimental/pilot scale to industrial scale, as the extraction/separation and purification steps are crucial to the economic competitiveness of microalgae-based pigments, while minimizing as much as possible the environmental impacts, as energy and solvent consumption.

Currently it is consensual that the development of a pigment production unit from microalgae must consider the pigment life cycle, in a "cradle-to-gate" perspective. All life cycle steps are interconnected and must be considered together in a systematic approach. This way it will be possible to have an integrated view of the whole process, allowing full optimization of the economic and environmental impacts. Only by doing this, microalgae-based pigments will be able to compete with synthetic or other natural-based pigments. The pigment life cycle comprises the following stages:

- Acquisition of the raw materials/nutrients needed to grow the microalgae, including in particular the water, air and nutrients (e.g., nitrogen, carbon, oligo-elements, among others) necessary for the microalgae growth;
- Cultivation of the microalgae, should be done in a way that optimizes the pigment yield;
- Microalgae harvesting that may involve water removal and biomass separation from the culture medium;
- Microalgae cell wall disruption, for example, to facilitate the diffusion of the solvents inside the solid matrix promoting the pigment solubilization;
- Separation/extraction and purification of the pigments extracted, in which the operating conditions determined at the experimental and/or pilot unit scale are relevant.

Also, linked to the previous pigment production life cycle stages, there is the process units' construction and implementation. This includes in particular the cultivation system that should ensure an optimized microalgae growth, both in terms of pigment yield and robustness, for example, avoiding as much as possible contamination and simplifying the nutrients feeding and microalgae harvesting.

From an economic and environmental point of view, pigment production must be coupled with the production of other products, trying to maximize the utilization of microalgae biomass generated, following a biorefinery approach. In practice, as pigments are high value compounds and are present in the biomass in small quantities, they should be the first microalgae components to be obtained. After the extraction/separation and purification of pigments, or of other compounds of high interest present in small amounts, a large quantity of biomass will be generated. Currently microalgae pigments are products aimed at niche markets, for customers who are willing to pay a premium price for products made from natural ingredients with potential bioactive and/or health effects.

Microalgae biomass contains many components, in particular proteins, lipids, and carbohydrates, each one with its specific values and useful for various applications (Morais Junior et al., 2020; Sathasivam et al., 2019). For instance, microalgal carbohydrates can be used to obtain ethanol (Chen et al., 2013; Seon et al.,

2020), an energy source or a chemical feedstock; proteins can be used for human and animal feed (Amorim et al., 2020) or for other novel applications, such as stabilizing agents of emulsions, fluid interfaces, and foams (Bertsch et al., 2021), and lipids can be used to produce biodiesel (Chhandama et al., 2021; Mata et al., 2010) or serve as monomers to obtain bio-based polymers (Hidalgo et al., 2019). Besides minimizing the losses of microalgae biomass and the costs of properly treating and processing it, obtaining more products improves the overall process viability, reducing both the production costs and operational risk (Chew et al., 2017; Préat et al., 2020; Tejada Carbajal et al., 2020).

The main challenges facing nowadays pigment production from microalgae lies in the development of cultivation and biomass processing systems that allow high productivity while maintaining the costs as low as possible. Only through the integration of the various process units, in a life cycle perspective, and process optimization and intensification, will be possible to achieve that goal (Jacob-Lopes et al., 2020). In the following subsections, some key aspects of the various process units for microalgae pigment production are discussed, highlighting relevant aspects and some issues that still need to be solved, including examples from practical cases of the production of microalgal pigments.

5.6.1 Microalgae Cultivation Systems

To be able to obtain significant quantities of microalgae pigments for commercialization, large-scale cultivation systems are necessary. Two types of cultivation systems are normally used for this purpose: open (raceways and shallow ponds) and closed systems (tubular photobioreactors). The first type is the most used option due to the low investment and operational costs. However, open raceways have some significant drawbacks, as they are more sensitive to contaminations, and the cultivation grow conditions are harder to control effectively, and significantly depend on the environmental conditions, potentially limiting the biomass and/or pigment productivity. Hence, open cultivation systems should be used only for robust microalgae species (e.g., *Chlorella*, *Scenedesmus*, and *Nannochloropsis*) (Martins et al., 2016), or to those that grow in very specific and/or harsh conditions (e.g., *Dunaliella salina* and *Arthrospira sp.*) that limit the contamination potential (Kratzer and Murkovic, 2021; Mata et al., 2016).

As pigments are considered high-value compounds and their yield is strongly dependent on the environmental conditions, in particular light, temperature and nutrients availability, closed cultivation systems are also seen as a good choice. Closed systems allow one to avoid culture contamination and to better control the growth conditions, in terms of light and temperature. Thus, a higher biomass quality can be achieved in closed cultivation systems, ensuring also a more regular supply of biomass, a key aspect for an efficient operation of the harvesting and separation/ extraction and purification systems (Borowitzka, 2013b).

Besides the cultivation system, other issues must be taken into account. Continuous or batch mode systems can be used, being the former preferred as they allow a more continuous supply of microalgae biomass, reducing the cultivation space requirements for the same production capacity. The cultivation mode also depends on the location, land costs and availability, potential of using local sources of nutrients (as waste streams), and climatic conditions (e.g., solar radiation) are also other factors that must be taken into account.

In practice, both types of systems can be used, either one of them, or combined. The selection of the most adequate type depends not only on the process economics, but also on the cultivation requirements. Thus, a proper selection of the microalgae species or strain is of the utmost importance. Some critical requirements include the ability to grow rapidly, being able to withstand variations in cultivation conditions, and the capacity to accumulate high quantities of the desired pigment. The accumulation of pigments may also depend on other factors, as specific nutrients deprivation, such as nitrogen, or the microalgae life stage, that have to be taken into account in developing an industrial production system. This information is obtained at experimental and/or at pilot unit scale, and an economically viable process depends on it, as well as the minimization of the environmental impacts.

An interesting example of the aforementioned pigments is astaxanthin. Currently, there are only two biologic sources that can compete with the synthetic version: the yeast *Phaffia rhodozyma* and the microalga *Haematococcus pluvialis*; with an average astaxanthin content of 8 mg/g and 60 mg/g, respectively. Hence, *H. pluvialis* is the preferred natural primary source of astaxanthin (Lim et al., 2018). This species of microalgae can be grown either in autotrophic or mixotrophic modes, and, as the astaxanthin biosynthesis process requires light, higher concentrations of astaxanthin are achieved in photoautotrophic cultures (Kang et al., 2005). Moreover, nitrogen deprivation, strong light, and high temperature and salinity also promote the production and accumulation of astaxanthin (Han et al., 2013).

Concerning the cultivation process, taking into account the *Haematococcus pluvialis* life cycle, a continuous culture system is not adequate. This species of microalgae has three life cycle stages. The first one corresponds to a biflagellate form, in which for favorable conditions the cells will divide rapidly until reaching high density. Then, under stress, by removing the favorable cultivation conditions, the cells lose their flagella and grow, becoming haematocysts if the stress is maintained, in which stage they accumulate large amounts of astaxanthin, as evidenced by the red color that the microalgae acquire in that stage (Ma et al., 2018). Therefore, a sequential cultivation strategy is a better option. Initially, *Haematococcus* is grown in small, closed photobioreactors under favorable conditions until they reach the desired cell density. Then, they are transferred to larger reactors, which could be closed or open, in which the cells are subjected to stress to induce the accumulation of astaxanthin. Usually, high light intensity, nutrient deprivation (in particular of nitrogen), and/or salt addition are the stresses applied. When the cells become red, they are harvested (Shah et al., 2016).

For other pigments, which are produced from other microalgae species, other cultivation strategies may be considered. In particular, for β-carotene from *Dunaliella salina* a similar strategy was used as for astaxanthin (Ben-Amotz, 1995; Raja et al., 2007). For phycocyanin from *A. platensis* open raceways are normally used (Pagels et al., 2019), as this species of microalgae grows in extreme conditions of high water temperature and pH values, thus limiting the potential contaminations. Although closed systems for *A.*

platensis show a better performance, the increase in costs makes it an uncompetitive option when compared with open ponds (Eriksen, 2008; Soni et al., 2017).

5.6.2 Harvesting and Biomass Pre-Processing

The harvesting process depends on the cultivation strategy, in particular if it is batch or continuous. Sedimentation followed by centrifugation can be accomplished, as it is done with *H. pluvialis*. For β-carotene from *Dunaliella salina* and phycocyanin from *A. platensis*, centrifugation coupled with or without filtration is adequate. The water removed from the harvested biomass can be recycled back to the cultivation media. Depending on the microalgae species and the separation/extraction process, flocculants and/or other adjuvants may be added to facilitate the separation (Caetano et al., 2019).

Usually the biomass paste is pre-processed, to ensure that the extraction/separation and purification processes are accomplished adequately. This may involve the breakage of the cell walls, to enable a better contact between solvents and the biomass components to be extracted. For example, for *H. pluvialis* a cracking mill can be used, and other options are available depending on the characteristics of the microalgae cell walls.

Afterward, the paste may be dried, to remove the water, which when non-polar solvents are used limits the process efficiency and leads to higher solvent losses and larger quantities of wastewater to process with the corresponding costs. This can be done, for example, in a spray dryer, as it is done sometimes for phycocyanin from *A. platensis* (da Silva et al., 2019). However, when the pigments are produced for low-value applications or products, as animal feed, biomass drying should be avoided due to the high energy consumption in the process. Hence, it is desirable in this case to perform the extraction process in the wet biomass. Even though a lower selectivity is expected in this case, the further refinement steps needed may be cheaper and use fewer resources than drying the biomass. Likewise, in a biorefinery context, the cruder extraction may simplify the production of other compounds (e.g., lipids) in an integrated fashion, reducing process cost (Branco-Vieira et al., 2020).

5.6.3 Extraction/Separation and Purification Processes

Various factors must be taken into account when choosing the adequate extraction/separation and purification process. The most important is the ability to extract the pigments in an efficient manner, which requires a high selectivity to the compound of interest and high yield. Experimental data and other information obtained at the lab and/or pilot-scale units is crucial to make that selection, and also to support the process scale-up to industrial production (Branco-Vieira et al., 2020).

The final application, market niche, or segment in which the pigments will be commercialized is the other fundamental aspect that must be considered. For human consumption, which includes pigments used as food colorants or for pharmaceutical and nutraceutical applications (food supplements), quality requirements (safety and purity) must be fulfilled. Thus, in this case, the selection of processes that use less toxic solvents, or solvents considered safe (e.g., ethanol), or even no solvents, is desirable. For other applications, as animal feed, the need to reduce costs leads to utilization of more volatile solvents, as it reduces energy consumption and facilitates solvent recovery (Branco-Vieira et al., 2020).

Process scalability is also influenced by the desired market niche or application intended, as the process involving lower value products will tend to have larger capacities, to be able to reduce costs via economies of scale. Moreover, process intensification could also be important, as the ability to combine different process units can contribute to reducing production costs. An interesting example that may be extended to pigment extraction is the ability to use hexane as a solvent to extract lipids from wet biomass for biodiesel production (Martin, 2016). As hexane selectively extracts non-polar compounds, and many of the pigments that can be potentially extracted from microalgae biomass are non-polar, this represents an interesting option.

Regarding further purification and polishing of extracted pigments, the final application is also an important factor. In some cases, as animal feed, there is no need to purify the final extract, as the quality requirements are lower and further purification will increase the costs and the final product price, normally no further processing is done. However, for human consumption and other applications that require higher purities (e.g., pharmaceutical applications), the initial extraction process must be coupled with further purification and refinement processes, to fulfill quality requirements. Various options exist and are proposed in the literature based on different physical or chemical separation methods, such as: precipitation, chromatography, esterification, among others (Benavides and Rito-Palomares, 2006; Chang and Sang, 2007; Fujii, 2012; Gantar et al., 2012; Rodrigues et al., 2020; Sørensen et al., 2013). However, most of the studies have focused on the experimental scale systems, with no consideration of the industrial scalability.

The various extraction/separation and purification processes available are described in the previous section 5.5. Considering astaxanthin, supercritical carbon dioxide extraction is a good option, and is often used in practice (Fujii, 2012). The same process, or solvent extraction, can be considered for β-carotene, even though for β-carotene powder stabilization is many times enough, as is the case of its utilization in animal feed.

5.7 IMPORTANCE AND COMMERCIAL APPLICATIONS OF MICROALGAE PIGMENTS

In the previous sections, the nature and properties of pigments were demonstrated, followed by a description of the microalgal biomass cultivation and downstream processing conditions needed to recover the compounds of interest. In this section, it will be presented some commercial applications already in practice, that are transforming microalgae cultivation and processing a highly growing and attractive business (Guenard, 2021; Herrador, 2016; VentureRadar, 2021).

5.7.1 NUTRACEUTICAL AND PHARMACEUTICAL USES

Several authors have reviewed the great potential of microalgae metabolites for nutraceutical and pharmaceutical applications due to their therapeutic properties (e.g., antioxidants, antimicrobials, anti-tumoral, anti-cancer, anticoagulants, among others) and high nutritional value of their biomasses (Braune et al., 2021; Ferdous and Yusof, 2021; Kiran and Venkata Mohan, 2021; Mehariya et al., 2021; Nicoletti, 2016).

Grover et al. (2021) performed *in vivo* assays in mice to evaluate the toxicity and pharmaceutical properties of C-phycocyanin (phycocyanin from cyanobacteria, C-PC) extracted from *Spirulina platensis*. This study showed that certain concentrations of C-PC (500 and 1000 mg·kg^{-1}) can enhance superoxide dismutase (SOD) and catalase (CAT) activity, both important enzymes that act against oxidative stress, and reduce the synthesis of interferon-γ (IFN-γ) and tumor necrosis factor-α (TNF-α), pro-inflammatory cytokines, without compromising the production of anti-inflammatory cytokines. Furthermore, the authors observed non-toxic effect in mice with C-PC dosage up to 2,000 mg·kg^{-1} body weight.

Jayappriyan et al. (2013) observed 80% inhibition of the proliferative activity of PC-3 human prostate adenocarcinoma in 24 h in vivo assays using 50 μM β-carotene from *Dunaliella salina* EU5891199. Moreover, the potential anti-cancer activity of several other microalgae metabolites has been addressed in the literature. Zhang, J. et al. (2019), reported potential anti-tumor activity against human colon cancer of extracellular polymeric substances (EPS), also known as exopolysaccharides, produced by three microalgae (*Chlorella pyrenoidosa* FACHB-9, *Scenedesmus* sp., and *Chlorococcum* sp.) and Sanjeewa et al. (2016) investigated the anti-inflammatory and anti-cancer activities of sterols from *Nannochloropsis oculata*.

Currently, *Spirulina* (*Arthrospira*) and other microalgae species (e.g., *Chlorella, Dunaliella, Haematococcus, Schizochytrium*, among others) are classified as Generally Recognized as Safe (GRAS) by FDA (US Food and Drug Administration). In particular, *Chlorella* and *Spirulina* are the most commercialized microalgae, especially as whole-cell, due to their rich biomass, including high protein content (García et al., 2017). However, several microalgae-based products have been successfully commercialized as well (Table 5.1).

Recently, Ratha et al. (2021) discussed the potential use of *Spirulina*-based nutraceuticals to enhance immunity and also slow the progress of COVID-19 infection. This study, together with several other microalgae-based studies, reinforces the potential of microalgae as a rich source of bioactive compounds, especially regarding human health, and the importance of continuous research and development in this area.

5.7.1.1 Cosmetics and Cosmeceutical Use

The Federal Food, Drug, and Cosmetic Act (FD&C Act), as well as the European Commission, define cosmetics as *products related to beauty, attractiveness, appearance, and skin care*. The term "cosmeceutical" is commonly used by the cosmetic industry to cosmetics that claim medicinal benefits; however, this term is not recognized by FDA (U.S. Food and Drug Administration), where this type of product can only be defined as drugs or as a combination of cosmetic and drugs (US Food & Drug Administration, 2020). Unlike cosmetics definition, drugs are defined as *products able to treat, mitigate, prevent, or cure a disease or affect in some way the structure or function of the human body*. In addition to products with drug-like benefits, the cosmetic industry has also increasingly sought to innovate in natural products to replace chemical substances in their formulations, due to their potential side effects, such as toxicity and hypersensitivity (Yarkent et al., 2020). In this

TABLE 5.1
Commercial Microalgae-based Products with Nutraceutical and Pharmaceutical Claims

Product	Claim	Source	Company
ASTAPURE® Astaxanthin	Skin, eye, brain, and cardiovascular health, fertility and immunity	*Haematococcus pluvialis*	Alga Technologies Ltd.[1]
AstaPure® Arava (astaxanthin, polysaccharides, vitamins, dietary fibers, fatty acids, and proteins)	Potential antioxidant and nutraceutical		
FucoVital™ (Fucoxanthin)	Liver health benefits and in metabolic syndrome, anti-inflammatory properties	*Phaeodactylum tricornutum*	
Natural Astaxanthin	Powerful antioxidant, skin, cardiovascular, brain, gastric, vision and eye health, muscle endurance and immune function	*H. pluvialis*	AstaReal®[2]
BioAstin® Hawaiian Astaxanthin®	Joint, tendon, skin, cardiovascular, eye and brain health and recovery after physical exercise	*H. pluvialis*	Cyanotech® Corporation[3]
Phycocyanin (powder)	Immunity improval and anti-cancer properties	*Spirulina*	Tianjin Norland Biotech Co.[4]
Astaxanthin (powder/oil)	Antioxidant, anti-tumor, and anti-cancer properties, also improve immunity and eyesight	*H. pluvialis*	
ZANTHIN® Natural Astaxanthin	Healthy aging (immunomodulatory and anti-inflammatory properties)	*H. pluvialis*	Valensa International[5]

Notes
[1] https://www.algatech.com/
[2] http://www.astareal.se/
[3] https://www.cyanotech.com/
[4] http://www.norlandbiotech.com/
[5] https://valensa.com/
Websites:

sense, the potential uses of microalgae metabolites in cosmetics have been extensively studied. In addition to natural colorants in make-up products (e.g., astaxanthin and phycobiliproteins), microalgae metabolites can be incorporated in cosmetics such as:

- Sunscreen – Microalgae synthesize various pigments to protect themselves against solar ultraviolet radiation (UVR) exposure (200–400nm) and oxidative stress. Carotenoids, phycobiliproteins, but especially mycosporine-like amino acids (MAAs) and scytonemin (only produced by cyanobacteria) are potential natural molecules to be incorporated into sunscreen formulations to overcome the disadvantages of conventional sunscreens concerning human health and environmental issues (Vega et al., 2020). MAAs are water-soluble secondary metabolites with low molecular weight (<400 Da) and high molar extinction coefficient (ε = 28,100 to 50,000 L·mol^{-1}·cm^{-1}) in UV-A (315–400nm) and UV-B (280–315nm) regions. These molecules are capable of absorbing UV radiation without forming reactive oxygen species (ROS), by releasing the absorbed UV energy as heat (Wada et al., 2013). The species *Aphanizomenon*, *Chlorogloeopsis* spp., *Lyngbya purpurem*, and *Nostoc commune* are the main sources of MAAs from cyanobacteria (Abu-Ghosh et al., 2021). Scytonemin is a lipid-soluble pigment (yellow-brown), with molar extinction coefficient (ε) equal to 16,200 L·mol^{-1}·cm^{-1}, mainly absorbing radiation in UV-A region (Amador-Castro et al., 2020). *Scytonema* sp. and *Nostoc* sp. are the main sources of scytonemin (Abu-Ghosh et al., 2021), although several other species have been reported to produce scytonemin (e.g., *Anabaena* sp., *Calothrix* sp., *Gloeocapsa* sp., *Lyngbya* sp., *Nostoc punctiforme*, *Rivularia* sp., among others) (Pathak et al., 2020; Rastogi et al., 2013; Rastogi and Incharoensakdi, 2014). In addition to the photoprotective effect, MAAs and scytonemin also present antioxidant, anti-inflammatory, and anti-aging properties. However, there is a lack of studies regarding the biocompatibility of these compounds in sunscreens formulations, and also their biosynthetic pathways are not completely elucidated (Amador-Castro et al., 2020).

- Skin whitening – Melanin is a protein that, in addition to defining the color of the skin, helps to protect it from the impacts of radiation. However, external factors can stimulate abnormal melanin production, causing the skin to darken. Active compounds with skin-whitening properties have been evaluated for their ability to inhibit the action of tyrosinase, the key enzyme in melanin biosynthesis (Aslam et al., 2021). Sahin (2018) evaluated the potential tyronase inhibition of aqueous and ethanolic extract of *Arthrospira platensis*. The results showed that ethanolic extracts present higher tyrosinase inhibitory capacity (IC$_{50}$:1.4 × 10^{-3} g·mL^{-1}) than in aqueous (IC$_{50}$:7.2 × 10^{-3} g·mL^{-1}), probably due to the presence of specific phenolic compounds, such as vanillic, caffeic, and ferulic acid. Lee et al. (2018) also observed inhibition of melanogenesis in B16F10 Melanoma Cells and Artificial Human Skin Equivalents, promoted by ethanolic extract of *Chlamydomonas reinhardtii* rich carotenoids (e.g., β-carotene, lutein, neoxanthin and violaxanthin) and phenolic compounds.

- Anti-aging, antioxidant, and anti-inflammatory agent – Skin aging is a natural process, but it can also be induced by several external factors that stimulate the formation of ROS and, consequently, the breakdown and reduction of collagen production, leading to the appearance of wrinkles (Nowruzi et al., 2020). Microalgae carotenoids, such as β-carotene, lycopene, astaxanthin, and lutein have been reported as powerful antioxidants with significant anti-aging effects (Pangestuti et al., 2020). Gunes et al. (2017) reported the promising application of *Spirulina platensis* extract as an antioxidant and skin healer. In vitro experiments showed that skin cream incorporated with 1.125% of crude extract after 5–10 days of the application resulted in 74.9% of wound closure against 33.9% of the cream control (cream without microalgal extract) and 23% without any cream. Additionally, de Melo et al. (2019) observed anti-inflammatory, antioxidant, antibacterial, and wound healing properties in *Chlorella vulgaris* extract incorporated in topical hydrogel.

- Moisturizing agent – Moisturizers are molecules capable of reducing water loss and retaining skin hydration, maintaining the skin's elasticity and healthy appearance. In this sense, some microalgae (e.g., *Chroococcidiopsis*, *Phormidium,* and *Nostoc* species), to survive in saline and desert habits, produce large amounts of polysaccharides and extracellular polymeric substances (EPS) to enhance the water retention (Nowruzi et al., 2020).

Among other applications, *Nannochloropsis gaditana* biomass has been used for thalassoteraphy (Mourelle et al., 2017). According to Pangestuti et al. (2020), Chlorophyll *a* and carotenoids are the potential molecules to be used for this purpose. Additionally, chlorophyll can be used in deodorants to mask odors (Stoyneva-Gärtner et al., 2020). Table 5.2 shows some examples of commercial microalgae-based cosmetics and ingredients available for cosmetic formulations.

The variety of microalgae species, as well as the great diversity of bioactive compounds in their biomass, make them a potential source of inputs for the cosmetic industry. In addition to the possibility of synergistic effects, given the richness of microalgae extracts, the replacement of chemical substances by biological ones is of extreme importance, especially concerning human health and environmental impact. However, studies are still needed to elucidate the biosynthesis of some microalgae compounds, to enhance the stability of these molecules in cosmetics and to make their large-scale production economically viable.

5.8 USES OF MICROALGAL PIGMENTS BOOSTING BIOECONOMY AND CIRCULAR ECONOMY

The number of applications of microalgal pigments is already expressive. Nevertheless, as research advances, other applications are being discovered or recovered.

The fluorescence characteristics of some microalgal pigments, makes their use in flow cytometry extremely important (Pagels et al., 2020). The increasing concern with environmental pollution has led younger generations of consumers to demand more environmentally friendly products. Thus, besides their search for natural food and cosmeceutical products, there is now a growing trend for products incorporating natural dyes. To address these new trends, the textile, plastic, paint, paper, and printing industries have driven their attention to naturally occurring pigments, of

TABLE 5.2

Commercial Microalgae-based Cosmetics and Ingredients for Cosmetic Formulations

Product	Claims/application	Microalgae	Company
Iraya algae body serum	Moisturizer, skin balance and toxins removal	*Spirulina* extract	Nykaa[1]
Cosmetic line containing Plasmarine Active Complex™	Anti-aging and anti-wrinkle, skin hydration, balance, repairing and protection	Concentrate of two marine microalgae	Bluevert™[2]
Intensive hair repair mask	Hair strength and revitalization	Cyanobacteria	Aubrey® Organics[3]
REVEAL color-correcting eye serum concealer	Cover up skin imperfections	Extracts of *Dunaliella salina*, *Haematococcus pluvialis* and *Aphanizomenon flos-aquae*	Algenist®[4]
Dermochlorella®[*]	Increases elastin and collagen production, prevents and treats stretch marks and reduces body flaccidity	Hiperconcentrated extract of *Chlorella vulgaris*	CODIF Technologie Naturelle[5]
PEPHA®-TIGHT[*]	Long-term skin firming effect after multiple applications	*Nannochloropsis oculata* extract	Royal DSM N.V.[6]
PEPHA®-CTIVE[*]	Increases ATP levels (mitochondria stimulation and protection), skin care	*Dunaliella salina* extract	
AlgAllure® AlgaRiche® Skin Care Collection	Anti-aging and antioxidant properties	Extract of red marine microalgae	AlgAllure®[7]
ALGAKTIV (cosmetics line)	Several applications, including skin whitening, anti-wrinkle agent, elasticity, protection and care for sensitive skin (induction of elastin and collagen production), and skin shield against urban pollution	Extracts of *Chlamydomonas reinhardtii*, *Chlorella emersonii*, *Chlorella vulgaris*, *Dicrateria rotunda*, *Haematococcus pluvialis*, *Phaeodactylum tricornutum*, *Porphyridium cruentum*, *Ruttnera lamellosa*, *Skeletonema costatum*, *Spirulina maxima* and *Synechococcus elongatus*	Greenaltech[8]

Notes

[*] Ingredient for skin care cosmetics.

Websites:

[1] http://www.nykaa.com/

[2] https://bluevert.com/

[3] https://aubreyorganics.com/

[4] https://www.algenist.com/

[5] http://www.codif-tn.com/

[6] https://www.dsm.com/

[7] https://algallure.com/

[8] http://algaktiv.com/

which those produced by microalgae represent an important source (Aruldass et al., 2018). In this context, the industry has been actively participating in European funded projects – ex. SeaColors, (Life (EU), 2016), and GreenColors (EU (ERDF), 2017), among others – while the younger designers pave their way, using micro-algae pigments to color fabric (Blond & Bieber, 2020).

Furthermore, extending this concept to energy harvesting using photovoltaic technology with architectonic features, the use of biocompatible, non-toxic, natural pigments, that are also cost-effective, represents a potential environmental friendly alternative for dye-sensitized solar cells, that has been the object of recent research with promising results (Orona-Navar et al., 2021).

Additionally, microalgae cultivation can be performed with the purpose of wastewater phycoremediation (Kalra et al., 2021). In the end, besides yielding purified water, microalgal biomass can be used for several applications, within a biorefinery concept, including the recovery of valuable pigments, as discussed before. This intertwining of applications and uses of microalgae biomass is boosting the circular economy of microalgae as a powerful renewable nature feedstock.

The number of industrial applications of microalgal pigments and microalgal biomass is growing, ranging from water recycling and nutrient recovery, to biofuels production, including food and feed, nutraceuticals, cosmeceuticals, and many other valuable uses (VentureRadar, 2021).

5.9   CONCLUSIONS

Microalgae pigments are naturally occurring valuable compounds, mainly bio-compatible and non-toxic, that can provide a handful of colors. Their use in many daily applications as food colorants, textile and medical dyes, or cosmeceutical, nutraceutical, and pharmaceutical compounds, as well as in energy harvesting systems such as sensitized dye cells, has been gaining terrain due to their avail-ability, relatively low production cost, and especially their environmentally friendly nature. Yet, the industrial scale-up of some of these applications is still facing a few hurdles, particularly for the most energy-intensive processes. Using only the pig-ments fraction of microalgal biomass has proved often costly. Thus, the most consensual approach to produce pigments from microalgae is within a biorefinery concept, where several fractions of the microalgal biomass are used, covering a wide range of applications, making the whole process cost-effective, boosting the microalgae bioeconomy, and increasing the circularity of this valuable feedstock.

ACKNOWLEDGMENTS

This work was financially supported by Base Funding – UIDB/00511/2020 of Laboratory for Process Engineering, Environment, Biotechnology and Energy – LEPABE – funded by National funds through the FCT/MCTES (PIDDAC); Base Funding – UIDB/04730/2020 of Center for Innovation in Engineering and Industrial Technology, CIETI – funded by national funds through the FCT/MCTES (PIDDAC); project IF/01093/2014/CP1249/CT0003 funded by national funds through FCT/MCTES; project "EXTRATOTECA – Microalgae Extracts for High Value Products" –

POCI-01–0247-FEDER-033784, funded by FEDER funds through COMPETE2020 – Programa Operacional Competitividade e Internacionalização (POCI) and by national funds (PIDDAC) through FCT/MCTES. António Martins thanks FCT (Fundação para a Ciência e Tecnologia) for funding through program DL 57/2016 – Norma transitória. Teresa Mata thanks the funding of Project NORTE-06-3559-FSE-000107, cofinanced by Programa Operacional Regional do Norte (NORTE2020), through Fundo Social Europeu (FSE). Wilson Júnior thanks European Union's Horizon 2020 research and innovation programme under the Marie Skłodowska-Curie grant agreement number 867473.

REFERENCES

Aberoumand, A., 2011. A review article on edible pigments properties and sources as natural biocolorants in foodstuff and food industry. *J. Dairy Food Sci.* 6, 71–78.

Abu-Ghosh, S., Dubinsky, Z., Verdelho, V., Iluz, D., 2021. Unconventional high-value products from microalgae: A review. *Bioresour. Technol.* 329, 124895. 10.1016/j.biortech.2021.124895

Amador-Castro, F., Rodriguez-Martinez, V., Carrillo-Nieves, D., 2020. Robust natural ultraviolet filters from marine ecosystems for the formulation of environmental friendlier bio-sunscreens. *Sci. Total Environ.* 749, 141576. 10.1016/j.scitotenv.2020.141576

Amarante, M.C.A. de, Corrêa Júnior, L.C.S., Sala, L., Kalil, S.J., 2020. Analytical grade C-phycocyanin obtained by a single-step purification process. *Process Biochem.* 90, 215–222. 10.1016/j.procbio.2019.11.020

Amorim, M.L., Soares, J., Coimbra, J.S. dos R., Leite, M. de O., Albino, L.F.T., Martins, M.A., 2020. Microalgae proteins: Production, separation, isolation, quantification, and application in food and feed. *Crit. Rev. Food Sci. Nutr.* 61, 1976–2002. 10.1080/10408398.2020.1768046

Arashiro, L.T., Boto-Ordóñez, M., Van Hulle, S.W.H., Ferrer, I., Garfí, M., Rousseau, D.P.L., 2020. Natural pigments from microalgae grown in industrial wastewater. *Bioresour. Technol.* 303, 122894. 10.1016/j.biortech.2020.122894

Aruldass, C.A., Dufossé, L., Ahmad, W.A., 2018. Current perspective of yellowish-orange pigments from microorganisms – A review. *J. Clean. Prod.* 180, 168–182. 10.1016/J.JCLEPRO.2018.01.093

Aslam, A., Bahadar, A., Liaquat, R., Saleem, M., Waqas, A., Zwawi, M., 2021. Algae as an attractive source for cosmetics to counter environmental stress. *Sci. Total Environ.* 772, 144905. 10.1016/j.scitotenv.2020.144905

Barkia, I., Saari, N., Manning, S.R., 2019. Microalgae for high-value products towards human health and nutrition. *Mar. Drugs*, 17, 304. 10.3390/MD17050304

Begum, H., Yusoff, F.M., Banerjee, S., Khatoon, H., Shariff, M., 2015. Availability and utilization of pigments from microalgae. *Crit. Rev. Food Sci. Nutr.* 56, 2209–2222. 10.1080/10408398.2013.764841

Ben-Amotz, A., 1995. New mode of *Dunaliella* biotechnology: Two-phase growth for β-carotene production. *J. Appl. Phycol.* 7, 65–68. 10.1007/BF00003552

Ben-Amotz, A., Shaish, A., Avron, M., 1991. The biotechnology of cultivating *Dunaliella* for production of β-carotene rich algae. *Bioresour. Technol.* 38, 233–235. 10.1016/0960-8524(91)90160-L

Benavente-Valdés, J.R., Aguilar, C., Contreras-Esquivel, J.C., Méndez-Zavala, A., Montañez, J., 2016. Strategies to enhance the production of photosynthetic pigments and lipids in chlorophycae species. *Biotechnol. Reports*. 10, 117–125. 10.1016/j.btre.2 016.04.001

Benavides, J., Rito-Palomares, M., 2006. Simplified two-stage method to B-phycoerythrin recovery from *Porphyridium cruentum. J. Chromatogr. B Anal. Technol. Biomed. Life Sci.* 844, 39–44. 10.1016/j.jchromb.2006.06.029

Benemann, J., 2016. The solar microalgae industry: Then, now, and coming. *Algae Europe*, Madrid. 10, 117–125.

Bertsch, P., Böcker, L., Mathys, A., Fischer, P., 2021. Proteins from microalgae for the stabilization of fluid interfaces, emulsions, and foams. *Trends Food Sci. Technol.* 108, 326–342. 10.1016/J.TIFS.2020.12.014

Blond & Bieber, 2020. Natural fabric dyeing using algae: Algaemy [WWW Document]. URL https://startupfashion.com/natural-fabric-dyeing-algaemy/ (accessed 8.19.21).

Borowitzka, L., Borowitzka, M., 1990. Commercial production of β-carotene by *Dunaliella salina* in open ponds. *Bull. Mar. Sci.* 47, 244–252.

Borowitzka, M., 2019. Microalgal metabolism and their utilisation, in: Malcata, F.X., Sousa Pinto, I., Guedes, A.C. (Eds.), *Marine Macro- and Microalgae: An Overview*. CRC Press, Taylor & Francis Group, Boca Raton, Florida, pp. 43–62.

Borowitzka, M.A., 2013a. Energy from microalgae: A short history. *Proceedings of Algae Europe*, Madrid. 1–15. 13–15 December 2016. 10.1007/978-94-007-5479-9_1

Borowitzka, M.A., 2013b. High-value products from microalgae-their development and commercialisation. *J. Appl. Phycol.* 25, 743–756. 10.1007/s10811-013-9983-9

Borowitzka, M.A., Huisman, J.M., Osborn, A., 1991. Culture of the astaxanthin-producing green alga *Haematococcus pluvialis* 1. Effects of nutrients on growth and cell type. *J. Appl. Phycol.* 3, 295–304. 10.1007/bf00026091

Branco-Vieira, M., Costa, D.M.B., Mata, T.M., Martins, A.A., Freitas, M.A.V., Caetano, N.S., 2020a. Environmental assessment of industrial production of microalgal biodiesel in central-south Chile. *J. Clean. Prod.* 266, 121756. 10.1016/j.jclepro.2020.121756

Branco-Vieira, M., Mata, T.M., Martins, A.A., Freitas, M.A.V., Caetano, N.S., 2020b. Economic analysis of microalgae biodiesel production in a small-scale facility. *Energy Reports* 6, 325–332. 10.1016/j.egyr.2020.11.156

Braune, S., Krüger-Genge, A., Kammerer, S., Jung, F., Küpper, J.H., 2021. Phycocyanin from *Arthrospira platensis* as potential anti-cancer drug: Review of in vitro and in vivo studies. *Life* 11, 1–14. 10.3390/life11020091

Burlew, J.S. (Ed.), 1953. *Algal culture: From laboratory to pilot plant, Carnegie Institution of Washington Publication 600*. Carnegie Institution of Washington Publication 600, Washington, D.C. 10.1016/0016-0032(54)90075-7

Caetano, N.S., Martins, A.A., Gorgich, M., Gutiérrez, D.M., Ribeiro, T.J., Mata, T.M., 2019. Flocculation of *Arthrospira maxima* for improved harvesting. *Energy Reports* 6(1), 423–428. 10.1016/j.egyr.2019.08.083

Camacho, F., Macedo, A., Malcata, F., 2019. Potential industrial applications and commercialization of microalgae in the functional food and feed industries: A short review. *Mar. Drugs* 17, 312. 10.3390/MD17060312

Chang, D.K., Sang, J.S., 2007. Selective extraction of free astaxanthin from *Haematococcus* culture using a tandem organic solvent system. *Biotechnol. Prog.* 23, 866–871. 10.1 021/bp0700354

Chekanov, K., Lobakova, E., Selyakh, I., Semenova, L., Sidorov, R., Solovchenko, A., 2014. Accumulation of astaxanthin by a new *Haematococcus pluvialis* strain BM1 from the white sea coastal rocks (Russia). *Mar. Drugs* 12, 4504–4520. 10.3390/md12084504

Chen, C.Y., Zhao, X.Q., Yen, H.W., Ho, S.H., Cheng, C.L., Lee, D.J., Bai, F.W., Chang, J.S., 2013. Microalgae-based carbohydrates for biofuel production. *Biochem. Eng. J.* 78, 1–10. 10.1016/J.BEJ.2013.03.006

Chen, M., 2014. Chlorophyll modifications and their spectral extension in oxygenic photosynthesis. *Ann. Rev. Biochem.* 83, 317–340. 10.1146/ANNUREV-BIOCHEM-072711-162943

Chen, M., Blankenship, R.E., 2011. Expanding the solar spectrum used by photosynthesis. *Trends Plant Sci.* 16, 427–431. 10.1016/J.TPLANTS.2011.03.011

Cheng, X., Shah, M., 2021. Bioextraction of astaxanthin adopting varied techniques and down-stream processing methodologies, in: Ravishankar, G.A. and Rao, A.R. *Global Perspectives on Astaxanthin.* Elsevier Inc., pp. 313–339. 10.1016/B978-0-12-823304-7.00028-3

Chew, K.W., Yap, J.Y., Show, P.L., Suan, N.H., Juan, J.C., Ling, T.C., Lee, D.J., Chang, J.S., 2017. Microalgae biorefinery: High value products perspectives. *Bioresour. Technol.* 229, 53–62. 10.1016/J.BIORTECH.2017.01.006

Chhandama, M.V.L., Satyan, K.B., Changmai, B., Vanlalveni, C., Rokhum, S.L., 2021. Microalgae as a feedstock for the production of biodiesel: A review. *Bioresour. Technol. Reports* 15, 100771. 10.1016/J.BITEB.2021.100771

Corrêa, P.S., Morais Júnior, W.G., Martins, A.A., Caetano, N.S., Mata, T.M., 2020. Microalgae biomolecules: Extraction separation and purification methods, *Processes* 9, 10.10.3390/pr9010010

Cuellar-Bermudez, S.P., Aguilar-Hernandez, I., Cardenas-Chavez, D.L., Ornelas-Soto, N., Romero-Ogawa, M.A., Parra-Saldivar, R., 2015. Extraction and purification of high-value metabolites from microalgae: Essential lipids, astaxanthin and phycobiliproteins. *Microb. Biotechnol.* 8, 190–209. 10.1111/1751-7915.12167

Cysewski, G.R., 2011. *Microalgae Production and their Use in Animal Feeds.* Cyanotech Corporation, Hawaii, U.S.A.

da Silva, S.C., Fernandes, I.P., Barros, L., Fernandes, Â., José Alves, M., Calhelha, R.C., Pereira, C., Barreira, J.C.M., Manrique, Y., Colla, E., Ferreira, I.C.F.R., Filomena Barreiro, M., 2019. Spray-dried *Spirulina platensis* as an effective ingredient to improve yogurt formulations: Testing different encapsulating solutions. *J. Funct. Foods.* 60, 103427. 10.1016/j.jff.2019.103427

Del Campo, J.A., García-González, M., Guerrero, M.G., 2007. Outdoor cultivation of microalgae for carotenoid production: Current state and perspectives. *Appl. Microbiol. Biotechnol* 74, 1163–1174. 10.1007/s00253-007-0844-9

Del Campo, J.A., Moreno, J., Rodríguez, H., Angeles Vargas, M., Rivas, J., Guerrero, M.G., 2000. Carotenoid content of chlorophycean microalgae: factors determining lutein accumulation in *Muriellopsis* sp. (Chlorophyta). *J. Biotechnol.* 76, 51–59. 10.1016/S01 68-1656(99)00178-9

Del Campo, J.A., Rodríguez, H., Moreno, J., Vargas, M.Á., Rivas, J., Guerrero, M.G., 2004. Accumulation of astaxanthin and lutein in *Chlorella zofingiensis* (Chlorophyta). *Appl. Microbiol. Biotechnol.* 64, 848–854. 10.1007/s00253-003-1510-5

Del Mondo, A., Smerilli, A., Sané, E., Sansone, C., Brunet, C., 2020. Challenging microalgal vitamins for human health. *Microb. Cell Factories* 19, 1–23. 10.1186/S12934-020-01459-1

de Melo, R.G., de Andrade, A.F., Bezerra, R.P., Viana Marques, D. de A., da Silva, V.A., Paz, S.T., de Lima Filho, J.L., Porto, A.L.F., 2019. Hydrogel-based *Chlorella vulgaris* extracts: a new topical formulation for wound healing treatment. *J. Appl. Phycol.* 31, 3653–3663. 10.1007/s10811-019-01837-2

Dickinson, S., Mientus, M., Frey, D., Amini-Hajibashi, A., Ozturk, S., Shaikh, F., Sengupta, D., El-Halwagi, M.M., 2016. A review of biodiesel production from microalgae. *Clean Technol. Environ. Policy* 193(19), 637–668. 10.1007/S10098-016-1309-6

Dolganyuk, V., Belova, D., Babich, O., Prosekov, A., Ivanova, S., Katserov, D., Patyukov, N., Sukhikh, S., 2020. Microalgae: A promising source of valuable bioproducts. *Biomol.* 10, 1153. 10.3390/BIOM10081153

do Nascimento, T.C., Nass, P.P., Fernandes, A.S., Vieira, K.R., Wagner, R., Jacob-Lopes, E., Zepka, L.Q., 2020. Exploratory data of the microalgae compounds for food purposes. *Data Br.* 29, 105182. 10.1016/J.DIB.2020.105182

Elleuch, F., Hlima, H. Ben, Barkallah, M., Baril, P., Abdelkafi, S., Pichon, C., Fendri, I., 2019. Carotenoids overproduction in *Dunaliella* sp.: Transcriptional changes and new insights through lycopene cyclase regulation. *Appl. Sci.* 9(24), 5389. 10.3390/app9245389

Eloka-Eboka, A.C., Inambao, F.L., 2017. Effects of CO_2 sequestration on lipid and biomass productivity in microalgal biomass production. *Appl. Energy* 195, 1100–1111. 10.101 6/J.APENERGY.2017.03.071

Eriksen, N.T., 2008. Production of phycocyanin—A pigment with applications in biology, biotechnology, foods and medicine. *Appl. Microbiol. Biotechnol* 80(1), 1–14. 10.1007/ S00253-008-1542-Y

Esland, L., Larrea-Alvarez, M., Purton, S., 2018. Selectable markers and reporter genes for engineering the chloroplast of *Chlamydomonas reinhardtii*. *Biol.* 7(4), 46. 10.3390/ BIOLOGY7040046

EU (ERDF), 2017. Algae-based dyes; A New Alternative For Sustainable Development In The Textile Industry - Aitex [WWW Document]. https://www.aitex.es/colorantes-procedentes-de-algas-greencolor/?lang=en (accessed 8.19.21).

Fábregas, J., Domínguez, A., García Àlvarez, D., Lamela, T., Otero, A., 1998. Induction of astaxanthin accumulation by nitrogen and magnesium deficiencies in *Haematococcus pluvialis*. *Biotechnol. Lett.* 20, 623–626. 10.1023/A:1005322416796

Ferdous, U.T., Yusof, Z.N.B., 2021. Medicinal prospects of antioxidants from algal sources in cancer therapy. *Front. Pharmacol.* 12, 593116. 10.3389/FPHAR.2021.593116

Fernandes, B.D., Mota, A., Teixeira, J.A., Vicente, A.A., 2015. Continuous cultivation of photosynthetic microorganisms: Approaches, applications and future trends. *Biotechnol. Adv.* 33, 1228–1245. 10.1016/J.BIOTECHADV.2015.03.004

Ferrazzano, G.F., Papa, C., Pollio, A., Ingenito, A., Sangianantoni, G., Cantile, T., 2020. Cyanobacteria and microalgae as sources of functional foods to improve human general and oral health. *Mol.* 25, 5164. 10.3390/MOLECULES25215164

Flórez-Miranda, L., Cañizares-Villanueva, R.O., Melchy-Antonio, O., Martínez-Jerónimo, F., Flores-Ortíz, C.M., 2017. Two stage heterotrophy/photoinduction culture of Scenedesmus incrassatulus: potential for lutein production. *J. Biotechnol.* 262, 67–74. 10.1016/j.jbiotec.2017.09.002

Fu, W., Paglia, G., Magnúsdóttir, M., Steinarsdóttir, E.A., Gudmundsson, S., Palsson, B.Ø., Andrésson, Ó.S., Brynjólfsson, S., 2014. Effects of abiotic stressors on lutein production in the green microalga *Dunaliella salina*. *Microb. Cell Factories.* 13, 1–9. 10.1186/1475-2859-13-3

Fujii, K., 2012. Process integration of supercritical carbon dioxide extraction and acid treatment for astaxanthin extraction from a vegetative microalga. *Food Bioprod. Process.* 90, 762–766. 10.1016/j.fbp.2012.01.006

Fujita, Y., Tsujimoto, R., Aoki, R., 2015. Evolutionary aspects and regulation of tetrapyrrole biosynthesis in cyanobacteria under aerobic and anaerobic environments. *Life.* 5, 1172–1203. 10.3390/LIFE5021172

Furuki, T., Maeda, S., Imajo, S., Hiroi, T., Amaya, T., Hirokawa, T., Ito, K., Nozawa, H., 2003. Rapid and selective extraction of phycocyanin from *Spirulina platensis* with ultrasonic cell disruption. *J. Appl. Phycol.* 15, 319–324. 10.1023/A:1025118516888

Galarza, J.I., Gimpel, J.A., Rojas, V., Arredondo-Vega, B.O., Henríquez, V., 2018. Over-accumulation of astaxanthin in *Haematococcus pluvialis* through chloroplast genetic engineering. *Algal Res.* 31, 291–297. 10.1016/j.algal.2018.02.024

Galasso, C., Gentile, A., Orefice, I., Ianora, A., Bruno, A., Noonan, D.M., Sansone, C., Albini, A., Brunet, C., 2019. Microalgal derivatives as potential nutraceutical and food supplements for human health: A focus on cancer prevention and interception. *Nutr.* 11, 1226. 10.3390/NU11061226

Galetovic, A., Seura, F., Gallardo, V., Graves, R., Cortés, J., Valdivia, C., Nuñez, J., Tapia, C., Neira, I., Sanzana, S., Gómez-Silva, B., 2020. Use of phycobiliproteins from atacama cyanobacteria as food colorants in a dairy beverage prototype. *Foods*. 9, 1–13. 10.3390/foods9020244

Gantar, M., Simović, D., Djilas, S., Gonzalez, W.W., Miksovska, J., 2012. Isolation, characterization and antioxidative activity of C-phycocyanin from *Limnothrix* sp strain 37-2-1. *J. Biotechnol*. 159(1–2), 21–26. 10.1016/j.jbiotec.2012.02.004

García, J.L., Vicente, M. de Galán, B., 2017. Microalgae, old sustainable food and fashion nutraceuticals. *Microb. Biotechnol*. 10, 1017–1024. 10.1111/1751-7915.12800

Gimpel, J.A., Henríquez, V., Mayfield, S.P., 2015. In metabolic engineering of eukaryotic microalgae: Potential and challenges come with great diversity. *Front. Microbiol*. 6, 1376. 10.3389/FMICB.2015.01376

Glazer, A.N., 1994. Phycobiliproteins – A family of valuable, widely used fluorophores. *J. Appl. Phycol*. 6, 105–112. 10.1007/BF02186064

Goiris, K., Muylaert, K., Fraeye, I., Foubert, I., De Brabanter, J., De Cooman, L., 2012. Antioxidant potential of microalgae in relation to their phenolic and carotenoid content. *J. Appl. Phycol*. 24, 1477–1486. 10.1007/S10811-012-9804-6

Gómez, P.I., González, M.A., 2005. The effect of temparature and irradiance on the growth and carotenogenic capacity of seven strains of *Dunaliella salina* (Chlorophyta) cultivated under laboratory conditions. *Biol. Res*. 38, 151–162. 10.4067/s0716-97602005000200005

Gorgich, M., Passos, M.L.C., Mata, T.M., Martins, A.A., Saraiva, M.L.M.F.S., Caetano, N.S., 2020. Enhancing extraction and purification of phycocyanin from *Arthrospira* sp. with lower energy consumption. *Energy Reports*. 6, 312–318. 10.1016/J.EGYR.2020.11.151

Grover, P., Bhatnagar, A., Kumari, N., Narayan Bhatt, A., Kumar Nishad, D., Purkayastha, J., 2021. C-Phycocyanin-a novel protein from *Spirulina platensis* – In vivo toxicity, antioxidant and immunomodulatory studies *Saudi J. Biol. Sci*. 28, 1853–1859. 10.1016/j.sjbs.2020.12.037

Guenard, R., 2021. The green machine: Commercializing microalgae products. *Inf. Int. News Fats, Oils, Relat. Mater*. 32, 6–12. 10.21748/inform.02.2021.06

Guiry, M.D., 2012. How many species of algae are there? *J. Phycol*. 48, 1057–1063. 10.1111/J.1529-8817.2012.01222.X

Gunes, S., Tamburaci, S., Dalay, M.C., Deliloglu Gurhan, I., 2017. In vitro evaluation of *Spirulina platensis* extract incorporated skin cream with its wound healing and antioxidant activities. *Pharm. Biol*. 55, 1824–1832. 10.1080/13880209.2017.1331249

Gurev, A., Dragancea, V., Haritonov, S., 2020. Microalgae – non-traditional sources of nutrients and pigments for functional foods. *J. Eng. Sci*. XXVII, 75–98. 10.5281/zenodo.3713372

Gustiningtyas, A., Setyaningsih, I., Hardiningtyas, S.D., Susila, A.A.R., 2020. Improvement stability of phycocyanin from *Spirulina platensis* encapsulated by water soluble chitosan nanoparticles. *IOP Conf. Ser. Earth Environ. Sci*. 414, 012005. 10.1088/1755-1315/414/1/012005

Gutiérrez, S., Lauersen, K.J., 2021. Gene delivery technologies with applications in microalgal genetic engineering. *Biology (Basel)* 10(4), 265. 10.3390/biology10040265

Habib, M., Ahsan B., 2008. *A Review on Culture, Production, and Use of Spirulina as Food for Humans and Feeds for Domestic Animals and Fish*. Food and Agriculture Organization of the United Nations, Rome, Italy.

Halim, R., Hosikian, A., Lim, S., Danquah, M.K., 2010. Chlorophyll extraction from microalgae: A review on the process engineering aspects. *Int. J. Chem. Eng*. 2010, 391632. 10.1155/2010/391632

Han, D., Li, Y., Hu, Q., 2013. Astaxanthin in microalgae: Pathways, functions and biotechnological implications. *Algae*. 28, 131–147. 10.4490/algae.2013.28.2.131

Hemantkumar, J.N., Rahimbhai, M.I., 2019. Microalgae and its use in nutraceuticals and food supplements, in: Vítová, M. (Ed.), *Microalgae – From Physiol. to Appl.* IntechOpen. 10.5772/INTECHOPEN.90143

Herrador, M., 2016. The Microalgae-Biomass Industry in Japan – An Assessment of Cooperation and Business Potential With European Companies. *EU-Japan Cent. Ind. Coop.* 1–184.

Hidalgo, P., Navia, R., Hunter, R., Gonzalez, M.E., Echeverría, A., 2019. Development of novel bio-based epoxides from microalgae *Nannochloropsis gaditana* lipids. *Compos. Part B Eng.* 166, 653–662. 10.1016/J.COMPOSITESB.2019.02.049

Jacob-Lopes, E., Queiroz, M.I., Zepka, L.Q., 2020. *Pigments from Microalgae Handbook.* Springer Nature Switzerland AG. 10.1007/978-3-030-50971-2_14

Jayappriyan, K.R., Rajkumar, R., Venkatakrishnan, V., Nagaraj, S., Rengasamy, R., 2013. In vitro anticancer activity of natural β-carotene from *Dunaliella salina* EU5891199 in PC-3 cells. *Biomed. Prev. Nutr.* 3, 99–105. 10.1016/j.bionut.2012.08.003

Jiménez, C., Niell, F.X., 1991. Growth of *Dunaliella viridis* Teodoresco: effect of salinity, temperature and nitrogen concentration. *J. Appl. Phycol.* 3, 319–327. 10.1007/BF00026094

Jin, E., Lee, C.G., Polle, J.E.W., 2006. Secondary carotenoid accumulation in *Haematococcus* (chlorophyceae): Biosynthesis, regulation, and biotechnology. *J. Microbiol. Biotechnol.* 16(6), 821–831. https://www.koreascience.or.kr/article/JAKO200629734275172.page

Kalaji, H.M., Schansker, G., Brestic, M., Bussotti, F., Calatayud, A., Ferroni, L., Goltsev, V., Guidi, L., Jajoo, A., Li, P., Losciale, P., Mishra, V.K., Misra, A.N., Nebauer, S.G., Pancaldi, S., Penella, C., Pollastrini, M., Suresh, K., Tambussi, E., Yanniccari, M., Zivcak, M., Cetner, M.D., Samborska, I.A., Stirbet, A., Olsovska, K., Kunderlikova, K., Shelonzek, H., Rusinowski, S., Bąba, W., 2016. Frequently asked questions about chlorophyll fluorescence, the sequel. *Photosynth. Res.* 132, 13–66. 10.1007/S11120-016-0318-Y

Kalra, R., Gaur, S., Goel, M., 2021. Microalgae bioremediation: A perspective towards wastewater treatment along with industrial carotenoids production. *J. Water Process Eng.* 40, 101794. 10.1016/J.JWPE.2020.101794

Kang, C.D., Lee, J.S., Park, T.H., Sim, S.J., 2005. Comparison of heterotrophic and photoautotrophic induction on astaxanthin production by *Haematococcus pluvialis*. *Appl. Microbiol. Biotechnol.* 68, 237–241. 10.1007/s00253-005-1889-2

Kannaujiya, V.K., Kumar, D., Pathak, J., Sinha, R.P., 2019. Phycobiliproteins and their commercial significance, in: Mishra, A.K., Tiwari, D.N., and Rai, A.N. (Eds.), *Cyanobacteria From Basic Sci. to Appl.* Elsevier Inc., pp. 207–216. 10.1016/B978-0-12-814667-5.00010-6

Kannaujiya, V.K., Sinha, R.P., 2016. An efficient method for the separation and purification of phycobiliproteins from a rice-field cyanobacterium *Nostoc* sp. strain HKAR-11. *Chromatographia* 79, 335–343. 10.1007/s10337-016-3025-0

Kato, S., Shinomura, T., 2020. Carotenoid Synthesis and Accumulation in Microalgae Under Environmental Stress, in: Jacob-Lopes, E., Queiroz, M., Zepka, L. (Eds.), *Pigments from Microalgae Handbook.* Springer, Cham, pp. 69–80. 10.1007/978-3-030-5 0971-2_4

Kaur, G., Khattar, J.I., Singh, D.P., Singh, Y., Nadda, J., 2009. Microalgae: a Source of Natural Colours, in: Khattar, J.I.S., Singh, D.P., Kaur, G. (Eds.), *Algal Biology and Biotechnology.* I.K. International Publishing House, Pvt. Ltd., New Delhi, India, pp. 129–149.

Kaur, S., Khattar, J.I.S., Singh, Y., Singh, D.P., Ahluwalia, A.S., 2019. Extraction, purification and characterisation of phycocyanin from *Anabaena fertilissima* PUPCCC 410.5: as a natural and food grade stable pigment. *J. Appl. Phycol.* 31, 1685–1696. 10.1007/s10811-018-1722-9

Khanra, S., Mondal, M., Halder, G., Tiwari, O.N., Gayen, K., Bhowmick, T.K., 2018. Downstream processing of microalgae for pigments, protein and carbohydrate in industrial application: A review. *Food Bioprod. Process.* 110, 60–84. 10.1016/j.fbp.201 8.02.002

Kim, D.Y., Vijayan, D., Praveenkumar, R., Han, J.I., Lee, K., Park, J.Y., Chang, W.S., Lee, J.S., Oh, Y.K., 2016. Cell-wall disruption and lipid/astaxanthin extraction from microalgae: *Chlorella* and *Haematococcus*. *Bioresour. Technol.* 199, 300–310. 10.1016/ J.BIORTECH.2015.08.107

Kiran, B.R., Venkata Mohan, S., 2021. Microalgal cell biofactory—therapeutic, nutraceutical and functional food applications. *Plants.* 10(5), 836. 10.3390/plants10050836

Kratzer, R., Murkovic, M., 2021. Food Ingredients and Nutraceuticals from Microalgae: Main Product Classes and Biotechnological Production. *Foods.* 10, 1626. 10.3390/ foods10071626

Kumar, A., Bera, S., 2020. Revisiting nitrogen utilization in algae: A review on the process of regulation and assimilation. *Bioresour. Technol. Reports* 12, 100584. 10.1016/ j.biteb.2020.100584

Kumar, P.K., Krishna, S.V., Naidu, S.S., Verma, K., Bhagawan, D., Himabindu, V., 2019. Biomass production from microalgae Chlorella grown in sewage, kitchen wastewater using industrial CO_2 emissions: Comparative study. *Carbon Resour. Convers.* 2, 126–133. 10.1016/j.crcon.2019.06.002

Lamers, P.P., Laak, C.C.W. van de Kaasenbrood, P.S., Lorier, J., Janssen, M., Vos, R.C.H. De, Bino, R.J., Wijffels, R.H., 2010. Carotenoid and fatty acid metabolism in light-stressed *Dunaliella salina*. *Biotechnol. Bioeng.* 106, 638–648. 10.1002/BIT.22725

Larkum, A.W., 2016. Photosynthesis and Light Harvesting in Algae, in: Borowitzka, M., Beardall, J., Raven, J. (Eds.), *The Physiology of Microalgae. Developments in Applied Phycology*, vol 6. Springer, Cham. 10.1007/978-3-319-24945-2_3

Lee, A., Kim, J.Y., Heo, J., Cho, D.H., Kim, H.S., An, I.S., An, S., Bae, S., 2018. The inhibition of melanogenesis via the PKA and ERK signaling pathways by *Chlamydomonas reinhardtii* extract in B16f10 melanoma cells and artificial human skin equivalents. *J. Microbiol. Biotechnol.* 28, 2121–2132. 10.4014/jmb.1810.10008

Levasseur, W., Perré, P., Pozzobon, V., 2020. A review of high value-added molecules production by microalgae in light of the classification. *Biotechnol. Adv.* 41, 107545. 10.1016/J.BIOTECHADV.2020.107545

Li, S., Luo, S., Guo, R., 2013. Efficiency of CO_2 fixation by microalgae in a closed raceway pond. *Bioresour. Technol.* 136, 267–272. https://doi.org/10.1016/J.BIORTECH.2013.03.025

Li, Y., Chen, M., Li, Y., Chen, M., 2015. Novel chlorophylls and new directions in photosynthesis research. *Funct. Plant Biol.* 42, 493–501. 10.1071/FP14350

Li, Y., Scales, N., Blankenship, R.E., Willows, R.D., Chen, M., 2012. Extinction coefficient for red-shifted chlorophylls: Chlorophyll d and chlorophyll f. Biochim. *Biophys. Acta - Bioenerg.* 1817, 1292–1298. 10.1016/J.BBABIO.2012.02.026

Liang, M.-H., Hao, Y.-F., Li, Y.-M., Liang, Y.-J., Jiang, J.-G., 2016. Inhibiting lycopene cyclases to accumulate lycopene in high β-carotene-accumulating *Dunaliella bardawil*. *Food Bioprocess Technol.* 96(9), 1002–1009. 10.1007/S11947-016-1681-6

Life (EU), 2016. SeaColors [WWW Document]. http://www.seacolors.eu/index.php/en/ (accessed 8.19.21).

Lim, K.C., Yusoff, F.M., Shariff, M., Kamarudin, M.S., 2018. Astaxanthin as feed supplement in aquatic animals. *Rev. Aquac.* 10, 738–773. 10.1111/raq.12200

Liu, C., Hu, B., Cheng, Y., Guo, Y., Yao, W., Qian, H., 2021. Carotenoids from fungi and microalgae: A review on their recent production, extraction, and developments. *Bioresour. Technol.* 337, 125398. 10.1016/J.BIORTECH.2021.125398

Liu, J., Lu, Y., Hua, W., Last, R.L., 2019. A new light on photosystem II maintenance in oxygenic photosynthesis. *Front. Plant Sci.* 10, Article number 975. 10.3389/FPLS.2019.00975

Liyanaarachchi, V.C., Premaratne, M., Ariyadasa, T.U., Nimarshana, P.H.V., Malik, A., 2021. Two-stage cultivation of microalgae for production of high-value compounds and biofuels: A review. *Algal Res.* 57, 102353. 10.1016/j.algal.2021.102353

Luimstra, V.M., Schuurmans, J.M., de Carvalho, C.F.M., Matthijs, H.C.P., Hellingwerf, K.J., Huisman, J., 2019. Exploring the low photosynthetic efficiency of cyanobacteria in blue light using a mutant lacking phycobilisomes. *Photosynth. Res.* 141, 291–301. 10.1 007/s11120-019-00630-z

Ma, R., Thomas-Hall, S.R., Chua, E.T., Eltanahy, E., Netzel, M.E., Netzel, G., Lu, Y., Schenk, P.M., 2018. Blue light enhances astaxanthin biosynthesis metabolism and extraction efficiency in *Haematococcus pluvialis* by inducing haematocyst germination. *Algal Res.* 35, 215–222. 10.1016/j.algal.2018.08.023

Machado, F.R.S., Trevisol, T.C., Boschetto, D.L., Burkert, J.F.M., Ferreira, S.R.S., Oliveira, J.V., Burkert, C.A. V., 2016. Technological process for cell disruption, extraction and encapsulation of astaxanthin from *Haematococcus pluvialis*. *J. Biotechnol.* 218, 108–114. 10.1016/J.JBIOTEC.2015.12.004

Márquez-Escobar, V.A., Bañuelos-Hernández, B., Rosales-Mendoza, S., 2018. Expression of a Zika virus antigen in microalgae: Towards mucosal vaccine development. *J. Biotechnol.* 282, 86–91. 10.1016/J.JBIOTEC.2018.07.025

Martin, G.J.O., 2016. Energy requirements for wet solvent extraction of lipids from microalgal biomass. *Bioresour. Technol.* 205, 40–47. 10.1016/j.biortech.2016.01.017

Martins, A.A., Mata, T.M., Oliveira, O., Oliveira, S., Mendes, A.M., Caetano, N.S., 2016. Sustainability evaluation of biodiesel from *Arthrospira platensis* and *Chlorella vulgaris* under mixotrophic conditions and salinity stress. *Chem. Eng. Trans.* 49, 571–576. 10.3303/CET1649096

Masojídek, J., Torzillo, G., Koblížek, M., 2013. Photosynthesis in microalgae. , in: *Handbook of Microalgal Culture: Applied Phycology and Biotechnology*, second ed., Wiley-Blackwell, Chichester, West Sussex, pp. 21–36. 10.1002/9781118567166.CH2

Masuda, T., Fujita, Y., 2008. Regulation and evolution of chlorophyll metabolism. *Photochem. Photobiol. Sci.* 7, 1131–1149. 10.1039/B807210H

Mata, T.M., Martins, A.A., Caetano, N.S., 2010. Microalgae for biodiesel production and other applications: A review. *Renew. Sustain. Energy Rev.* 14, 217–232. 10.1016/j.rser.2009.07.020

Mata, T.M., Martins, A.A., Oliveira, O., Oliveira, S., Mendes, A.M., Caetano, N.S., 2016. Lipid content and productivity of *Arthrospira platensis* and *Chlorella vulgaris* under mixotrophic conditions and salinity stress. *Chem. Eng. Trans.* 49, 187–192. 10.3303/CET1649032

Mehariya, S., Goswami, R.K., Karthikeysan, O.P., Verma, P., 2021. Microalgae for high-value products: A way towards green nutraceutical and pharmaceutical compounds. *Chemosphere* 280, 130553. 10.1016/j.chemosphere.2021.130553

Mitchell, R., Goacher, P., 2017. *D-Factory: The microalgae biorefinery | Final report | Economic assessment of Dunaliella-based algae biorefinery concepts*. European Union's Seventh Framework Programme project under grant agreement no. 613870 CORDIS | European Commission. London.

Molino, A., Iovine, A., Casella, P., Mehariya, S., Chianese, S., Cerbone, A., Rimauro, J., Musmarra, D., 2018. Microalgae characterization for consolidated and new application in human food, animal feed and nutraceuticals. *Int. J. Environ. Res. Public Heal. 2018.* 15, 2436. 10.3390/IJERPH15112436

Morais Junior, W.G., Gorgich, M., Corrêa, P.S., Martins, A.A., Mata, T.M., Caetano, N.S., 2020. Microalgae for biotechnological applications: Cultivation, harvesting and biomass processing. *Aquaculture* 528, 735562. 10.1016/j.aquaculture.2020.735562

Morocho-Jácome, A.L., Ruscinc, N., Martinez, R.M., de Carvalho, J.C.M., Santos de Almeida, T., Rosado, C., Costa, J.G., Velasco, M.V.R., Baby, A.R., 2020.

(Bio)Technological aspects of microalgae pigments for cosmetics. *Appl. Microbiol. Biotechnol. 2020.* 104, 9513–9522. 10.1007/S00253-020-10936-X

Mourelle, M.L., Gómez, C.P., Legido, J.L., 2017. The potential use of marine microalgae and cyanobacteria in cosmetics and thalassotherapy. *Cosmetics* 4(4), 46. 10.3390/cosmetics4040046

Müh, F., Zouni, A., 2020. Structural basis of light-harvesting in the photosystem II core complex. *Protein Sci.* 29, 1090–1119. 10.1002/PRO.3841

Mulders, K.J.M., Janssen, J.H., Martens, D.E., Wijffels, R.H., Lamers, P.P., 2014a. Effect of biomass concentration on secondary carotenoids and triacylglycerol (TAG) accumulation in nitrogen-depleted *Chlorella zofingiensis. Algal Res.* 6, 8–16. 10.1016/j.algal.2014.08.006

Mulders, K.J.M., Lamers, P.P., Martens, D.E., Wijffels, R.H., 2014b. Phototrophic pigment production with microalgae: Biological constraints and opportunities. *J. Phycol.* 50, 229–242. /10.1111/jpy.12173

Nicoletti, M., 2016. Microalgae nutraceuticals. *Foods* 5, 1–13. 10.3390/foods5030054

Novoveská, L., Ross, M.E., Stanley, M.S., Pradelles, R., Wasiolek, V., Sassi, J.-F., 2019. Microalgal carotenoids: A review of production, current markets, regulations, and future direction. *Mar. Drugs 2019.* 17, 640. 10.3390/MD17110640

Nowruzi, B., Sarvari, G., Blanco, S., 2020. The cosmetic application of cyanobacterial secondary metabolites. *Algal Res.* 49, 101959. 10.1016/j.algal.2020.101959

Nwoba, E.G., Parlevliet, D.A., Laird, D.W., Alameh, K., Moheimani, N.R., 2019. Light management technologies for increasing algal photobioreactor efficiency. *Algal Res.* 39, 101433. 10.1016/J.ALGAL.2019.101433

Ogbonna, J.C., Masui, H., Tanaka, H., 1997. Sequential heterotrophic/autotrophic cultivation – An efficient method of producing *Chlorella* biomass for health food and animal feed. *J. Appl. Phycol.* 9, 359–366. 10.1023/A:1007981930676

Orona-Navar, A., Aguilar-Hernández, I., Nigam, K.D.P., Cerdán-Pasarán, A., Ornelas-Soto, N., 2021. Alternative sources of natural pigments for dye-sensitized solar cells: Algae, cyanobacteria, bacteria, archaea and fungi. *J. Biotechnol.* 332, 29–53. 10.1016/J.JBIOTEC.2021.03.013

Ota, S., Morita, A., Ohnuki, S., Hirata, A., Sekida, S., Okuda, K., Ohya, Y., Kawano, S., 2018. Carotenoid dynamics and lipid droplet containing astaxanthin in response to light in the green alga *Haematococcus pluvialis. Sci. Reports.* 8, 1–10. 10.1038/s41598-018-23854-w

Pagels, F., Guedes, A.C., Amaro, H.M., Kijjoa, A., Vasconcelos, V., 2019. Phycobiliproteins from cyanobacteria: Chemistry and biotechnological applications. *Biotechnol. Adv.* 37, 422–443. 10.1016/J.BIOTECHADV.2019.02.010

Pagels, F., Salvaterra, D., Amaro, H.M., Guedes, A.C., 2020. Pigments from microalgae, in: Jacob-Lopes, E., Maroneze, M.M., Queiroz, M.I., and Zepka, L.Q. (Eds.), *Handbook of Microalgae-Based Processes and Products.* Elsevier Inc. (Academic Press), London, UK. 10.1016/b978-0-12-818536-0.00018-x

Pareek, S., Sagar, N.A., Sharma, S., Kumar, V., Agarwal, T., González-Aguilar, G.A., Yahia, E.M., 2017. Chlorophylls: Chemistry and biological functions. *Fruit and Vegetable Phytochemicals: Chemistry and Human Health*, second ed., pp. 269–284. 10.1002/9781119158042.CH14

Pathak, J., Pandey, A., Maurya, P.K., Rajneesh, R., Sinha, R.P., Singh, S.P., 2020. Cyanobacterial secondary metabolite scytonemin: A potential photoprotective and pharmaceutical compound. *Proc. Natl. Acad. Sci. India Sect. B - Biol. Sci.* 90, 467–481. 10.1007/s40011-019-01134-5

Pereira, A.G., Otero, P., Echave, J., Carreira-Casais, A., Chamorro, F., Collazo, N., Jaboui, A., Lourenço-Lopes, C., Simal-Gandara, J., Prieto, M.A., 2021. Xanthophylls from the sea: Algae as source of bioactive carotenoids. *Mar. Drugs.* 19, 188. 10.3390/MD19040188

Pick, U., Zarka, A., Boussiba, S., Davidi, L., 2018. A hypothesis about the origin of carotenoid lipid droplets in the green algae *Dunaliella* and *Haematococcus*. *Planta.* 249, 31–47. 10.1007/S00425-018-3050-3

Pourhajibagher, M., Bahador, A., 2021. Antimicrobial properties, anti-virulence activities, and physico-mechanical characteristics of orthodontic adhesive containing C-phycocyanin: a promising application of natural products. *Folia Med. (Plovdiv).* 63, 113–121. 10.3897/folmed.63.e52756

Préat, N., Taelman, S.E., De Meester, S., Allais, F., Dewulf, J., 2020. Identification of microalgae biorefinery scenarios and development of mass and energy balance flow-sheets. *Algal Res.* 45, 101737. 10.1016/J.ALGAL.2019.101737

Rabbani, S., Beyer, P., Lintig, J. V., Hugueney, P., Kleinig, H., 1998. Induced β-carotene synthesis driven by triacylglycerol deposition in the unicellular Alga *Dunaliella bardawil*. *Plant Physiol.* 116, 1239–1248. 10.1104/pp.116.4.1239

Raja, R., Hemaiswarya, S., Rengasamy, R., 2007. Exploitation of *Dunaliella* for β-carotene production. *Appl. Microbiol. Biotechnol.* 74(3), 517–523. 10.1007/S00253-006-0777-8

Rammuni, M.N., Ariyadasa, T.U., Nimarshana, P.H.V., Attalage, R.A., 2019. Comparative assessment on the extraction of carotenoids from microalgal sources: Astaxanthin from *H. pluvialis* and β-carotene from *D. salina*. *Food Chem.* 277, 128–134. 10.1016/J.FOODCHEM.2018.10.066

Ramos-Romero, S., Torrella, J.R., Pagès, T., Viscor, G., Torres, J.L., 2021. Edible microalgae and their bioactive compounds in the prevention and treatment of metabolic alterations. *Nutr. 2021.* 13, 563. 10.3390/NU13020563

Rastogi, R.P., Incharoensakdi, A., 2014. UV radiation-induced biosynthesis, stability and antioxidant activity of mycosporine-like amino acids (MAAs) in a unicellular cyanobacterium Gloeocapsa sp. CU2556. *J. Photochem. Photobiol. B Biol.* 130, 287–292. 10.1016/J.JPHOTOBIOL.2013.12.001

Rastogi, R.P., Madamwar, D., Nakamoto, H., Incharoensakdi, A., 2020. Resilience and self-regulation processes of microalgae under UV radiation stress. *J. Photochem. Photobiol. C Photochem. Rev.* 43, 100322. 10.1016/j.jphotochemrev.2019.100322

Rastogi, R.P., Sinha, R.P., Incharoensakdi, A., 2013. Partial characterization, UV-induction and photoprotective function of sunscreen pigment, scytonemin from *Rivularia* sp. HKAR-4. *Chemosphere* 93, 1874–1878. 10.1016/J.CHEMOSPHERE.2013.06.057

Ratha, S.K., Renuka, N., Rawat, I., Bux, F., 2021. Prospective options of algae-derived nutraceuticals as supplements to combat COVID-19 and human coronavirus diseases. *Nutrition* 83, 111089. 10.1016/j.nut.2020.111089

Rochaix, J.-D., 2002. The three genomes of *Chlamydomonas*. *Photosynth. Res.* 73, 285–293. 10.1023/A:1020484105601

Rodrigues, R.D.P., Silva, A.S. e., Carlos, T.A.V., Bastos, A.K.P., de Santiago-Aguiar, R.S., Rocha, M.V.P., 2020. Application of protic ionic liquids in the microwave-assisted extraction of phycobiliproteins from *Arthrospira platensis* with antioxidant activity. *Sep. Purif. Technol.* 252, 117448. 10.1016/j.seppur.2020.117448

Román, R.B., Alvárez-Pez, J.M., Fernández, F.G.A., Grima, E.M., 2002. Recovery of pure b-phycoerythrin from the microalga *Porphyridium cruentum*. *J. Biotechnol.* 93, 73–85. 10.1016/S0168-1656(01)00385-6

Rosales-Mendoza, S., García-Silva, I., González-Ortega, O., Sandoval-Vargas, J.M., Malla, A., Vimolmangkang, S., 2020. The potential of algal biotechnology to produce antiviral compounds and biopharmaceuticals. *Molecules* 25, 4049. 10.3390/MOLECULES25184049

Rubio, F.C., Camacho, F.G., Sevilla, J.M.F., Chisti, Y., Grima, E.M., 2003. A mechanistic model of photosynthesis in microalgae. *Biotechnol. Bioeng.* 81, 459–473. 10.1002/BIT.10492

Sahin, S.C., 2018. The potential of *Arthrospira platensis* extract as a tyrosinase inhibitor for pharmaceutical or cosmetic applications. *South African J. Bot.* 119, 236–243. 10.1016/j.sajb.2018.09.004

Sanjeewa, K.K.A., Fernando, I.P.S., Samarakoon, K.W., Lakmal, H.H.C., Kim, E.A., Kwon, O.N., Dilshara, M.G., Lee, J.B., Jeon, Y.J., 2016. Anti-inflammatory and anti-cancer activities of sterol rich fraction of cultured marine microalga *Nannochloropsis oculata*. *Algae* 31, 277–287. 10.4490/algae.2016.31.6.29

Santos-Sánchez, N.F., Salas-Coronado, R., Villanueva-Cañongo, C., Hernández-Carlos, B., 2019. Antioxidant compounds and their antioxidant mechanism, in: Shalaby, E. (Ed.), *Antioxidants*. IntechOpen. 10.5772/INTECHOPEN.85270

Sarkar, D., Shimizu, K., 2015. An overview on biofuel and biochemical production by photosynthetic microorganisms with understanding of the metabolism and by metabolic engineering together with efficient cultivation and downstream processing. *Bioresour. Bioprocess* 2, 1–19. 10.1186/S40643-015-0045-9

Sathasivam, R., Ki, J.-S., 2018. A review of the biological activities of microalgal carotenoids and their potential use in healthcare and cosmetic industries. *Mar. Drugs.* 16, 26. 10.3390/MD16010026

Sathasivam, R., Radhakrishnan, R., Hashem, A., Abd_Allah, E.F., 2019. Microalgae metabolites: A rich source for food and medicine. *Saudi J. Biol. Sci.* 26, 709–722. 10.1016/j.sjbs.2017.11.003

Schliep, M., Cavigliasso, G., Quinnell, R.G., Stranger, R., Larkum, A.W.D., 2013. Formyl group modification of chlorophyll a: A major evolutionary mechanism in oxygenic photosynthesis. *Plant. Cell Environ.* 36, 521–527. 10.1111/PCE.12000

Schoefs, B., 2002. Chlorophyll and carotenoid analysis in food products. Properties of the pigments and methods of analysis. *Trends Food Sci. Technol.* 13, 361–371. 10.1016/S0924-2244(02)00182-6

Seon, G., Kim, H.S., Cho, J.M., Kim, M., Park, W.-K., Chang, Y.K., 2020. Effect of post-treatment process of microalgal hydrolysate on bioethanol production. *Sci. Reports.* 10, 1–12. 10.1038/s41598-020-73816-4

Seyfabadi, J., Ramezanpour, Z., Khoeyi, Z.A., 2011. Protein, fatty acid, and pigment content of *Chlorella vulgaris* under different light regimes. *J. Appl. Phycol.* 23, 721–726. 10.1007/s10811-010-9569-8

Shah, M.M.R., Liang, Y., Cheng, J.J., Daroch, M., 2016. Astaxanthin-producing green microalga *Haematococcus pluvialis*: From single cell to high value commercial products. *Front. Plant Sci.* 7, 531. 10.3389/FPLS.2016.00531

Shen, G., Canniffe, D.P., Ho, M.-Y., Kurashov, V., van der Est, A., Golbeck, J.H., Bryant, D.A., 2019. Characterization of chlorophyll f synthase heterologously produced in *Synechococcus* sp. PCC 7002. *Photosynth. Res.* 140, 77–92. 10.1007/S11120-018-00610-9

Shi, T.Q., Wang, L.R., Zhang, Z.X., Sun, X.M., Huang, H., 2020. Stresses as first-line tools for enhancing lipid and carotenoid production in microalgae. *Front. Bioeng. Biotechnol.* 8, 610. 10.3389/fbioe.2020.00610

Silva, J.C. da, Lombardi, A.T., 2020. Chlorophylls in microalgae: Occurrence, distribution, and biosynthesis, In: *Pigments from Microalgae Handbook*. Springer, Cham, pp. 1–18. 10.1007/978-3-030-50971-2_1

Soni, R.A., Sudhakar, K., Rana, R.S., 2017. Spirulina – From growth to nutritional product: A review. *Trends Food Sci. Technol.* 69, 157–171. 10.1016/j.tifs.2017.09.010

Sørensen, L., Hantke, A., Eriksen, N.T., 2013. Purification of the photosynthetic pigment C-phycocyanin from heterotrophic *Galdieria sulphuraria*. *J. Sci. Food Agric.* 93, 2933–2938. 10.1002/jsfa.6116

Stahl, W., Sies, H., 2012. β-Carotene and other carotenoids in protection from sunlight. *Am. J. Clin. Nutr.* 96, 1179S–1184S. 10.3945/AJCN.112.034819

Stirbet, A., Lazár, D., Guo, Y., Govindjee, G., 2020. Photosynthesis: Basics, history and modelling. *Ann. Bot.* 126, 511–537. 10.1093/AOB/MCZ171

Stoyneva-Gärtner, M., Uzunov, B., Gärtner, G., 2020. Enigmatic microalgae from aero-terrestrial and extreme habitats in cosmetics: The potential of the untapped natural sources. *Cosmetics* 7, 1–22. 10.3390/cosmetics7020027

Sun, L., Wang, S., Chen, L., Gong, X., 2003. Promising fluorescent probes from phycobiliproteins. *IEEE J. Sel. Top Quantum Electron.* 9, 177–188. 10.1109/JSTQE.2003.812499

Takaichi, S., 2020. Carotenoids in phototrophic microalgae: Distributions and biosynthesis, in: *Pigments from Microalgae Handbook.* Springer, Cham, pp. 19–41. 10.1007/978-3-030-50971-2_2

Tejada Carbajal, E.M., Martínez Hernández, E., Fernández Linares, L., Novelo Maldonado, E., Limas Ballesteros, R., 2020. Techno-economic analysis of *Scenedesmus dimorphus* microalgae biorefinery scenarios for biodiesel production and glycerol valorization. *Bioresour. Technol. Reports.* 12, 100605. 10.1016/J.BITEB.2020.100605

Torres-Tiji, Y., Fields, F.J., Mayfield, S.P., 2020. Microalgae as a future food source. *Biotechnol. Adv.* 41, 107536. 10.1016/J.BIOTECHADV.2020.107536

Tudor, C., Gherasim, E.C., Dulf, F.V., Pintea, A., 2021. In vitro bioaccessibility of macular xanthophylls from commercial microalgal powders of *Arthrospira platensis* and *Chlorella pyrenoidosa.* *Food Sci. Nutr.* 9, 1896–1906. 10.1002/FSN3.2150

Urreta, I., Ikaran, Z., Janices, I., Ibañez, E., Castro-Puyana, M., Castañón, S., Suárez-Alvarez, S., 2014. Revalorization of *Neochloris oleoabundans* biomass as source of biodiesel by concurrent production of lipids and carotenoids. *Algal Res.* 5, 16–22. 10.1016/j.algal.2014.05.001

US Food & Drug Administration, 2020. "Cosmeceutical" | FDA [WWW Document]. *US Food Drug Adm.* URL https://www.fda.gov/cosmetics/cosmetics-labeling-claims/cosmeceutical (accessed 8.15.21).

Vaquero, I., Mogedas, B., Ruiz-Domínguez, M.C., Vega, J.M., Vílchez, C., 2014. Light-mediated lutein enrichment of an acid environment microalga. *Algal Res.* 6, 70–77. 10.1016/j.algal.2014.09.005

Vega, J., Bonomi-Barufi, J., Gómez-Pinchetti, J.L., Figueroa, F.L., 2020. Cyanobacteria and red macroalgae as potential sources of antioxidants and UV radiation-absorbing compounds for cosmeceutical applications. *Mar. Drugs.* 18(12), 659. 10.3390/md18120659

VentureRadar, 2021. Top Microalgae companies | VentureRadar [WWW Document]. https://www.ventureradar.com/keyword/Microalgae (accessed 8.19.21).

Vonshak, A., Richmond, A., 1988. Mass production of the blue-green alga *Spirulina*: An overview. *Biomass.* 15, 233–247. 10.1016/0144-4565(88)90059-5

Wada, N., Sakamoto, T., Matsugo, S., 2013. Multiple roles of photosynthetic and sunscreen pigments in cyanobacteria focusing on the oxidative stress. *Metabolites.* 3, 463–483. 10.3390/metabo3020463

Wan, M., Zhang, Z., Wang, J., Huang, J., Fan, J., Yu, A., Wang, W., Li, Y., 2015. Sequential heterotrophy–dilution–photoinduction cultivation of *Haematococcus pluvialis* for efficient production of astaxanthin. *Bioresour. Technol.* 198, 557–563. 10.1016/J.BIORTECH.2015.09.031

Wang, C., Zhao, Y., Wang, L., Pan, S., Liu, Y., Li, S., Wang, D., 2021. C-phycocyanin mitigates cognitive impairment in doxorubicin-induced chemobrain: impact on neuroinflammation, oxidative stress, and brain mitochondrial and synaptic alterations. *Neurochem. Res.* 46, 149–158. 10.1007/s11064-020-03164-2

Wang, P., Grimm, B., 2021. Connecting chlorophyll metabolism with accumulation of the photosynthetic apparatus. *Trends Plant Sci.* 26, 484–495. 10.1016/J.TPLANTS.2020.12.005

Willows, R.D., 2007. Chlorophyll Synthesis, in: Wise, R.R., Hoober, J.K. (Eds.), *The Structure and Function of Plastids. Advances in Photosynthesis and Respiration*, vol 23, Springer, Dordrecht, pp. 295–313. 10.1007/978-1-4020-4061-0_15

Xu, Y., Ibrahim, I.M., Harvey, P.J., 2016. The influence of photoperiod and light intensity on the growth and photosynthesis of *Dunaliella salina* (chlorophyta) CCAP 19/30. *Plant Physiol. Biochem.* 106, 305–315. 10.1016/J.PLAPHY.2016.05.021

Yaakob, M.A., Mohamed, R.M.S.R., Al-Gheethi, A., Aswathnarayana Gokare, R., Ambati, R.R., 2021. Influence of nitrogen and phosphorus on microalgal growth, biomass, lipid, and fatty acid production: An overview. *Cells.* 10, 393. 10.3390/cells10020393

Yarkent, Ç., Gürlek, C., Oncel, S.S., 2020. Potential of microalgal compounds in trending natural cosmetics: A review. *Sustain. Chem. Pharm.* 17, 100304. 10.1016/j.scp.2020.1 00304

Yarnold, J.E., 2016. *Photosynthesis of Microalgae in Outdoor Mass Cultures and Modelling its Effects on Biomass Productivity for Fuels, Feeds and Chemicals*. University of Queensland, Australia.

Yeager, D.G., Lim, H.W., 2019. What's new in photoprotection: A review of new concepts and controversies. *Dermatol. Clin.* 37, 149–157. 10.1016/J.DET.2018.11.003

Zhang, J., Liu, L., Ren, Y., Chen, F., 2019. Characterization of exopolysaccharides produced by microalgae with antitumor activity on human colon cancer cells. *Int. J. Biol. Macromol.* 128, 761–767. 10.1016/j.ijbiomac.2019.02.009

Zhang, Y., Shi, M., Mao, X., Kou, Y., Liu, J., 2019. Time-resolved carotenoid profiling and transcriptomic analysis reveal mechanism of carotenogenesis for astaxanthin synthesis in the oleaginous green alga *Chromochloris zofingiensis. Biotechnol. Biofuels* 12, 1–19. 10.1186/S13068-019-1626-1

Zhao, Y., Yue, C., Ding, W., Li, T., Xu, J.W., Zhao, P., Ma, H., Yu, X., 2018. Butylated hydroxytoluene induces astaxanthin and lipid production in *Haematococcus pluvialis* under high-light and nitrogen-deficiency conditions. *Bioresour. Technol.* 266, 315–321. 10.1016/j.biortech.2018.06.111

Zilinskas, B.A., Greenwald, L.S., 1986. Phycobilisome structure and function. *Photosynth. Res.* 10, 7–35. 10.1007/BF00024183

6 Microalgae for Animal and Fish Feed

Margarida Costa, Joana G. Fonseca, and Joana L. Silva
Allmicroalgae – Natural Products, Pataias, Portugal

Jean-Yves Berthon
Greentech SA, Saint Beauzire, France

Edith Filaire
University Clermont Auvergne, UNH (Human Nutrition Unity), ECREIN Team, Clermont-Ferrand, France

CONTENTS

6.1 Introduction: Background and Driving Forces..177
6.2 Definition of Microalgae...178
6.3 Microalgal Biochemical and Nutritional Value 180
    6.3.1 Proteins 182
    6.3.2 Lipids...182
    6.3.3 Carbohydrates..183
    6.3.4 Pigments 184
    6.3.5 Vitamins...185
6.4 Industrial Applications of Microalgae in the Functional Aquafeed Industry..185
6.5 Industrial Applications of Microalgae in the Functional Terrestrial Feed Industry 189
6.6 Commercialized Formulated Aqua and Terrestrial Feed...........................191
6.7 Challenges and Future Perspectives 203
6.8 Summary..203
References...204

6.1 INTRODUCTION: BACKGROUND AND DRIVING FORCES

In applied phycology, the term *microalgae* refers to microscopic algae and photosynthetic bacteria, such as cyanobacteria. Both classes are considered a potential source of energy, fuel, food, and many other interesting commercial products. Microalgae are the primary producers found in several ecosystems, both marine and freshwater. Seventy-two thousand five hundred species and 16 classes have been listed. They have a unique biochemical composition and produce primary and secondary metabolites with interesting biological properties. In particular, these compounds have shown potential as antioxidant, anti-inflammatory, and anti-cancer agents. Microalgae have been used for centuries as human food or animal feed. However, large-scale industrial cultivation is relatively recent technology. Recent articles review state of the art biotechnological production and the use of microalgae. However, this chapter aims to provide an overview of microalgae properties and to examine its potential in use as a feed product.

6.2 DEFINITION OF MICROALGAE

The estimated number of microalgae species is between 30,000–1,000,000. While there is vast application potential due to their unique metabolism, chemical diversity, and ability to tolerate wastewater for culture, certain limitations and challenges remain for the commercial development of microalgae and microalgae-related products. The most complex and crucial issues are related to science, technology, legislation, administration, and marketing gaps.

The concept of *algae* is associated with a diversity of micro- and macro-organisms that can proliferate through photosynthesis (Patras et al., 2019). Both macro- and microalgae can grow in freshwater and marine environments. Microalgae include eukaryotic microalgae and prokaryotic cyanobacteria (Sathasivam et al., 2019), which are ubiquitous and can be found in the most diverse environments, including extreme ecosystems experiencing high or low temperatures, light intensity, pH, and salinity (Martínez-Francés and Escudero-Oñate, 2018).

Eucaryotic microalgae have traditionally been classified according to their pigments. The current classification systems consider other criteria, including the chemical nature of photosynthetic storage products, the organization of photosynthetic membranes, and other morphological features (Remize et al., 2021). Blue-green algae (*Cyanophyceae*), green algae (*Chlorophyceae*), *Bacillariophyceae* (including diatoms), and *Chrysophyceae* (including golden algae) have been described as the most abundant microalgal phyla (García et al., 2017).

Environmental diversity results in the synthesis and secretion of primary and secondary metabolites, which have industrial applications and principal health benefits. (De Morais et al., 2015). Microalgal biomass has the potential to be used as a feedstock for an alternative and innovative source of nutrients, in cosmetics, pharmaceuticals, and nutraceuticals (Rumin et al., 2020). It is important to note that the interest of all those markets in microalgae has been bolstered by the growing concerns concerning environmental sustainability and regulatory issues linked to synthetic chemicals found in food and feed and in beauty products and health care.

Microalgae cultivation can take place in three principal ways, namely photoautotrophic, heterotrophic, and mixotrophic. Most microalgae are designated as photoautotrophs once they can generate energy directly from light, CO_2, and water, typically through photosynthesis, producing organic compounds. Autotrophic microalgae use inorganic carbon forms, such as CO_2 or bicarbonate (HCO_3^-) (Borowitzka, 2018). Additionally, some microalgae species are heterotrophic using organic substrates, such as acetate or glucose, both as energy and carbon sources, in the absence of light (Perez-Garcia et al., 2015). According to the conditions, several of those organisms have a plastic metabolism that allows switching between auto- and heterotrophy. When the microalgae combine the two metabolisms, they are defined as mixotrophs (Pires, 2015). Those organisms can metabolize, simulta- neously, both carbon forms in the presence of light.

Compared with autotrophic microalgal growth, heterotrophic and mixotrophic cultivation are more productive in terms of biomass yield (Pires, 2015), as het- erotrophic production, the cultivation mode achieving higher productivities. Besides higher productivity, heterotrophy reveals many other advantages. Heterotrophic cultivation occurs in closed fermenters, significantly reducing the crop area used and allowing strict parameter control (Hu et al., 2018), such as pH, temperature, oxygen levels, and nutrients. The contaminations and consequent culture collapse are, moreover, better avoided due to this strict control and sterilization. Therefore, heterotrophic cultivation is economically advantageous once the cost per produced biomass is significantly lower (Hu et al., 2018).

Chen (1996) and Borowitzka (1999) had long recognized that the cell density of a heterotrophic culture could vary between 20 and 100 g.L^{-1} (Chen, 1996; Borowitzka, 1999). However, cell densities of 174.5 and 255 g.L^{-1} had already been reported for *Chlorella vulgaris* (Barros et al., 2019) and *Chlorella protothecoides* (Ghidossi et al., 2017), respectively. Compared to autotrophic cultivation, heterotrophy had allowed enhancing *C. vulgaris* final biomass up to 137.4 times (Barros et al., 2019). However, the chlorophyl content had demonstrated to be lower when *C. vulgaris* was cultivated heterotrophically, increasing from 5 mg g^{-1} DW to 24 mg g^{-1} DW when the culture was exposed to autotrophic conditions (Barros et al., 2019). The same changes in- duced a 70% increase in protein content in this microalga (Barros et al., 2019). Therefore, the biochemical composition of the final biomass achieved depends on the trophic state of the cultures. The lipid content of a heterotrophic *C. protothecoides* culture was demonstrated to be 55.2% (Xu et al., 2006), compared to 37.5% in the same autotrophic species (Krzemińska et al., 2015). Similar patterns were reported for different microalgae: autotrophic *C. sorokiniana* presented 12–18% lipid content, compared to 24–31% in the heterotrophic biomass (Rosenberg et al., 2014); auto- trophic *Chlorella minutissima* had shown 20.2% lipids, compared to 36.2% of het- erotrophic biomass (Dubey et al., 2015).

Several limitations are associated besides hetero- and mixotrophic cultivation of microalgae, undoubtedly offering higher productivities and cell densities. The in- ability to produce light-induced metabolites, primarily related to antioxidant ac- tivities, represents a great constraint. Besides, there is also the indirect need for arable land to produce the organic carbon sources used during those cultivations (Borowitzka, 1999), suppressing one of the most significant microalgae cultivation advantages. It is also important to note that only a limited number of microalgae species can metabolize organic carbon sources (Perez-Garcia et al., 2015).

Generally speaking, selecting a suitable microalgae cultivation system depends on the species and added-value compounds intended, as well as their final appli- cation. Therefore, several parameters should be considered for cultivation, such as light efficiency, pH and temperature control, species/biocompounds productivity, hydrodynamic stress, the need for an axenic culture and harvesting, and ease of scale-up (Guedes and Malcata, 2012). The primary decision when cultivating microalgae is whether to use closed or open systems. The open systems require low investment and operational costs, and they are also easy to build and operate. However, contamination issues are more common and difficult to control and high cell densities are not reached due to self-shading effects (Wu et al., 2017). Most of the species commercialized and produced industrially on open systems, like *Spirulina* sp. and *Dunaliella salina*, grow in extreme environments, suitable for cultivation in open pounds avoiding the most common contaminations.

On the other hand, closed systems have shown to be more efficient than open ones, allowing better control of the culture parameters. Close bioreactors include tubular configurations, flat plate reactors, and fermenters (Perez-Garcia et al., 2015). The first two configurations are ideal for the autotrophic cultivation of microalgae and the fermenters are used for heterotrophic growth.

Despite a wide range of advantages of using closed systems for microalgae cultivations, some constraints also arise. Those systems are commonly associated with fouling and overheating, cleaning is more time- and resource-consuming and oxygen accumulation (in the autotrophic reactors) results in a decrease in culture growth. Additionally, the closed systems require relatively high investment and operational costs.

Thus, there are many opportunities to cultivate microalgae, which are dependent on the species aimed to produce and the application of the final biomass. Besides this, the investment and operational costs are key factors when selecting the trophic mode of cultivation and the suitable bioreactor.

6.3 MICROALGAL BIOCHEMICAL AND NUTRITIONAL VALUE

The bioactive compounds produced by microalgae and the cellular content are strain-specific and respond to biotic and abiotic factors, e.g., growth phase, light intensity, etc. (Lafarga, 2020). Therefore, these factors should be considered when cultivating microalgae to manipulate their cell composition and biomass productivity, cell metabolism pathway, and final bioactivity (Matos et al., 2017). Most metabolites accumulate intracellularly, even if, in the case of exometabolites, their excretion to the culture medium takes place.

Microalgae compounds can be grouped into proteins/enzymes, acids, pigments, and vitamins, derived from primary metabolism. Secondary bioactive compounds can also be synthesized (Levasseur et al., 2020). Figure 6.1 summarizes the nutritional components that can be acquired from microalgae biomass.

Even though the biochemical composition appears species-, and even strain-dependent, protein is typically the principal organic constituent (12–35%), usually followed by lipids (7–23%) and carbohydrates (5–23%) (Vieira et al., 2020). It is important to note that these proportions may change depending on the culturing conditions (Vieira et al., 2020).

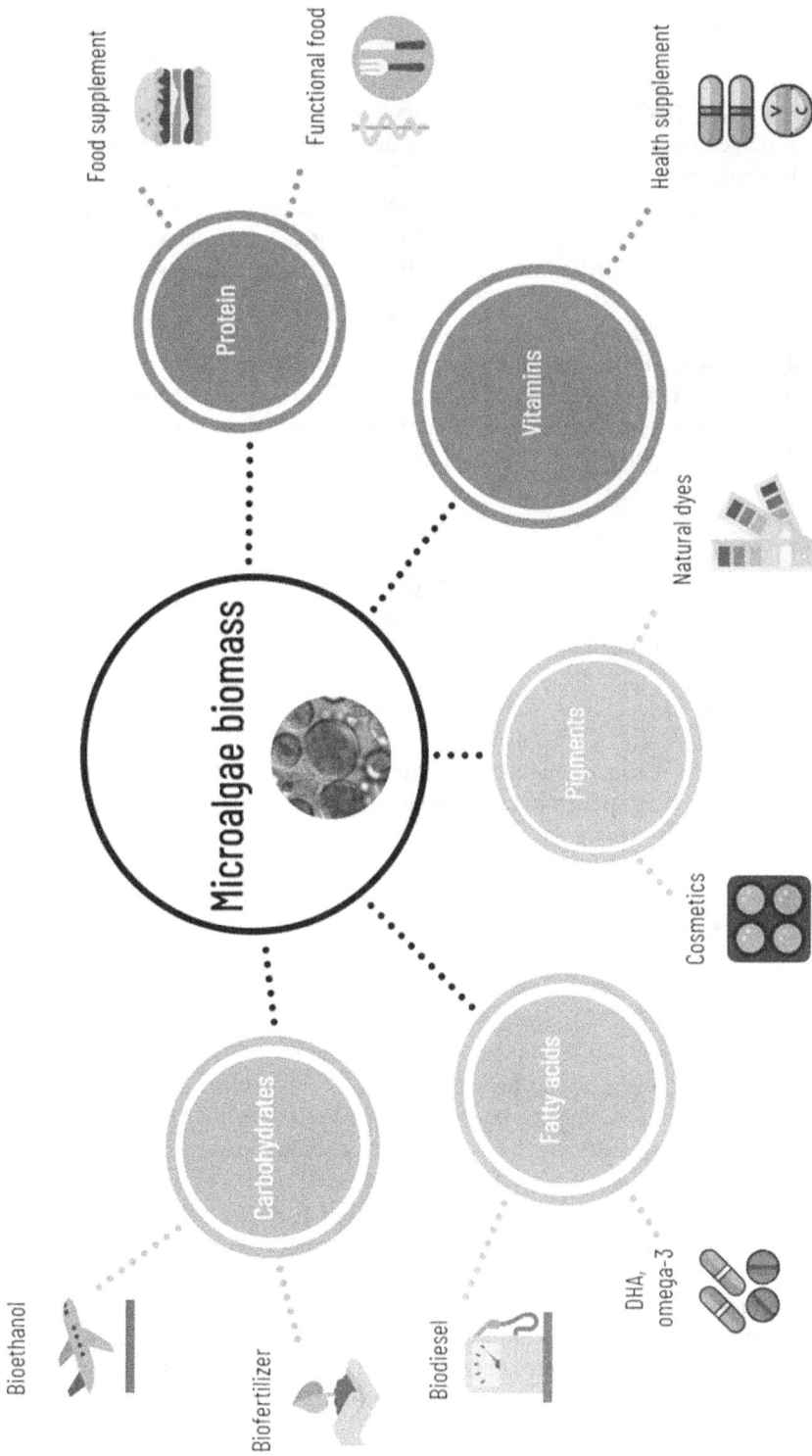

FIGURE 6.1 Algal biomass and their applications.

6.3.1 PROTEINS

Microalgae have been recognized as a protein source since the 1950s (Venkataraman, 1997). Depending on the species and culturing factors, the protein content ranges from 42% to over 70% on a dry weight basis, reported for specific cyanobacteria species (Barkia et al., 2019).

In terms of quality, microalgae have a rich and varied amino acid composition, containing all the essential amino acids (EEAs) humans are incapable of synthesizing (Barkia et al., 2019). In addition, this group of organisms contains non-essential amino acids (NEAAs), such as arginine, aspartic, proline, glutamic acid, glycine, cysteine. NEAAs of *C. vulga*ris and *Haematococcus pluvialis* appear around 51.03 and 48.50% of the total amino acid profile, respectively. Biological properties in immunity, gene expression regulation, redox homeostasis, and cell communication that can be used in the nutritional, cosmetics, and pharmaceuticals fields have been reported for those compounds (Jacob-Lopes et al., 2019).

6.3.2 LIPIDS

Particular attention has been given recently to microalgae lipids. These compounds can make up to 74% of microalgae's total biochemical composition (Sathasivam et al., 2019). Polyunsaturated fatty acids (PUFAs) have recently been the subject of attention because of their health benefits. Based on 12 to 24 carbon atom fatty acids, these molecules include the n-3 and n-6 polyunsaturated fatty acid families. Among PUFAs, the bioactivity of docosahexaenoic acid (DHA, 22:6) and eicosapentaenoic acid (EPA, 20:5) is usually highlighted. For example, brain function, more specifically, short- and long-term memory, can be improved by DHA. This molecule also has a positive effect on cognitive decline, depression, bipolar symptoms, and mood swings. EPA and/or DHA have also been associated with a preventive role in cardiovascular diseases (CVDs), and cancer. Generally speaking, the bioactive lipid properties are expected to be of central interest in developing novel ingredients or compounds for various sectors, including the feed and food industry, with high commercial value (Calder, 2006).

Fish such as salmon, tuna, mackerel, sardines, anchovies, herring, or pollock is a good source of EPA and DHA for human consumption. However, marine fish cannot give the global demand for n-3 fatty acids due to their limited stocks and some undesirable contaminants, encouraging industries to find novel alternative sources (Oliver et al., 2020). In this sense, the feasibility of using microalgae as an alternative has been long-explored (Remize et al., 2021). However, it is essential to note that the content of n-3 PUFAs varies within species according to environmental factors, given that it is naturally relatively low (Sahu et al., 2013). Thus, cell EPA and DHA productivity improvement strategies have been investigated, including manipulating culturing conditions, partial or complete deprivation of nitrogen, or using two-phase culturing approaches (Wang et al., 2019). Acting directly on the microalgae metabolism and applying the heterotrophic mode is another way to stimulate lipid production (Remize et al., 2021). Thus, under optimal cultivation conditions, several species, especially those belonging to the genera *Chlorella*, *Nannochloropsis*, and *Dunaliella* are described as showing exceptionally high amounts of lipids in their cell mass.

Other essential fatty acids such as alpha-linolenic acid (ALA), gamma-linolenic acid (GLA), linoleic acid (LA), and arachidonic acid (ARA) are produced by microalgae. They can be applied by the feed and food markets. They have also proven to be active on healing and wound repair, as well as having anti-microbial properties (Choopani et al., 2016).

6.3.3 CARBOHYDRATES

Besides lipids, microalgal biomass can be a source of carbohydrates with great industrial applications. They are represented by poly- or oligosaccharides present in vacuoles and cell walls. Cellulose, and amylose are some main polysaccharides present in microalgae, which also excrete exopolysaccharides (EPS). These carbohydrates have different biological roles linked to 1) energy reservation, 2) formation of the cell wall and 3) cell communication.

Tetraselmis sp., *Isochrysis* sp., *Porphyridium cruentum*, *Porphyridium purpureum*, *Chlorella* sp. and *Rhodella reticulata* are the main genera and species used for the production of polysaccharides. More particularly, *Chlorella* sp. is described as having high carbohydrate content, with *C. vulgaris* able to accumulate 37–55% dry weight (Illman et al., 2000). *Chlamydomonas reinhardtii* and *Scenedesmus obliquus* have been indicated as suitable for biofuel feedstock due to their 45–60% carbohydrate content. With appropriate cultivation conditions, such as under three-day nitrogen starvation, the production of microalgal carbohydrates can be improved. The biomass concentration and carbohydrate content of *S. obliquus* reached 4.96 g.L^{-1} and 51.8%, respectively under this stress condition (Ho et al., 2012).

The polysaccharides derived from *Porphyridium* sp. are described as having many techno-functional and biological properties (Delattre et al., 2016). *Porphyridium cruentum* is a fast and flexibly grown eukaryotic marine microalga. In addition, its cells are encapsulated in a mucilaginous sheath, representing a cell's structural polysaccharide. Some of these polysaccharides are excreted in this medium in the form of exopolysaccharides (EPSs). The exact physiological function of EPS remains ambiguous, even if the literature reports it as the prevention against desiccation and protection against many environmental conditions such as pH, temperature, salinity, and irradiance (Ramus, 1972). Many factors such as culture media, mode of cultivation (batch, continuous), illumination, and salinity impact the soluble EPSs (Ramus, 1972).

The most studied red microalgae exopolysaccharides were produced and extracted from *Rhodella reticulata*, *Porphyridium* sp., *P. aerugineum*, and *P. cruentum*. *Porphyridium cruentum* is described as producing a high amount of exopolysaccharides (0.1–0.7 g.L^{-1}), such as *Arthrospira platensis* (0.37 g.L^{-1}), *Botryococcus braunii* (0.25–1 g.L^{-1}) and *Dunaliella salina* (0.94 g.L^{-1}) (Trabelsi et al., 2013; Casadevall et al., 1985; Mishra and Jha, 2009).

6.3.4 PIGMENTS

Microalgae appear macroscopically in different colors due to the presence of pigments, which absorb visible light and have a fundamental role in photosynthetic cell metabolism. The main classes of pigments found in microalgae include chlorophylls (0.5–1.0% DW), carotenoids (0.1–0.2% DW, but some species can achieve up to 14%), and phycobiliproteins (up to 8% DW) (D'Alessandro and Filho, 2016). For several years, pigments have acquired crucial importance in various fields, namely the pharmaceutical, medical, and food fields (Santiago-Santos et al., 2004), because of their health benefits, including antioxidant, anti-cancer, and anti-inflammatory properties. They are also used in cosmetics because they can replace artificial colors (Rodrigues et al., 2015).

Chlorophylls are green pigments with polycyclic planar structures esterified by a phytol side chain. According to their structural features and wavelength absorption, different chlorophylls (a, b, c, d, e) have been identified in microalgae. Chlorophyll a is present in all photoautotrophic algae and the only found in Cyanobacteria and Rhodophyta. Chlorophyll b is found, generally, in green algae (Stengel et al., 2011). Chlorophylls c, d and e can be found in diverse marine microalgae and freshwater diatoms. When isolated, the fraction containing chlorophyll represents 0.5–1.0% DW (Vieira et al., 2020).

Carotenoids are another class of pigments found in microalgae in abundance, ranging from red and brown to orange and yellow colors. Their biological properties related to redox homeostasis, anti-cancer, regulation of triglycerides and HDL (high-density lipoprotein) (Tanaka et al., 2012; Berthon et al., 2017) allow this group of compounds to accumulate several industrial applications in cosmetics, food and feed.

Carotenoids are classified as α, β, ε, and γ or oxygen-containing compounds, named xanthophylls (Guedes and Malcata, 2012). The last group involves lutein, violaxanthin, spirilloxanthin, neoxanthin, and fucoxanthin. Their distribution patterns vary depending on the species, with more than 600 different carotenoids identified.

Chlorophyceae class represents the primary source of carotenoids within the microalgae group of organisms. Those organisms can produce carotenes (β-carotene, lycopene) and xanthophylls (astaxanthin, violaxanthin, antheraxanthin, zeaxanthin, neoxanthin and lutein, among others). Other pigments such as fucoxanthin, diatoxanthin and diadinoxanthin are produced by different microalgae phyla (Berthon et al., 2017). *Dunaliella salina, Dunaliella bardawil, Dunaliella tertiolecta,* and *Scenedesmus almeriensis* demonstrated their capability in producing a significant proportion of β-carotene when compared to the other pigments (Berthon et al., 2017).

Another carotenoid that is industrially exploited not only in the cosmetic field, but also in aquaculture, is astaxanthin. However, the production cost of this pigment, mainly produced by *Haematococcus* sp., is very high (Canales-Gómez et al., 2010). In aquaculture feed, it has been used to culture salmon, shrimp, ornamental fish and sea bream. 3 g.kg^{-1} feed of *Haematococcus pluvialis* administration effectively enhanced the oxidative status and other biochemical parameters in rainbow trout (Sheikhzadeh et al., 2012).

Despite also being produced by *Chlorella zofingiensis*, *Chlorococcum* sp., *Scenedesmus* sp. and yeast *Xanthophyllomyces dendrorhous* (Yuan et al., 2011), *Haematococcus pluvialis* is the main source of this pigment once it can synthesize up to 81% astaxanthin out of its total carotenoids (Rammuni et al., 2019).

Phycobiliproteins represent the last class of pigments found in red algae, cyanobacteria, cryptophytes, and glaucocystrophytes. According to their absorption spectra, they are classified into four major subgroups: 1) red phycoerytrin, 2) magenta phycoerythrocyanin, 3) blue phycocyanin, and 4) light blue allophycocyanin. In the medical field, they are used as markers in flow cytometry, microscopy, and DNA tests linked to their highly sensitive fluorescent properties. The natural blue pigment phycocyanin, the major microalgae phycobiliprotein, is used for multiple applications in the pharmaceutical field due to its bioactive properties such as antioxidant, anti-inflammatory, and anti-cancer activities. The cosmetic industry also seeks it out for application into lipsticks, eyeliners, and so on (Bingula et al., 2016). Moreover, since 2013, the Food and Drug Administration has accepted its use in food matrices as a coloring additive.

6.3.5 Vitamins

Microalgae represent a source of vitamins which is more complete than terrestrial plants. These microorganisms produce vitamin A, B1, B2, B6, B12, C, E, K, niacin, nicotinate, biotin, and folic acid. The concentrations of vitamin A, B_1, B_2, E, and niacin can achieve those found in vegetables by some microalgae genus, such as *Arthrospira* sp., *Chlorella* sp., and *Scenedesmus* sp. (Del Mondo et al., 2020). *Dunaliella* is also known to highly accumulate vitamins B_2, B_{12}, B_9, B_3, C, and E. More specifically, Hernández-Carmona et al., (2009) reported that *Eisenia arborea*, a species of brown microalgae, contains 3.44 mg.g^{-1} of vitamin C, which is similar to that of mandarin oranges.

Therefore, microalgae are essential sources of macro- and micro-nutrients with interesting market possibilities in nutraceuticals and feed. However, due to food safety regulations, especially in Europe, few of them are currently marketed.

6.4 INDUSTRIAL APPLICATIONS OF MICROALGAE IN THE FUNCTIONAL AQUAFEED INDUSTRY

Up to the present day, microalgae have diverse industrial applications, including food formulation, cosmetics, health products, and fertilizers. Due to their high nutritional and functional value, microalgae have a high potential for application in the feed industry, being incorporated as a feed supplement, enhancing animal performance. Moreover, the increased capacity of microalgae to produce natural anti-microbial compounds promoted the use of some species for immunostimulant applications. Moreover, microalgae cultivation has already been tested for wastewater treatment and biofuel production (Rumin et al., 2020).

The increasing global population and increasing demand for protein will promote feed ingredient soybean production. (FAO, 2018). This process will involve using large areas of arable land, causing the destruction of several terrestrial ecosystems. In addition, several marine ecosystems are being destroyed due to anthropogenic activities, diminishing their natural resources and decreasing the fish and seafood stocks available for human consumption. Nowadays, aquaculture has been demonstrated to be the best way to mitigate lack of natural resources and obtain edible protein. The aquafeed market has been growing faster than other food sectors (Napier et al., 2019) and it is projected to reach 73.15 million tonnes of feeds by 2025, representing a growth of 8–10% (FAO 2018). Feed inputs for aquaculture production represent 40–75% of the global production costs, representing the most relevant driver in this market (FAO, 2018). The increasing demand for meat by a rising population will become particularly dramatic in the coming decades, because dedicated soybean food crops, the conventional feedstuff for animal feeding, will need to occupy an increasing fraction of arable land (FAO, 2018). Aquaculture is currently the world's most efficient producer of edible protein, and continues to grow faster than any other major food sector in the world in response to the rapidly increasing global demand for fish and seafood (Napier et al., 2019). Feed inputs for aquaculture production represent 40–75% of aquaculture production costs and are a key market driver for aquaculture production (FAO, 2018). The aquafeed market is expected to grow 8–10% per annum and is the production of compound feeds is projected to reach 73.15 million tons in 2025 (FAO, 2018).

Recent studies have shown that animal feed supplemented with microalgae, either as a live feed or as an additive, present higher quality and performance, enhance antiviral and antibacterial protection, and improve the immune system, and consequently the disease resistance stress (Remize et al., 2021). Moreover, microalgae-enriched feed has been proven to contribute positively to animal physiology, improving protein turnover, gut function, and stress tolerance (Rehberg-Haas et al., 2015) which helps to achieve a final product with high quality and performance. Despite all advantages of microalgae as livestock feed, it is essential to define a feeding objective to adapt the use of different microalgae according to their biochemical characterization, such as protein, carbohydrates, lipids, vitamins, and pigments composition.

Astaxanthin is one of the pigments used in the aquaculture industry, and its benefits are well documented. Lim et al., (2018) showed that this pigment enhances the immune system, increasing the resistance to infectious diseases in farmed fish. Moreover, due to its high antioxidant activity, astaxanthin could improve the reproductive performance of aquatic animals, increasing the quality of eggs, the growth rate of larvae, and their survival (Lim et al., 2018). Recently, Jaseera and co-workers noted that 2% *Aurantiochytrium* sp. incorporation can significantly improve the growth, survival, nutritional quality of giant tiger prawn, *Penaeus monodon* postlarvae, and increase the tolerance to stress (Jaseera et al., 2021).

Fish-derived fishmeal (FM) and fish oil (FO) used in aquafeeds have long been identified as non-sustainable resources. The use of grain and oilseed crops, such as soy or corn, to substitute FM and FO have been debated in the past. However, it faces critical concerns once terrestrial plant ingredients appear to have low digestibility, anti-nutritional factors, and a deficient amino acid profile, as well as lacking the bioactive long-chain omega-3 EPA and DHA (Borowitzka, 2018, Li et al., 2009).

The use of microalgae blends in fish feed can be found in the literature from the perspective of allowing fish to benefit from a combination of the biochemical compositions of several microalgae species. A blend of *Nannochloropsis* sp. and *Isochrysis* sp. had proven to be a good substitute of 15% fishmeal of juvenile Atlantic cod, *Gadus morhua*, not interfering with fish survival and feed conversion ratio (Walker and Berlinsky, 2011). Sarker et al., (2020) combined two commercially available microalgae, *Nannochloropsis oculata* and *Schizochytrium* sp., to a fish-free feed for Nile tilapia (*Oreochromis niloticus*) that would promote its health performance. Compared to the control diet, Nile tilapia fed with microalgae showed higher growth, weight gain, and specific growth rate. Higher fillet lipid and DHA content, the highest degree of *in vitro* protein hydrolysis and protein digestibility were also obtained on fish fed with the experimental diet (Sarker et al., 2020).

Several studies exploring the possibility of substituting fishmeal with *Arthrospira platensis*, *Chlorella* sp., *Scenedesmus* sp., *Nannofrustulum* sp. and *Tetraselmis suecica* for diverse fish species are reported in the literature (Shah et al., 2018). A 5% replacement of fishmeal with *Spirulina pacifica* significantly increased weight gain, protein efficiency ratio and feed intake of Parrot Fish, *Oplegnathus fasciatus* (Kim et al., 2013). The same study reported that a 15% supplementation resulted in higher muscle protein and lower whole-body lipid (Kim et al., 2013). Hajiahmadian et al., (2012) reported that *Spirulina platensis* substitution of fishmeal up to 20% led to weight gain and increased specific growth rate and feed conversion ratio of golden barb fish, *Puntius gelius* (Hajiahmadian et al., 2012). The growth of silver seabream, *Rhabdosargus sarba*, was also not affected by a substitution up to 50% *Spirulina* sp. (El-Sayed, 1994). Neither were red tilapia fingerlings, *Oreochromis* sp., by 30% substitution with *Arthrospira maxima* (Rincón et al., 2012). Juvenile Nile tilapia presented a higher feed conversion ratio when 50% fishmeal was substituted by *Spirulina* sp. (Hussein et al., 2013), despite Velasquez and co-workers having found that 30% inclusion is the optimal level of replacement (Velasquez et al., 2016).

Twenty percent *Desmodesmus* sp. was included as a fishmeal substitute in Atlantic salmon, *Salmo salar*, feed without compromising the animal growth indexes (Kiron et al., 2016). Sørensen et al., (2016) verified they were able to use *Phaeodactyum tricornutum* as a substitute for 6% of fishmeal in Atlantic salmon, not compromising the nutrient digestibility and growth performance (Sørensen et al., 2016). *T. suecica* was described as able to replace 20% of fish protein, not interfering with the animal growth and quality of European juvenile sea bass meat (Tulli et al., 2012). Also, the complete substitution of fishmeal by *Chlorella* sp. resulted in an increase in the final weight, improvement in total cholesterol, LDL, triglyceride levels, as well as the reproductive performance of zebrafish, *Danio rerio* (Carneiro et al., 2020).

Aquaculture of mollusks is currently the market utilizing most of the feed-destined produced microalgae. The most recent numbers go back to the 1990s, but it was calculated that in 1999, 62% of the aquaculture-predetermined microalgae were used for mollusks. While 21% were used for crustaceans, only 16% were applied to fish aquaculture. Due to their richness in n-3 PUFAs, *Isochrysis* affinis *lutea* (T-iso), *Pavlova lutheri*, and *Chaetoceros* sp. represent the classic microalgae used in shellfish hatcheries (Packer et al., 2016). When released into the cultivation tanks, microalgae are quickly and efficiently filtered from the water. n-3 PUFAs are also crucial for cultivating crustaceans, which can feed directly on microalgae (Spolaore et al., 2006).

Other microalgae had been included in aquafeed due to their immunostimulant properties. 6–8% *Chlorella vulgaris* feed supplementation enhanced prophenolox-idase activity and total hemocyte counts and resistance of giant freshwater prawn *Macrobrachium rosenbergii* postlarvae against *Aeromonas hydrophila* infection (Maliwat et al., 2017). Other *Chlorella* sp. supplementation in the diet of rainbow trout *Oncorhynchus mykiss* fingerlings improved the animal growth and physiolo-gical parameters and stimulated the resistance to bacterial infection (Quico et al., 2021). *Labeo rohita* fingerlings demonstrated to be immunostimulated by 0.5 g *Euglena virilis* kg^{-1} dry diet, with increased levels of superoxide anion production, lysozyme, serum bactericidal activity, serum protein, and albumin, as well as in-creased resistance to *A. hydrophila* (Das et al., 2009). Watanuki et al., (2006) de-monstrated that 5–10% of *Spirulina platensis* has the same effect on carp, *Cyprinus carpio* (Watanuki et al., 2006).

As illustrated, microalgae can also be used as probiotics. *Chaetoceros* sp., *Pavlova* sp., and *Isochrysis* sp. have been shown to improve pearl oyster resistance to bacterial pathogens when added to their diets (Shah et al., 2018). Inulin, galactooligosaccharides, xylooligosaccharides, agarose-derived oligosaccharides, neoagaro-oligosaccharides, alginate-derived oligosaccharides, arabinoxylans, galactans, and β-glucans are a few examples of microalgal compounds with prebiotic activities. The proliferation of leu-kocytes as monocyte-macrophages and neutrophils, as well as phagocytic activity and secretion of immune mediators (e.g., cytokines), are associated with this class of bioactive compounds (Vetvicka et al., 2021). Paramylon is a linear β-1,3 polymer of glucose initially isolated from *Euglena gracilis* and represents an immunostimulant intensively used in aquaculture. These prebiotic compounds are incorporated as feed supplements, enhancing the immune performance of species such as mussels and Atlantic salmon (Kiron et al., 2016).

Beta-glucan derived from yeast, mainly *Saccharomyces cerevisiae*, has to date been the most successful prebiotic on the market. Some other products can also be found on the market, such as WellMune^{TM}, by Biothera Corporation (Eagan, MN, USA), BetaGlucans, by BioTec Pharmacon (Tromsø, Norway), and Macrogard^{TM} by Immunocorp (Werkendam, Netherlands). Paramylon, available commercially as Algamune™ by Algal Scientific Corporation (Plymouth, MI, USA). This β-1,3-glucan yield can go up to 90% DW when *Euglena gracilis* is grown heterotrophically (Barsanti et al., 2001). Despite many compounds already being commercially available, further research in this topic appears to be necessary. The probability of finding compounds with prebiotic activities into marine microalgae species is enormous. However, their complex polymer-derived structures are a puzzle for researchers.

Direct utilization of microalgae as feed is also practiced in aquaculture. Heterotrophic protists and small zooplankton, such as *Brachionus* sp. or *Artemia salina*, are critical players in supplying the microalgae nutrients and functionality to higher trophic levels (Camacho et al., 2019). The heterotrophic dinoflagellate *Crypthecodinium cohnii* can be easily found in the market traded as a FO substitute due to its DHA content for seab-ream, *Sparus aurata*, microdiets (Ganuza et al., 2008). These characteristics seem to be of great importance for seabream larvae and resulted in similar performances compared to classical fisheries diets (Bec et al., 2006).

Nannochloropsis sp., *Isochrysis galbana* and *Schizochytrium mangrovei* are also widely used due to their EPA and DHA composition for rotifer production. Those microalgae were demonstrated to increase rotifer survival, productivity, the efficiency of feed assimilation, and biochemical composition (Ferreira et al., 2009; Ferreira et al., 2008).

Therefore, the potential of using microalgae in aquafeed is tremendous. Besides being a reliable and unconventional source of nutrients, microalgae biomass is rich in bioactive metabolites, which confer cultivation advantages to cultured animals.

6.5 INDUSTRIAL APPLICATIONS OF MICROALGAE IN THE FUNCTIONAL TERRESTRIAL FEED INDUSTRY

The incorporation of microalgae in non-aqua feed, while not being as common, can be seen in the literature and the market. Several products using *Spirulina* sp. as a pet feed supplement are commercialized, such as Phycon® pastes, from Phycon (The Netherlands) or Allvitae®, from Allmicroalgae (Portugal). Besides *Spirulina* sp. PUFA-rich microalgae are most commonly found in pet food – for cats and dogs, once essential to keep the animals' health intact, especially during their growth and reproduction. Together with these products, the literature supports the supplementation of *Schizochytrium* sp. into canine diets as a source of n-3 LC-PUFA DHA (Hadley et al., 2017). Souza et al., (2019) described that a 0.4% *Schizochytrium* sp. diet supplementation is pleasant to dogs' palate while increasing the metabolizable energy and nutrients digestibility. Also, it stimulates phagocytic cells and the phagocytosis of monocytes, while not affecting the animals' fecal characteristics, biochemical profile, and blood hemogram (Souza et al., 2019).

Many experiments have been applied to farm animals showing that dietary microalgae can improve their growth performance and health condition. Poultry feeding assays are reported in the literature, demonstrating the physiological and functional effects of microalgae supplementation on those animals. Waldenstedt et al., (2003) verified that *Haematococcus pluvialis* supplementation could reduce caecal colonization of *Clostridium perfringens* on female broiler chickens (Waldenstedt et al., 2003). Specific-pathogen-free chicken fed with 1–2 g.kg^{-1} *Spirulina* sp. of ration exhibited optimum immune response, increased protection against heterologous virus strains, and reduced viral shedding (Abotaleb et al., 2020). The immune stimulation conferred by *Spirulina* sp. in chickens was also confirmed by Mirzaie et al. (2018). Moreover, *Spirulina* sp. supplementation decreased stress hormones and serum lipid parameters and elevated antioxidant status, while not affecting animal performance characteristics (Mirzaie et al., 2018). Kang et al., (2013) noted that 1% fresh liquid *Chlorella vulgaris* supplementation improved body weight gain, immune factors, and the production of *Lactobacillus* bacteria in the intestinal microflora of broiler chickens (Kang et al., 2013). The beneficial effects of this microalga have also been reported for ducks. Oh et al., (2015) found that the animal body weight was increased in line with the supplementation of heterotrophic *C. vulgaris,* and the meat quality was positively affected (Oh et al., 2015).

Studies on pig feeding using microalgae are scarce when compared with other animals. Furbeyre et al., (2017) demonstrated that 1% *Arthrospira platensis* and *C. vulgaris* supplementation is responsible for improving the intestine mucosal architecture and nutrient digestibility of weaned piglets. Moreover, *C. vulgaris* was demonstrated to play a role in regulating digestive disorders after weaning, avoiding diarrhea (Furbeyre et al., 2017). In growing pigs, the inclusion of 0.1% heterotrophic *C. vulgaris* improved the growth performance, nutrient digestibility, microbial shedding (decreased *Escherichia coli* and higher *Lactobacillus* sp.), and decrease the fecal noxious gas emissions (ammonia and hydrogen sulphide) (Yan et al., 2012). Coelho et al., (2020) showed that the growth performance, carcass, and meat quality traits remained stable when finishing pigs were supplemented with 5% *C. vulgaris* diets. This inclusion increased the levels of some lipid-soluble antioxidant pigments and n-3 PUFA, and decreased the n-6:n-3 fatty acid ratio, improving the nutritional value of pork fat (Coelho et al., 2020).

A limited number of ruminant feeding assays are reported in the literature. The reason is mainly attributed to the amount of microalgae necessary to perform those assays being enormous when compared to other animals (Becker, 2004), as well as the technical details necessary to conduct them. Nevertheless, one of the first microalgae feeding assays, performed by Hintz et al., (1966) demonstrated that when weaning lambs were fed with supplementation of a microalgae mixture, containing *Chlorella* spp., *Scenedesmus obliquus,* and *Scenedesmus quadricauda*, significant weight gain occurred compared to the other rations (Hintz et al., 1966). Alves et al. (2018) used sheep ruminal fluid, hypothesizing the protection of *Nannochloropsis oceanica* cell walls to EPA. These authors demonstrated that *N. oceanica* resultant EPA metabolism was remarkably reduced, demonstrating, for the first time in ruminants, the kinetics of EPA biohydrogenation class products and the formation of 20:0 fatty acids (Alves et al., 2018).

The supplementation of 10%, but not 20%, *A. platensis* to weaned lambs led to an increase in body weight and score, and body condition score when compared to the non-supplemented control group (Holman, 2012).

Kulpys et al., (2009) evaluated the effect of *A. platensis* when supplemented to lactating cows, during a 90-day trial. It was verified that a 200g daily supplementation of *Spirulina* sp. led to an increase in cow bodyweight of 8.5–11% and 21% of milk when compared to the control group. The authors even concluded that this supplementation is economically sustainable (Kulpys et al., 2009).

Although the microalgal species was not revealed, a study performed by Elzinga et al., (2019) demonstrated that a DHA-rich microalgae supplementation given to horses with equine metabolic syndrome can modulate their metabolic condition and reduce inflammation. This study is especially relevant since a big portion of the equine population is predisposed to develop this metabolic syndrome (Elzinga et al., 2019).

Several other microalgae have been experimented on in lab animals, supporting the potential of the use of this group of microorganisms in animal diets (Navarro et al., 2016). Therefore, besides minor effects being found in terms of meat quality, the inclusion of different species of microalgae in animal diets can improve their productivity by increasing growth performance parameters and stimulating their immune response. The scientific community has demonstrated the role of several microalgal strains in the most diverse terrestrial animals, increasing opportunities for the establishment of robust and scientifically supported feed products.

6.6 COMMERCIALIZED FORMULATED AQUA AND TERRESTRIAL FEED

While the potential of microalgae as a feed supplement for the most diverse animals has long been recognized, the products available on the market are still limited to a few formulations. During the 1960s, protein-rich microalgae were the main product available. The commercialization of *Dunaliella* sp. and *Haematococcus* sp. aimed at the functionality and color conferred by their pigments, β-carotene, and astaxanthin, appeared and had their boom during the 1980s. The 1990s brought the lipid-rich microalgae and EPA and DHA became a trend in feed products (Camacho et al., 2019). Nowadays, the market is populated by diverse microalgae-based products commercialized by several companies, some microalgae producers, and other feed formulators, as demonstrated in Table 6.1.

Allmicroalgae Natural Products, located in Portugal, and Necton, located in the same country, are two examples of companies whose core activity is microalgae production and which have a brand of aquafeed products in the market, Allvitae® and Phytobloom®, respectively.

In the aqua market, two significant categories of microalgae-derived feed products can be found: microalgae cultures, usually concentrated, mostly used for larval fish, shrimp, and bivalves, and formulated rations, supplemented with microalgae. Algagen LLC, located in Florida (USA), supplying concentrated microalgae cultures for the most diverse aqua cultivation. However, in Belgium Proviron commercialized their microalgae for feed application in powder form. Others, like Sera in Germany formulated products on the market, instead of pure microalgae.

Most of the commercialized microalgae for feed applications are being used fresh, either as the only component or as an additive. Some suppliers produce microalgae seed cultures so that the aquaculture farmers can have an on-site microalgae cultivation for further utilization. Algagen LLC is one company with a diverse offer of seed microalgae species. There is also a large number of options in terms of species. As an example, GreenSea, located in France, has more than ten species available as a product.

The market of microalgae-based feed for terrestrial animals is not as developed as that of aquatic animals. The products only started to appear on the market late in the 20th century, and the majority are nutritional supplements. *Chlorella* sp. and *Spirulina* sp. are the predominant microalgae in these products, such as pâtés and biscuits for cats and dogs commercialized by Yarrah, in the Netherlands, or Equialgae, the ration sold by NeoAlgae for horses. However, other species appear supplemented in the animal feed, such as *Tetraselmis chui*, *Nannochloropsis oceanica* and *Scenedesmus obliquus* used in Allvitae® by Allmicroalgae Natural Products, as is demonstrated in Table 6.2.

TABLE 6.1
Description of Some Aqua-feed Microalgae-based Products Available in the Market

Product	Composition	Microalgae Species	Company	Price	Applications
Instant Algae® Nanno 3600	Microalgae concentrate	Nannochloropsis sp.	Reed Mariculture Inc. (California, USA)	62.50€ (1L)	Finfish and shellfish hatchery
Instant Algae® Tetraselmis 3600	concentrate	Tetraselmis sp.		70.83€ (1L)	Finfish, shellfish and shrimp hatchery
Instant Algae® Iso 1800		T- Isochrysis		45.81€ (1L)	Finfish, shellfish and shrimp hatchery
Instant Algae® Pavlova 1800		Pavlova sp.		45.81€ (1L)	Finfish, shellfish and shrimp hatchery
Instant Algae® TW 1800		Thalassiosira weissflogii and T. pseudonana		30.83€ (1L)	Finfish, shellfish and shrimp hatchery
Rotifer Diet®		Nannochloropsis sp. and Tetraselmis sp.		62.50€ (1L)	Finfish hatchery
RotiGrow® OneStep	Liquid	High-yield microalgal blend		59.38€ (1L)	Finfish hatchery
RotiGrow® Plus		Omega algal blend		59.38€ (1L)	Finfish and shellfish hatchery
Rotigrow® Nanno		Nannochloropsis sp.		59.38€ (1L)	Finfish and shellfish hatchery
Chlorella V12	Live microalgae	Chlorella sp.		n.d	Finfish hatchery
N-Rich® High Pro N-Rich® PL Plus N-Rich® Ultra PL	Liquid	Concentrated blend of microalgae		45.81€ (1L)	Finfish hatchery
RotiGreen® Nanno		Nannochloropsis sp.		62.50€ (1L)	Finfish hatchery
RotiGreen® Iso		T- Isochrysis		45.81€ (1L)	
Rotigreen® Omega		Blend of Nannochloropsis sp. and other DHA-producing microalgae (n.d.)		45.81€ (1L)	
Shellfish Diet® 1800	Microalgae concentrate	Blend of Isochrysis sp., Pavlova sp., Tetraselmis sp., Thalassiosira weissflogii and T. pseudonana		44.05€ (1L)	Shellfish hatchery

(Continued)

TABLE 6.1 (Continued)
Description of Some Aqua-feed Microalgae-based Products Available in the Market

Product	Composition	Microalgae Species	Company	Price	Applications
LPB Frozen Shellfish Diet®		Blend of *Tetraselmis* sp., *Thalassiosira weissflogii* and *T. pseudonana*			Shellfish hatchery
TDO Chroma BOOST™		*Haematococcus* sp.		22.92 to 31.25€ (1Kg)	Finfish hatchery
Phytobloom Green Formula	Microalgae concentrate	*Nannochloropsis* sp.	Necton, S.A. (Algarve, Portugal)	n.d	Finfish, shellfish and shrimp hatchery
		Isochrysis sp.		n.d	Shellfish hatchery
		Tetraselmis sp.		n.d	n.d.
Phytobloom Green Ice	Frozen paste	*Nannochloropsis* sp.		n.d	Finfish, shellfish and shrimp hatchery
		Isochrysis sp.		n.d	Finfish and shrimp hatchery
		Phaeodactylum sp.		n.d	Shellfish hatchery
		Tetraselmis sp.		n.d	Shrimp hatchery
Phytobloom Green Prof	Microalgae powder	*Nannochloropsis* sp.		n.d	Finfish, shellfish and shrimp hatchery
		Isochrysis sp.		n.d	Finfish, shellfish and shrimp hatchery
		Phaeodactylum sp.		n.d	Shellfish hatchery
		Tetraselmis sp.		n.d	Shellfish and shrimp hatchery
AlgaGenPods™ Tisbe	Liquid	Live *Tisbe biminiensis* and microalgae (n.d.)	Algagen LLC (Florida, USA)	85€ (1Kg)	Seahorse, mandarins, wrasses, invert larvae
AlgaGenPods™ Apocylops		*Apocylops panamensis* and microalgae (n.d.)		85€ (1Kg)	Corals and marine-fish-larvae
AlgaGenPods™ Pseudodiaptomus		*Pseudodiaptomus pelagicus* and microalgae (n.d.)		15€ (1Kg)	Leafy seadragon, fish, reidi seahorse, clown and mandarin larval fish, seahorses, sessile invertebrates, and small reef fish

(Continued)

TABLE 6.1 (Continued)
Description of Some Aqua-feed Microalgae-based Products Available in the Market

Product	Composition	Microalgae Species	Company	Price	Applications
AlgaGenPods™ Parvocalanus		Parvocalanus crassirostris and microalgae (n.d.)		103€ (1Kg)	Sea tang, sergeant major, anthias, several larval fish, sessile invertebrates, corals and small reef fish
AlgaGenPods™ Acaria		Acartia tonsa and microalgae (n.d.)		85€ (1Kg)	Aquatic toxicology; ideal for fish breeding efforts
Phyto-Plasm™ Brown		Three types of brown phytoplankton/microalgae and zooxanthellae		45.83€ (1L)	Most pods, clams, flame scallops, shrimp and corals
Phyto-Plasm™ Green		Two types of green phytoplankton/microalgae and zooxanthellae			Amphipods, copepods, shrimp larvae, sponges, feather-duster, gorgonians and soft corals.
Phyto-Plasm™ Zooxanthellae		Blend of two species of symbiotic microalgae			Corals
Phyto-Plasm™ Gallon		Blend of microalgae algae either green or brown and zooxanthellae			Corals
PhycoPure™ Reefblend		Blend of microalgae/phytoplankton		55€ (1Kg)	Packed and diverse reef tanks, including zooxanthellae
PhycoPure™ Copepod Blend				n.d	Clams and corals
PhycoPure™ Greenwater				n.d	Rotifers, feather dusters and leathers
PhycoPure™ Zooxanthellae				132€ (1Kg)	Clams and corals

(Continued)

TABLE 6.1 (Continued)

Description of Some Aqua-feed Microalgae-based Products Available in the Market

Product	Composition	Microalgae Species	Company	Price	Applications
Proviron ChaetoPrime	Microalgae powder	Chaetoceros muelleri	Proviron (Hemiksem, Belgium)	270€ (1Kg)	Bivalves and shrimp or prawn larviculture
Proviron NannoPrime		Nannochloropis sp.		240€ (1Kg)	Finfish and shrimp larviculture
Proviron IsoPrime		Isochrysis aff. galbana T-ISO		270€ (1Kg)	Finfish, shrimp and prawn larviculture
Proviron TetraPrime S		Tetraselmis suecica		240€ (1Kg)	Bivalve, shrimp and prawn larviculture
Proviron TetraPrime C		Tetraselmis chui		240€ (1Kg)	Bivalve, shrimp and prawn larviculture
Proviron ThalaPrime W		Thalassiosira weissflogii		270€ (1Kg)	Bivalve and shrimp larviculture
Phycom® Algae flakes	Flakes of different sizes	Chlorella vulgaris	Phycom® (Veenendaal, Netherlands)	n.d	Freshwater fish
Phycom® Algae flakes		Chlorella sorokiniana		n.d	Freshwater fish
Phycom® Algae powder	Microalgae powder	Chlorella vulgaris		n.d	Freshwater fish
Phycom® Algae powder		Chlorella sorokiniana		n.d	Freshwater fish
Phycom® Algae paste		Chlorella vulgaris		n.d	Freshwater fish
Phycom® Algae paste	Microalgae paste	Chlorella sorokiniana		n.d	Freshwater fish
Phycom® Algae pellets (IQF)	Soluble pellets	Chlorella vulgaris		n.d	Freshwater fish
Phycom® Algae pellets (IQF)		Chlorella sorokiniana		n.d	Freshwater fish
Sera Crabs Nature	Sinking loops	Blend of Spirulina sp. and Haematococcus sp.	Sera (Heinsberg, Germany)	n.d	Crustaceans
Sera Micron Nature	Powder	Zooplankton, Spirulina sp. (51%) and Haematococcus sp.		n.d	Fish, amphibians and Artemia nauplii

(Continued)

TABLE 6.1 (Continued)
Description of Some Aqua-feed Microalgae-based Products Available in the Market

Product	Composition	Microalgae Species	Company	Price	Applications
Sera Plankton Tabs Nature	Sinking tablets	Zooplankton, *Spirulina* sp. (24%) and *Haematococcus* sp.		n.d	Fish that eat at the bottom and invertebrates
Sera Shrimps Nature	Granules	*Spirulina* sp. (5%) and *Haematococcus* sp.		n.d	Shrimp
Sera Catfish Chips Nature	Sinking tablets	*Spirulina* sp. (4%) and *Haematococcus* sp.		n.d	Rasping and suckermouth bottom fish
Sera Herbs'n'Loops Nature	Powder loops	*Spirulina* sp. and *Haematococcus* sp.		n.d	Tortoises and other herbivorous reptiles
Sera Guppy Gran Nature	Soft granules	*Spirulina* sp. (4%) and *Haematococcus* sp.		n.d	Herbivorous fish that mainly eat in the middle water layers, such as guppies
Sera KOI Professional Spirulina Color Food	Granules	*Spirulina* sp. (6.3%) and *Haematococcus* sp.		n.d	n.d.
Sera Spirulina Tabs Nature	Attaching tablets	*Spirulina* sp. (24%) and *Haematococcus* sp.		n.d	Herbivorous fish and invertebrates
Sera Cichlid Green XL Nature	Floating granules	*Spirulina* sp. (10%) and *Haematococcus* sp.		n.d	Herbivorous fish
Sera Discus Color Nature	Soft granules	*Spirulina* sp. and *Haematococcus* sp.		n.d	Discus fish
Sera Flora Nature	Flakes	*Spirulina* sp. (7%) and *Haematococcus* sp.		n.d	Herbivorous ornamental fish
Hagen Nutrafin MAX Spirulina Flakes	Flakes	*Spirulina* sp.	Hagen (Quebec, Canada)	n.d	Ornamental fish

(Continued)

TABLE 6.1 (Continued)
Description of Some Aqua-feed Microalgae-based Products Available in the Market

Product	Composition	Microalgae Species	Company	Price	Applications
Allvitae® Aqua	Microalgae paste	Nannochloropsis sp., Chlorella sp. Phaeodactylum tricornutum, Tetraselmis sp.	Allmicroalgae Natural Products S.A. (Pataias, Portugal)	n.d.	Finfish, shellfish and shrimp hatchery
	Powder	Nannochloropsis sp., Chlorella sp. Phaeodactylum tricornutum, Tetraselmis sp.			
Algamune™	Powder	Euglena gracilis	Algal Scientific Corporation (Michigan, USA)	n.d.	Shrimp
Green feed Nanno CLA101	Microalgae concentrate	Nannochloropsis oculata	GreenSea (Mèze, France)	n.d.	Finfish hatchery
Green Feed Chaeto CLA110-G		Chaetoceros gracilis		n.d.	Shellfish and shrimp hatchery
Green Feed Chaeto CLA110-C		Chaetoceros calcitrans		n.d.	Shellfish and shrimp hatchery
Green Feed Tetra CLA106-S		Tetraselmis suecica		n.d.	n.d.
Green Feed Tetra CLA106-C		Tetraselmis chui		n.d.	Shellfish and shrimp hatchery
Green Feed Iso CLA112		Tisochrysis lutea		n.d.	Finfish hatchery
Green Feed Phaeo CLA102		Phaeodactylum tricornutum		n.d.	Shellfish and shrimp hatchery
Green Feed Duna CA114	Live culture	Dunaliella tertiolecta		n.d.	n.d.
Green Feed Porphy CLA103	Microalgae concentrate	Porphyridium cruentum		n.d.	n.d.

(Continued)

TABLE 6.1 (Continued)
Description of Some Aqua-feed Microalgae-based Products Available in the Market

Product	Composition	Microalgae Species	Company	Price	Applications
Nutritional mix CLA006		Tetraselmis sp., Porphyridium sp., Nannochloropsis sp., Phaeodactylum sp. blend		n.d.	Shellfish hatchery, invertebrates and corals
Nutritional mix CLA007		Tetraselmis sp., Porphyridium sp., Nannochloropsis sp., Phaeodactylum sp. blend		n.d.	Shellfish hatchery, invertebrates and corals
Green feed Nanno CLA1011		Nannochloropsis oculata		n.d.	Finfish hatchery
BIOMIN PA101	Powder	Tetraselmis chui		n.d.	Shellfish and shrimp hatchery
BIOMIN PA102		Phaeodactylum tricornutum		n.d.	Shellfish and shrimp hatchery
BIOMIN PA103		Dunaliella spp.		n.d.	Fish feed formulation
BIOMIN PA106		Spirulina sp.		n.d.	Fish feed formulation
BIOMIN PA109		Chlorella sp.		n.d.	Fish feed formulation
BIOMIN PA118		Haematococcus sp.		n.d.	Shrimp hatchery and fish feed formulation
Nanno CLA101F20	Frozen paste	Nannochloropsis sp.		n.d.	Finfish hatchery
CA104	Microalgae concentrate	Chlorella vulgaris		n.d.	Fresh water organisms
CA105		Selenastrum capricornutum		n.d.	Fresh water organisms
CA111		Rhodomonas salina		n.d.	Copepods
CA117		Scenedesmus subspicatus		n.d.	Fresh water organisms

n.d. – not described

TABLE 6.2

Description of Some Terrestrial Feed Microalgae-based Products Available in the Market

Product	Composition	Microalgae Species	Company	Price	Applications
Phycom® Algae flakes	Brittle flakes	*Chlorella vulgaris*	Phycom (Veenendaal, Netherlands)	n.d.	Poultry and piglets, horses, dogs and cats, freshwater fish
Phycom® Algae flakes		*Chlorella sorokiniana*		n.d.	
Phycom® Algae powder	Microalgae powder	*Chlorella vulgaris*		n.d.	
Phycom® Algae powder		*Chlorella sorokiniana*		n.d.	
Phycom® Algae paste	Microalgae paste	*Chlorella vulgaris*		n.d.	
Phycom® Algae paste		*Chlorella sorokiniana*		n.d.	
Phycom® Algae pellets (IQF)	Pellets	*Chlorella vulgaris*		n.d.	
Phycom® Algae pellets (IQF)		*Chlorella sorokiniana*		n.d.	
ALLVITAE Mature Pets (7+)	Enriched with microalgae powder	*Chlorella vulgaris, Tetraselmis chui, Nannochloropsis oceanica* and/or *Scenedesmus obliquus*	Allmicroalgae Natural Products S.A. (Pataias, Portugal)	n.d.	Cats and dogs
ALLVITAE Junior Pets		*Chlorella vulgaris, Tetraselmis chui, Nannochloropsis oceanica* and/or *Scenedesmus obliquus*			Cats and dogs
ALLVITAE Piglets		*Chlorella vulgaris, Tetraselmis chui, Nannochloropsis oceanica* and/or *Scenedesmus obliquus*			Swine

(Continued)

TABLE 6.2 (Continued)
Description of Some Terrestrial Feed Microalgae-based Products Available in the Market

Product	Composition	Microalgae Species	Company	Price	Applications
ALLVITAE Performance		Chlorella vulgaris, Tetraselmis chui, Nannochloropsis oceanica and/or Scenedesmus obliquus			Swine
ALLVITAE Organic Eggs		Chlorella vulgaris, Tetraselmis chui, Nannochloropsis oceanica and/or Scenedesmus obliquus			Aviary, game birds
ALLVITAE Game Birds		Chlorella vulgaris, Tetraselmis chui, Nannochloropsis oceanica and/or Scenedesmus obliquus			Aviary, game birds
Yarrah Organic dog food pâté with beef and chicken	Microalgae-enriched pâté	Spirulina sp.	Yarrah (Harderwijk, Netherlands)	9.70€ (1Kg)	Dogs
Yarrah Organic Sensitive dry dog food	Microalgae-enriched pâté	Spirulina sp.		10€ (1Kg)	Dogs with stomach or digestion problems
Yarrah Organic dog food pâté with chicken	Microalgae-enriched pâté	Spirulina sp.		6.70€ (1Kg)	Dogs
Yarrah Organic cat food chunks with fish	Microalgae-enriched chunks	Spirulina sp.		11.50€ (1Kg)	Cats
Yarrah Organic cat food pâté with fish	Microalgae-enriched pâté	Spirulina sp.		6.70€ - (1Kg)	Cats

(Continued)

TABLE 6.2 (Continued)
Description of Some Terrestrial Feed Microalgae-based Products Available in the Market

Product	Composition	Microalgae Species	Company	Price	Applications
Yarrah Organic vegan dog biscuits for smaller dogs	Microalgae-enriched biscuits	Spirulina sp.		14.80€ (1Kg)	Dogs
Dr. Mercola SpiruGreen	Tablets (supplement)	Blend of Arthrospira platensis (97%) and Haematococcus pluvialis (3%)	Cape Dr. Mercola (Cape Coral, USA)	n.d.	Cats and dogs
Algae-to-Omega – Horse Omega-3 Supplement	Enriched with microalgae powder	n.d.	Equi-Force Equine Products, LLC (Kentucky, USA)	64.88€ (1Kg)	Horses
Equialgae – Horses	Enriched with microalgae powder	Blend of Spirulina sp. and Chlorella sp.	NeoAlgae (Gijón, Espanha)	27.30€ (1Kg)	Horses
Arenus Animal Health-Aleira	Enriched with microalgae powder	Algae-derived DHA	Arenus Animal Health (Fort Collins, USA)	106.62€ (1Kg)	Horses
Arenus Animal Health-Releira	Enriched with microalgae powder	Algae-derived DHA		55.19€ (1Kg)	
AlgenPower CHLORELLA – Premium Vitties	Microalgae-enriched biscuits	Enriched with 2% Chlorella sp.	AlgenPower (Hausmannstätten, Austria)	49.76€ (1Kg)	Dogs
AlgenPower CHLORELLA – Premium Vitties	Microalgae-enriched biscuits	Enriched with 2% Chlorella sp.		49.76€ (1Kg)	Horses
AlgEnerg Pvt Ltd - Animal Feed Supplement	Microalgae powder	Enriched with Spirulina sp.	AlgEnerg Pvt Ltd (New Delhi, India)	4.99€ (1Kg)	Cows, buffalo, calves, sheeps, goats, horses and poultry

(Continued)

TABLE 6.2 (Continued)
Description of Some Terrestrial Feed Microalgae-based Products Available in the Market

Product	Composition	Microalgae Species	Company	Price	Applications
SmartPak Equine - Smart & Simple™ Spirulina Pellets	Pellets	Enriched with *Spirulina* sp.	SmartPak Equine (Plymouth, USA)	26.20€ (1Kg)	Horses
SmartPak Equine - SmartBreathe® Ultra Pellets	Pellets			21.20€ (1Kg)	
SmartPak Equine - SmartItch-Ease™	n.d.			54.90€ (1Kg)	
Crypto Lina Crypto Aero Metabolism Crypto Aero Plus	Microalgae powder	Enriched with *Spirulina* sp.	Crypto Aero Wholefood Horse Feed (Wellington, Florida)	22.75€ (1Kg) 83.44€ (1Kg) 100.82€ (1Kg)	Horses
Air-Way EQ (Pellets)	Microalgae powder	Enriched with *Spirulina* sp.	Med-Vet Pharmaceuticals (Eden Prairie, USA)	928.81€ (1Kg)	Horses
Wholistic Spirulina	Microalgae powder	Enriched with *Spirulina* sp.	Wholistic Pet Organics (Bedford, USA)	148.07€ (1Kg)	Dogs
Algamune™	Powder	*Euglena gracilis*	Algal Scientific Corporation (Michigan, USA)	n.d.	Pigs and poultry

n.d. – not described

The market perspective for microalgae-based feed products is positive for the coming years. Besides representing an excellent alternative to the classical ingredients used for aquafeed, research and development support the evidence of the health-promoting effects induced by this group of organisms. There is also a growing need to diminish the number of antibiotics used during fish production. It is then expected that the offer on products supplemented with microalgae shall increase significantly in the next few years.

6.7   CHALLENGES AND FUTURE PERSPECTIVES

The recent rapid evolution in the microalgae biotechnology field led to an increase in the algal bioeconomy applied to the feed industry. Microalgae researchers have been focusing on increasing biomass productivity while reducing production costs. Significant achievements have been reported for bioreactor design, harvesting techniques, strain development, adaptation, and genetic and metabolic engineering, allowing improvements in biomass and added-value compounds productivity. Culturing conditions have also been manipulated to increase the biomass content in carbohydrates, proteins, lipids, pigments, and other metabolites of interest. All those improvements allowed for the achievement of a cell factory undergoing continuous improvement and becoming more effective in carbon capture and more suitable for market applications, such as feed. In spite of all these improvements, the price of microalgal feed is still higher when compared to the traditional ingredients, while conferring to the product functionalities that are not seen in the other crop ingredients. Another constraint in microalgae marketing relates to aquaculture demand for live biomass instead of dried powder. The sourcing and transport of concentrated and frozen biomass inevitably increase the product's final cost.

The balanced nutritional profile and functionality of the biomass are critical factors for the use of these microorganisms in animal feed. Further studies are needed, especially for product formulation, but the future seems promising for this market.

6.8 SUMMARY

Scientifically validated evidence showing that microalgal metabolites develop functional roles when incorporated in feed has been increasing in recent times. Biotechnological research in this field is promising. New bioactive metabolites relevant in the most diverse areas, from human to animal health, are likely to be found in the near future. The microalgae market is also a likely to become a growing sector once consumers become more aware of its nutritional and functional properties and are consequently willing to pay to benefit from that. The possible use of microalgae is then destined to become a prominent reality, intending to promote the growth and health of the fed animal. Beyond microalgae functionality being well documented, further studies are needed in the field of product formulation.

REFERENCES

Abotaleb, M.M., Mourad, A., Abousenna, M.A., Helal, A.M., Nassif, S.A., Elsafty, M.M., 2020. The effect of spirulina algae on the immune response of spf chickens to commercial inactivated Newcastle vaccine in poultry. *VacciMonitor.* 29, 2.

Alves, S.P., Mendonça, S.H., Silva, J.L., Bessa, R.J.B., 2018. *Nannochloropsis oceanica*, a novel natural source of rumen-protected eicosapentaenoic acid (EPA) for ruminants. *Sci. Reports.* 8, 1.

Barkia, I., Saari, N., Manning, S.R., 2019. Microalgae for high-value products towards human health and nutrition. *Mar. Drugs* 24, 304.

Barros, A., Pereira, H., Campos, J., Marques, A., Varela, J., Silva. J., 2019. Heterotrophy as a tool to overcome the long and costly autotrophic scale-up process for large scale production of microalgae. *Sci. Reports* 9, 13935.

Barsanti, L., Vismara, R., Passarelli, V., Gualtieri, P., 2001. Paramylon (β-1,3-Glucan) content in wild type and WZSL mutant of *Euglena gracilis*. Effects of growth conditions. *J. Appl. Phycol.* 13, 59–65.

Bec, A., Martin-Creuzburg, D., Von Elert., E., 2006. Trophic upgrading of autotrophic picoplankton by the heterotrophic nanoflagellate *Paraphysomonas* sp. *Limnol. Oceanog.* 51, 1699–1797.

Becker, W., 2004. Microalgae in human and animal nutrition, in: Richmond, A. (Ed.), *Handbook of Microalgal Culture: Biotechnology and Applied Phycology*, p. 18. Blackwell Science, London.

Berthon, J.Y., Nachat-Kappes, R., Bey, M. Cadoret, J.P., Renimel, I., Filaire, E. 2017. Marine algae as attractive source to skin care. *Free Radical Res.* 510, 555–567.

Bingula, R., Dupuis, C., Pichon, C., Berthon, J.Y., Filaire, M., Pigeon, L., Filaire, E., 2016. Study of the effects of betaine and/or C-phycocyanin on the growth of lung cancer A549 cells in vitro and in vivo. *J. Oncol.* 2016, 8162952.

Borowitzka, M.A., 1999. Commercial production of microalgae: Ponds, tanks, tubes and fermenters. *Prog. Ind. Microbiol.* 35, 313–321.

Borowitzka, M.A., 2018. Biology of microalgae, in: *Microalgae in Health and Disease Prevention*, pp. 23–72.

Calder, P.C., 2006. N-3 polyunsaturated fatty acids, inflammation, and inflammatory diseases. *Am. J. Clin. Nutr.* 83, 1505S–1519S.

Camacho, F. Macedo, A., Malcata, F., 2019. Potential industrial applications and commercialization of microalgae in the functional food and feed industries: A short review. *Mar. Drugs.* 17, 312.

Canales-Gómez, E., Correa, G., Teresa Viana, M., 2010. Effect of commercial carotene pigments (Astaxanthin, Cantaxanthin and β-Carotene) in Juvenile abalone *Haliotis rufescens* diets on the color of the shell or nacre. *Vet. Mex.* 41, 3.

Carneiro, W.F., Dias Castro, T.F., Orlando, T.M., Meurer, F., Paula, D.A.J., Virote, B.C.R., Vianna, A.R.C.B., Murgas, L.D.S., 2020. Replacing fish meal by *Chlorella* sp. meal: Effects on zebrafish growth, reproductive performance, biochemical parameters and digestive enzymes. *Aquaculture* 528, 735612.

Casadevall, E., Dif, D., Largeau, C., Gudin, D., Chaumont, D., Desanti, O., 1985. Studies on batch and continuous cultures of *Botryococcus braunii*: Hydrocarbon production in relation to physiological state, cell ultrastructure, and phosphate nutrition. *Biotechnol. Bioeng.* 27, 286–295.

Chen, F., 1996. High cell density culture of microalgae in heterotrophic growth. *Trends Biotech.* 14, 21–426.

Choopani, A., Poorsoltan, M., Fazilati, M., Mohammad Latifi, A., Salavati, H., 2016. Spirulina: A source of gamma-linoleic acid and its applications. *J. Appl. Biotech. Rep.* 3, 483–488.

Coelho, D., Pestana J., Almeida, J.M., Alfaia, C.M., Fontes, C.M.G.A., Moreira, O., Prates, J.A.M., 2020. A high dietary incorporation level of *Chlorella vulgaris* improves the nutritional value of pork fat without impairing the performance of finishing pigs. *Animals* 10, 12.

D'Alessandro, E.B., Filho, N.R.A., 2016. Concepts and studies on lipid and pigments of microalgae: A review. *Renew. Sustain. Energy Rev.* 58, 832–841.

Das, B.K., Pradhan, J., Sahu, S., 2009. The effect of *Euglena viridis* on immune response of Rohu, Labeo Rohita (Ham.). *Fish Shellfish Immunol.* 26, 871–876.

Delattre, C., Guillaume P., Laroche, C., Michaud, P., 2016. Production, extraction and characterization of microalgal and cyanobacterial exopolysaccharides. *Biotech. Adv.* 15, 1159–1179.

Del Mondo, A., Smerilli, A., Sané, E., Sansone, C., Brunet, C., 2020. Challenging microalgal vitamins for human health. *Microbial Cell Factories.* 201.

De Morais, M.G., Vaz, B.D.S., De Morais, E.G., Costa, J.A.V., 2015. Biologically active metabolites synthesized by microalgae. *BioMed. Res. Int.* 215.

Dubey, K.K., Kumar, S., Dixit, D., Kumar, P., Kumar, D., Jawed, A., Haque, S., 2015. Implication of industrial waste for biomass and lipid production in *Chlorella minutissima* under autotrophic, heterotrophic, and mixotrophic grown conditions. *Appl. Biochem. Biotech.* 176, 1581–1595.

El-Sayed, A.F.M., 1994. Evaluation of soybean meal, spirulina meal and chicken offal meal as protein sources for silver seabream (*Rhabdosargus sarba*) fingerlings. *Aquaculture.* 127, 169–176.

Elzinga, S.E., Betancourt, A., Stewart, J.C., Altman, M.H., Barker, V.D., Muholland, M., Bailey, S., Brennan, K.M., Adams, A.A., 2019. Effects of Docosahexaenoic acid–rich microalgae supplementation on metabolic and inflammatory parameters in horses with equine metabolic syndrome. *J. Equine Veterinary Sci.* 83, 102811.

FAO, 2018. *The State of Fisheries and Aquaculture in the World 2018 – Meeting the sustainable development goals.* Rome. Licence: CC BY-NC-SA 3.0 IGO.

Ferreira, M., Coutinho P., Seixas, P., Fábregas, J., Otero, A., 2009. Enriching rotifers with 'premium' microalgae. *Nannochloropsis gaditana. Mar. Biotechol.* 22, 3.

Ferreira, M., Maseda, A., Fábregas, J., Otero, A., 2008. Enriching rotifers with "premium" microalgae. Isochrysis aff. galbana clone T-ISO. *Aquaculture* 279(2008), 126–130.

Furbeyre, H.J., Van Milgen, T., Mener Gloaguen, M., Labussière, E., 2017. Effects of dietary supplementation with freshwater microalgae on growth performance, nutrient digestibility and gut health in weaned piglets. *Animal.* 11, 183–192.

Ganuza, E., T. Benítez-Santana, E. Atalah, O. Vega-Orellana, R. Ganga, and M. S. Izquierdo., 2008. *Crypthecodinium cohnii* and *Schizochytrium* sp. as Potential substitutes to fisheries-derived oils from seabream *(Sparus aurata)* microdiets. *Aquaculture.* 277, 109–116.

García, José L., de Vicente, M., Galán., B., 2017. Microalgae, old sustainable food and fashion nutraceuticals. *Microbio. Biotech.* 10, 1017–1024.

Ghidossi, T., Marison, I., Devery, R., Gaffney, D., Forde. C., 2017. Characterization and optimization of a fermentation process for the production of high cell densities and lipids using heterotrophic cultivation of *Chlorella protothecoides*. *Industr. Biotech.* 13.

Guedes, A.C., Malcata, F.X., 2012. Nutritional Value and Uses of Microalgae in Aquaculture. In Muchlisin, Z. (Ed.), *Aquaculture.* In Tech, Rijeke, Croatia.

Hadley, K.B., Bauer, J., Milgram, N.W., 2017. The oil-rich alga *Schizochytrium* sp. as a dietary source of docosahexaenoic acid improves shape discrimination learning associated with visual processing in a canine model of senescence. *Prostaglandins Leukot. Essent. Fatty Acids.* 118, 10–18.

Hajiahmadian, M., Vajargah, M.F., Farsani, H.G., Mohammad., M., 2012. Effect of *Spirulina platensis* meal as feed additive on growth performance and survival rate in golden barb fish, *Puntius gelius* (Hamilton, 1822). *J. Fisheries Int.* 7, 61–64.

Hernández-Carmona, G., Carrillo-Domínguez, S., Arvizu-Higuera, D.L., Rodríguez-Montesinos, Y.E., Murillo-Álvarez, J.I., Muñoz-Ochoa, M., María Castillo-Domínguez, R., 2009. Monthly variation in the chemical composition of *Eisenia arborea* J.E. Areschoug. *J. Appl. Phycol.* 21, 607–616.

Hintz, H. F., Heitman, Weir, H.W.C., Torell, D.T., Meyer., J.H., 1966. Nutritive Value of Algae Grown on Sewage2. *J. Anim. Sci.* 25.

Ho, S.H., Chen, C.Y., Chang, J.S., 2012. Effect of Light Intensity and Nitrogen Starvation on CO_2 Fixation and Lipid/Carbohydrate Production of an Indigenous Microalga *Scenedesmus obliquus* CNW-N. Biores. *Technol.* 113, 244–252.

Holman, B., 2012. Growth and Body Conformation Responses of Genetically Divergent Australian Sheep to Spirulina (*Arthrospira platensis*) Supplementation. *Am. J. Exper. Agriculture.* 2, 2.

Hu, J., Nagarajan, D., Zhang, Q., Chang, J.S., Lee, D.J., 2018. Heterotrophic Cultivation of Microalgae for Pigment Production: A Review. *Biotectnol. Adv.* 36, 54–67.

Hussein, E.E.S., Dabrowski, K., El-Saidy, M.S.D., Lee, B.J., 2013. Enhancing the Growth of Nile Tilapia Larvae/Juveniles by Replacing Plant (Gluten) Protein with Algae Protein. *Aquaculture Res.* 44, 1365–2109.

Illman, A. M., Scragg, A.H., Shales, S.W., 2000. Increase in Chlorella Strains Calorific Values When Grown in Low Nitrogen Medium. *Enzyme Microb.Technol.* 27, 631–635.

Jacob-Lopes, E., Maroneze, M.M., Deprá, M.C., Sartori, R.B., Dias, R.R., Zepka, L.Q., 2019. Bioactive Food Compounds from Microalgae: An Innovative Framework on Industrial Biorefineries. *Curr. Opinion Food Sci.* 25, 1–7.

Jaseera, K.V., Ebeneezar, S., Sayooj P., Nair, A.V., Kaladharan, P., 2021. Dietary Supplementation of Microalgae, *Aurantiochytrium* sp. and Co-Feeding with Artemia Enhances the Growth, Stress Tolerance and Survival in *Penaeus monodon* (Fabricius, 1798) Post Larvae. *Aquaculture.* 533, 1–12.

Kang, H. K., Salim, H.M., Akter Kim, N.D.W., Kim, J.H., Bang, H.T., Kim, M.J., Na, J.C., Choi, H.C., Suh, O.S., 2013. Effect of various forms of dietary Chlorella supplementation on growth performance, immune characteristics, and intestinal microflora population of broiler chickens. *J. Appl. Poultry Res.* 22, 100–108.

Kim, S.S., Rahimnejad, S., Kim, S.K.W., Lee, K.J., 2013. Partial replacement of fish meal with *Spirulina pacifica* in diets for parrot fish (Oplegnathus Fasciatus). *Turk. J. Fisheries Aquatic Sci.* 13, 197–204.

Kiron, V., Kulkarni, A., Dahle, D., Vasanth, G., Lokesh, J., Elvebo, O., 2016. Recognition of purified beta 1,3/1,6 glucan and molecular signalling in the intestine of Atlantic Salmon. *Dev. Comp. Immunol.* 56, 57–65.

Kiron, V., Sørensen, M., Huntley, M., Vasanth, G.K., Gong, Y., Dahle, D., Palihawadana, A.M., 2016. Defatted biomass of the Microalga, *Desmodesmus* sp., can replace fish-meal in the feeds for Atlantic Salmon. *Front. Mar. Sci.* 17 (May). 10.3389/fmars.2016.00067

Krzemińska, I., Piasecka, Nosalewicz, A.A., Simionato, D., Wawrzykowski, J., 2015. Alterations of the lipid content and fatty acid profile of *Chlorella protothecoides* under different light intensities. *Bioresour. Technol.* 196, 72–77.

Kulpys, J., Paulauskas, E., Pilipavicius, V., Stankevicius, R., 2009. Influence of Cyanobacteria *Arthrospira* (Spirulina) *platensis* biomass additives towards the body condition of lactation cows and biochemical milk indexes. *Agronomy Res.* 7, 823–835.

Lafarga, T., 2020. Cultured microalgae and compounds derived thereof for food applications: strain selection and cultivation, drying, and processing strategies. *Food Rev. Int.* 36, 559–583.

Levasseur, W., Perré, P., Pozzobon, V., 2020. A review of high value-added molecules production by microalgae in light of the classification. *Biotech Adv.* 41, 107547.

Li, P., Mai, K., Trushenski, J., Wu, G., 2009. New developments in fish amino acid nutrition: towards functional and environmentally oriented aquafeeds. *Amino Acids.* 37, 43–53.

Lim, K.C., Yusoff, F.M., Shariff, M., Kamarudin, M.S., 2018. Astaxanthin as feed supplement in aquatic animals. *Rev. Aquac.* 10(3), 738–773.

Maliwat, Gian Carlo, Stephanie Velasquez, Jan Lorie Robil, Merab Chan, Rex Ferdinand Tayamen, T.M., Ragaza, J.A., 2017. Growth and immune response of giant freshwater prawn *Macrobrachium rosenbergii* (De Man) postlarvae fed diets containing Chlorella vulgaris (Beijerinck). *Aquaculture Res.* 48, 1666–1676.

Martínez-Francés, E., Escudero-Oñate, C., 2018. Cyanobacteria and microalgae in the production of valuable bioactive compounds, in: Martínez-Francés, E. and Escudero-Oñate, C. (Eds.), *Microalgal Biotechnol.* 6, 104–128.

Matos, J., Cardoso, C., Bandarra, N.M., Afonso, C., 2017. Microalgae as healthy ingredients for functional food: A review. *Food Funct.* 1, 2672–2685.

Mirzaie, S., Zirak-Khattab, F., Hosseini, A.H., Donyaei-Darian, H., 2018. Effects of dietary spirulina on antioxidant status, lipid profile, immune response and performance characteristics of broiler chickens reared under high ambient temperature *Asian-Australasian J. Anim. Sci.* 31, 556–563.

Mishra, A., Jha, B., 2009. Isolation and characterization of extracellular polymeric substances from micro-algae *Dunaliella salina* under Salt Stress. *Bioresource Technol.* 100, 3382–3386.

Napier, J.A., Olsen, R.E., Tocher, D.R., 2019. Update on GM canola crops as novel sources of omega-3 fish oils. *Plant Biotechnol. J.* 17, 703–705.

Navarro, F., Forján, E., Vázquez, M., Montero, Z., Bermejo, E., Castaño, M.A., Alberto Toimil, A., et al., 2016. Microalgae as a safe food source for animals: Nutritional characteristics of the acidophilic microalga *Coccomyxa onubensis. Food Nutr. Res.* 60, 30472.

Oh, S. T., Zheng, L., Kwon, H.J., Choo, Y.K., Lee, K.W., C. W. Kang, An, B.K., 2015. Effects of dietary fermented *Chlorella vulgaris* (CBT®) on growth performance, relative organ weights, cecal microflora, tibia bone characteristics, and meat qualities in pekin ducks. *Asian-Australasian J. Anim. Sci.* 28, 95–101.

Oliver, L., Dietrich, T., Marañón, I., Villarán, M.C., Barrio, R.J., 2020. Producing omega-3 polyunsaturated fatty acids: A review of sustainable sources and future trends for the EPA and DHA market. *Resources*, 9, 148.

Packer, M.A., Harris, G.C., Adams, S.L., 2016. Food and feed applications of algae, in: *Algae Biotechnology, Green Energy and Technology.* Springer, Cham.

Patras, D., Moradu, C.V., Socaciu, C., 2019. Bioactive ingredients from microalgae: food and feed applications. *Bull. Univ. Agr. Sci. Vet. Med. Cluj-Napoca Food Sci. Techno.* 76, 1.

Perez-Garcia, O., Bashan, Y., Bashan, Y., Bashan, Y., 2015. Microalgal heterotrophic and mixotrophic culturing for bio-refining: From metabolic routes to techno-economics, in: *Algal Biorefineries: Vol. 2: Products and Refinery Design.* Springer International Publishing, Switzerland.

Pires, J.C.M., 2015. Mass production of microalgae, in: *Handbook of Marine Microalgae: Biotechnology Advances*, pp. 55–68, Elsevier, Amsterdam.

Quico, A.C., Astocondor, M.M., Ortega, R.A., 2021. Dietary supplementation with *Chlorella peruviana* improve the growth and innate immune response of rainbow trout *Oncorhynchus mykiss* fingerlings. *Aquaculture.* 533, 736117.

Rammuni, M.N., Ariyadasa, T.U.A., Nimarshana, P.H.V., Attalage, R.A.A., 2019. Comparative assessment on the extraction of carotenoids from microalgal sources: Astaxanthin from H. pluvialis and β-carotene from D. salina. *Food Chem.* 30, 128–134.

Ramus, J., 1972. The production of extracellular polysaccharide by the unicellular red alga Porphyridium aerugineum. *J. Phycol.* 8, 97–111.

Rehberg-Haas, S., Meyer, S., Tielmann, M., Lippemeier, S., Vadstein, O., Bakke, I., Kjørsvik, E., Evjemo, J.O., Schulz, C., 2015. Use of the microalga Pavlova viridis as enrichment product for the feeding of Atlantic cod larvae (*Gadus morhua*). *Aquaculture.* 438(141), 150.

Remize, M., Brunel, Y., Silva, J.L.S., Berthon, J.Y., Filaire, E., 2021. Microalgae n-3 PUFAs production and use in food and feed industries. *Mar. Drugs.* 19, 113.

Rincón, D., Velásquez, D.H.A., Dávila, M.J., Semprun, A.M., Morales, E.D., Hernández, J.L., 2012. Substitution levels of fish meal by *Arthrospira* (=Spirulina) *maxima* meal in experimental diets for red tilapia fingerlings (*Oreochromis* sp.). *Revista Colombiana de Ciencias Pecuarias.* 25, 3.

Rodrigues, D.B., Menezes, C.R., Mercadante, A.Z., Jacob-Lopes, E., Zepka, L.Q., 2015. Bioactive pigments from microalgae *Phormidium autumnale*. *Food Res. Int.* 77, 273–279.

Rosenberg, J.N., Kobayashi, N., Barnes, A., Noel, E.A., Betenbaugh, M.J., Oyler., G.A., 2014. Comparative analyses of three chlorella species in response to light and sugar reveal distinctive lipid accumulation patterns in the microalga *C. sorokiniana*, in: Duhalt, R.V. (Ed.), *PLoS One.* 9, e92460.

Rumin, J., Nicolau, E., Gonçalves de Oliveira, R., Fuentes-Grünewald, C., Picot, L., 2020. Analysis of scientific research driving microalgae market opportunities in Europe. *Marine Drugs.* 18, 264.

Sahu, A., Pancha, I., Jain, D., Paliwal, C., Ghosh, T., Patidar, S., Bhattacharya, S., Mishra, S., 2013. Fatty acids as biomarkers of microalgae. *Phytochem.* 89, 53–58.

Santiago-Santos, MaC, Ponce-Noyola, T., Olvera-Ramirez, R., Ortega-Lopez, J., Ca~nizares-Villanueva, R. O. 2004. Extraction and purification of phyco-cyanin from Calothrix sp. *Process. Biochem.* 39, 2047–2052.

Sarker, C., Kapuscinski, P.K., McKuin, A.R., Fitzgerald, B., Nash, D.S.F., Greenwood, H.M., 2020. Microalgae-blend tilapia feed eliminates fishmeal and fish oil, improves growth, and is cost viable. *Sci. Rep.* 10, 19326.

Sathasivam, R., Radhakrishnan, R., Hashem, A., Allah, E.F.A., 2019. Microalgae metabolites: A rich source for food and medicine *Saudi J. Biol. Sci*, 26, 709–722.

Shah, M.R., Lutzu, G.A., Alam, A., Sarker, P., Chowdhury, M.A.K., Parsaeimehr, A., Liang, Y., Daroch., M., 2018. Microalgae in aquafeeds for a sustainable aquaculture industry. *J. Appl. Phycol.* 30, 197–213.

Sheikhzadeh, N., Tayefi-Nasrabadi, H., Oushani, A.K., Enferadi, M.H.N., 2012. Effects of *Haematococcus pluvialis* supplementation on antioxidant system and metabolism in rainbow trout (*Oncorhynchus mykiss*). *Fish Physiol. Biochem.* 38, 413–419

Sørensen, M., Berge, G.M., Reitan, K.I., Ruyter, B., 2016. Microalga *Phaeodactylum tricornutum* in feed for Atlantic Salmon (*Salmo salar*) – Effect on nutrient digestibility, growth and utilization of feed. *Aquaculture.* 460, 116–123.

Souza, C.M.M., De Lima, D.C., Bastos, D.S., de Oliveira, S.G., Beirão, B.C.B., Félix, A.P., 2019. Microalgae *Schizochytrium* sp. as a source of docosahexaenoic acid (DHA): Effects on diet digestibility, oxidation and palatability and on immunity and inflammatory indices in dogs. *Anim. Sci. J.* 90, 1567–1574.

Spolaore, P., Joannis-Cassan, C., Duran, E., Isambert, A., 2006. Commercial applications of microalgae. *J.Biosc. Bioeng.* 101, 87–96.

Stengel, D. B., Connnan, S., Popper, Z. A. 2011. Algal chemodiversity and bioactivity: sources of natural variability and implications for commercial application. *Biotechnol. Adv.* 29, 483–501.

Tanaka, T., Shnimizu, M., Moriwaki, I., 2012. Cancer chemoprevention by carotenoids. *Molecules*, 14, 3202–3240.

Trabelsi, L., Ouada, H.B., Zili, F., Mazhoud, N., Ammar, J., 2013. Evaluation of *Arthrospira platensis* extracellular polymeric substances production in photoautotrophic, heterotrophic and mixotrophic conditions. *Folia Microbiol*. 58, 39–45.

Tulli, F., Zittelli, G.C., Giorgi, G. Poli, B.M., Tibaldi, E., Tredici, M.R., 2012. Effect of the inclusion of dried *Tetraselmis suecica* on growth, feed utilization, and fillet composition of European sea bass Juveniles fed organic diets. *J. Aquatic Food Product Technol*. 21, 188.

Velasquez, S.F., Chan, M.A., Abisado, R.G., Traifalgar, R.F.M., Tayamen, M.M., Maliwat, G.C.F., Ragaza, J.A., 2016. Dietary spirulina (*Arthrospira platensis*) replacement enhances performance of juvenile Nile Tilapia (*Oreochromis niloticus*). *J. Appl. Phycol*. 28, 1023–1030.

Venkataraman, L.V., 1997. *Spirulina platensis* (Arthrospira): Physiology, cell biology and biotechnology, in: Vonshak, A. (Ed.), *J. Appl. Phycol*. 9, 295–296.

Vetvicka, V., Teplyakova, T.V., Shintyapina, A.B., Korolenko, A., 2021. Effects of medicinal fungi-derived β-Glucan on tumor progression. *J. Fungi*. 7, 250.

Vieira, M.V., Pastrana, L.M., Fuciños, P., 2020. Microalgae encapsulation systems for food, pharmaceutical and cosmetics applications. *Mar. Drugs* 18, 644.

Waldenstedt, L., Inborr, J., Hansson, I., Elwinger., K., 2003. Effects of astaxanthin-rich algal meal (*Haematococcus pluvalis*) on growth performance, caecal campylobacter and clostridial counts and tissue astaxanthin concentration of broiler chickens. *Anim. Feed Sci. Technol*. 108, 119–132.

Walker, A.B., Berlinsky, D.L., 2011. Effects of partial replacement of fish meal protein by microalgae on growth, feed intake, and body composition of Atlantic Cod. *North Am. J. Aquaculture*. 73, 76–83.

Wang, Xi, Fosse, H.K., Li, K., Chauton, M.S., Vadstein, O., Reitan, K.I., 2019. Influence of nitrogen limitation on lipid accumulation and EPA and DHA content in four marine microalgae for possible use in aquafeed. *Front. Mar. Sci*. 6, 95. doi: 10.3389/fmars.2019.00095

Watanuki, H., Ota, K., Tassakka, A.C.M.A.R., Kato, T., Sakai, M., 2006. Immunostimulant Effects of Dietary *Spirulina platensis* on Carp, *Cyprinus carpio*. *Aquaculture* 258, 1–4.

Wu, Z., Dejtisakdi, W., Kermanee, P., Ma, C., Arirob, W., Sathasivam, R., Juntawong, N., 2017. Outdoor cultivation of *Dunaliella salina* KU 11 using brine and saline lake water with raceway ponds in Northeastern Thailand. *Biotechnol. Appl. Biochem*. 64, 938–943.

Xu, H., Miao, X., Wu, Q., 2006. High quality biodiesel production from a microalga *Chlorella protothecoides* by heterotrophic growth in fermenters. *J. Biotech*. 126, 499–507.

Yan, L., Lim, S.U., H. Kim, I.H., 2012. Effect of fermented Chlorella supplementation on growth performance, nutrient digestibility, blood characteristics, fecal microbial and fecal noxious gas content in growing pigs. *Asian-Australas J. Anim. Sci*. 25, 1742–1747.

Yuan, J.P., Peng, J., Yin, K., Wang., J.H., 2011. Potential health-promoting effects of Astaxanthin: A high-value carotenoid mostly from microalgae. *Mol. Nutr. Food Res*. 55, 150–165.

7 Algal-Sourced Biostimulants and Biofertilizer for Sustainable Agriculture and Soil Enrichment
Algae for Fertilizers and Soil Conditioners

P. Muthukumaran
Department of Biotechnology, Kumaraguru College of Technology, Coimbatore, Tamilnadu, India

J. Arvind
Dhirajlal Gandhi College of Technology, Omalur, Tamil Nadu, India

M. Kamaraj
Department of Biotechnology, College of Biological and Chemical Engineering, Addis Ababa Science and Technology University, Addis Ababa, Ethiopia

A. Manikandan
Department of Industrial Biotechnology, Bharath Institute of Higher Education and Research, Chennai, Tamilnadu, India

CONTENTS

7.1 Introduction...212
7.2 Microalgae Production Costs 214
7.3 Difficulties as GHG Emissions and Resources Utilization.......................215
7.4 Algal Fertilizer Benefits 216
    7.4.1 N and P Demand..216

DOI: 10.1201/9781003195405-7

7.5   Soil Quality Improvements 217
7.6   Details about Algal Bacterial Interactions in Soil 219
       7.6.1   Mutualism 221
       7.6.2   Commensalism 221
       7.6.3   Parasitism 221
7.7   Quoram Sensing – A Molecular Approach to Understand Plant–Algae
       Interaction 222
       7.7.1   Quorum Sensing 222
       7.7.2   Plant-Algae Interaction 223
7.8   Benefits and Detrimental Effects as Soil Conditioner:
       As Microbial Inoculant to the Soil – Alter the Native
       Microbiome/Affect the PGPR – MORE of Antagonism Synergism 223
       7.8.1   Carbon Fixation in Soil 225
       7.8.2   Nutrient Mobilization 225
7.9   Conclusion and Future Directions 225
       7.9.1   A Few Future Directions 227
References 227

7.1   INTRODUCTION

In the era of modern agriculture, the development of sustainable and eco-friendly agriculture systems is being established by monitoring environmental pollutions and ecological disasters in particular ecosystems (Massah and Azadegan, 2016). For example, environmental and ecological problems, which also encompass the inadequacy in soil fertility, biodiversity depreciation, and underground water contamination, were noticed because of excess use of synthetic fertilizers (Bai et al., 2019; Srivastav, 2020). To address these major issues, organic fertilizers from biological resources, such as eukaryotic microalgae or prokaryotic cyanobacteria, and algal fertilizers can fix carbon dioxide and nitrogen in the atmosphere and improve soil fertility (Singh et al., 2019; Chojnacka et al., 2020). In addition, algae can be used as a promising biological resource to recover nutrients from wastewater, and similarly, literature reports revealed that algae could be cultivated in wastewater to mitigate environmental pollution (Papadopoulos et al., 2020; Hu et al., 2020). Due to its versatile applications and uses, algal biomass is being used as biofertilizer to create eco-friendly agriculture (Solovchenko et al., 2016; Renuka et al., 2018).

In modern days of agriculture, due to intensive farming, excessive fertilizers were applied to increase crop yield, production, and productivity. The overuse of synthetic chemicals and fertilizers led to a serious ecologic imbalance in agricultural ecosystems, and even a tremendous reduction in the quality of crops. To address this, agricultural practices are being switched over to organic farming to overcome health issues and meet consumer standards (Kramer et al., 2006). When we switch over to the use of biofertilizers, it leads to an increase in plant growth and yield, and we can overcome the adverse effects of chemical fertilizers. Biofertilizers enhance crop productivity by nitrogen fixation, phosphate solubilization, and production of phytohormones (Pereira and Verlecar, 2005). A few microorganisms, such as *Rhizobium, Azotobacter spp., Bacillus megatherium*, arbuscular mycorrhiza, blue-green algae, seaweed, and earthworms, were used as biofertilizers (Karthikeyan et al., 2008). Among several marine renewable resources, seaweed is one of the promising sources that can be used as food, feed, fodder, fertilizer, agar, alginate, carrageenan, and a source of various useful chemicals (Sahoo, 2000). In recent days, seaweed has been used as fertilizer to replace synthetic fertilizer (Crouch and van Staden, 1993) (Figure 7.1).

To meet our current demand, that is, ecological and economic benefits, various research communities started to establish three major representations to exploit algal biofertilizers in agronomy: i) algal biomass, which could be transformed and utilized by the soil microflora, assisting as a slow-release fertilizer to continually deliver nutrients to plants (Dineshkumar et al., 2018; Saadaoui et al., 2019); ii) live algal cells could be augmented into the soil to promote the microbial community, advance the

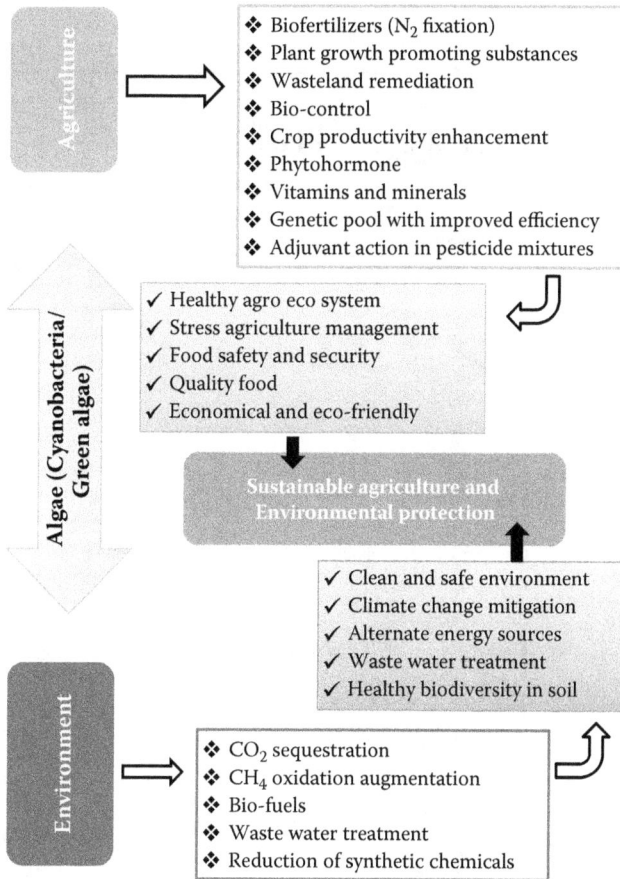

FIGURE 7.1 Overview of the role of algae in agriculture and the environment.

soil fertility, and regulate soil moisture (Sepehr et al., 2019); and iii) algae extract encompassing amino acids and minerals can be capitalized on as liquid bio-fertilizer, sprayed onto the plant leaves' surface (Deepika and Mubarakali, 2020) (Figure 7.1).

For example, *Nannochloropsis* sp., *Chlorella* sp., and *Spirulina* sp. can be used as slow-releasing biofertilizers (Dineshkumar et al., 2018). Similarly, liquid biofertilizers chiefly supply the algal extricates supplemented with the nutrients essential to plants' growth. Vijayakumar et al. (2019) demonstrated the effects of liquid biofertilizer enriched with amino acids and essential minerals on *Capsicum annum*: enhanced the root length, total dry weight, leaf area, and the number of branches and pods. Various plants and algae used as biofertilizers and biostimulants are listed in Tables 7.1 and 7.2.

7.2 MICROALGAE PRODUCTION COSTS

Mass production of microalgae by outdoor cultivation requires approximately 23 € per kg of biomass (Coppens et al., 2016); the microalgal harnessing cost was assessed at production, equivalent to 289 € per kg of nitrogen. Synthetic chemical fertilizers with a

TABLE 7.1
Major Agricultural Crop and Algae for Experimental Studies

Sl. No.	Crops	Algae Used	References
1	Rice	*Anabaena* and *Nostoc*	(Venkataraman, 1981)
2		*Calothrix elenkinii*	(Priya et al., 2015; Ranjan et al., 2016)
3		*Anabaena*	(Prasanna et al., 2013)
4	Wheat	*Cylindrospermum sphaerica*	(Nisha et al., 2007)
5		*Anabaena*	(Prasanna et al., 2016)
6		*Calothrix ghosei, Hapalosiphon intricatus, Nostoc sp*	(Karthikeyan et al., 2007)
7		Cyanobacterial consortia and their consortia with eubacteria	(Rana et al., 2012; Rana et al., 2015)
8		Consortia of cyanobacteria and green algae	(Renuka et al., 2017)
9		*Chlorella vulgaris*	(Uysal et al., 2015)
10	Maize	*Anabaena*	(Prasanna et al., 2016)
11		Nostoc strain 2S9B and Anabaena strains LC2, C5	(Svircev et al., 1997)
12		*Chlorella-Bacteria*	(Yilmaz and Sönmez, 2017)
13		*Chlorella vulgaris*	(Uysal et al., 2015)
14	Tomato	*Acutodesmus dimorphus*	(Gomiero et al., 2011)
15		*Nannochloropsis*	(Coppens et al., 2016)
16	Pearl millet	*Anabaena doliolum HH-209*	(Nisha et al., 2007)
17	*Beta vulgaris*	*Nostoc strain 2S9B and Anabaena strains LC2, C5*	(Svircev et al., 1997)

TABLE 7.2
Microalgal and Cyanobacteria and Their Applications

Sl. No.	Algal Species	Used As	References
1	Cyanobacteria *Spirulina platensis*	Foliar treatment for brinjal	(Dias et al., 2016)
3	*Consortium* ZOB1	–	(Zayadan et al., 2014)
4	*Anabaena sp. Aulosira sp. Cylindrospermum sp. Nostoc sp. Tolypothrix sp*	Rice cultivation	(Ashok et al., 2017)
6	Nitrogen-fixing cyanobacteria	Rice cultivation	(Padhy et al., 2016)
7	*Frankia Hsli*10	–	(Srivastava and Mishra, 2014)
8	Microalgae *Chlorella sp*	Soil mulching	(Marks et al., 2017)
9	*Chlorella vulgaris*	Soil application and foliar spray	(Ozdemir et al., 2016)
10	*Chlorella pyrenoidosa*	Rice seed treatment	(Yadavalli and Heggers, 2013)
11	*Acutodesmus dimorphus*	Soil application and foliar spray	(Garcia-Gonzalez and Sommerfeld, 2016)
12	*Microcystis aeruginosa* MKR 0105 *Anabaena* PCC 7120, *Chlorella sp*	Foliar spray	(Grzesik et al., 2017)

maximum of 14% N, 7% P, 15% K, similarly, slow-releasing organic fertilizers with 4% N, 2% P, 5% K with a market value of 7.9 € and 11 € per kg of nitrogen, respectively, the cost established via literature survey, the existing microalgae cultivation systems cost 5 €/kg of microalgae biomass production in an open pond; in case of bioreactors, it goes up to 50 €/kg (Fernández et al., 2019). Similarly, Acién et al. (2012) reported that the total cost of algal biomass reduced from 69 €/kg to 12.6 €/kg by scaling up studies. In dry algal biomass, the nitrogen content is around 5.87%; for that, the unit cost of nitrogen will be approximately 214.65 €/kg. By comparing all those literature, other than the advantages of algal fertilizer, it was noticed that the cost of microalgae-based biofertilizer is always greater than synthetic fertilizer (Oostlander et al., 2020).

7.3 DIFFICULTIES AS GHG EMISSIONS AND RESOURCES UTILIZATION

The elevated levels of Green House Gases (GHGs) in the atmosphere care ausing tremendous climatic change, which includes temperature escalation, fluctuations in rainfall patterns, increases in the sea levels, floods, droughts, and increased occurrences of extreme climatic phenomena, etc., which is nothing but "global warming" (Dawson et al., 2011). Microalgae mitigate atmospheric CO_2 levels by biological CO_2 fixation (Brennan and Owende, 2010; Mutanda et al., 2011). For example, 1 ton of microalgal biomass can fix CO_2 up to 1.83 tonnes, and it is 10–50 times more efficient than photosynthetic plants (Wang et al., 2008). Due to modern agricultural practices, agricultural product yields have increased, which has lead to increased food security and limited agricultural land-usage and greenhouse gas emissions (Burney et al., 2010). The use of agro-waste as feedstocks for biofuel production is insufficient to meet our food production (Searchinger et al., 2015); to overcome this, microalgae are one of the promising biological resources to increase productivity yields of crops (Moody et al., 2014). Several works of literature revealed that using algal biomass as a source of animal or human food reduces emissions and other environmental impacts to the ecosystem.

7.4 ALGAL FERTILIZER BENEFITS

7.4.1 N AND P DEMAND

In general, both nitrogen (N) and phosphorus (P) in soil determine agricultural crop yields. At the same time, excessive anthropogenic N and P may have a tremendous impact on natural environments, and it leads to ecological and evolutionary consequences; it may start from individual species to entire ecosystems. For example, (i) reduced biodiversity (Lambers et al., 2010), and indirect effects on biodiversity, (ii) increased local extinction by competitive species, (iii) altered plant community and ecosystem (Rohr et al., 2016), (iv) reduction in the functional traits of the population, in particular communities and ecosystems (Díaz et al., 2006), and (v) ecosystem service structural alteration.

N and P are essential for photosynthetic processes, cell growth, metabolism, and protein synthesis (Chapin et al., 2011). For more than 50 years, industrial nitrogen was widely produced by the Haber–Bosch process. It leads to increased deposition of oxidized and reduced nitrogen freshwater and coastal areas (Galloway et al., 2013). Similarly, anthropogenic nitrogen deposition also takes place via fossil fuel combustion and atmospheric nitrogen-fixing by legumes (Ciais et al., 2014). It leads to elevated nitrogen deposition in an ecosystem, and it shifts ecosystems from N to P constraint with unfavorable ecosystem impacts (Elser et al., 2010). Similarly, deposition of phosphorus from phosphorus-rich dust from sand and agricultural soils shifts ecosystems from P to N limitation (Brahney et al., 2015).

During the 1970s to 1980s, due to disproportionate use of fertilizer and manure phosphorus in industrially developed nations, low P use efficiency was noticed, which led to huge amounts of surplus phosphorus and residuals reported in soils (Syers 2006). It led to industrialized nations taking action to sort out issues and increase their phosphorus use efficiency by several approaches, such as reduction of input, mining of the residual phosphorus in the soil, and agricultural management practices, to improve uptake of phosphorus by crop plants (Sattari et al., 2012). Similarly, a lot of published data revealed an increase in crop yields (Bouwman et al., 2017). In upcoming years, phosphorus usage will determine sustainable production of foods to meet world population growth from 7.3 billion in 2015 to 9.7 billion in 2050. Based on the predicted assessment, regional phosphorus production costs are expected in the range of 20–60% resource depletion by 2100 (Van Vuuren et al., 2010).

Microalgae is one of the main feedstocks that are widely used as main sources of products, such as bioactive compounds, biofuels, biofertilizers, carbon sequestration, and wastewater treatment (Gupta et al., 2013; Renuka et al., 2015). Excessive use of synthetic agrochemicals leads to adverse ecological degradation of soil fertility, eutrophication, and loss of biodiversity (Chagnon et al., 2015; Hallmann et al., 2014; Van der Sluijs et al., 2015). To address this issue, a microalgal biofertilizer is the best remedy to retain the ecological balance of the soil ecosystem. Very particularly, it can be cultivated in wastewater; we can recover excess nutrients from the agricultural runoffs and reuse water, microalgae can be used in the sequestration of carbon dioxide and nitrous oxides in atmospheric air, as well as to reduce greenhouse gas emissions (Brennan and Owende, 2010). Based on literature availability, it was revealed that microalgae biomass was widely used as a biofertilizer (Wuang et al., 2016).

Any form of biofertilizer mainly contains living microorganisms, which enhance the seed germination and growth of plants. Mostly they facilitate the uptake of soil nutrients, such as nitrogen, phosphate, potassium, and other mineral nutrients, by converting them from insoluble to soluble by the production of organic substances (Reddy and Saravanan, 2013). In general, biofertilizers are classified based on the nature of microorganisms and application: nitrogen-fixators; phosphates- and potassium solubilizing biofertilizers; phosphorus-mobilizing biofertilizers; and biofertilizers for secondary macronutrients, zinc and iron solubilizers, plant-growth-promoting rhizobacteria (PGPR), etc. (Bhattacharjee and Dey, 2014). The application of microalgae as a biofertilizer will enhance productivity and crop quality by improving soil fertility (Pérez-Montaño et al., 2014). Few studies report the utilization of microalgae/cyanobacteria as biofertilizers, plant growth promoters, and bio-control agents for various crops (Chaudhary et al., 2012). For wheat crops, these microalgae increase the yield and micronutrient concentration (Rana et al., 2012; Prasanna et al., 2013). Similarly, microalgal consortia along with N/P/K fertilizer will save up to 25% N utilization for wheat cultivation (Renuka et al., 2016). Several studies revealed that microalgal biofertilizers enhance greater nutrient uptake, higher biomass accumulation, and greater crop yields (Faheed and Abd-El Fattah, 2008) (Table 7.3).

7.5 SOIL QUALITY IMPROVEMENTS

As of now, different approaches for soil quality improvement have been proposed, which includes biological nitrogen fixation, application of biofertilizer, soil surface residue reclamation, animal manure recycling, and conservation agriculture (Mupangwa et al., 2012). Soil biota, especially soil microorganisms, play a vital role in the biogeochemical cycle of macro- and micronutrients, plant growth enhancement, enhanced soil fertility, and improved agroecosystem productivity (Lian et al., 2010; Prasanna et al., 2012). The size and number of soil microbial communities in a particular soil will indicate soil fertility and health as well as quality; these parameters are essential for the development of sustainable agriculture (Sparling, 1997) (Figure 7.2).

When we use microalgae as biofertilizers, apart from a supplement of N, P, and potassium (K), it improves soil organic matter, soil structure, and soil moisture and

TABLE 7.3

Microalgal and Cyanobacterial Isolates as Biostimulants and Purpose of Investigation

Sl. No.	Microalgae	Purpose of Investigation	Plant/ Crop Used	References
1	*Acutodesmus dimorphus*	Aqueous extract: seed primer, foliar spray, and biofertilizer to promote seed germination, plant growth enhancement, and fruit production	Tomato	(Garcia-Gonzalez and Sommerfeld, 2016)
2	*Chlorella spp*	Co-cultivation of Chlorella spp and tomato in a hydroponic system	Tomato	(Zhang et al., 2017)
3	Cyanobacterial strains- BF1 *Anabaena torulosa* BF2, *Nostoc carneum* BF3, *Nostoc piscinale* BF4, *Anabaena doliolum, Anabaena sp.* (CW1), *Anabaena sp.* (CR1)	Soil microbial activities and the compositional changes in the microbial communities	Maize	(Prasanna et al., 2016)
4	*Calothrix elenkinii* (RPC1)	Rhizosphere microbiome of rice	Rice	(Ranjan et al., 2016)
5	*Sargassum johnstonii*	Aqueous extract: Used to study growth and yield of tomato plants and to improve lycopene and vitamin C content	Tomato	(Kumari et al., 2011)
6	*Ulva lactuca, Caulerpa sertularioides, Padina gymnospora,* and *Sargassum liebmannii*	Aqueous extract as biostimulants on the germination and growth of tomato plants	Tomato	(Hernández-Herrera et al., 2014)
7	*Sargassum wightii*	Aqueous extract on growth and yield of wheat	Wheat	(Kumar and Sahoo, 2011)
8	Wastewater grew microalgal consortia-cyanobacteria (Cyanophyta) and green algae (Chlorophyta)	Micronutrient enrichment in grains and facilitating enhanced micronutrients (Zn, Fe, Cu, and Mn)	Wheat	(Renuka et al., 2017)

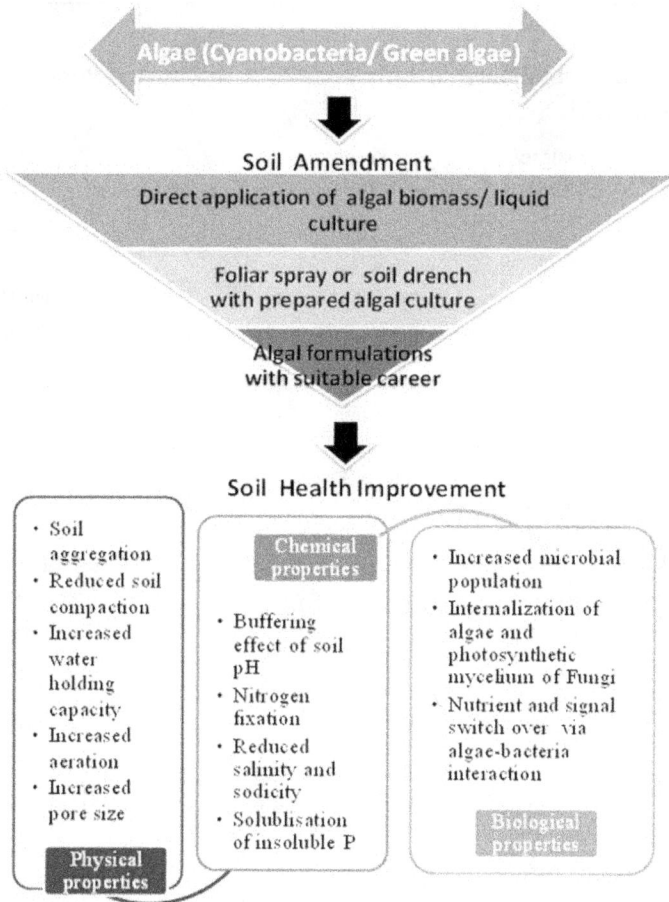

FIGURE 7.2 Overview of benefits of algae application in soil health.

enhances and stimulates microbial colonization and growth with maximal enzymatic activities (Lupatini et al., 2016; Rahman et al., 2017; Sarma et al., 2017), even at very short periods (Suleiman et al., 2018). Several literature reports revealed that different types of algae are widely distributed in soil, which include green (Chlorophyta), blue-green (cyanobacteria), yellow-green (Xanthophyta), and diatoms (Bacillariophyta). This group of algal species can produce polysaccharides, and it possesses the ability of nitrogen fixation (Figure 7.2 and Table 7.4).

7.6 DETAILS ABOUT ALGAL BACTERIAL INTERACTIONS IN SOIL

Most of the algae live in symbiosis with other microorganisms throughout their lifespan in one way or another (Dittami et al., 2014). Organic substances facilitate signaling molecules to nurture bacterial populations in the phycosphere. For example, *Phaeodactylum tricornutum* and *Thalassiasira pseudonana* are involved in

TABLE 7.4

Microalgal and Cyanobacterial Isolates as Biofertilizer and Purpose of Investigation

Sl. No.	Microalgae/ Cyanobacteria	Purpose of Investigation	Plant/ Crop Used	References
1	*Acutodesmus dimorphus*	Aqueous extract: seed primer, foliar spray, and biofertilizer to evaluate seed germination, plant growth, and fruit production	Tomato	(Garcia-Gonzalez and Sommerfeld, 2016)
2	Dry biomass of *Nannochloropsis spp., Ulothrix spp.,* and *Klebsormidium spp.*	Quality and economic value of tomato fruits	Tomato	(Coppens et al., 2016)
3	Dry biomass of *Nannochloropsis spp., Ulothrix spp.* and *Klebsormidium spp.*	Improving carotenoids in petals of the rose	Rose	(Lachman et al., 2001)
4	*C. vulgaris*	Assessing effects of agronomic and physiological responses on lettuce	Lettuce	(Faheed and Abd-El Fattah, 2008)
5	Aquaculture wastewater and a marine culture of *Nannochloropsis* spp	N mineralization study		(Stadler et al., 2006)
6	*Nostoc spp., Hapalosiphon spp.,*	Seed germination, root and shoot growth, the weight of rice grains, and protein content of rice	Rice	(Misra and Kaushik, 1989)
7	*Aulosira fertilissima*	Seed germination, root and shoot growth, the weight of rice grains, and protein content of rice	Rice	(Singh and Trehan, 1973)
8	*Azotobacter chroococcum* (W5), *Mesorhizobium ciceri* (F 75), *Pseudomonas striata* (P27), and *Serratia marcescens* (L11)	Inoculation of wheat seeds with cyanobacterial bacterial biofilm	Wheat	(Swarnalakshmi et al., 2013)

nitrogen and organic carbon utilization and cell wall assembly, co-occurring these algae with bacteria around 200 million years (Bowler et al., 2008). Similarly, bacteria enhance algal growth, spore germination, morphogenesis, and pathogen resistance (Amin et al., 2015; Ramanan et al., 2016). Algae–bacteria interactions may facilitate all possible symbiotic relationships, though, and these types of interactions are explored in the planktonic zone but not yet completely (Ramanan et al., 2016). Algal and bacterial interactions in the soil as well as any ecosystem are broadly classified into three types: i) Mutualism, ii) Commensalism, and iii) Parasitism.

7.6.1 MUTUALISM

Mutual interactions are extremely rare in algal as well as bacterial species. Mostly these kinds of interactions occur in very rare events that are species specific. The main motto of this interaction between two species mostly exchanges nutritional requirements from algae to bacteria and/or vice versa. For example, micronutrients like vitamins (Teplitski and Rajamani, 2011) and macronutrients such as nitrogen and carbon (Kazamia et al., 2012; Kim et al., 2014) are mostly exchanged between by algae and bacteria. For these kinds of mutual interactions, they alter their metabolism.

7.6.2 COMMENSALISM

In this interaction, any organism, either bacteria or algae, benefits. This interaction is considered between non-interacting partners (Zapalski, 2011). For example, *Chlamydomonas reinhardtii* and heterotrophic bacteria widely occurred as commensal. This alga *C. reinhardtii* uses vitamin B_{12} from heterotrophic bacteria, but bacteria do not utilize organic carbon, which was released by *C. reinhardtii*. Another study revealed vitamin B_{12}-dependent microalga *Lobomonas rostrata* and the bacterium *Rhizobium loti* (Kazamia et al., 2012). Villa et al. (2014) demonstrated co-cultures of microalgae with *Azotobacter vinelandii*. In this study, microalgal species *Neochloris oleoabundans* and *Scenedesmus* sp. BA032 used *A. vinelandii* siderophores as a nitrogen source; this is another example to understand commensalism between algae and bacteria. In most cases, alga gain benefits from bacterial species.

7.6.3 PARASITISM

Parasitism is one of the best examples of detrimental interaction, where only one species benefits and the another one is affected. In most cases, smaller organisms act as a parasite, particularly algae-bacterial interactions; in most of the cases, microalgae and cyanobacterial blooms dominate the growth of other microorganisms (Wang et al., 2010). Around 10% of known algae (red algae) are parasitic (Hancock et al., 2010). Secreting glucosidases, chitinases, cellulases, and other enzymes by bacteria showed parasitism against algal species by rupturing of the algal cell wall (Wang et al., 2010). For example, most of the parasitic bacteria are located near the algal cell wall to degrade the algal cell wall (Wang et al., 2010), and these parasites, in general,

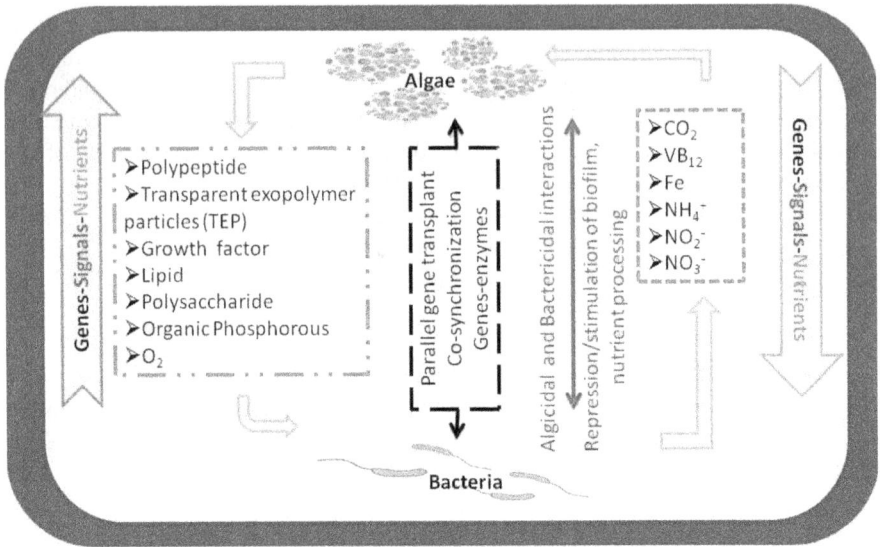

FIGURE 7.3   Algal-bacterial interactions.

have huge industrial demands and applications. Enzymes, such as cellulases, hemi-cellulases, and pectinases, are obtained from these bacterial parasites and used for various industries like food, brewery, wine, textile, and paper industries (Bhat, 2000). Similarly, chitinolytic enzymes from bacterial parasites, used for the preparation of pharmaceutical oligosaccharides, are used to prevent the spread of pathogenic transmission (Dahiya et al., 2006) (Figure 7.3).

In general, a variety of interactions between algae and bacteria have been described in various studies, which can range from beneficial to detrimental toward algal growth. By understanding, interaction and control of certain interactions may serve as a highly useful study to enhance low volume, high-value products from algae species, or we can control algal blooms and their adverse effects on ecosystems, or to enhance biomass production with low production cost by co-culture of bacterial consortia.

7.7   QUORAM SENSING – A MOLECULAR APPROACH TO UNDERSTAND PLANT–ALGAE INTERACTION

7.7.1   QUORUM SENSING

Quorum sensing is the best well-known gene regulatory system studied in bacteria that directly contributes to biofilm formation (Pasmore and Costerton, 2003; Wolfe et al., 2003; Stanley and Lazazzera, 2004; Dobretsov et al., 2009). Acyl homoserine lactones (AHLs) are the best example of bacterial quorum sensing signals. For example, inside bacterial cells, LuxR-like regulators and LuxR–AHL complex bind within the promoters region of the genes (Zhang et al., 2002; Koch et al., 2005). The role of quorum sensing has limitless uses and applications. A few roles are (i) it is involved in biofilm formation and surface attachments that facilitate colonization of bacteria; (ii) it may favor nutrient acquisition by bacteria; (iii) it regulates microbial population dynamics by the production of antimicrobial compounds; and (iv) it induces phenotypic heterogeneity in unpredicted environments (Rolland et al., 2016).

An array of published literature states that bacterial quorum sensing is altered by compounds from algae and vascular plants (Koch et al., 2005; Gao et al., 2007; Skindersoe et al., 2008). One of the pioneering researches revealed the occurrence of AHLs in cyanobacterial blooms (Bachofen and Schenk, 1998). A famous freshwater alga *Chlamydomonas reinhardii* can synthesis at least a dozen chemical compounds that mimic bacterial AHL activity, and can promote diverse quorum sensing receptors. Similarly, both *Chlamydononas reinhardtii* and *Chlorella sp.* can stimulate quorum sensing reliant luminescence of *Vibrio harveyi* (Teplitski et al., 2004).

7.7.2 PLANT-ALGAE INTERACTION

Several studies revealed that various group of cyanobacteria colonized the rhizosphere region and various plant parts (Gantar and Elhai, 1999; Krings et al., 2009), and also, a few studies reported that, cyanobacteria shows symbiotic association with algae, fungi, gymnosperms, pteridophytes, and vascular plants (Meeks and Elhai, 2002; Santi et al., 2013). Similarly, *Calothrix elenkinii* could colonize in root and shoot tissues of rice, and it enhances N fixing and phosphate solubilizing microbial populations (Priya et al., 2015). Gantar et al. (1993) demonstrated by wheat plant with two cyanobacterial isolates, namely *Nostoc* 2S9B and Anabaena CS. *Nostoc* 2S9B showed association and colonization in wheat plant roots, and no colonization of Anabaena CS with plant roots was noticed. Experiments using the wheat plant as a model plant, when radiolabelled N, elucidated that the cyanobacteria commit to the plant N, regardless of their colonization/affiliation with the host plant (Gantar et al., 1993). Similarly, Nostoc 2S9B colonize between intercellular spaces, and the epidermal and cortex cell region of wheat root tissue (Gantar, 2000; Gantar and Elhai, 1999; Nilsson et al., 2002).

Table 7.5 shows various types of associations with plants by algae, especially cyanobacteria.

7.8 BENEFITS AND DETRIMENTAL EFFECTS AS SOIL CONDITIONER: AS MICROBIAL INOCULANT TO THE SOIL – ALTER THE NATIVE MICROBIOME/AFFECT THE PGPR – MORE OF ANTAGONISM SYNERGISM

Literature reports revealed that various studies that support both essential research and functional research, the beneficial attributes of algal bio-fertilizers are huge, such as it improves and facilitates plant growth, increases crop yield, promotes association with soil microorganisms, and improves the nutritional value of fruits and seed germination, etc. (Renuka et al., 2018; Vijayakumar et al., 2019; Deepika and Mubarak Ali, 2020).

For example, in the soil, algal biomass gets assimilated, and the nutrients release are effectively catalyzed by the soil microorganisms. Proteobacteria (35–42%), Acidobacteria (12–15%), and Bacteroidetes (8–10%) are three major microbial species that play a vital role in the cucumber rhizosphere (Lv et al., 2020).

TABLE 7.5

Symbiotic Association of Algae with Vascular and Non-vascular Plants

Sl. No.	Plant	Algae	Nature of Association	References
1	Wheat	*Anabaena laxa* and *Calothrix* sp.	Colonisation of roots and stems	(Karthikeyan et al., 2009; Babu et al., 2015)
		Nostoc sp. strain 2S9B	Establishment of para-nodules	(Gantar and Elhai, 1999)
		Nostoc sp. strain 2S9B	Colonization in roots	(Gantar et al., 1993)
		Nostoc sp. strain 2S9B	Colonization of roots	(Gantar, 2000)
2	Rice	*Anabaena laxa* and *Calothrix* sp.	Colonisation of roots and stems	(Bidyarani et al., 2015)
		Nostoc strains	Colonisation in roots- surface and intracellular surfaces	(Nilsson et al., 2002)
3	Cotton	Biofilms of *Anabaena* with Trichoderma/ Azotobacter	Colonisation of roots	(Triveni et al., 2015)
4	Cycads (Maccrozamia, Lepidozamia, Bowenia and Cycas)	*Nostoc* spp., *Tolypothrix* sp. and *Leptolyngbya* sp.	Endosymbiont	(Cuddy et al., 2012)

Similarly, after the application of cyanobacteria as biofertilizer, apart from nitrogen fixation, cyanobacteria release certain extracellular polymeric substances and plant growth-promoting (PGP); these substances alter soil quality and favour the plant's growth (Li et al., 2019). Some cyanobacteria minimize the frequency of agricultural diseases by arresting the growth of *Fusarium wilt, Rhizoctonia solani, Sclerotinia sclerotiorum,* and*Alternaria porri* (Roberti et al., 2015; Prasanna et al., 2015).

Apart from this, some species of algae, especially cyanobacteria, showed a potential threat to the environmental ecosystem. For example, cyanobacteria are normally applied to soil as biofertilizer nitrogen, but one of the potential threats is a cyanobacterial toxin and biological invasion (Do Nascimento et al., 2019; Schaefer et al., 2020). A few of them are microcystins, nodularin, aplysiatoxins, cylin-drospermopsin, saxitoxins, anatoxin, and lyngbyatoxin (Van Apeldoorn et al., 2007). Similarly, living cyanobacteria wildly used in aquatic ecosystems migrate to the aquatics near the agricultural lands, culminating in algal bloom and biological invasion in a particular ecosystem (Chen et al., 2006). Moreover, some other unfavorable factors include residual nutrients in the particular ecosystem after algae intensive cultivation, difficulties in transportation and storage of liquid algal bio-fertilizer, and uncontrolled rate of degradation of algal biomass when we use algae as extended-release bio-fertilizer; the negligible rate in consumer-preference compared as an alternative synthetic fertilizer, poses as the major challenge in modern agriculture by using algal biofertilizer as a biological resource.

7.8.1 CARBON FIXATION IN SOIL

Every year, about 0.05% of this global energy is captured by photosynthetic plants by photosynthesis (Zhu et al., 2008). Literature reports revealed that microbial growth, multiplication, and metabolic activity of microorganisms in the soil eco-system help to improve the carbon content in the soil by releasing nutrients from an unavailable state to an available state (Docampo et al., 2010; Manjunath et al., 2011; Malik et al., 2013; Malik et al., 2013). Similarly, soil microorganisms facilitate crop improvement, nutrient cycling, and improved crop productivity (Singh et al., 2019). About 35% of terrestrial carbon comprises 6–9% of the total earth's surface. The estimated carbon in wetlands is rates up to 3 $gCm^{-2}d^{-1}$ (Bonneville et al., 2008; Zhou et al., 2009) (Figure 7.4).

7.8.2 NUTRIENT MOBILIZATION

Technically, micronutrients play a vital role in the metabolism of plants, and the physiological function of plants and their plant products' quality depends on available nutrients in the soil (Rana et al., 2012; Prasanna et al., 2013). In developing nations like Southeast Asia and sub-Saharan Africa alone, around 2 billion people suffered from micronutrient deficiency, compared to the rest of the world (Ramakrishnan, 2002). To address these issues, a lot of agricultural practices are implemented to increase yields; along with this, biofortification plays an additional role in enhancing the nutritional status of edible plant parts. Apart from this, due to global demands, excessive application of synthetic fertilizers take place, and it leads to decreased accumulation of important micronutrients, such as Zn, Fe, Cu, and Mn in cereal grains, and these micronutrients are essential for the human diet (Zhang et al., 2012).

7.9 CONCLUSION AND FUTURE DIRECTIONS

Microalgae and their products, either in the form of stimulants and/or biofertilizers, gain greater attention for crop productions to improve the quality as well as productivity of various crops. Based on literature survey, it was noticed that different formulations are readily available for almost all categories of soil ecosystems and crops. By comparing synthetic chemicals, cost of production, and other advantages and particularly their eco-friendly nature, microalgae and their products emerge as an area for various industrial applications, which include food, pharma, agricultural, environmental, and health sectors. These products enhance the growth, yield of agricultural crops by mobilizing, solubilizing, or uptake of nutrients by plants. However, there are several scopes that are there to enhance the application of

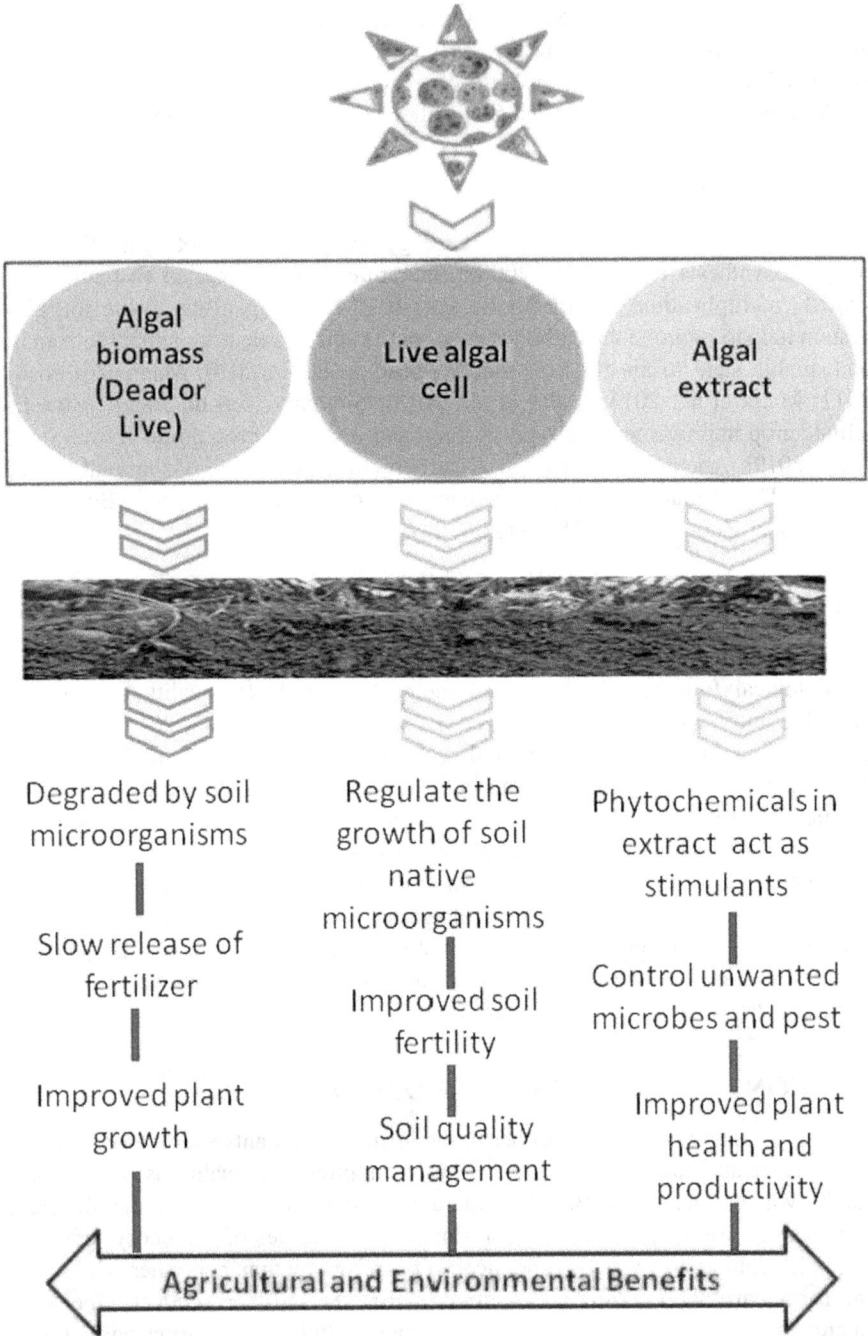

FIGURE 7.4 Benefits of algae to agriculture and the environment.

microalgae in agricultural sectors, particularly how this microalgal and their algal-based products can be mass-produced to meet our industrial demand.

7.9.1 A FEW FUTURE DIRECTIONS

1. Microalgae isolates are commercially available, but the application of algae and algae products in the field of agriculture is limited. We need to fill the gap.
2. At the time of microalgae production, several abiotic factors can influence the biochemical composition of microalgae, which may alter the properties of final products (biofertilizer and biostimulants). We can find an alternative sustainable solution for the production of algal-based products.
3. Crop cultivation, a different agronomic practice, may influence crop growth and yield. It is highly challengeable and needs to be addressed based on the performance of biofertilizers and biostimulants from algal sources.
4. Various biotic and abiotic factors may influence the growth of the plant as well as algae. We need to identify an appropriate alternative solution to manage/control both biotic and abiotic factors in soil and the association of algae with soil, soil biota, plants, etc.
5. Potential threats of cyanobacteria to the environment are due to toxins in certain species of cyanobacteria that may harm a particular ecosystem. Care should be taken before the selection of algal saucers as biofertilizers and/ or biostimulants.
6. The high cost of algal biomass is another challenge in algal research. For this, we need to identify cost-effective and eco-friendly cultivation practices for mass production of algal biomass.

REFERENCES

Acién, F.G., Fernández, J.M., Magán, J.J., Molina, E., 2012. Production cost of a real microalgae production plant and strategies to reduce it. *Biotechnol Adv.* 30(6):1344–1353.

Amin, S. A., Hmelo, L. R., van Tol, H. M., Durham, B. P., Carlson, L. T., Heal, K. R., Morales, R. L., Berthiaume, C. T., Parker, M. S., Djunaedi, B., Ingalls, A. E., Parsek, M. R., Moran, M. A., Armbrust, E. V., 2015. Interaction and signalling between a cosmopolitan phytoplankton and associated bacteria. *Nature* 522(7554), 98–101.

Ashok, A.K., Ravi, V., Saravanan, R., 2017. Influence of cyanobacterial auxin on sprouting of taro (Colocasia esculenta var. antiquorum) and corm yield. Ind. *J. Agric. Sci.* 87(11), 1437–1444.

Babu, S., Prasanna, R., Bidyarani, N., Singh, R., 2015. Analysing the colonisation of inoculated cyanobacteria in wheat plants using biochemical and molecular tools. *J. App. Phycol.* 27(1), 327–338.

Bachofen, R., Schenk, A., 1998. Quorum sensing autoinducers: Do they play a role in natural microbial habitats? *Microbiol. Res.* 153(1), 61–63.

Bai, X., Wang, Y., Huo, X., Salim, R., Bloch, H., Zhang, H., 2019. Assessing fertilizer use efficiency and its determinants for apple production in China. *Ecol. Indic.* 104, 268–278.

Bhat, M.K., 2000. Cellulases and related enzymes in biotechnology. *Biotechnol. Adv.* 18(5), 355–383.

Bidyarani, N., Prasanna, R., Chawla, G., Babu, S., Singh, R., 2015. Deciphering the factors associated with the colonization of rice plants by cyanobacteria. *J. Basic Microbiol.* 55(4), 407–419.

Bonneville, M.C., Strachan, I.B., Humphreys, E.R., Roulet, N.T., 2008. Net ecosystem CO_2 exchange in a temperate cattail marsh in relation to biophysical properties. *Agric. Forest Meteorol.* 148(1), 69–81.

Bouwman, A.F., Beusen, A.H.W., Lassaletta, L., Van Apeldoorn, D.F., Van Grinsven, H.J. M., Zhang, J., 2017. Lessons from temporal and spatial patterns in global use of N and P fertilizer on cropland. *Sci. Rep.* 7(1), 40366.

Bowler, C., Allen, A.E., Badger, J.H., Grimwood, J., Jabbari, K., Kuo, A., Maheswari, U., et al., 2008. The Phaeodactylum genome reveals the evolutionary history of diatom genomes. *Nature* 456(7219), 239–244.

Brahney, J., Mahowald, N., Ward, D.S., Ballantyne, A.P., Neff, J.C., 2015. Is atmospheric phosphorus pollution altering global alpine Lake stoichiometry? *Global Biogeochem. Cycles* 29(9), 1369–1383.

Brennan, L., Owende, P., 2010. Biofuels from microalgae—A review of technologies for production, processing, and extractions of biofuels and co-products. *Renew. Sustain. Energy Rev.* 14(2), 557–577.

Burney, J.A., Davis, S.J., Lobell, D.B., 2010. Greenhouse gas mitigation by agricultural intensification. *Proceedings of the National Academy of Sciences of the United States of America*, 107(26), 12052–12057.

Chagnon, M., Kreutzweiser, D., Mitchell, E.A.D., Morrissey, C.A., Noome, D.A., & Van der Sluijs, J.P., 2015. Risks of large-scale use of systemic insecticides to ecosystem functioning and services. Environ. *Sci. Poll. Res. Int.* 22(1), 119–134.

Chapin, F. S., Matson, P. A., Vitousek, P., 2011. *Principles of terrestrial ecosystem ecology.* Springer Science and Business Media.

Chaudhary, V., Prasanna, R., Nain, L., Dubey, S. C., Gupta, V., Singh, R., Jaggi, S., Bhatnagar, A.K., 2012. Bioefficacy of novel cyanobacteria-amended formulations in suppressing damping off disease in tomato seedlings. *World J. Microbiol. Biotechnol.* 28(12), 3301–3310.

Chen, W., Song, L., Gan, N., Li, L., 2006. Sorption, degradation and mobility of microcystins in Chinese agriculture soils: risk assessment for groundwater protection. *Environ. Pollut.* 144, 752–758.

Chojnacka, K., Moustakas, K., Witek-Krowiak, A., 2020. Bio-based fertilizers: A practical approach towards circular economy. *Bioresour. Technol.* 295, 122223.

Ciais, P., Sabine, C., Bala, G., Bopp, L., Brovkin, V., Canadell, J., Thornton, P., 2014. Carbon and other biogeochemical cycles. In Climate Change 2013: the physical science basis. *Contribution of Working Group I to the Fifth Assessment Report of the Intergovernmental Panel on Climate Change.* Cambridge University Press, pp. 465–570.

Coppens, J., Grunert, O., Van Den Hende, S., Vanhoutte, I., Boon, N., Haesaert, G., De Gelder, L., 2016. The use of microalgae as a high-value organic slow-release fertilizer results in tomatoes with increased carotenoid and sugar levels. *J. App. Phycol.* 28(4), 2367–2377.

Crouch, I.J., van Staden, J., 1993. Evidence for the presence of plant growth regulators in commercial seaweed products. *Plant Growth Regul.* 13(1), 21–29.

Cuddy, W.S., Neilan, B.A., Gehringer, M.M., 2012. Comparative analysis of cyanobacteria in the rhizosphere and as endosymbionts of cycads in drought-affected soils. *FEMS Microbiol. Ecol.* 80(1), 204–215.

Dahiya, N., Tewari, R., Hoondal, G.S., 2006. Biotechnological aspects of chitinolytic enzymes: A review. *Appl. Microbiol. Biotechnol.* 71(6), 773–782.

Dawson, T.P., Jackson, S.T., House, J.I., Prentice, I.C., Mace, G.M., 2011. Beyond predictions: Biodiversity conservation in a changing climate. *Science* 332(6025), 53–58.

Deepika, P., Mubarakali, D., 2020. Production, and assessment of microalgal liquid fertilizer for the enhanced growth of four crop plants. *Biocatal. Agric. Biotechnol.* 28, 101701.

Dias, G.A., Rocha, R.H.C., Araújo, J.L., De Lima, J.F., Guedes, W.A., 2016. Growth, yield, and postharvest quality in eggplant produced under different foliar fertilizer (*Spirulina platensis*) treatments. *Semina: Ciências Agrárias*, 37(6), 3893–3902.

Díaz, S., Fargione, J., Chapin III, F. S., Tilman, D., 2006. Biodiversity loss threatens human well-being. *PLOS Biol.* 4(8), e277.

Dineshkumar, R., Kumaravel, R., Gopalsamy, J., Sikder, M. N. A., Sampathkumar, P., 2018. Microalgae as bio-fertilizers for rice growth and seed yield productivity. *Waste Biomass Valor.* 9(5), 793–800.

Dittami, S.M., Eveillard, D., Tonon, T., 2014. A metabolic approach to study algal–bacterial interactions in changing environments. *Molecul. Ecol.* 23(7), 1656–1660.

Dobretsov, S., Teplitski, M., Paul, V., 2009. Mini-review: Quorum sensing in the marine environment and its relationship to biofouling. *Biofouling* 25(5), 413–427.

Docampo, R., Ulrich, P., Moreno, S.N.J., 2010. Evolution of acidocalcisomes and their role in polyphosphate storage and osmoregulation in eukaryotic microbes. *Philos. Trans. R. Soc. Lon. B Biol. Sci.* 365(1541), 775–784.

Do Nascimento, M., Battaglia, M. E., Sanchez Rizza, L., Ambrosio, R., Arruebarrena Di Palma, A., Curatti, L., 2019. Prospects of using biomass of N_2-fixing cyanobacteria as an organic fertilizer and soil conditioner. *Algal Res.* 43, 11.

Elser, J.J., Peace, A.L., Kyle, M., Wojewodzic, M., McCrackin, M. L., Andersen, T., Hessen, D.O., 2010. Atmospheric nitrogen deposition is associated with elevated phosphorus limitation of lake zooplankton. *Ecol. Lett.* 13(10), 1256–1261.

Faheed, F.A., Abd-El Fattah, Z., 2008. Effect of *Chlorella vulgaris* as biofertilizer on growth parameters and metabolic aspects of lettuce plant. *J. Agric. Soc. Sci.* 4, 165–169.

Fernández, F.G.A., José María Fernández, S., Grima, E.M., 2019. *Costs analysis of microalgae production.* Biofuels from algae (2nd ed) (pp. 551–566).

Galloway, J.N., Leach, A.M., Bleeker, A., & Erisman, J.W., 2013. A chronology of human understanding of the nitrogen cycle. *Philos. Trans. R. Soc. Lon. B Biol. Sci.* 368(1621), 20130120.

Gantar, M., 2000. Mechanical damage of roots provides enhanced colonization of the wheat endorhizosphere by the dinitrogen-fixing cyanobacterium *Nostoc* sp. strain 2S9B. *Biol. Fert. Soils* 32(3), 250–255.

Gantar, M., Elhai, J., 1999. Colonization of wheat para-nodules by the N2-fixing cyanobacterium *Nostoc* sp. strain 2S9B. *New Phytol.* 141(3), 373–379.

Gantar, M., Kerby, N.W., Rowell, P., 1993. Colonization of wheat (*Triticum vulgare* L.) by N_2-fixing cyanobacteria: III. The role of a hormogonia-promoting factor. *New Phytol.* 124(3), 505–513.

Gao, M., Chen, H., Eberhard, A., Gronquist, M.R., Robinson, J.B., Connolly, M., Teplitski, M., Rolfe, B.G., Bauer, W.D., 2007. Effects of AiiA-mediated quorum quenching in *Sinorhizobium meliloti* on quorum-sensing signals, proteome patterns, and symbiotic interactions. *Mol Plant Microbe Interact.* 20(7), 843–856.

Garcia-Gonzalez, J., Sommerfeld, M., 2016. Biofertilizer and biostimulant properties of the microalga *Acutodesmus dimorphus*. *J. Appl. Phycol.* 28(2), 1051–1061.

Gomiero, T., Pimentel, D., Paoletti, M.G., 2011. Environmental impact of different agricultural management practices: Conventional vs. organic agriculture. *Crit Rev Plant Sci.* 30(1–2), 95–124.

Grzesik, M., Romanowska-Duda, Z., Kalaji, H.M., 2017. Effectiveness of cyanobacteria and green algae in enhancing the photosynthetic performance and growth of willow (*Salix viminalis* L.) plants under limited synthetic fertilizers application. *Photosynthetica* 55(3), 510–521.

Gupta, V., Ratha, S.K., Sood, A., Chaudhary, V., Prasanna, R., 2013. New insights into the biodiversity and applications of cyanobacteria (blue-green algae)—Prospects and challenges. *Algal Res.* 2(2), 79–97.

Hallmann, C.A., Foppen, R.P.B., van Turnhout, C.A.M., de Kroon, H., Jongejans, E., 2014. Declines in insectivorous birds are associated with high neonicotinoid concentrations. *Nature* 511(7509), 341–343.

Hancock, L., Goff, L., Lane, C., 2010. Red algae lose key mitochondrial genes in response to becoming parasitic. *Genome Biol. Evol.* 2, 897–910.

Hernández-Herrera, R.M., Santacruz-Ruvalcaba, F., Ruiz-López, M.A., Norrie, J., Hernández-Carmona, G., 2014. Effect of liquid seaweed extracts on growth of tomato seedlings (*Solanum Lycopersicum* L.). *J. Appl. Phycol.* 26(1), 619–628.

Hu, X., Meneses, Y. E., Aly Hassan, A.A., 2020. Integration of sodium hypochlorite pre-treatment with co-immobilized microalgae/bacteria treatment of meat processing wastewater. *Bioresour. Technol.* 304.

Karthikeyan, A., Nagasathya, A., Shanthi, V., Priya, E., 2008. Hypersaline Cyanobacterium: A potential biofertilizer for *Vigna mungo*. L (Black Gram). Am. Euras. *J. Sustain. Agricu.* 2, 87–91.

Karthikeyan, N., Prasanna, R., Nain, L., Kaushik, B.D., 2007. Evaluating the potential of plant growth promoting cyanobacteria as inoculants for wheat. *Eur. J. Soil Biol.* 43(1), 23–30.

Karthikeyan, N., Prasanna, R., Sood, A., Jaiswal, P., Nayak, S., Kaushik, B.D., 2009. Physiological characterization and electron microscopic investigations of cyano-bacteria associated with wheat rhizosphere. *Folia Microbiologica* 54(1), 43–51.

Kazamia, E., Czesnick, H., Nguyen, T.T., Croft, M.T., Sherwood, E., Sasso, S., Hodson, S.J., Warren, M.J., Smith, A.G., 2012. Mutualistic interactions between vitamin B12-dependent algae and heterotrophic bacteria exhibit regulation. *Environ. Microbiol.* 14(6), 1466–1476.

Kim, B.H., Ramanan, R., Cho, D.H., Oh, H.M., Kim, H.S., 2014. Role of Rhizobium, a plant growth promoting bacterium, in enhancing algal biomass through mutualistic inter-action. *Biomass Bioener.* 69, 95–105.

Koch, B., Liljefors, T., Persson, T., Nielsen, J., Kjelleberg, S., Givskov, M., 2005. The LuxR receptor: The sites of interaction with quorum-sensing signals and inhibitors. *Microbiol.* 151(11), 3589–3602.

Kramer, S.B., Reganold, J.P., Glover, J.D., Bohannan, B.J.M., Mooney, H.A., 2006. Reduced nitrate leaching and enhanced denitrifier activity and efficiency in organically fertilized soils. *Proc. Natl. Acad. Sci. USA* 103(12), 4522–4527.

Krings, M., Hass, H., Kerp, H., Taylor, T.N., Agerer, R., Dotzler, N., 2009. Endophytic cyanobacteria in a 400-million-yr-old land plant: A scenario for the origin of a sym-biosis? *Rev. Palaeobot. Palyno.* 153(1–2), 62–69.

Kumar, G., Sahoo, D., 2011. Effect of seaweed liquid extract on growth and yield of *Triticum aestivum* var. Pusa Gold. *J. Appl. Phycol.* 23(2), 251–255.

Kumari, R., Kaur, I., Bhatnagar, A.K., 2011. Effect of aqueous extract of *Sargassum john-stonii* Setchell & Gardner on growth, yield and quality of *Lycopersicon esculentum* Mill. *J. Appl. Phycol.* 23(3), 623–633.

Lachman, J., Orsak, M., Pivec, V., Kratochvilova, D., 2001. Anthocyanins and carotenoids-major pigments of roses. *Hortic. Sci.* 28, 33–39.

Lambers, H., Brundrett, M.C., Raven, J.A., Hopper, S.D. 2010. Plant mineral nutrition in ancient landscapes: high plant species diversity on infertile soils is linked to functional diversity for nutritional strategies. *Plant Soil* 334, 11–31. 10.1007/s11104-010-0444-9

Li, H., Zhao, Q., Huang, H., 2019. Current states and challenges of salt-affected soil remediation by cyanobacteria. *Sci. Total Environ.* 669, 258–272.

Lian, B., Wang, B., Pan, M., Liu, C., Teng, H.H., 2010. Microbial release of potassium from K-bearing mineralsby thermophilic fungus *Aspergillus fumigatus*. *Geochimica et Cosmochimica Acta* 72, 87–98.

Lupatini, M., Korthals, G.W., de Hollander, M., Janssens, T.K., Kuramae, E.E., 2016. Soil microbiome is more heterogeneous in organic than in conventional farming system. *Front. Microbiol.* 7, 2064.

Lv, J., Liu, S., Feng, J., Liu, Q., Guo, J., Wang, L., Jiao, X., Xie, S., 2020. Effects of microalgal biomass as biofertilizer on the growth of cucumber and microbial communities in the cucumber rhizosphere. *Turk. J. Bot.* 44, 167.

Malik, M.A., Khan, K.S., Marschner, P., Ali, S., 2013. Organic amendments differ in their effect on microbial biomass and activity and on P pools in alkaline soils. *Biol. Fert. Soils.* 49(4), 415–425.

Manjunath, M., Prasanna, R., Sharma, P., Nain, L., Singh, R., 2011. Developing PGPR, consortia using novel genera Providencia and Alcaligenes along with cyanobacteria for wheat. *Arch. Agron. Soil Sci.* 57(8), 873–887.

Marks, E.A., Miñón, J., Rad, C. et al., 2017. Application of a microalgal slurry to soil stimulates heterotrophic activity and promotes bacterial growth. *Sci. Total Environ.* 605–606, 610–617.

Massah, J., Azadegan, B.J., 2016. Effect of chemical fertilizers on soil compaction and degradation. *AMA Agric. Mech. Asia Afr. Lat. Am.* 47, 44–50.

Meeks, J.C., Elhai, J., 2002. Regulation of cellular differentiation in filamentous cyanobacteria in free-living and plant-associated symbiotic growth states. *Microbiol. Mol. Biol. Rev.* 66(1), 94–121.

Misra, S., Kaushik, B.D., 1989. Growth promoting substances of cyanobacteria. In vitamins and their influence on rice plant. *Proc. Ind. Natl. Sci. Acad. B Biol. Sci.*, 55, 295–300.

Moody, J.W., McGinty, C.M., Quinn, J.C., 2014. Global evaluation of biofuel potential from microalgae. *Proc. Natl. Acad. Sci. USA.* 111(23), 8691–8696.

Mupangwa, W., Twomlow, S., Walker, S., 2012. Reduced tillage, mulching and rotational effects on maize (*Zea mays* L.): Cowpea (*Vigna unguiculata* (Walp.) L.) and Sorghum (*Sorghum bicolor* L. (Moench)) yields under semi-arid conditions. *Field Crops Res.*, 132, 139–148.

Mutanda, T., Ramesh, D., Karthikeyan, S., Kumari, S., Anandraj, A., Bux, F., 2011. Bioprospecting for hyper-lipid producing microalgal strains for sustainable biofuel production. *Bioresour. Technol.* 102(1), 57–70.

Nilsson, M., Bhattacharya, J., Rai, A.N., Bergman, B., 2002. Colonization of roots of rice (*Oryza sativa*) by symbiotic Nostoc strains. *New Phytol.* 156(3), 517–525.

Nisha, R., Kaushik, A., Kaushik, C.P., 2007. Effect of indigenous cyanobacterial application on structural stability and productivity of an organically poor semi-arid soil. *Geoderma.* 138(1–2), 49–56.

Oostlander, P.C., van Houcke, J., Wijffels, R.H., Barbosa, M.J., 2020. Microalgae production cost in aquaculture hatcheries. *Aquac.* 525, 735310.

Ozdemir, S., Sukatar, A., Oztekin, G.B., 2016. Production of *Chlorella vulgaris* and its effects on plant growth, yield and fruit quality of organic tomato grown in greenhouse as biofertilizer. Tarim Bilimleri Dergisi – *J. Agric. Sci.* 22(4), 596–605.

Padhy, R.N., Nayak, N., Dash-Mohini, R.R., Rath, S., Sahu, R.K., 2016. Growth, metabolism and yield of rice cultivated in soils amended with fly ash and cyanobacteria and metal loads in plant parts. *Rice Sci.* 23(1), 22–32.

Papadopoulos, K.P., Economou, C.N., Tekerlekopoulou, A.G., Vayenas, D.V., 2020. Two-step treatment of brewery wastewater using electrocoagulation and cyanobacteria-based cultivation. J. Environ. *Manag.* 265, 110543.

Pasmore, M., Costerton, J.W., 2003. Biofilms, bacterial signaling, and their ties to marine biology. *J. Ind. Microbiol. Biotechnol.* 30(7), 407–413.

Pereira, N., Verlecar, X.N., 2005. Role of marine algae in organic farming. *Curr. Sci.* 89, 593–594.

Pérez-Montaño, F., Alías-Villegas, C., Bellogín, R.A., del Cerro, P., Espuny, M.R., Jiménez-Guerrero, I., López-Baena, F.J., et al., 2014. Plant growth promotion in cereal and leguminous agricultural important plants: From microorganism capacities to crop production. *Microbiol. Res.* 169(5–6), 325–336.

Prasanna, R., Babu, S., Bidyarani, N., Kumar, A., Triveni, S., Monga, D., Mukherjee, A.K., et al., 2015. Prospecting cyanobacteria-fortified composts as plant growth promoting and biocontrol agents in cotton. *Exp. Agric.* 51(1), 42–65.

Prasanna, R., Babu, S., Rana, A., Kabi, S.R., Chaudhary, V., Gupta, V., Kumar, A., et al., 2013. Evaluating the establishment and agronomic proficiency of cyanobacterial consortia as organic options in wheat–rice cropping sequence. *Exp. Agric.* 49(3): 416–434.

Prasanna, R., Kanchan, A., Ramakrishnan, B., Ranjan, K., Venkatachalam, S., Hossain, F., Shivay, Y. S., et al., 2016. Cyanobacteria-based bioinoculants influence growth and yields by modulating the microbial communities favourably in the rhizospheres of maize hybrids. *Eur. J. Soil Biol.* 75, 15–23.

Prasanna, R., Rana, A., Chaudhary, V., Joshi, M., Nain, L., 2012. Cyanobacteria-PGPR interactions for effective nutrient and pest management strategies in agriculture, in: Satyanarayana, T., Johri, B.N., Prakash, A. (Eds.), *Microorganisms in Sustainable Agriculture and Biotechnology*. Springer, pp. 173–195.

Prasanna, R., Sharma, E., Sharma, P., Kumar, A., Kumar, R., Gupta, V., Pal, R.K., et al., 2013. Soil fertility and establishment potential of inoculated cyanobacteria in rice crop grown under non-flooded conditions. *Paddy Water Environ.* 11(1–4), 175–183.

Priya, H., Prasanna, R., Ramakrishnan, B., Bidyarani, N., Babu, S., Thapa, S., Renuka, N., 2015. Influence of cyanobacterial inoculation on the culturable microbiome and growth of rice. *Microbiol. Res.* 171, 78–89.

Rahman, M.T., Zhu, Q.H., Zhang, Z.B., Zhou, H., Peng, X., 2017. The roles of organic amendments and microbial community in the improvement of soil structure of a vertisol. *Appl. Soil Ecol.* 111, 84–93.

Ramakrishnan, U., 2002. Prevalence of micronutrient malnutrition worldwide. *Nutr. Rev.* 60, S46–S52.

Ramanan, R., Kim, B.H., Cho, D.H., Oh, H.M., Kim, H.S., 2016. Algae–bacteria interactions: Evolution, ecology and emerging applications. *Biotechnol. Adv.* 34(1), 14–29.

Rana, A., Joshi, M., Prasanna, R., Shivay, Y.S., Nain, L., 2012. Biofortification of wheat through inoculation of plant growth promoting rhizobacteria and cyanobacteria. *Eur. J. Soil Biol.* 50, 118–126.

Rana, A., Kabi, S.R., Verma, S., Adak, A., Pal, M., Shivay, Y.S., Prasanna, R., Nain, L., 2015. Prospecting plant growth promoting bacteria and cyanobacteria as options for enrichment of macro and micronutrients in grains in rice—Wheat cropping sequence. *Cogent. Food Agric.* 1(1), 1037379.

Ranjan, K., Priya, H., Ramakrishnan, B., Prasanna, R., Venkatachalam, S., Thapa, S., Tiwari, R., et al., 2016. Cyanobacterial inoculation modifies the rhizosphere microbiome of rice planted to a tropical alluvial soil. *Appl. Soil Ecol.* 108, 195–203.

Reddy, C.A., Saravanan, R.S., 2013. Polymicrobial multi-functional approach for enhancement of crop productivity. *Adv. Appl. Microbiol.* 82, 53–113.

Renuka, N., Guldhe, A., Prasanna, R., Singh, P., Bux, F., 2018. Microalgae as multi-functional options in modern agriculture: Current trends, prospects and challenges. *Biotechnol. Adv.* 36(4), 1255–1273.

Renuka, N., Prasanna, R., Sood, A., Ahluwalia, A. S., Bansal, R., Babu, S., Singh, R., et al., 2016. Exploring the efficacy of wastewater-grown microalgal biomass as a biofertilizer for wheat. *Environ. Sci. Poll. Res. Int.* 23(7), 6608–6620.

Renuka, N., Prasanna, R., Sood, A., Bansal, R., Bidyarani, N., Singh, R., Shivay, Y.S., et al., 2017. Wastewater grown microalgal biomass as inoculants for improving micronutrient availability in wheat. *Rhizosphere* 3, 150–159.

Renuka, N., Sood, A., Prasanna, R., Ahluwalia, A.S., 2015. Phycoremediation of wastewaters: A synergistic approach using microalgae for bioremediation and biomass generation. *Int. J. Environ. Sci. Tech.* 12(4), 1443–1460.

Roberti, R., Galletti, S., Burzi, P. L., Righini, H., Cetrullo, S., Perez, C., 2015. Induction of defence responses in zucchini (*Cucurbita pepo*) by *Anabaena* sp. water extract. *Biol. Control* 82, 61–68.

Rohr, R.P., Saavedra, S., Peralta, G., Frost, C.M., Bersier, L.F., Bascompte, J., Tylianakis, J.M., 2016. Persist or produce: A community trade-off tuned by species evenness. *Am. Natur.* 188(4), 411–422.

Rolland, J.L., Stien, D., Sanchez-Ferandin, S., Lami, R., 2016. Quorum sensing and quorum quenching in the phycosphere of phytoplankton: A case of chemical interactions in ecology. *J. Chem. Ecol.* 42(12), 1201–1211.

Saadaoui, I., Sedky, R., Rasheed, R., Bounnit, T., Almahmoud, A., Elshekh, A., Dalgamouni, T., et al., 2019. Assessment of the algae-based biofertilizer influence on date palm (*Phoenix dactylifera* L.) cultivation. *J. Appl. Phycol.* 31(1), 457–463.

Sahoo, D., 2000. *Farming the Ocean: Seaweeds Cultivation and Utilization.* Aravali, New Delhi.

Santi, C., Bogusz, D., Franche, C., 2013. Biological nitrogen fixation in non-legume plants. *Ann. Bot.* 111(5), 743–767.

Sarma, B., Borkotoki, B., Narzari, R., Kataki, R., Gogoi, N., 2017. Organic amendments: Effect on carbon mineralization and crop productivity in acidic soil. *J. Clean. Prod.* 152, 157–166.

Sattari, S.Z., Bouwman, A.F., Giller, K.E., van Ittersum, M.K., 2012. Residual soil phosphorus as the missing piece in the global phosphorus crisis puzzle. *Proc. Natl. Acad. Sci. USA.* 109(16), 6348–6353.

Schaefer, A.M., Yrastorza, L., Stockley, N., Harvey, K., Harris, N., Grady, R., Sullivan, J., et al., 2020. Exposure to microcystin among coastal residents during a cyanobacteria bloom in Florida. *Harmful Algae* 92, 101769.

Searchinger, T., Edwards, R., Mulligan, D., Heimlich, R., Plevin, R., 2015. Environmental Economics. Do biofuel policies seek to cut emissions by cutting food? *Science.* 347(6229), 1420–1422.

Sepehr, A., Hassanzadeh, M., Rodriguez-Caballero, E., 2019. The protective role of cyanobacteria on soil stability in two aridisols in northeastern Iran. *Geoderma Reg.* 16, e00201.

Singh, J.S., Kumar, A., Singh, M., 2019. Cyanobacteria: A sustainable and commercial bioresource of bio-fertilizer and bio-fuel from waste waters. *Environ. Sustain. Indic.* 3–4, 100008.

Singh, V.P., Trehan, K., 1973. Effect of extracellular products of *Aulosira fertilissima* on the growth of rice seedlings. *Plant Soil.* 38(2), 457–464.

Skindersoe, M.E., Ettinger-Epstein, P., Rasmussen, T.B., Bjarnholt, T., de Nys, R., Givskov, M., 2008. Quorum sensing antagonism from marine organisms. *Mar. Biotechnol.* 10(1), 56–63.

Solovchenko, A., Verschoor, A.M., Jablonowski, N.D., Nedbal, L., 2016. Phosphorus from wastewater to crops: An alternative path involving microalgae. *Biotechnol. Adv.* 34(5), 550–564.

Sparling, G.P., 1997. Soil microbial biomass, activity and nutrient cycling as indicators of soil health, in: Pankurst, C.E., Doube, B.N., Gupta, V.S.R. (Eds.), *Biological Indicators of Soil Health*, Vol. 4. CAB, New York, USA, pp. 97–119.

Srivastav, A.L., 2020. Chemical fertilizers and pesticides: Role in groundwater contamination, in: Prasad, M.N.V. (Ed.), *Agrochemicals Detection, Treatment, and Remediation*. Butterworth-Heinemann, Oxford, UK, pp. 143–159.

Srivastava, A., Mishra, A.K., 2014. Regulation of nitrogen metabolism in salt tolerant and salt sensitive Frankia strains. *Ind. J. Exp. Biol.* 52(4), 352–358.

Stadler, C., von Tucher, S., Schmidhalter, U., Gutser, R., Heuwinkel, H., 2006. Nitrogen release from plant-derived and industrially processed organic fertilisers used in organic horticulture. *J. Plant Nutr. Soil Sci.* 169(4), 549–556.

Stanley, N.R., Lazazzera, B.A., 2004. Environmental signals and regulatory pathways that influence biofilm formation. *Mol. Microbiol.* 52(4), 917–924.

Suleiman, A.K.A., Lourenço, K.S., Pitombo, L.M., Mendes, L.W., Roesch, L.F.W., Pijl, A., Carmo, J.B., et al., 2018. Recycling organic residues in agriculture impacts soil-borne microbial community structure, function and N_2O emissions. *Sci. Total Environ.* 631–632, 1089–1099.

Svircev, Z., Tamas, I., Nenin, P., Drobac, A., 1997. Co-cultivation of N_2-fixing cyanobacteria and some agriculturally important plants in liquid and sand cultures. *App. Soil Ecol.* 6(3), 301–308.

Swarnalakshmi, K., Prasanna, R., Kumar, A., Pattnaik, S., Chakravarty, K., Shivay, Y.S., Singh, R., Saxena, A.K., 2013. Evaluating the influence of novel cyanobacterial biofilmed biofertilizers on soil fertility and plant nutrition in wheat. *Eur. J. Soil Biol.* 55, 107–116.

Syers, J.K., 2008. Efficiency of soil and fertilizer phosphorus: Reconciling changing concepts of soil phosphorus chemistry with agronomic information. In *The 18th World Congress of Soil Science*. https://www.fao.org/3/a1595e/a1595e.pdf

Teplitski, M., Chen, H., Rajamani, S., Gao, M., Merighi, M., Sayre, R.T., Robinson, J. B., et al., 2004. *Chlamydomonas reinhardtii* secretes compounds that mimic bacterial signals and interfere with quorum sensing regulation in bacteria. *Plant Physiol.* 134(1), 137–146.

Teplitski, M., Rajamani, S., 2011. Signal and nutrient exchange in the interactions between soil algae and bacteria, in: Witzany, G. (Ed.), *Biocommunication in Soil Microorganisms*. Springer, Berlin, Germany, pp. 413–426.

Triveni, S., Prasanna, R., Kumar, A., Bidyarani, N., Singh, R., Saxena, A.K. 2015. Evaluating the promise of Trichoderma and Anabaena based biofilms as multifunctional agents in Macrophominaphaseolina-infected cotton crop. Biocont. *Sci. Technol.* 25(6), 656–670.

Uysal, O., Uysal, F. O., Ekinci, K., 2015. Evaluation of microalgae as microbial fertilizer. Eur. J. Sustain. *Develop.* 4(2), 77.

Van Apeldoorn, M.E., Van Egmond, H.P., Speijers, G.J., Bakker, G.J., 2007. Toxins of cyanobacteria. *Mol. Nutr. Food Res.* 51(1), 7–60.

Van der Sluijs, J.P., Amaral-Rogers, V., Belzunces, L.P., Bijleveld van Lexmond, M.F., Bonmatin, J.M., Chagnon, M., et al., 2015. Conclusions of the Worldwide Integrated Assessment on the risks of neonicotinoids and fipronil to biodiversity and ecosystem functioning. *Environ. Sci. Poll. Res. Int.* 22(1), 148–154.

Van Vuuren, D.P., Bouwman, A.F., Beusen, A.H.W., 2010. Phosphorus demand for the 1970–2100 period: A scenario analysis of resource depletion. *Glob. Environ. Change.* 20(3), 428–439.

Venkataraman, G.S., 1981. Blue–green algae for rice production, in: F.A.O. Soils (Ed.) [Bulletin], 46. Rome, Italy.

Vijayakumar, S., Durgadevi, S., Arulmozhi, P., Rajalakshmi, S., Gopalakrishnan, T., Parameswari, N., 2019. Effect of seaweed liquid fertilizer on yield and quality of *Capsicum annum* L. *Acta Ecologica Sinica*. 39, 406–410.

Villa, J.A., Ray, E.E., Barney, B.M., 2014. *Azotobacter vinelandii* siderophore can provide nitrogen to support the culture of the green algae *Neochloris oleoabundans* and *Scenedesmus* sp. BA032. *FEMS Microbiol. Lett.* 351(1), 70–77.

Wang, B., Li, Y., Wu, N., Lan, C.Q., 2008. CO_2 bio-mitigation using microalgae. *Appl. Microbiol. Biotechnol.* 79(5), 707–718.

Wang, X., Li, Z., Su, J., Tian, Y., Ning, X., Hong, H., Zheng, T., 2010. Lysis of a red-tide causing alga, *Alexandrium tamarense*, caused by bacteria from its phycosphere. *Biol. Control.* 52(2), 123–130.

Wolfe, A.J., Chang, D.E., Walker, J.D., Seitz-Partridge, J.E., Vidaurri, M.D., Lange, C.F., Prüss, B.M., Henk, M.C., Larkin, J.C., Conway, T., 2003. Evidence that acetyl phosphate functions as a global signal during biofilm development. *Mol. Microbiol.* 48(4), 977–988.

Wuang, S.C., Khin, M.C., Chua, P.Q.D., Luo, Y.D., 2016. Use of Spirulina biomass produced from treatment of aquaculture wastewater as agricultural fertilizers. *Algal Res.* 15, 59–64.

Yadavalli, R., Heggers, G.R.V.N., 2013. Two stage treatment of dairy effluent using immobilized *Chlorella pyrenoidosa*. *J. Environ. Health Sci. Eng.* 11(1), 36.

Yilmaz, E., Sönmez, M., 2017. The role of organic/bio–fertilizer amendment on aggregate stability and organic carbon content in different aggregate scales. *Soil Till. Res.* 168, 118–124.

Zapalski, M.K., 2011. Is absence of proof a proof of absence? Comments on commensalism. *Palaeogeogr. Palaeoclimatol. Palaeoecol.* 302(3–4), 484–488.

Zayadan, B.K., Matorin, D.N., Baimakhanova, G.B., Bolathan, K., Oraz, G.D., Sadanov, A.K., 2014. Promising microbial consortia for producing biofertilizers for rice fields. *Microbiol.* 83(4), 391–397.

Zhang, J., Wang, X., Zhou, Q., 2017. Co-cultivation of *Chlorella* spp and tomato in a hydroponic system. *Biomass Bioener.* 97, 132–138.

Zhang, R.G., Pappas, K.M., Brace, J.L., Miller, P.C., Oulmassov, T., Molyneaux, J.M., Anderson, J.C., et al., 2002. Structure of a bacterial quorum-sensing transcription factor complexed with pheromone and DNA. *Nature.* 417(6892), 971–974.

Zhang, Y.-Q., Deng, Y., Chen, R.-Y., Cui, Z.-L., Chen, X.-P., Yost, R., Zhang, F.-S., Zou, C.-Q., 2012. The reduction in zinc concentration of wheat grain upon increased phosphorus-fertilization and its mitigation by foliar zinc application. *Plant Soil.* 361(1–2), 143–152.

Zhou, L., Zhou, G., Jia, Q., 2009. Annual cycle of CO_2 exchange over a reed (*Phragmites australis*) wetland in Northeast China. *Aquat. Bot.* 91(2), 91–98.

Zhu, X.G., Long, S.P., Ort, D.R., 2008. What is the maximum efficiency with which photosynthesis can convert solar energy into biomass? *Curr. Opin. Biotechnol.* 19(2), 153–159.

8 Recent Trends in Microalgal Refinery for Sustainable Biopolymer Production

Menghour Huy, Ann Kristin Vatland, Reza Zarei, and Gopalakrishnan Kumar
Department of Chemistry, Bioscience, and Environmental Engineering, Faculty of Science and Technology, University of Stavanger, Stavanger, Norway

CONTENTS

8.1 Introduction..238
8.2 Parameters Affecting Microalgae Biomass Cultivation.............................239
    8.2.1 Temperature..239
    8.2.2 pH 240
    8.2.3 Nutrients 240
    8.2.4 CO_2 Aeration in Photoautotrophic Microalgae 240
8.3 Biopolymer Production from Microalgae Biomass....................................241
    8.3.1 Polyhydroxyalkanoates (PHA)..242
    8.3.2 Polylactic Acid or Polylactide (PLA)...243
    8.3.3 Polysaccharides...243
    8.3.4 Protein..244
8.4 Biopolymer's Extraction Technologies 245
    8.4.1 Pretreatment/Cell Wall Disruption...245
    8.4.2 Extraction Technology 246
        8.4.2.1 Solvent Extractions...246
        8.4.2.2 Supercritical Fluids Extraction......................................248
        8.4.2.3 Subcritical Fluid Extraction 249
8.5 Future Prospects and Research Needs...250
8.6 Conclusions...251
Acknowledgment 252
References...252

DOI: 10.1201/9781003195405-8

8.1   INTRODUCTION

Plastic material has served in many sectors, such as rigid packaging, building and construction, textile induction, electronics, consumers goods, etc. (European Bioplastics, 2019; Geyer et al., 2017), where the global demand for plastic has raised from 1.5 to 368 million tons between 1950 and 2019 (Plastics Europe, 2020). Less awareness and poor management of plastic waste from the olden days have left a serious scar on the planet. Footprints of ocean dumping, river dumping, and non-degradable material dumped in the landfills have given examples of why this synthetic plastic should be replaced by an alternative or more biodegradable material. The use of synthetic plastic has been around for only the past seven decades, and hundreds of millions of tons of plastic waste generation has been annually reported for this past decade (Costa et al., 2019). In the United States alone, about 35 million tons of plastic wastes were deposed in 2018. However, 80.5% were dumped in the landfill, and only 6.8% were recycled; 12.7% were used for combustion (EPA, 2020). The concerns of plastic disposal have shifted the plastic production market toward finding more green, sustainable, bio-based material to replace this synthetic material.

European countries as well as the rest of the world have given more strict policies against single-use plastic to slow down the trench of plastic consumption. Plastic waste management (recycling and landfill dumping) has improved significantly over this past decade (Plastics Europe, 2020). The trench of plastic consumption has slightly slowed down; yet, over 300 million tons of plastic still need to be supplied annually (Plastics Europe, 2020). For these demands with the strict rule, the bioplastic market has grown dynamically, where about 2.11 million tons were produced in 2019 and aim to hit about 2.43 million tons in 2024 (European Bioplastics, 2019).

Bioplastic can be classified into two main categories bio-based/non-biodegradable and biodegradable plastic. Polyethylene (PE), polyethylene terephthalate (PET), polyamides (PA), polypropylene (PP), polyethylene furanoate (PEF), and poly (trimethylene terephthalate) (PTT) are in the bio-based/non-biodegradable group, where biodegradable plastic consists of polybutylene adipate terephthalate (PBAT), polyhydroxyalkanoate (PHA), polylactic acid (PLA), polybutylene succinate (PBS), and starch blends (European Bioplastics, 2019). PHAs and PLA products are known biopolymers, which can be derived from microalgae biomass. The current market (2021) trench of bio-based plastic consists of 44.5% bio-based/non-degradable and 55% biodegradable products (European Bioplastics, 2019). Increased biodegradable products will help reduce plastic footprints, therefore satisfying the strict policy applied in Europe.

The diversity of microalgal biorefinery technologies have given many routes to solve environmental issues, such as supplying feed for animal, supplying feedstock for biofuel, serving as biomaterial, etc. Microalgae biomass is packed with valuable compounds and can accumulate highly targeted composition with components such as proteins, carbohydrates, and lipids for the final refinery of the products. Polyhydroxyalkanoate (PHA), for instance, is known to be obtained from microalgae cells synthesized from lipid in microalgae biomass (Devadas et al., 2021) or derived from other microbial sugar fermentation using microalgae biomass as feedstock (Costa et al., 2019). Therefore, sugar fermentation is predicted to be the largest production method in the PHAs market between 2020 and 2025 (Research and Markets, 2021). Despite the uses of biopolymers for bioplastic production, polysaccharides from microalgae can also be utilized for cosmetic applications, as well as protein derived from microalgae to be used in animal feed.

From an animal feed point of view, commercial aquafeed production was estimated to be increased over two-fold (80 million tons) by 2025 compared to 2014. According to the numbers reported in 2019, 3–6 million tons of low-value fish are being fed directly to fish and shrimp (Kim et al., 2019). This raises the concern of the meat quality of fish and shrimp, where microalgae biomass is believed to be a potential alternative to replace aquafeed partly or fully in the near future.

Comparing with crops as a source of feedstock, microalgae biomass can be cultivated and harvested rapidly regardless of the season with minimal land uses (Wicker et al., 2021). However, current technology producing biopolymers from microalgae biomass is very expensive, which is why it is still challenging to compete with synthetic plastic and replacing feed in the market. Research in finding the alternative feeding source in the upstream process of the production (microalgae cultivation) in order to optimize parameters and product extraction efficiency in the downstream process is in the progress to make the final products more economically favorable to the market. In addition to this, rapid nutrient consumption capabilities have foreseen microalgae technology in wastewater treatment promising a feasible cost-effective feedstock for the industrial-scale operation. This potential of microalgae biotechnology in wastewater has led towards better energy/resource management in the circular bioeconomy. Besides the nutrient recovery capability of microalgae biomass and producing biodegradable plastic, the recent investigation also pointed toward the potential of using biopolymers derived from microalgae to aggregate nano- and micro-plastic from wastewater (Cunha et al., 2020).

8.2 PARAMETERS AFFECTING MICROALGAE BIOMASS CULTIVATION

8.2.1 TEMPERATURE

Temperature is one of the most essential factors for efficient microalgae growth. Polyunsaturated fatty acids (PUFA) can be accumulated and act as a barrier for the microalgae nutrient uptake ability in lower temperature growing. Only the strains with targeted PUFA in the downstream processing will be selected and cultivated in the cold temperature. The increase of temperature to the optimum condition induces greater biomass production. However, growing biomass above the optimum could inhibit the productivity as well as the quality of the biomass. Generally, the temperature should be maintained between 25 to 35 °C (Renaud et al., 2002). For instance, *Chlorella sp.* can grow generally at 5 °C to 30 °C but the optimum was found at 25 °C (Senge and Senger, 1990), 37 °C for *Chlorella sorokiniana* (Zheng et al., 2013), 25 to 30 °C for *Botryocuccus braunii* (Ruangsomboon, 2012), and 28 °C *Haematococcus pluvialis* (Hong et al., 2016). In general, microalgae will only rapidly produce biomass within their optimal temperature. Above its maximum temperature, heat stress will slow down their growth drastically or dead cells will start accumulating in the biomass (Hong et al., 2016).

8.2.2 pH

pH is also importantly affected enzyme activity, and it may also influence the cell growth and metabolism of microalgae. In general, preferable pH values for microalgae cultivation are close to pH 7.0 (Sakarika and Kornaros, 2016) but depending on species, the optimum condition for pH values can be different. Each species has its specific pH tolerance. For instance, the optimum pH value of *Dunaliella salina* was found to be at pH 11, *Dunaliella acidophila* at pH 3 (Varshney et al., 2015). *Scenedesmus spp.* preferred pH ranging from 7.0 to 9.0 for their optimal biomass productivity (Difusa et al., 2015). To sum up, selecting microalgae strains with the optimal culturing pH is crucial and essentially leads to better mass cultivation.

8.2.3 Nutrients

Selecting nutrients is one of the crucial parts of efficient cultivation in order to obtain a favorable growth rate. In general, nitrogen and phosphorus have been marked as most needed for efficiently culturing microalgae. To obtain optimum biomass productivity of microalgae, N:P ratio is a key factor that needs to be considered for the experimental settings. Mostly, microalgae biomass consists of 3–12% of nitrogen and 0.3–3% of phosphorus (Reynolds, 2006). In general, the red field ratio of carbon, nitrogen, and phosphorous concentration is a biological constant inherent to the fundamental protein-to-RNA ratio and has N:P in 16:1 ratio across the living entities on Earth. A more recent, comprehensive revision of ocean organic particulates reported a global median N:P ratio of 22:1 (Monfet and Unc, 2017). For instance, one study carried out various N:P ratios and found out that *Nanochloropsis oculata* and *Tisochrysis lutea* provided higher growth rates in 20:1 N:P ratios (Rasdi and Qin, 2015). In general wastewater, the effluent consists of a complex mixture of chemical composition, in which most forms of nitrogen and phosphorus are represented. Most microalgae can only directly consume inorganic nitrogen for their growth, but some microalgae can uptake organic forms of nitrogen, especially amino acids and urea or purines in the wastewater (Monfet and Unc, 2017). All of the nitrogens must be considered with respect to phosphorus ratio in order to optimize the mass production of microalgae biomass. N:P ratios of wastewater can be seen in Table 8.1 below.

8.2.4 CO$_2$ Aeration in Photoautotrophic Microalgae

Microalgae are photosynthetic organisms that need CO$_2$ for respiration. Due to their potential of capturing CO$_2$ by respiration, microalgae have gained wide attention as promising in the utilization and fixing of CO$_2$. In addition, these organisms have other promising applications that will be discussed further and hence will improve

TABLE 8.1

Reported N:P Molar Ratio of Wastewater

Wastewater	Characteristic			Reference
	N (mg/L)	P (mg/L)	N:P (molar ratio)	
Domestic	40	8	11	(Tchobanoglous and Burton, 1991)
Beef cattle feedlot	63	14	10	(Bradford et al., 2008)
Dairy	185	30	14	(Bradford et al., 2008)
Poultry feedlot	802	50	36	(Bradford et al., 2008)
Paper mill	11	0.6	41	(Pokhrel and Viraraghavan, 2004)
Textile	90	18	11	(Fongsatitkul et al., 2004)
Urban wastewater effluent	27.4	11.8	5	(Martínez, 2000)
Secondarily treated sewage	15.5	0.5	66	(Xin et al., 2010)
Municipal (centrate) wastewater	116.1	212.0	1	(Li Y. et al., 2011)
Primary clarifier effluent	51	2.1	52	(Woertz et al., 2009)

the production of cleaner and more environmentally friendly bioproducts considering various aspects (Li F. et al., 2011; Mohsenpour and Willoughby, 2016). Depending on the CO_2 concentration, microalgae rapidly produce biomass correlated to the percentage CO_2 in the gas and aeration rate system. These microorganisms can convert CO_2 into biomass through photosynthesis with an efficiency several times higher than terrestrial plants (Gonçalves et al., 2014). The investigation on the various concentrations of CO_2 in microalgae cultivation has exhibited higher biomass productivities with the higher CO_2 concentration in the flue gas (Ji et al., 2016). Some microalgae are capable of tolerating a very high percentage of CO_2 through their respiration, e.g., up to 15% for *Scenedesmus obliquus* (Kaewkannetra et al., 2012), 16–34% for *Haematococcus pluvialis* (Huntley and Redalje, 2007), 40% for *Chlorella sp.* (Sakai et al., 1995), 60% for *Chlorococcum lttorale* (Chihara et al., 1994), and 70% for *Chlorella KR-1* (Sung et al., 1998). However, the interest component of microalgae must be carefully considered if the selected microalgae strain is subjected to high CO_2 concentration.

8.3 BIOPOLYMER PRODUCTION FROM MICROALGAE BIOMASS

Biopolymer production from microalgae biomass is getting more interesting due to its degradability, high biomass productivity, capability of growing in most environments (outdoor and indoor), cost-effectiveness, etc. Microalgae can synthesize polymeric compounds that enable them to be used for numerous useful applications.

Examples of biopolymeric compounds synthesized are PHAs/PLAs, poly-saccharides, and proteins. For instance, synthesized PHAs and PLAs production can be extracted then utilized as biodegradable plastic for packaging as well as construction material. Polysaccharides and protein are more interesting in the pharmaceutical/nutraceutical industries (Lutzu et al., 2021).

8.3.1 Polyhydroxyalkanoates (PHA)

PHAs are one of the alternative biopolymers to replace petroleum-based products. They are biodegradable plastic, which is naturally synthesized and accumulated by various microorganisms. The interests of PHAs caught by researchers in biopolymers production are mainly due to their structural properties, which can be processed (extrusion and injection molding) and are very close to polypropylene among biopolymers production (Costa et al., 2019). PHAs are classified as short-chain-length and medium-chain-length depending on the carbon atoms in the chain of polymers (Anjum et al., 2016).

Depending on strains and growth conditions, microalgae can store very high PHA content in their cell, e.g., 69% of PHA content was reported in freshwater microalgae *Nostoc muscorum* with the optimized condition (Bhati and Mallick, 2015; Costa et al., 2019). However, this high production yield could only be achieved by a proper chemical growth medium responding to their favorable growth condition. Strategies used to enhance PHAs content are to manipulate the growth condition in the upstream processing, such as nutrient starvation, high salinity, and gas exchange limitation. One of the most promising factors enhancing PHA accumulation in microalgae biomass is nutrient depletion. For instance, the study of Costa et al. (2018) addressed two microalgae strains; *Synechococcus subsalsus* and *Spirulina sp.* LEB-18 were capable of enhancing 3-folds and 4-folds by the reduction of nitrogen availability during cultivation, reaching a maximum PHAs of about 16%, and 12%, respectively. Without any nitrogen depletion strategies, this same author reported that production can yield only less than 3% (Costa et al., 2018). Additionally, Rueda et al. (2020) conducted a study on cultivating cyanobacteria in an agricultural runoff by three full-scale semi-closed photobioreactors outdoors, to examine the bioremediation and biopolymers production. The same author reported that nutrient (nitrogen and phosphorus) removal rate on an average over 95% in the first photobioreactor with the addition of $NaHCO_3$, improved PHB accumulation as maximum 4.5%$_{Volatile\ Suspended\ Solids\ (VSS)}$ of PHB was achieved (1.7%$_{VSS}$ without $NaHCO_3$) in the second and third photobioreactor. In Rueda et al. (2020), the authors addressed this production yield was close to the lab-scale operation (6.5%$_{dried\ cell\ weight\ (dcw)}$ of PHB), reported by the other study of using cyanobacteria and secondary urban wastewater as feedstock (Arias et al., 2018). Regardless of low PHB concentration, this research displayed the feasibility of utilizing agricultural runoff as a source for biopolymer production in an outdoor full-scale operation (Rueda et al., 2020). Despite direct PHB extraction from microalgae biomass, this microalgae biomass can be used as feedstock for microbial fermentation, which was reported by Rahman et al. (2015); the PHB production yield maximum was 31% in *Escherichia coli* (*E. coli*) fermentation. According to both authors, microalgae biomass may have potential in wastewater treatment bioeconomy in regards to PHB production, which can be obtained either with a direct extraction treatment facility with low production yield or by serving as feedstock in microbial fermentation to enrich the production (Rahman et al., 2015).

8.3.2 Polylactic Acid or Polylactide (PLA)

PLA is one of the most popular alternatives to replace petroleum-based products. It is non-toxic, renewable, and biodegradable. Therefore, PLA can serve in various applications, such as biomedical products, filament for 3D printing, food packaging, textile, building blocks, etc. (Karan et al., 2019; Lutzu et al., 2021). PLA is a biopolymer produced by several microbial fermentation processes that polymerize lactic acids into PLA (Lutzu et al., 2021). Some biofuel fermentation methods could also potentially produce lactic acids as byproducts, which could further be utilized as cost-effective feedstock for PLA as the co-products of the main targeted fuel production. The structural properties of PLA vary and therefore can be found in three forms: poly-L-Lactic acid (PLLA), poly-D-Lactic acid (PDLA), and poly-D,L-Lactic acid (PDLLA). For all these forms, algal biomass has recently been shown as one of the most promising resources. Synthesis of PLA from microalgae biomass involving step-processes consist of (1) hydrolyzing microalgae polysaccharides to simple sugars to enhance microbes bioavailability, (2) using microbial fermentation turning them into ethanol and lactic-acids, and (3) polymerized lactic acids into PLA by using direct a chemical or fermentation process. In a recent development, Santos Assunção et al. (2021) reported the use of PHB/PLA with nanoparticles of carotenoids extracted from microalgae biomass as encapsulants; encapsulation techniques are known to provide the enhancements of bioavailability, water solubility, and stability of hydrophobic carotenoids. The results of this study concluded that the encapsulation achieved efficiency levels higher than 80%, good morphology, and high thermal resistance, which can withstand general temperatures used (<185 °C) in the food processing industry; the lowest temperature of the formulation in the study indicated the endothermic event at 186 °C (Santos Assunção et al., 2021). Since they are both biodegradable materials, they can be envisioned as a promising alternative option to be used in the food industry.

8.3.3 Polysaccharides

Polysaccharides are natural biopolymers considered biodegradable and nontoxic. Algal polysaccharides serve in numerous applications, such as in agricultural, biomedical, pharmaceutical industries, etc (El-Naggar et al., 2020). Microalgae can synthesize polysaccharides such as cellulose and hemicellulose as a functional cell wall response to cultivation environmental, and starch responses to nutrient depletion and dark\light illumination cycle. These polysaccharides can vary greatly depending on the strains, and their culturing environment (Lutzu et al., 2021). For instance, *Chlorella zofingiensis* can accumulate 9.7% starch (45.9% carbohydrate in total) with nitrogen repletion medium, where 43.4% starch (of 66.9% carbohydrate in total) was found in nitrogen depletion medium (Zhu et al., 2014). In another study with two different types of nutrient starvation strategies, *Chlamydomonas reihardtii* grown in nitrogen-deficient liquid tris-acetate-phosphate (TAP) medium contained about 35% of starch (Koo et al., 2017), whereas 49% of starch content was achieved with sulfur starved TAP medium (Mathiot et al., 2019). In the same study of Mathiot et al. (2019), it was also reported that *Chlamydomonas reihardtii (11–32 A)* outperformed the other nine strains, including *Ankistrodesmus falcatus*, *Chlamydomonas reihardtii (DW15)*, *Chlorella sorokiniana*, *Chlorella variabilis*, *Chlorella vulgaris*, *Parachlorella kessleri*, *Scenedesmus acutus*, *Scenedesmus obliquus*, and *Scenedesmus sp.* in terms of starch content. Starch contents in algal biomass were assessed on cell lysate with the assistance of enzymatic digestion of extracted starch into glucose or simple sugar. This proved to be an alternative method of using microalgae as a feedstock for starch-based bioplastic. Even though starch content in microalgae can be much higher than for most terrestrial plants, starch extraction from microalgae can be complicated and affected by high cell wall to cell volume ratio, and the thick rigid cell wall caused by starch accumulation strategy. Currently, microalgae starch-based technology is expensive. Therefore, in order to make the methods more feasible considering industrial scale, most researchers attempt to extract multiple products aiming to satisfy economic requirements. For instance, due to high starch contents, microalgae biomass is getting interesting for bioenergy sectors for use as an alternative feedstock for bioethanol/biohydrogen fermentation. Then, byproduct resultant from bioethanol fermentation (lactic acid) can be polymerized into PLAs as a co-product of bioethanol.

8.3.4 Protein

Currently, utilization of microalgae protein is getting more and more interesting, with the potential of fully or partially replacing current protein sources in animal feed production. Microalgae can accumulate very high protein content, with concentrations that may reach over 60%. Some microalgae synthesize protein in the form of biopolymer-binding amino acids to form glycoprotein as its cell wall. This cell wall can be rigid and limits the bioavailability to the consumers as well as lowers the protein extraction yield if biomass is designated to the extraction process. The rigidity and structure of microalgae protein vary from one another based on their isolation environment as well as cultivation process. Regarding their biodigestibility, three major types of microalgae biomass are being used in the animal feed industries, such as a whole, disrupted cells, and defatted biomass. Microalgae strains such as *Desmodesmus sp.* and *Nannochloropsis sp.* can score as high as soybean meal to be used for Atlantic salmon, whereas *Scenedesmus sp.* compared with pea, ranked equal for common carp in terms of apparent digestibility coefficients (Kim et al., 2019). The purpose of the preprocess resulting as disrupted cells or defatted biomass is to improve the bioavailability of microalgae constituents to the consumers. Protein monomer forms (amino acids) of microalgae biomass can be obtained by physical/chemical/biological methods that break down crude protein into amino acids. *Haematococcus pluvialis*'s amino acid composition reported by Zhu et al. (2020) and Safi et al. (2013) shows significantly high levels, even compared to fresh krill meal reported by Sørensen et al. (2011) as fish feed ingredient (Safi et al., 2013; Sørensen et al., 2011; Zhu et al., 2020). Apart from the use in feed production, microalgae protein-based products are as well foreseen as useful biodegradable bioplastic fibers in other applications such as building blocks material, food packaging, and biomedical products (Lutzu et al., 2021).

8.4 BIOPOLYMER'S EXTRACTION TECHNOLOGIES

The complexity and efficiency of biopolymer extraction from microalgae biomass can be affected by upstream processing, and therefore culturing of microalgae biomass should be thoughtfully considered. The unfavorable growth environment of microalgae cultivation can lead cells to develop thick and rigid cell walls to protect themself from disruption in the multiplication stage (Domozych et al., 2012). Depending on microalgae strains and their structure, cultivated biomass is normally examined in the deep consideration process as whole biomass or disrupted biomass. Hence, insight into parameters influencing cell wall thickness development, and applying them in the upstream processing stage, can improve later downstream processing extraction cost.

8.4.1 PRETREATMENT/CELL WALL DISRUPTION

The typical cell wall can be classified into four groups: simple cell membrane, cell membrane with extracellular material, cell membrane with intracellular material, and cell membrane with intracellular and extracellular material (Gonzalez-Fernandez and Muñoz, 2017). Disruption of these cell walls can be done by various methods, such as applying physical/mechanical, chemical, and biological force.

Physical/mechanical cell disruptions use physical force to break down or separate cells from the cell wall, such as pressuring, clashing, grinding, etc. Various physical/mechanical applications, such as freeze-fracture, freeze-drying, mild temperature, autoclaving, stream explosive, bead milling, expeller press, high-speed homogenization, high-pressure homogenization, microwave, ultrasonication, and pulsed electric field treatment are already established (Gonzalez-Fernandez and Muñoz, 2017; Günerken et al., 2015; Wicker et al., 2021). The strong advantage of using physical/mechanical cell disruption compared to chemical and biological methods is due to its effectiveness without any additional chemicals, which may lead to lesser unwanted byproducts (Günerken 2015). If the final products are required to be strictly impurity-free, and in order to be used as health supplements/feed in pharmaceutical or nutraceutical industries, physical/mechanical techniques have a great advantage. However, the drawback of physical/mechanical is that they require a massive amount of energy input, which sets them aside from chemical and biological treatments (Sankaran et al., 2020).

In terms of treatment cost, chemical cell disruption is the most selected technique compared to physical/mechanical and biological techniques (Velazquez-Lucio et al., 2018). This is due to a variety of compounds that can be used, such as oxidative, acid, alkaline agents/solvents, etc., to disrupt the cell wall (Mendes-Pinto et al., 2001). Operational parameters such as temperature, time, biomass ratio, and type of solvents used correspond to microalgae strains/cell walls must be properly controlled to ensure that the optimal extraction is reached with minimal loss and is free from unwanted byproducts (Günerken et al., 2015). As mentioned, unlike physical/mechanical cell disruption, a chemical compound used can leave un-pleasant constituents resulting from the product treatment. Depending on the che-mical solvents used, final product purifications/refinement of chemical cell disruption can also be complicated and costly if the final products are designated to the pharmaceutical or nutraceutical industry (D'Hondt et al., 2017). Alternatively, pretreatment by chemical agents is commonly used in bioenergy and biomaterial sectors, due to their relatively low energy usage and since byproducts are con-sidered minor (Velazquez-Lucio et al., 2018).

The biological method is done by using the enzyme to digest sugar/protein-bound in cell walls, which results in disruption of the cell wall. This method is also called enzymatic hydrolysis, which converts complex biopolymers into simpler monomeric forms, carbohydrate to monomeric sugars, and/or proteins to amino acids. Enzymatic hydrolysis is often selected because the operation condition is milder compared to most chemical methods (Velazquez-Lucio et al., 2018). Temperature, time, stirring, and biomass/enzyme ratio are highly influenced/af-fected by upstream processing/environmental conditions of cultivated microalgae biomass. Since most biosynthesis of biopolymers requires stress conditions to ac-cumulate biopolymers in microalgae cells, the cell wall of microalgae also thickens in correlation to the applied conditions. Hence, the scaling-up process of enzymatic hydrolysis in itself causes challenges that need considerations, especially due to the related enhanced needs for high-cost enzymes (Yap et al., 2016).

8.4.2 Extraction Technology

8.4.2.1 Solvent Extractions

Normally, with/without pretreatment, solvent extraction of biopolymers requires many steps of washing. These techniques consist of a series steps of solvent mixing, precipitating, rinsing, washing, drying, and separation, which leads to many frac-tional losses of the component in the test method (Gonzalez-Fernandez and Muñoz, 2017). Therefore, as mentioned above, with different types of solvents and mi-croalgae strains, optimization of operating parameters must be studied individually to enhance the extraction/separation efficiency. This method is still commonly used in downstream processing, as a variety of solvents can be selected proprietary to the production yield responding with minimum operational cost (Velazquez-Lucio et al., 2018). Like chemical cell disruption pretreatment, single solvents such as methanol, acetone, ethanol, ethyl acetate, isopropanol, and hexane are the most common, as well as combined solvents such as chloroform:methanol, etc. (Gonzalez-Fernandez and Muñoz, 2017). From the nutraceutical point of view, a comparative study of safety solvents used to extract bioactive astaxanthin, di-chloromethane, and acetone were examined with Soxhlet extraction accompanied by mechanical ball-milling cell disruption. The result of this study (43 wt % ex-traction yield, 30.4 mg/g dry weight of *Haematococcus pluvialis*) concluded that with slightly harsh condition (500 rpm, 60 mins) dichloromethane outperformed acetone for astaxanthin extractability, but the antioxidant activity is lower compared to acetone (Irshad et al., 2019a). This same author also developed a simpler combined process where mechanical cell disruption and solvent extraction were combined in one-pot aiming to avoid the need for the drying process of the biomass after cell disruption. The bioactive astaxanthin extraction yields of this method reached over 31.6 mg/g dry weight of *Haematococcus pluvialis* (47 wt %) by using ethanol with the extraction time of 30 mins at a mild condition of 200 rpm, room temperature, and atmospheric pressure (Irshad et al., 2019b) (Figure 8.1).

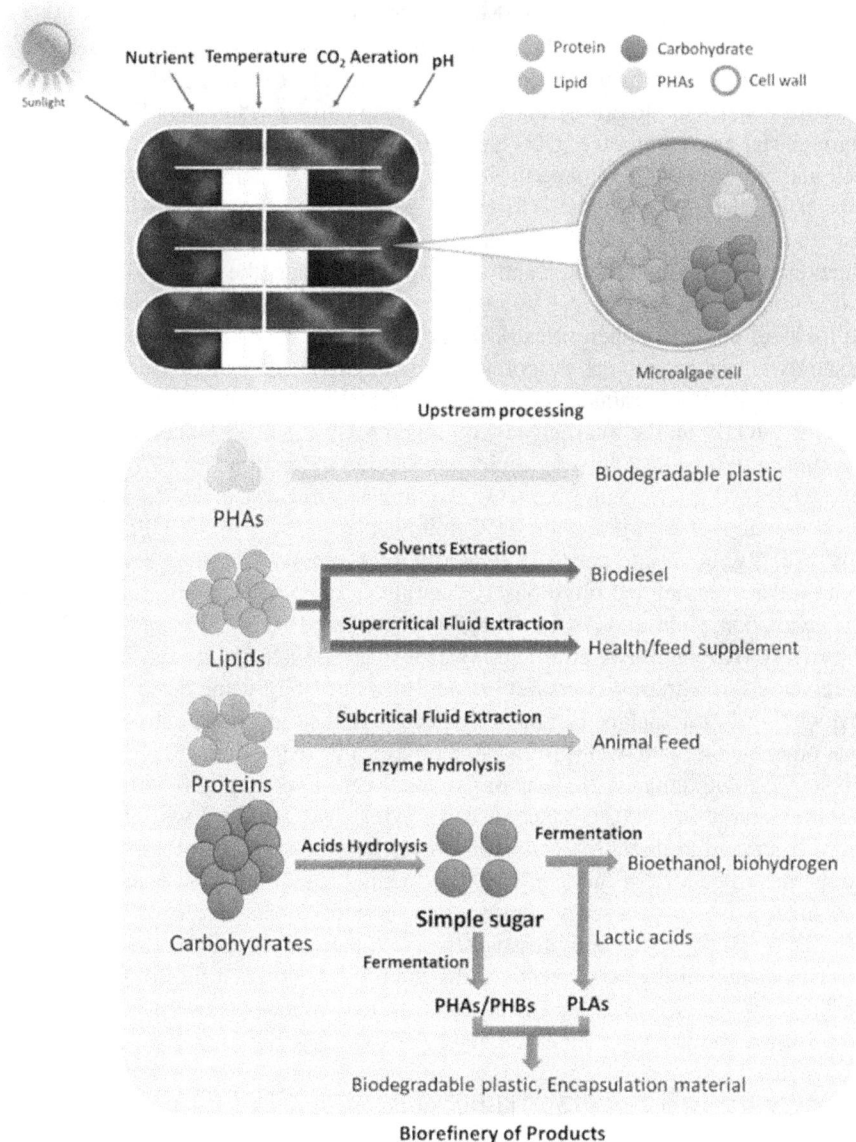

FIGURE 8.1 Microalgae biorefinery from its targeted composition.

8.4.2.2   Supercritical Fluids Extraction

In the recent development of extraction technology, supercritical fluid extraction overtook most of the nutraceutical and pharmaceutical application industries. This was due to its rapid extraction process and effectiveness, and since there were no byproducts, it resulted in an ultra-pure final product. Depending on the targeted component, fluids such as CO_2, ethanol, water, and the combination of CO_2 with other solvents are being used. Since different fluids have their independent properties, temperature, pressure, and critical density must be performed correspondingly to hit the optimal efficiency (Marcus, 2018). Regarding the properties of CO_2, it is the most popular among the other fluids in this technology. Using the chemical and physical properties of CO_2, this selected extractant is then performed in a supercritical state (gas to critical state) to permeate into the whole cell. Since it travels through the cell membrane, CO_2 can bind with targeted compounds of the whole cell and extract specific compounds from the whole cell before turning back into a gas state. Then, CO_2 evaporates from the extracted components, leaving almost no trace of residue solvents in the final production. With this method, the final production purification is significantly less complex than the conventional solvent extraction method. This is due to the properties of CO_2 at the atmospheric pressure in the final phase of supercritical CO_2 extraction (SC-CO_2) process, where CO_2 is instantly turned back to gas evaporate completely from the final product. Depending on the microalgae strains, without cell disruption, SC-CO_2 process can be performed directly on the microalgae with high pressure. For instance, in the study of Krichnavaruk et al. (2008), by direct astaxanthin extraction from *Haematococcus pluvialis* (feed grade) with pure CO_2, the highest yield was achieved about 34% at high pressure of 50 MPa with 70 °C followed by 25.4% at 40 MPa with 70 °C. Interestingly, in an attempt to facilitate the high pressure, by the presence of soybean (10% volume) and olive oil (10% volume) as co-solvents mixture with CO_2, the extraction yield at 40 MPa with 70 °C increased to 36.36% and 51.03%, respectively (Krichnavaruk et al., 2008). However, the need to design optimal pressure vessels, integrating also safety issues, this method of using direct extraction of the whole-cell microalgae can be less selective in the industrial-scale operation. On the other hand, with well-grounded biomass, the extraction yield was reported above 97% with ethanol (9.4% mass) as cosolvent at 300 bars (30 MPa) with 60 °C (Valderrama et al., 2003); hence, a higher yield was achieved with milder conditions compared to direct extraction. However, since the cost is increased in relation to the increased size of the extractor, this method is still the most energy extensive technology due to the energy consumption required to heat and cool the vessel, as well as pressurizing the compressor vessel to the optimal condition reported by the previous study in the larger scale of the industry. Significant improvement of this technology was achieved by Cheng et al. (2018) using ethanol (20% volume) as cosolvents, almost complete extraction (~98%) at low pressure of 8 MPa (slightly above the critical point), and 55 °C with the disrupted cell of *Haematococcus pluvialis* (Cheng et al., 2018). This might be due to the proper ruptured cell provided by hydrothermal cell wall disruption (at the temperature of 200 °C with 8 MPa pressure for 10 min) accommodated this extraction yield. (Table 8.2)

TABLE 8.2
Critical Properties of the Common Solvent

Solvent	Critical Temperature (K)	Critical Pressure (MPa)
CO_2	304.1	7.38
Ethanol	513.9	6.14
Methanol	512.6	8.09
Water	647.096	22.064
Acetone	508.1	4.70

8.4.2.3 Subcritical Fluid Extraction

Subcritical fluid extraction is using the same fluids as the supercritical fluid extraction method, in order to extract targeted components from the biomass. However, the major difference between these two methods is that the entire process takes a longer time and is less efficient in general. However, it is still one of the most popular technologies due to its milder process (below critical state) concerning the applied pressure. Subcritical water extraction (SWE) is one of the most used techniques to extract protein and carbohydrates from microalgae biomass. SWE is regarded as a clean, green, and efficient extraction method as only water is used to extract the constituent from the biomass (Álvarez-Viñas et al., 2021). The risk of hazardous contamination of the final product that may be caused by other fluids such as methanol could be cut down by this process. Further, since this technology is water-based, wet biomass can be directly used in the process itself, which is beneficial to cut down the need for the drying process in the extraction process, which is normally required for other methods. Moreover, SWE provides better protein extraction yield with a shorter extraction time compared to conventional alkali, solvents, or enzymatic hydrolysis (Álvarez-Viñas et al., 2021). Due to the properties of water, operating pressure is normally milder in terms of pressure (under 22 MPa); however, it requires high temperature (from 100 °C to the critical temperature of water). It is also confirmed in an optimization study with central composite design (CCD), by extracting biochemical compounds from *Chlorella vulgaris*, that temperature is the most critical parameter in this technology (Awaluddin et al., 2016). Temperature, extraction time, and biomass loading highly affect the carbohydrate and protein yield. In the study of Garcia-Moscoso et al. (2013), protein was extracted directly from wet *Scenedesmus sp.* biomass using flash hydrolysis in subcritical water extraction. This method resulted in 30–66% mainly water-soluble peptides and free amino acids in the hydrolysate with 280 °C temperature, demonstrating the potential application for a microalgal biorefinery feedstock (Garcia-Moscoso et al., 2013). In addition, Passos et al. (2015), using pretreated mixed microalgae culture (mainly *Stigeodonium sp.*, *Monoraphidium sp.*, *Nitzschia sp.* and *Navicula sp.*), found that hydrothermal pretreatment improved 9-fold, 23-fold, 11-fold, and 31-fold for soluble organic matter, proteins, carbohydrates, and lipids, respectively. Thistreatment enhanced 28% of methane production

TABLE 8.3
Extraction Technology

Extraction Technology	Pros	Cons
Solvent extraction	Cost-effective technologyA variety of solvents can be selected	Harsh operationUnwanted byproductsCan be complexed in the purification process
Supercritical fluid extraction	Effective, rapid extractionHigh yieldUltrapure final productLess complicated in the purification process	High pressureSafety concernsHigh capita investment
Subcritical fluid extraction	Milder operationApplicable for wet biomassHigher protein and carbohydrate yield	High temperatureLong extraction time

compared to the non-pretreated biomass with the optimal condition of 130 °C and 15 mins extraction time (Passos et al., 2015). This improved methane production was due to the increased soluble organic matters that provide a better digestibility to bacterial consortium. (Table 8.3)

8.5 FUTURE PROSPECTS AND RESEARCH NEEDS

The recent development of microalgae technology in wastewater treatment facilities has potentially given many alternative routes to solve the environmental issue as well as bioeconomy. In terms of bioremediation, microalgae can fix carbon from flue gas as well as the atmosphere, and nutrients such as nitrogen and phosphorus from the waste stream by turning them into value-added biomass. A recent study by Cunha et al. (2020) showed that biopolymers derived from microalgae can potentially remediate micro/nano plastic from wastewater treatment. Moreover, Rueda et al. (2020) exhibited the extra benefit besides bioremediation by efficiently utilizing microalgae biomass into PHB production in full-scale operation. However, PHB content was still low since the study was using mixed consortia with wastewater. Regarding the use of wastewater sources, the highest PHA content shown was 31% in *E. coli* fermentation of microalgae biomass cultivated with wastewater according to Costa et al. (2019). Life cycle assessment should also be used in order to analyze which is more feasible and eco-friendly, comparing direct extraction from biomass or the enrichment of production with microbial fermentation. Alternatively, examination of the high potential pure strain *Nostoc muscorum* with wastewater as a growth medium could be significantly beneficial in the improvement of the sustainability of microalgae eco-biotechnology. Therefore, balancing the biopolymer content yield with a respectively low ratio of the cell wall to cell volume could potentially improve the downstream processing. Insightful research on optimizing cell wall to cell volume ratio prior to extraction method should be given more focus.

Therefore, if the cell wall to cell volume ratio is properly optimized in the upstream processing itself, the operation cost can be significantly improved by giving milder cell wall disruption or considering removing cell wall disruption pretreatment from the process chain itself. Or, if cell wall disruption is required, a process combining cell wall pretreatment and extraction suggested by Irshad et al. (2019b), could improve the operation cost by removing the required drying process before the extraction. Followed by this, both the risk of getting contamination by unpleasant byproducts and the costs due to time-consuming drying processes can be excluded here. This suggested method is foreseen in the continuous rapid extraction applied in the nutraceutical industry.

Selecting SC-CO_2 extraction technology could be seen into two options, which direct extraction from microalgae whole biomass, or by extract of the disrupted biomass with significantly low pressure. Direct extraction could be done by optimizing extraction efficiency using cyst germination (*Haematococcus pluvialis*) reported by Choi et al. (2015) with the upstream processing itself. By understanding the cell wall development of selected strains, this idea could also be implemented to predict how other strains respond to parameters that might influence their cell wall development. Results from the idea may lead to tremendously milder operations in the extraction technology. However, structural properties of the cell wall should draw more focus using this technique, before going into the industrial scale. The disrupted cells of microalgae biomass could be operated in a significantly lower operating pressure. However, one increase in the process chain will significantly reflect the overall cost. The overall operation cost itself can't be justified by only extraction yield alone if this is taken into the industry. It also requires reactor vessel design, biomass loading rate, CO_2 flow rate, the extraction process, operation times, labors, etc, in which life cycle assessment is considered for each step of the critical analysis. In the use of microalgae as a food/feeding source, instead of optimizing the production yield, byproducts resulted from the extraction should not be overlooked; however, it needs to be deeply investigated to make the methods more economically and environmentally feasible, upscaling the technology to an industrial level.

8.6 CONCLUSIONS

All in all, the versatility of microalgal biomass has served in various applications, such as nutritional foods/health supplements for humans, animal feed, bioenergy production, biodegradable plastic, etc. The rapid production with valuable compositions of microalgae biomass has proven to be more selective over crops and terrestrial plants for the past few decades. However, microalgal biorefinery technology is still expensive due to the resulted rigid biomass from the optimization strategies. The choice of strains and optimizing strategies concerning the rigidity of its cell wall structure should be carefully considered to make it viable for later processes. Therefe, downstream processing/extraction technology must be carefully selected, especially for the products designated for nutraceutical/pharmaceutical uses.

ACKNOWLEDGMENT

The authors would like to thankfully acknowledge the Norwegian Ministry of Education and Research through the university (University of Stavanger) granted program.

REFERENCES

Álvarez-Viñas, M., Rodríguez-Seoane, P., Flórez-Fernández, N., Torres, M., Díaz-Reinoso, B., Moure, A., Domínguez, H., 2021. Subcritical water for the extraction and hydrolysis of protein and other fractions in biorxefineries from agro-food wastes and algae: A review. *Food Bioprocess Technol.* 14, 373–387. 10.1007/s11947-020-02536-4

Anjum, A., Zuber, M., Zia, K., Noreen, A., Anjum, M., Tabasum, S., 2016. Microbial production of polyhydroxyalkanoates (PHAs) and its copolymers: A review of recent advancements. *Int. J. Biol. Macromol.* 89, 161–174. 10.1016/j.ijbiomac.2016.04.069

Arias, D., Uggetti, E., García-Galán, M., García, J., 2018. Production of polyhydroxybutyrates and carbohydrates in a mixed cyanobacterial culture: Effect of nutrients limitation and photoperiods. *N. Biotechnol.* 42, 1–11. 10.1016/j.nbt.2018.01.001

Assunção, L., Bezerra, P., Poletto, V., Rios, A., Ramos, I., Ribeiro, C., Machado, B., Izabel Druzian, J., Costa, J., Nunes, I., 2021. Combination of carotenoids from Spirulina and PLA/PLGA or PHB: New options to obtain bioactive nanoparticles. *Food Chem.* 346. 10.1016/j.foodchem.2020.128742

Awaluddin, S., Izhar, S., Hiroyuki, Y., Danquah, M., Harun, R., 2016. Sub-critical water technology for enhance extraction of bioactive compound from microalgae. *J. Eng. Sci. Technol.* 11, 63–72. 10.1155/2016/5816974

Bhati, R., Mallick, N., 2015. Poly(3-hydroxybutyrate-co-3-hydroxyvalerate) copolymer production by the diazotrophic cyanobacterium *Nostoc muscorum* Agardh: Process optimization and polymer characterization. *Algal Res.* 7, 78–85. 10.1016/j.algal.2014.12.003

Bradford, S., Segal, E., Zheng, W., Wang, Q., Hutchins, S., 2008. Reuse of concentrated animal feeding operation wastewater on agricultural lands. *J. Environ. Qual.* 37, S97–S115. 10.2134/jeq2007.0393

Cheng, X., Qi, Z., Burdyny, T., Kong, T., Sinton, D., 2018. Low pressure supercritical CO_2 extraction of astaxanthin from *Haematococcus pluvialis* demonstrated on a microfluidic chip. *Bioresour. Technol.* 250, 481–485. 10.1016/j.biortech.2017.11.070

Chihara, M., Nakayama, T., Inouye, I., Kodama, M., 1994. Chlorococcum littorale, a new marine green Coccoid Alga (Chlorococcales, Chlorophyceae). *Arch. für Protistenkd.* 144, 227–235. 10.1016/S0003-9365(11)80133-8

Choi, Y., Hong, M., Sim, S., 2015. Enhanced astaxanthin extraction efficiency from *Haematococcus pluvialis* via the cyst germination in outdoor culture systems. *Process Biochem.* 50, 2275–2280. 10.1016/j.procbio.2015.09.008

Costa, S., Miranda, A., Andrade, B., Assis, D., Souza, C., Morais, M., Costa, J., Druzian, J., 2018. Influence of nitrogen on growth, biomass composition, production, and properties of polyhydroxyalkanoates (PHAs) by microalgae. *Int. J. Biol. Macromol.* 116, 552–562. 10.1016/j.ijbiomac.2018.05.064

Costa, S., Miranda, A., Morais, M., Costa, J., Druzian, J., 2019. Microalgae as source of polyhydroxyalkanoates (PHAs) — A review. *Int. J. Biol. Macromol.* 131, 536–547. 10.1016/j.ijbiomac.2019.03.099

Cunha, C., Silva, L., Paulo, J., Faria, M., Nogueira, N., Cordeiro, N., 2020. Microalgal-based biopolymer for nano- and microplastic removal: A possible biosolution for wastewater treatment. *Environ. Pollut.* 263. 10.1016/j.envpol.2020.114385

Devadas, V., Khoo, K., Chia, W., Chew, K., Munawaroh, H., Lam, M., Lim, J., Ho, Y., Lee, K., Show, P.L., 2021. Algae biopolymer towards sustainable circular economy. *Bioresour. Technol.* 325, 124702. 10.1016/j.biortech.2021.124702

D'Hondt, E., Martín-Juárez, J., Bolado, S., Kasperoviciene, J., Koreiviene, J., Sulcius, S., Elst, K., Bastiaens, L., 2017. 6 - Cell disruption technologies, in: Gonzalez-Fernandez, C., Muñoz, R. (Eds.), *Microalgae-Based Biofuels and Bioproducts, Woodhead Publishing Series in Energy.* Cambridge: Woodhead Publishing, pp. 133–154. 10.101 6/B978-0-08-101023-5.00006-6

Difusa, A., Talukdar, J., Kalita, M., Mohanty, K., Goud, V., 2015. Effect of light intensity and pH condition on the growth, biomass and lipid content of microalgae *Scenedesmus* species. *Biofuels* 6, 37–44. 10.1080/17597269.2015.1045274

Domozych, D., Ciancia, M., Fangel, J., Mikkelsen, M., Ulvskov, P., Willats, W., 2012. The cell walls of green algae: A journey through evolution and diversity. *Front. Plant Sci.* 3, 1–7. 10.3389/fpls.2012.00082

El-Naggar, N., Hussein, M., Shaaban-Dessuuki, S., Dalal, S., 2020. Production, extraction and characterization of Chlorella vulgaris soluble polysaccharides and their applications in AgNPs biosynthesis and biostimulation of plant growth. *Sci. Rep.* 10, 1–19. 10.1038/s41598-020-59945-w

EPA - United States Enviromental Protection Agency, 2020. Advancing Sustainable Materials Management: Facts 2018 Tables and Figures. Assessing Trends in Matrials Generation and Management in the United States. Adv. Sustain. Mater. Manag. Facts Fig. 2018.___(online 2021.12.06) https://www.epa.gov/sites/default/files/2021-01/ documents/2018_ff_fact_sheet_dec_2020_fnl_508.pdf

European Bioplastics, 2019. Bioplastics Market Development Update 2019. 14th European Bioplastics Conference.

Fongsatitkul, P., Elefsiniotis, P., Yamasmit, A., Yamasmit, N., 2004. Use of sequencing batch reactors and Fenton's reagent to treat a wastewater from a textile industry. *Biochem. Eng. J.* 21, 213–220. 10.1016/j.bej.2004.06.009

Garcia-Moscoso, J., Obeid, W., Kumar, S., Hatcher, P., 2013. Flash hydrolysis of microalgae (Scenedesmus sp.) for protein extraction and production of biofuels intermediates. *J. Supercrit. Fluids* 82, 183–190. 10.1016/j.supflu.2013.07.012

Geyer, R., Jambeck, J., Law, K., 2017. Production, use, and fate of all plastics ever made. *Sci. Adv.* 3, 25–29. 10.1126/sciadv.1700782

Gonçalves, A., Simões, M., Pires, J., 2014. The effect of light supply on microalgal growth, CO_2 uptake and nutrient removal from wastewater. *Energy Convers. Manag.* 85, 530–536. 10.1016/j.enconman.2014.05.085

Gonzalez-Fernandez, C., Muñoz, R., 2017. Microalgae-based biofuels and bioproducts, in: *From Feedstock Cultivation to End-products*, 1st ed. Cambridge: Woodhead Publishing. 10.1016/C2015-0-05935-4

Günerken, E., D'Hondt, E., Eppink, M., Garcia-Gonzalez, L., Elst, K., Wijffels, R., 2015. Cell disruption for microalgae biorefineries. *Biotechnol. Adv.* 33, 243–260. 10.1016/ j.biotechadv.2015.01.008

Hong, M., Choi, Y., Sim, S., 2016. Effect of red cyst cell inoculation and iron(II) supplementation on autotrophic astaxanthin production by *Haematococcus pluvialis* under outdoor summer conditions. *J. Biotechnol.* 218, 25–33. 10.1016/j.jbiotec.2015.11.019

Huntley, M., Redalje, D., 2007. CO2 mitigation and renewable oil from photosynthetic microbes: A new appraisal, *Mitigation and Adaptation Strategies for Global Change.* 10.1007/s11027-006-7304-1

Irshad, M., Hong, M., Myint, A., Kim, J., Sim, S., 2019a. Safe and complete extraction of Astaxanthin from *Haematococcus pluvialis* by efficient mechanical disruption of cyst cell wall. *Int. J. Food Eng.* 1–16. 10.1515/ijfe-2019-0128

Irshad, M., Myint, A., Hong, M., Kim, J., Sim, S., 2019b. One-pot, simultaneous cell wall disruption and complete extraction of Astaxanthin from *Haematococcus pluvialis* at room temperature. *ACS Sustain. Chem. Eng.* 7, 13898–13910. 10.1021/acssuschemeng.9b02089

Ji, M.-K., Yun, H.-S., Hwang, J.-H., Salama, E.-S., Jeon, B.-H.,Choi, J. 2016. Effect of flue gas CO_2 on the growth, carbohydrate and fatty acid composition of a green microalga Scenedesmus obliquus for biofuel production. *Environ. Technol.* 38, 2085–2092. 10.1080/09593330.2016.1246145

Kaewkannetra, P., Enmak, P., Chiu, T., 2012. The effect of CO2 and salinity on the cultivation of *Scenedesmus obliquus* for biodiesel production. *Biotechnol. Bioprocess Eng.* 17, 591–597. 10.1007/s12257-011-0533-5

Karan, H., Funk, C., Grabert, M., Oey, M., Hankamer, B., 2019. Green bioplastics as part of a circular bioeconomy. *Trends Plant Sci.* 24, 237–249. 10.1016/j.tplants.2018.11.010

Kim, S., Less, J., Wang, L., Yan, T., Kiron, V., Kaushik, S., Lei, X., 2019. Meeting global feed protein demand: Challenge, opportunity, and strategy. *Annu. Rev. Anim. Biosci.* 7, 221–243. 10.1146/annurev-animal-030117-014838

Koo, K., Jung, S., Lee, B., Kim, J., Jo, Y., Choi, H., Kang, S., Chung, G., Jeong, W., Ahn, J., 2017. The mechanism of starch over-accumulation in *Chlamydomonas reinhardtii* high-starch mutants identified by comparative transcriptome analysis. *Front. Microbiol.* 8, 1–12. 10.3389/fmicb.2017.00858

Krichnavaruk, S., Shotipruk, A., Goto, M., Pavasant, P., 2008. Supercritical carbon dioxide extraction of astaxanthin from *Haematococcus pluvialis* with vegetable oils as co-solvent. *Bioresour. Technol.* 99, 5556–5560. 10.1016/j.biortech.2007.10.049

Li, F., Yang, Z., Zeng, R., Yang, G., Chang, X., Yan, J., Hou, Y., 2011. Microalgae capture of CO_2 from actual flue gas discharged from a combustion chamber. *Ind. Eng. Chem. Res.* 50, 6496–6502. 10.1021/ie200040q

Li, Y., Chen, Y., Chen, P., Min, M., Zhou, W., Martinez, B., Zhu, J., Ruan, R., 2011. Characterization of a microalga *Chlorella sp.* well adapted to highly concentrated municipal wastewater for nutrient removal and biodiesel production. *Bioresour. Technol.* 102, 5138–5144. 10.1016/j.biortech.2011.01.091

Lutzu, G., Ciurli, A., Chiellini, C., Di Caprio, F., Concas, A., Dunford, N., 2021. Latest developments in wastewater treatment and biopolymer production by microalgae. *J. Environ. Chem. Eng.* 9, 104926. 10.1016/j.jece.2020.104926

Marcus, Y., 2018. Extraction by subcritical and supercriticalwater, methanol, ethanol and their mixtures. *Separations* 5(1). 10.3390/separations5010004

Martínez, M., 2000. Nitrogen and phosphorus removal from urban wastewater by the microalga *Scenedesmus obliquus*. *Bioresour. Technol.* 73, 263–272. 10.1016/S0960-8524(99)00121-2

Mathiot, C., Ponge, P., Gallard, B., Sassi, J., Delrue, F., Le Moigne, N., 2019. Microalgae starch-based bioplastics: Screening of ten strains and plasticization of unfractionated microalgae by extrusion. *Carbohydr. Polym.* 208, 142–151. 10.1016/j.carbpol.2018.12.057

Mendes-Pinto, M., Raposo, M., Bowen, J., Young, A., Morais, R., 2001. Evaluation of different cell disruption processes on encysted cells of Haematococcus pluvialis: Effects on astaxanthin recovery and implications for bio-availability. *J. Appl. Phycol.* 13, 19–24. 10.1023/A:1008183429747

Mohsenpour, S., Willoughby, N., 2016. Effect of CO_2 aeration on cultivation of microalgae in luminescent photobioreactors. *Biomass Bioenerg.* 85, 168–177. 10.1016/j.biombioe.2015.12.002

Monfet, E., Unc, A., 2017. Defining wastewaters used for cultivation of algae. *Algal Res.* 24, 520–526. 10.1016/j.algal.2016.12.008

Passos, F., Carretero, J., Ferrer, I., 2015. Comparing pretreatment methods for improving microalgae anaerobic digestion: Thermal, hydrothermal, microwave and ultrasound. *Chem. Eng. J.* 279, 667–672. 10.1016/j.cej.2015.05.065

Plastics Europe, 2020. An analysis of European plastics production, demand and waste data, in: *Plastics – the Facts 2020*. (online, 2021.12.06) https://plasticseurope.org/knowledge-hub/plastics-the-facts-2020/

Pokhrel, D., Viraraghavan, T., 2004. Treatment of pulp and paper mill wastewater - A review. *Sci. Total Environ.* 333, 37–58. 10.1016/j.scitotenv.2004.05.017

Rahman, A., Putman, R., Inan, K., Sal, F., Sathish, A., Smith, T., Nielsen, C., Sims, R., Miller, C., 2015. Polyhydroxybutyrate production using a wastewater microalgae based media. *Algal Res.* 8, 95–98. 10.1016/j.algal.2015.01.009

Rasdi, N., Qin, J., 2015. Effect of N:P ratio on growth and chemical composition of *Nannochloropsis oculata* and *Tisochrysis lutea*. *J. Appl. Phycol.* 27, 2221–2230. 10.1007/s10811-014-0495-z

Renaud, S., Thinh, L., Lambrinidis, G., Parry, D., 2002. Effect of temperature on growth, chemical composition and fatty acid composition of tropical Australian microalgae grown in batch cultures. *Aquaculture*. 211, 195–214. 10.1016/S0044-8486(01)00875-4

Research and Markets, 2021. Global Polyhydroxyalkanoate (PHA) Market by Type (Short Chain Length, Medium Chain Length), Production Method (Sugar Fermentation, Vegetable Oil Fermentation, Methane Fermentation), Application, and Region - Forecast to 2025, Markets and Markets.

Reynolds, C., 2006. *The Ecology of Phytoplankton (Ecology, Biodiversity and Conservation)*. Cambridge: Cambridge University Press. doi:10.1017/CBO9780511542145

Ruangsomboon, S., 2012. Effect of light, nutrient, cultivation time and salinity on lipid production of newly isolated strain of the green microalga, *Botryococcus braunii* KMITL 2. *Bioresour. Technol.* 109, 261–265. 10.1016/j.biortech.2011.07.025

Rueda, E., García-Galán, M., Ortiz, A., Uggetti, E., Carretero, J., García, J., Díez-Montero, R., 2020. Bioremediation of agricultural runoff and biopolymers production from cyanobacteria cultured in demonstrative full-scale photobioreactors. *Process Saf. Environ. Prot.* 139, 241–250. 10.1016/j.psep.2020.03.035

Safi, C., Charton, M., Pignolet, O., Silvestre, F., Vaca-Garcia, C., Pontalier, P., 2013. Influence of microalgae cell wall characteristics on protein extractability and determination of nitrogen-to-protein conversion factors. *J. Appl. Phycol.* 25, 523–529. 10.1007/s10811-012-9886-1

Sakai, N., Sakamoto, Y., Kishimoto, N., Chihara, M., Karube, I., 1995. *Chlorella* strains from hot springs tolerant to high temperature and high CO_2. *Energy Convers. Manag.* 36, 693–696. /10.1016/0196-8904(95)00100-R

Sakarika, M., Kornaros, M., 2016. Effect of pH on growth and lipid accumulation kinetics of the microalga *Chlorella vulgaris* grown heterotrophically under sulfur limitation. *Bioresour. Technol.* 219, 694–701. 10.1016/j.biortech.2016.08.033

Sankaran, R., Parra Cruz, R., Pakalapati, H., Show, P., Ling, T., Chen, W., Tao, Y., 2020. Recent advances in the pretreatment of microalgal and lignocellulosic biomass: A comprehensive review. *Bioresour. Technol.* 298, 122476. 10.1016/j.biortech.2019.122476

Senge, M., Senger, H., 1990. Response of the photosynthetic apparatus during adaptation of *Chlorella* and *Ankistrodesmus* to irradiance changes. *J. Plant Physiol.* 136, 675–679. 10.1016/S0176-1617(11)81343-X

Sørensen, M., Berge, G., Thomassen, M., Ruyter, B., Hatlen, B., Ytrestøyl, T., Aas, T., Åsgård, T., 2011. Today's and tomorrow's feed ingredients in Norwegian aquaculture, ISBN: 9788272519437.

Sung, K., Lee, J., Shin, C., Park, S., 1998. Isolation of a new highly CO2 tolerant fresh water microalga *Chlorella sp. KR-1. Renew. Energy* 16, 1019–1022. 10.1016/S0960-1481 (98)00362-0

Tchobanoglous, G., Burton, F., Metcalf & Eddy, 1991. *Wastewater Engineering : Treatment, Disposal, and Reuse*, 3rd ed. New York: McGraw-Hill.

Valderrama, J., Perrut, M., Majewski, W., 2003. Extraction of Astaxantine and phycocyanine from microalgae with supercritical carbon dioxide. *J. Chem. Eng. Data* 48, 827–830. 10.1021/je020128r

Varshney, P., Mikulic, P., Vonshak, A., Beardall, J., Wangikar, P., 2015. Extremophilic micro-algae and their potential contribution in biotechnology. *Bioresour. Technol.* 184, 363–372. 10.1016/j.biortech.2014.11.040

Velazquez-Lucio, J., Rodríguez-Jasso, R., Colla, L., Sáenz-Galindo, A., Cervantes-Cisneros, D.E., Aguilar, C.N., Fernandes, B.D., Ruiz, H.A., 2018. Microalgal biomass pre-treatment for bioethanol production: A review. *Biofuel Res. J.* 5, 780–791. 10.18331/ BRJ2018.5.1.5

Wicker, R., Kumar, G., Khan, E., Bhatnagar, A., 2021. Emergent green technologies for cost-effective valorization of microalgal biomass to renewable fuel products under a bior-efinery scheme. *Chem. Eng. J.* 415, 128932. 10.1016/j.cej.2021.128932

Woertz, I., Feffer, A., Lundquist, T., Nelson, Y., 2009. Algae grown on dairy and municipal wastewater for simultaneous nutrient removal and lipid production for biofuel feed-stock. *J. Environ. Eng.* 135, 1115–1122. 10.1061/(asce)ee.1943-7870.0000129

Xin, L., Hong-Ying, H., Jia, Y., 2010. Lipid accumulation and nutrient removal properties of a newly isolated freshwater microalga, *Scenedesmus sp. LX1*, growing in secondary effluent. *N. Biotechnol.* 27, 59–63. 10.1016/j.nbt.2009.11.006

Yap, B., Crawford, S., Dagastine, R., Scales, P., Martin, G., 2016. Nitrogen deprivation of microalgae: effect on cell size, cell wall thickness, cell strength, and resistance to mechanical disruption. *J. Ind. Microbiol. Biotechnol.* 43, 1671–1680. 10.1007/s10295-016-1848-1

Zheng, Y., Li, T., Yu, X., Bates, P., Dong, T., Chen, S., 2013. High-density fed-batch culture of a thermotolerant microalga *Chlorella sorokiniana* for biofuel production. *Appl. Energy* 108, 281–287. 10.1016/j.apenergy.2013.02.059

Zhu, S., Wang, Y., Huang, W., Xu, J., Wang, Z., Xu, J., Yuan, Z., 2014. Enhanced Accumulation of Carbohydrate and Starch in *Chlorella zofingiensis* Induced by Nitrogen Starvation. *Appl. Biochem. Biotechnol.* 174, 2435–2445. 10.1007/s12010-014-1183-9

Zhu, Y., Zhao, X., Zhang, X., Liu, H., Ao, Q., 2020. Amino acid, structure and antioxidant properties of *Haematococcus pluvialis* protein hydrolysates produced by different proteases. *Int. J. Food Sci. Technol.* 1–11. 10.1111/ijfs.14618

9 Algae as a Source of Polysaccharides and Potential Applications

Sonal Tiwari
Plant Cell Biotechnology Department (PCBT), CSIR-Central Food Technological Research Institute (CFTRI), Mysuru, Karnataka, India

E Amala Claret and Vikas S. Chauhan
Plant Cell Biotechnology Department (PCBT), CSIR-Central Food Technological Research Institute (CFTRI), Mysuru, Karnataka, India
and
Academy of Scientific and Innovative Research (AcSIR), Ghaziabad, India

CONTENTS

9.1 Introduction...258
9.2 Source and Chemical Structure of Macroalgal Polysaccharides................260
    9.2.1 Alginate...260
        9.2.1.1 Source..260
        9.2.1.2 Structure 260
    9.2.2 Carrageenans...263
        9.2.2.1 Source..263
        9.2.2.2 Structure 263
    9.2.3 Chitin and Chitosan..265
        9.2.3.1 Source..265
        9.2.3.2 Structure 265
    9.2.4 Fucoidans..266
        9.2.4.1 Source..266
        9.2.4.2 Structure 266
    9.2.5 Laminarin..267
        9.2.5.1 Source..267
        9.2.5.2 Structure 267
    9.2.6 Agar 268
        9.2.6.1 Source..268

DOI: 10.1201/9781003195405-9

             9.2.6.2 Structure 268
      9.2.7 Ulvan...269
             9.2.7.1 Source...269
             9.2.7.2 Structure 269
9.3   Microalgal Polysaccharides...270
9.4   Extraction and Purification of Polysaccharides...................................272
9.5   Potential Applications of Algal Polysaccharides 275
      9.5.1 Biological Activity of Algal Polysaccharides for
             Pharmaceutical Applications..275
             9.5.1.1 Anti-inflammatory and Immunomodulatory Activities 275
             9.5.1.2 Antioxidant Activity...276
             9.5.1.3 Anticancer and Antitumor Activities..............................276
             9.5.1.4 Antiviral Activity..277
             9.5.1.5 Antibacterial Activity 277
             9.5.1.6 Anticoagulant and Antithrombotic Activities..................277
             9.5.1.7 Antilipidemic and Antihepatotoxic Activities 278
             9.5.1.8 Neuroprotective Activity 278
      9.5.2 Cosmetic Applications 278
      9.5.3 Biomedical Applications 279
      9.5.4 Applications in Food Biotechnology..280
      9.5.5 Applications in Agriculture...281
      9.5.6 Applications in Bioenergy 282
9.6   Future Prospects 282
9.7   Conclusion..286
References..286

9.1 INTRODUCTION

Polysaccharides are polymers of simple carbohydrates (monosaccharides) linked together by glycosidic linkages. Polysaccharides are either homogenous or heterogeneous based on monomer composition. The hydrolysis of the former results in its oligosaccharides or different types of monosaccharide constituents (e.g., Pectin), and the latter will give the same sugar units (e.g., glycogen) (Yalpani, 2013; Khalil et al., 2018; Chen et al., 2020, Lahaye, 2001). In the living system, polysaccharides play a vital role as a functional unit, structural unit, and storage form of energy; hence, they are the most abundant biopolymer on earth. Due to their ubiquitous nature and beneficial biological activity, polysaccharides are extensively used in biomedical and food production sectors for their applications as biofilms, emulsifiers, stabilizers, drug delivery biomolecules, antimicrobial agents, anticoagulants, biofuels, and so on (Kraan, 2012; Zhang et al., 2010; Raposo et al., 2013). Until recently, however, our knowledge of polysaccharides from a natural source was limited due to the complexity of their structure, lack of understanding of their biosynthetic pathway, and extraction principles. This has created a need for exploring a wide variety of natural polysaccharides from readily available sources in significant quantity and quality to study their potential applications. The structural complexity of polysaccharides is more diverse than that of other biopolymers, such as proteins and nucleic acids (Chen et al., 2020; Kraan, 2012; Armisen and Gaiatas, 2009). Consider having a disaccharide with two same hexose sugars; it can form 12 different disaccharides. These two monosaccharides can link with two possible linkages, such as α and β, at six different points. Furthermore, polysaccharides are either branched or linear and, in part, contain a broad range of functional groups (sulfate, acetate, amino, or methyl groups) on one or more monomers (Arnosti et al., 2021). Polysaccharide bioactivity is determined by molecular characteristics, such as molecular sizes, types, and ratios of monosaccharide constituents and type of glycosidic linkages (Kong and Mooney, 2005; Kraan, 2012). Ideal active polysaccharides are primarily neutral or anionic polymers with 4–25 degrees of polymerization. They are very rarely obtained from a natural source; therefore, large polymers are treated by chemical (oxidizing agents), enzymatical (enzymes), or physical (microwave-assisted hydrolysis) methods during extraction. Recent development in these techniques made it possible to control the degree of depolymerization and to get the desired polysaccharide by adding or deleting functional groups (Hernandez-Carmona et al., 2013).

Algae is one of the major producers and commercial sources of a variety of polysaccharides, including sulfated polysaccharides as in complex cell walls (cellulose, sulfated galactan, xylan, or mannan fibrils), mucopolysaccharides as in mucilage, and storage polysaccharides (Kraan, 2012). They are simple photosynthetic organisms that are distributed throughout the earth, from hot springs to cold poles. They are important primary producers and also habitat-structuring species in their ecosystem. Earlier studies suggest that there are somewhere between 30,000 to 1 million species of algae globally, yet only a few hundred are explored for their commercially exploitable metabolites (Guiry, 2012, Ferdouse et al., 2018). Algae exist as unicellular or multicellular species of varying sizes. Typical unicellular microalgae measure between 2–20μm, while multicellular seaweeds can reach more than 70 cm in length with a growth rate of 2–50 cm per day, and kelp (brown seaweeds) covers 25% of the global coastline, forming an underwater forest (Wernberg et al., 2019, Ferdouse et al., 2018). The rapid growth rate of algae, as well as their ability to grow utilizing CO_2 as a sole carbon source and light as energy, may be able to meet the continuous supply of biomass necessary on a big scale for the manufacturing process (Wernberg et al., 2019). Algae have long been used as a delicacy and functional food by aboriginal peoples of the Far East and Asian countries, e.g., jelly, Dulse, Kombu. In Western countries, like Hawaii in the USA, nearly 40 species of seaweed are eaten raw, pickled, or along with food items. There is a growing body of literature that recognizes the importance of algae as a rich source of a wide variety of useful metabolites, so they are undergoing a revolution in terms of making large-scale healthier food products and purified metabolites (Sartal et al., 2011; Ścieszka and Klewicka, 2019). It is estimated that there are about 221 species of seaweed used worldwide for their commercial value (Ferdouse et al., 2018). Some of the prominently cultivated seaweeds for commercial purposes are *Saccharina japanoica* (33%), *Eucheuma spp.* (17%), *Kappaphycus alvarezii, Undaria pinnatifida,* and *Porphyra spp.* Nearly 32 million metric tons of seaweed valuing about 12 billion US$ were produced in 2017, 70% of which was used as food, and the balance was used as the source of hydrocolloids (carrageenan, alginate, agar), cattle feed, agriculture fertilizers, and metabolites. Globally 50 countries are involved in seaweed cultivation; of those, Asian countries, such as China (17.5 million tonnes), Indonesia (9.7 million tonnes), and Korea (1.7 million tons) are the largest producers of commercial seaweed. They contribute 99% of the global production of seaweed (Ferdouse et al., 2018).

Like other biochemical components of algae, their polysaccharide content shows extreme variation based on external condition, stage of development, and season. The total carbohydrate content of microalgae ranges from 21% to 50% or more, while in macroalgae, it is 4% to 76% of their total dry weight, as seen in Tables 9.1 and 9.2.

According to Ferdouse et al. (2018), nearly 40% of the world's hydrocolloids are manufactured from seaweed extracts such as agar, alginate, and carrageenans. In fact, the remarkable rise in Southeast Asia seaweed cultivation owes to the increasing demand for raw material in the Phycocolloids industries. Different species of algae have their unique polysaccharide content. Most of the well-known major polysaccharides of macroalgae and some microalgae are structurally characterized. Algin, Fucoidan, and Fucin are anionic polysaccharides of brown algae (Phaeophyceae). The average weight ratio of Alginates, fucoidans, and cellulose in brown algae is 3:1:1, respectively. Red algae (Rhodophyceae) contain agars and carrageenans, xylans, mannans, porphyran, cellulose, and floridean starch, while green algae (Chlorophyceae) contain ulvan, sulfatedgalactan, cellulose, xylans, starch, and β-D-mannans (Verma et al., 2015).

9.2 SOURCE AND CHEMICAL STRUCTURE OF MACROALGAL POLYSACCHARIDES

9.2.1 ALGINATE

9.2.1.1 Source

The cell wall of Brown algae (Class Phaeophyceae) majorly contains a linear water-soluble heteropolysaccharide, Alginate. E. C. C. Stanford, a British chemical scientist, elucidated the structure of alginates in the 1880s (Draget and Taylor, 2011). Algae belonging to genus *Laminaria, Alaria, Macrocystis*, and *Ascophyllum* are the major source of alginate for large-scale production. Other sources include members of *Durvillaea, Lessonia, Sargassum,* and *Ecklonia. Turbinari, Laminaria* and *Alaria* are farmed, while some species are harvested from natural habitats for alginate production since offshore farming of seaweed is an expensive process (Draget and Taylor, 2011; Chee et al., 2011). Alginate with high l-guluronic acid content is found in the outer cortex of matured stipes of *Laminaria hyperborean,* while its leaves have high mannuronic acid, which gives the flexibility to float (Draget and Taylor, 2011).

9.2.1.2 Structure

Alginate is a copolymer of guluronic acid (G) and mannuronic acid (M) with G α1–4 and M β1–4 linkages in the form of a homopolymeric (MM/GG) and heteropolymeric (MG/GM) chain (Figure 9.1). The mannuronic (M) and guluronic (G) acid residues of the polymer are arranged in an irregular, blockwise fashion along the chain. The physical property and reactivity of alginate depend on the proposition

TABLE 9.1

Algal Polysaccharides, Their Chemical Component, Type of Glycosidic Linkage and Source

Polysaccharide Name	Major Chemical Component	Minor Chemical Component	Type of Glycosidic Linkage	Algal Source	Reference
Alginate	guluronic acid, mannuronic acid	–	α-(1–4), β-(1–4)	*Laminaria, Alaria, Macrocystis, Ascophyllum*	Hasnain et al., 2020
Carrageenans	D galactopyranose, 3,6 anhydrogalactose	–	α-(1–3), β-(1–4)	*Kappaphycus alvarezii, Eucheuma denticulate, Chondrus crispus*	Youssouf et al., 2017
Chitin	N-acetyl-d-glucosamine	–	β-(1,4)	*Thalassiosira, Cyclotella*	Durkin et al., 2009
Chitosan	d-glucosamine, N-acetyl-d-glucosamine	–	β-(1,4)		
Fucoidans	l-fucose	mannose, glucose, arabinose, xylose, uronic acid	α-(1–3), α- (1–4)	*Fucus sp, Chorda filum, Laminaria digitata, Ascophyllum nodosum, Cladosiphon okamuranus*	Chizhov et al., 1999, Chevolot et al., 2001
Laminarin	d-glucopyranose,	uronic acid, mannitol	β-(1–3), β-(1–6)	*Laminaria digitata,Saccharina latissima, Fucus vesiculosus, Ascophyllum nodosum, Eisenia bicyclis, Sargassum henslowianum*	Kadam et al., 2015, Ermakova et al., 2013, Cui et al., 2021
Agrose (Agar)	galactose, 3,6-anhydrogalactose	–	α-(1–4), β- (1–3)	*Gelidium, Gracilaria, Acantkopeltis, Ceramium, Pterocladia*	Armisen and Gaiatas 2009
Agropectin (Agar)	galactose	pyruvate, β-d-xylopyranose, 4-o-Me-a-l-galactopyranose	β-(1,4), a-(1,3)		
Ulvan	L-rhamnose, D-xylose, D-glucuronic acid, L-iduronic acid	–	α-(1,4), β-(1,4)	*Ulva*	Lahaye and Robic 2007

TABLE 9.2
Polysaccharide Content (% dry wt.) of Different Seaweeds with Sulfate and Uronic Acid Composition

Species	Group	Total Carbohydrate (%)	Sulfate (%)	Uronic Acids (%)	References
Ulva pertusa	Chlorophyta	33.4	19.9	19.2	Zhang et al., 2010
Enteromorpha linza	Chlorophyta	47.9	16.2	11.9	Zhang et al., 2010
Bryopsis plumose	Chlorophyta	41.8	7.56	7.73	Zhang et al., 2010
Capsosiphon fulvescens	Chlorophyta	64	15.4	13.2	Synytsya et al., 2015
Laminaria japonica	Phaeophyceae	66.7	25.9	8.45	Zhang et al., 2010
Sargassum vulgare	Phaeophyceae	16–63.1	3.1–22.6	11.1–89.4	Dore et al., 2013
Porphyra haitanensis	Rhodophyta	78.9	17.7	0	Zhang et al., 2010
Pterocladia capillacea	Rhodophyta	35–57	18.7–28	1.27–5.87	Fleita et al., 2015
Gloiopeltis furcata	Rhodophyta	66.3–71.9	22.7–31.2	–	Yu et al., 2010

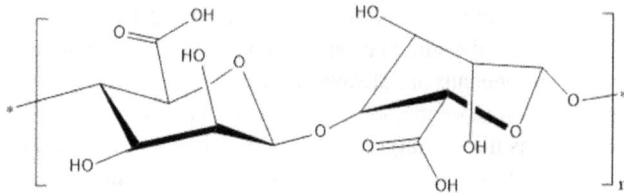

FIGURE 9.1 Structure of Alginate.

and conformation of these blocks. Due to varying linkage positions, MM blocks form a flat ribbon; GG blocks form a buckled ribbon and a helix-like structure for the MG/GM blocks. The monomer distribution and the sequential structure of different alginate are very difficult to figure out from the diverse species of algae (Draget and Taylor, 2011; Lee and Mooney, 2012). Alginate exists as calcium, magnesium, and sodium salt of alginic acid, and it constitutes 20–47% of the total dry weight of algal biomass, while its acid form, alginic acid, consists only of linear polyuronic acid (Pathak et al., 2008). Alginate is capable of absorbing water about 200–300 times its weight because of its ionotropic ability to form hydrogels by inter- and intra-molecular crosslink with polyvalent cations or polyamines in a colloidal solution (LeRoux et al., 1999).

9.2.2 CARRAGEENANS

9.2.2.1 Source

Carrageenans form the major part of the cell wall and intercellular matrix of the red algae belonging to Gigartinaceae, Hypneaceae, Solieriaceae, Phyllophoraceae, and Furcellariaceae families. Farmed seaweeds can produce up to 60% carrageenans of their total dry weight and are in ideal condition; this can reach about 70–80%. The carrageenans content varies among red algae families and also following their reproductive phase. It was first produced from *Chondrus crispus*. The commercial source of Iota (l) Carrageenans is farm *Eucheuma denticulatum*, and kappa (κ) Carrageenans is farm *Kappaphycus alvarezii*. The carbohydrate content of *Eucheuma denticulatum* predominantly contains 87% of l Carrageenans, and *Kappaphycus alvarezii* is composed of 68% of κ carrageenans and 32% of l Carrageenans (Youssouf et al., 2017). Species like *Chondrus crispus* and *Sarcothalia crispate* contain evenly proportionate κ and lambda (λ) Carrageenans. About 90% of the global production of carrageenans is obtained from *Eucheuma* and *Kappaphycus* (Campbell and Hotchkiss, 2017; Tuvikene, 2021). The Philippines, Indonesia, and East Africa are the major producers of these warm-water cultivated species (Ferdouse et al., 2018).

9.2.2.2 Structure

Carrageenans have high molecular weight; they are a linear, hydrophilic polysaccharide that is made of repeated disaccharides of alternating sulfated (15–40%) or non-sulfated D galactopyranose and 3,6 anhydrogalactose units bound by α-1–3 and β-1–4 glycosidic linkages (Tuvikene, 2021). The sulfate groups are found esterified to hydroxyl groups of carbon C6 or C2 atoms. The structure of the polysaccharide can be disrupted by ionic exchange and is greatly influenced by temperature, ph, and cation concentration since the sulfate esters are highly acid, which are always ionized. This is the reason carrageenans are always anionic, even at lower pH. On hydrolysis, carrageenans give rise to three major fractions as κ (kappa), ι (iota), and λ (lambda), and four minor fractions like μ (mu), v (nu), θ (theta), and ε (xi) (Cui, 2005). Based on their composition, there are nearly 15 different types of carrageenan, among which three have gained importance commercially, namely κ, ι, and λ carrageenans. These three types differ in the properties of esterified sulphate groups. The kappa-carrageenan is a polymer with repeated units of altering (1,3)-β-galactopyranose-4-sulfate and (1,4)-α-3,6-anhydrogalactopyranose residues, as shown in Figure 9.2. Iota-carrageenans (Figure 9.3) and lambda carrageenans (Figure 9.4) differ from κ-carrageenan in having a sulfate group at carbon C2 atom in former and carbon C2 as well as C6 in latter. The 3,6-anhydro-D-galactopyranosyl residues are responsible for the gelling property of carrageenans in an aqueous solution. Kappa carrageenans form strong and stiff gels when added to a solution containing potassium chloride than other salts. The iota-carrageenan does not show such specificity and always produces weaker elastic gels than κ carrageenans. The reason behind this is κ carrageenans

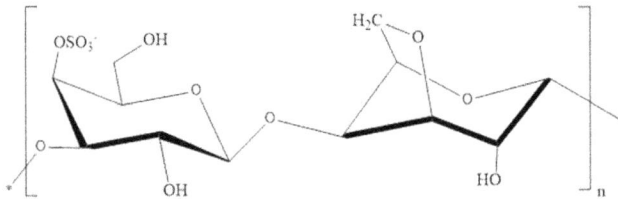

FIGURE 9.2   Structure of κ-carrageenan.

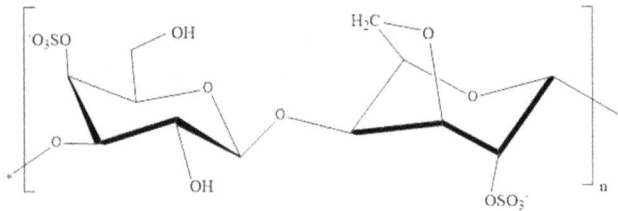

FIGURE 9.3   Structure of ι-carrageenan.

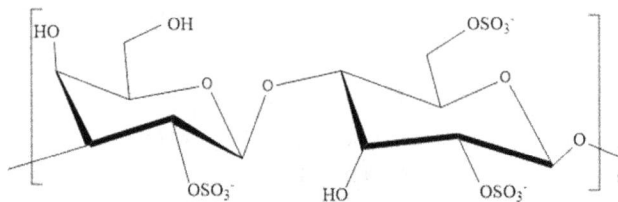

FIGURE 9.4   Structure of λ-carrageenans.

naturally have lower sulfate residues, about 20%, whereas ι and λ carrageenans contain 33% and 41% sulphate group (Youssouf et al., 2017).

9.2.3 CHITIN AND CHITOSAN

9.2.3.1 Source

Chitin and chitosan are the second most abundant rigid, elastic structural polysaccharides after cellulose and lignin. They were first discovered in 1884. These highly organized microfibrils give immense strength to the cell wall and exoskeleton of arthropods, fungi, bacteria, and few algal species. From 1965 onwards, the presence of chitin in a few species of algae has been reported. Diatoms such as *Thalassiosira* and *Cyclotella* of class *Mediophyceae* are a significant producer of chitin (Durkin et al., 2009). They produce long, thin β chitin fibers that extend from theca of the cell to the siliceous diatoms wall through fultoportulae. *Thalassiosira,* during the reproductive phase secrets chitin strands that connect newly divided daughter cells to form a chain that supports their buoyancy in its aquatic ecosystem (Walsby and Xypolyta, 1977). Cell walls of *crustose corallines* (red algae), *Clathromorphum compactu,* and green algae, such as *Pithophora oedogonium* and *Chlorella vulgaris*, contain a small fraction of Chitin. It gives protection to algae from ocean acidification and grazing by herbivorous organisms (Danyal et al., 2013).

9.2.3.2 Structure

Chitin and chitosan are linear biopolymers of high molecular weight (≥ 106 Da). The major composition of chitin is repeating units of *N*-acetyl-d-glucosamine bound by β-(1,4)-glycosidic linkage (Figure 9.5). Chitin is a structural and functional analog of cellulose in plants where the acetamide group of former is replaced with a hydroxyl group at the C2 position of glucose residue (Durkin et al., 2009). Cationic heteropolysaccharides and chitosan are analogs of chitin with varying degrees of deacetylation and depolymerisation (Figure 9.6). They are a copolymer of d-glucosamine and *N*-acetyl-d-glucosamine attached by β-(1,4)-glycosidic linkage. The presence of a free amino group in its structure makes chitosan susceptible to acid digestion, whereas chitin is insoluble at any pH. Chitosan is prepared from α-chitin by deacetylation, with 40–50% aqueous alkali solution at boiling temperature for a few hours (Mohammed et al., 2013). Chitin and chitosan are always

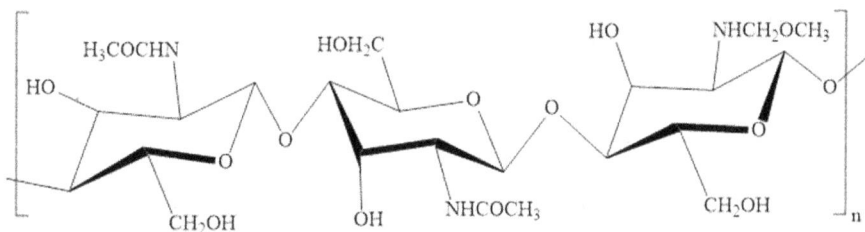

FIGURE 9.5   Structure of Chitin.

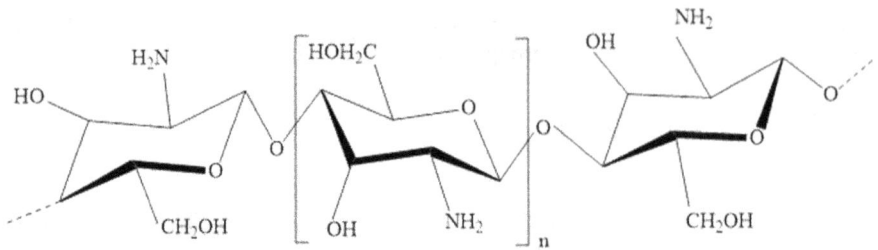

FIGURE 9.6 Structure of Chitosan.

complexed with other macromolecules and minerals. Chitin exists in two allomorphs, namely α, β, and rarely γ form, depending on its source. Though α-Chitin is the abundant allomorph in nature, algae contain only β-chitin, which shows a parallel arrangement of chitin chains in contrast to α-chitin (Durkin et al., 2009).

9.2.4 FUCOIDANS

9.2.4.1 Source

Fucoidans are structural and storage polysaccharides of the cell wall and intercellular mucilage of brown algae (Phaeophyceae). They are also known as fucan, fucosan, fuconoids, or sulphated fucan. The first fucoidan was extracted and studied by Kylin in 1913 from different species of brown algae (Verma et al., 2015). The main source of fucoidan is *Fucus vesiculosus*, more than 20% of its dry weight. The molecular weight of fucoidan ranges from 43–1550 kDa (Rioux et al., 2007). Fucoidan of *Cladosiphon okamuranus* contained 70.13% fucose and 15.16% sulphate (Lim et al., 2019).

9.2.4.2 Structure

Fucoidans are anionic complex branched polysaccharides with high l-fucose residues linked by 1,3 or 1,3 and 1,4 α linkages and sulfate esters (Cui et al., 2018). The complexity of its structure is because of the presence of about 10% of other monomers apart from fucose, such as mannose, glucose, arabinose, and xylose, along with acetyl/sulfate groups, uronic acid, and protein, as shown in Figure 9.7.

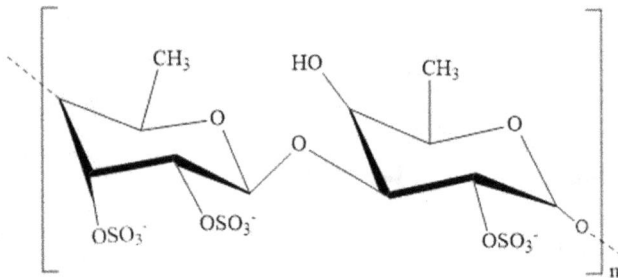

FIGURE 9.7 Structure of Fucoidan.

The commercial fucoidan of *Fucus vesiculosus* consisted of fucose, galactose, glucose, mannose, xylose, uranic acid, and sulfate in the molar ratio of 100: 3: trace: 2: 4: 20: 120 (Nishino et al., 1994). Based on its core region, fucoidan is grouped into two variants. One with a chain of 1,3-linked α-l-fucopyranose units as in fucoidan of *Chorda filum, Laminaria digitata,* and *Cladosiphon okamuranus* (Chizhov et al., 1999). Another with a chain of both 1,3- and 1,4-linked α-l fucopyranose units as found in fucoidan of *Ascophyllum nodosum, Fucus distichus,* and *Fucus evanescence* (Chevolot et al., 2001). Based on the type of core chain, this polymer is sulfonate and/or acetylated, either at O-4 or O-2, sometime at both positions of fucose residue. The branch point is created by fucose residue at every 2–3 monomers within the chain.

9.2.5 LAMINARIN

9.2.5.1 Source
Laminarin is otherwise known as laminaran, or leucosin, which is the reserved food material of brown algae belonging to the family Laminariaceae and was first described by Schmiedeberg in 1885. The *Laminarin digitata* is the commercial source of Laminarin, which constitutes up to 35% of its dry weight (Sanjeewa et al., 2017). Species other than those belonging to the family Laminariaceae also produce Laminarin, such as *Saccharina latissima* (33%), *Fucus vesiculosus,* and *Ascophyllum nodosum* (4.5%) (Kadam et al., 2015). Most of the laminarins are low molecular weight (~5 kDa) glucan except the one isolated from *Eisenia bicyclis,* which has the highest MW of 19–27 kDa (Ermakova et al., 2013). A laminarin-type polysaccharide with a molecular weight of 8.4 kDa is reported in *Sargassum henslowianum* (Cui et al., 2021).

9.2.5.2 Structure
Laminarin is a branched, hydrophilic β-glucan. The chemical structure of laminarin consists of the main chain made of 20–33 units of d-glucopyranose residues with β-(1–3) linkage along with 6-*O*-branching and β-(1–6)-interchain links (Rioux et al., 2007) (Figure 9.8). They also have some amount of uronic acid and mannitol (2.4 to 3.7%). The laminarin of *Eiseniabi cyclis* is a linear chain of (1,3)-β-d-glucanswith (1,3) and (1,6) linkage in a ratio of 2.6:1 and 13.2% sulfate (Ermakova et al., 2013). The solubility of Laminarin is affected by the amount of branching. The highly branched polymer is soluble even in cold water, while the laminarins with lesser branches are only soluble in water at high temperatures. Based on the terminal sugar residue of the glycan chain, Laminarin is either M or G type (Rioux et al., 2007). In the case of M type, terminal sugar is a non-reducing 1-linked D-mannitol, and in G type, it is a reducing glucose residue. Generally, only a small portion of the polysaccharide will have G chains, and the majority are M chains (Read et al., 1996). The ratio of M:D in glucan is species dependant, *Laminaria digitata* has M and G in the ratio of 3:1, while in *Cystoseira barbata* and *C. crinite* M chain is absent. There are some exceptional laminarins found in *Cystoseira* species with N-acetylhexosamine-terminated chains in smaller quantity (Sellimi et al., 2018).

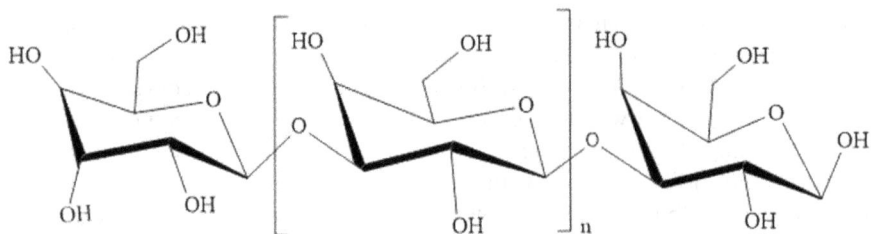

FIGURE 9.8 Structure of Laminarian.

9.2.6 AGAR

9.2.6.1 Source

Agar is a linear polysaccharide and gelling hydrocolloid found in the cell wall of algal species predominantly belonging to the genera *Gelidium, Gracilaria, Acantkopeltis, Ceramium,* and *Pterocladia* (Armisen and Gaiatas, 2009). Agar is known as Kanten (frozen sky) in Japan, where its commercial production started in the 17th century from *Gelidium* (Armisen and Gaiatas, 2009). Other red algae species like *Gracilaria, Ahnfeltia,* and *Pterocladiella* are also used for agar production in countries like India, Portugal, New Zealand, and Egypt (Nishinari and Fang, 2017). The molecular weight of agar is species-dependent and also varies according to extraction methods. Commercial agar has an MW of 106.4 to 243.5 kDa (Mitsuiki et al., 1999), while agar from wild seaweed has MW in the range of 540 kDa (Rodríguez et al., 2009). The agar content of red algae varies between species, *Gelidium* 20–40%, *Gracilaria* 15–39%, *Pterocladiella* 32%, and *Laurencia* 25% (Lee et al., 2017).

9.2.6.2 Structure

Agar is a large complex heteropolysaccharide made of two subunits, namely agarose and agaropectin. Overall, it is a polymer of repeating disaccharides called agarobiose composed of galactose and 3,6- anhydrogalactose residues (Figure 9.9). Agarose is a neutral, high molecular weight (120 kDa) linear polymer of alternating (1–4)-linked 3,6 anhydro-α-L-galactopyranose and (1–3)-linked β-D-galactose residues (Rodríguez et al., 2009), as shown in Figure 9.10. Due to this property, its concentration in agar affects the strength of the gel. Agar with a high concentration of agarose and less sulphate (0.5%) makes a stronger gel. Generally, a typical agar is composed of 70% agarose, and the rest is a branched agaropectin chain.

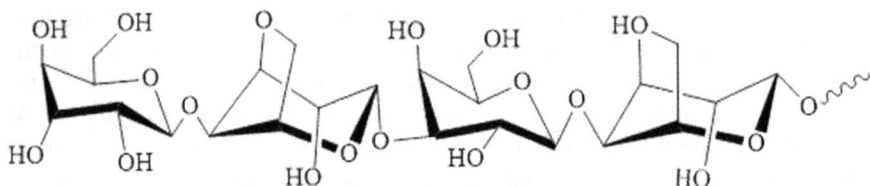

FIGURE 9.9 Structure of Agar.

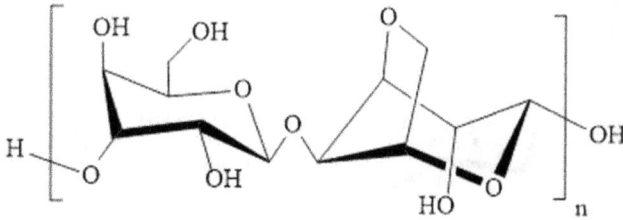

FIGURE 9.10   Structure of Agarose.

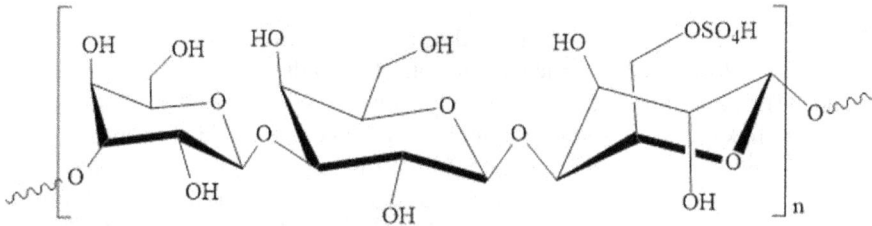

FIGURE 9.11   Structure of Agaropectin.

Agaropectin contains calcium, magnesium, potassium, and sodium sulfate esters at the 6th position of the β-(1,4)-galactose or in the 4th and 6th positions of a-(1,3)-galactose unit. Additionally, methyl groups, pyruvate, β-d-xylopyranose, and 4-o-Me-a-l-galactopyranose residues can occur in its structure, as shown in Figure 9.11. Since agaropectin contains more anionic groups (5–8%) and high molecular weight (12.6 kDa), it constitutes the non-gelling portion of agar. Agar gel formation can occur at a low concentration of 0.2% of agar at 80°C, followed by cooling at 40°C (Lahaye and Rochas, 1991; Asadova et al., 2020).

9.2.7   Ulvan

9.2.7.1 Source

Ulvans are charged water-soluble sulphated heteropolysaccharide fiber found in the cell wall of green marine algae (Chlorophyta) of genus *Ulva*, sea lettuce. Kylin was originally named ulvanulvin and ulvacin. The total ulvan content of *Ulva* is about 9–36% of its dry weight, which is about 38–54% of total cell wall polysaccharide (Lahaye and Robic, 2007).

9.2.7.2 Structure

The predominated monomeric units of ulvan are α- and β-(1,4)-linked mono-saccharides, such as L-rhamnose, D-xylose, D-glucuronic acid, and L-iduronic acid (Lahaye and Robic, 2007). Ulvan resembles mammalian glycosaminoglycans by having a central backbone made mostly of two disaccharides units, namely ulva-nobiuronic acid A (β-d-glucuronosyluronic acid-(1,4)-α-l-rhamnose 3-sulfate) and B (α-l-iduronopyranosic acid – (1,4)-α-l-rhamnose 3-sulfate), hardly with aldo-bioses/Ulvanobioses type U (β-D-xylose (1,4)-linked to α- L-rhamnose 3-sulfate).

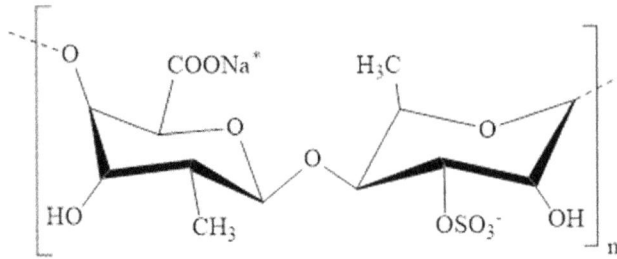

FIGURE 9.12 Structure of Ulvan.

These uronic acids are linked to sulfated neutral sugars; few are attached to the 2nd position of xylose residue, while most are attached to the 3rd position of rhamnose residue. The glycosidic linkage is predominantly (1,4) type, rarely (1,2)- and (1,3)-glycosidic linkages as in the ulvanobiuronic acid (glucuronic acid (1,2)-linked to the rhamnose residue), as shown in Figure 9.12 (Kidgell et al., 2019). The cell wall of Ulva also contains two other polysaccharides, namely xyloglucan and glucuronan, but the only ulvan has both rhamnose and iduronic acid, which constitute 45% of total polysaccharides (Lahaye and Robic 2007). Ulvan molecular weights are in the range of 189 to 8200 kDa, and its rheological properties are greatly influenced by pH (Bobin-Dubigeon et al., 1997).

9.3 MICROALGAL POLYSACCHARIDES

Prokaryotic cyanobacteria (mostly studied along with microalgae) and eukaryotic microalgae produce a wide variety of polysaccharides. These polysaccharides are part of polymeric reserve food material, cell wall, and complex extracellular secretions of the cell. Extracellular polysaccharides (EPS) have gained more recognition due to their industrial applications, such as gums, biostimulants, bioflocculants, etc. Polysaccharides with high molecular weights of 5×10^5 to 2×10^6 Da account for 40–95% of the total extracellular polymeric substance of microalgae, and their composition is species-specific, as seen in Table 9.3. They are mostly secreted to protect the cells from a stressful environment and for cell adhesion. Among different microalgae, diatoms are the largest producer of EPSs and share a similarity in structure. They all (1) are sulfated heteropolysaccharides; (2) contain glucose, galactose, arabinose, rhamnose, mannose, and xylose; and (3) have Mannan associated with their cell wall (a universal feature of diatoms) (Chiovitti et al., 2005). A major difference between cyanobacterial and microalgae EPS is the absence of fructose in the latter, except in *Dunaliella salina*. About 150 cyanobacterial strains are reported to produce EPS with xylose, arabinose, fucose, rhamnose, and galactose, along with uronic acids as their major composition (Pereira et al., 2009). Red microalgae cell wall contains sulfated anionic polysaccharides with major residues, such as glucose, galactose, and xylose, along with a small number of mannose, fucose, ribose, arabinose, rhamnose, and other methylated sugars. Their EPS majorly contains glucose, xylose, galactose, and glucuronic acid as methylated and sulfated residues. EPSs of marine green microalgae,

TABLE 9.3

Polysaccharides Content (% dry wt.) of Extracellular Secretion of Microalgae and Cyanobacteria with Sulfate and Uronic Acid Composition

Species	Group	Total Carbohydrate (%)	Sulfate (%)	Uronic Acids (%)	References
Prophyridium sp.	Rhodophyta	67	7.6–14.6	7.8–10	Raposo et al., 2013, Balti et al., 2018
Rhodella sp.	Rhodophyta	54.4	8	5–7.8	Raposo et al., 2013, Villay et al., 2013
Spirulina platensis	Cyanobacteria	13	5–20	7	Raposo et al., 2013, Trabelsi et al., 2009
Chlorella sp	Chlorophyta	68.4	7.8–9.4	3.7–9.0	Raposo et al., 2013, Zhou et al., 2016
Phaeodactylum tricornutum	Bacillariophyceae	70	7.5–13.3	1.4–6.3	Raposo et al., 2013, Ford and Percival, 1965
Cylindrotheca closterium	Bacillariophyceae	–	0–10.9	4.8–21.0	Raposo et al., 2013
Nacicula salinarum	Bacillariophyceae	–	6.3–11.5	7.7–8.0	Raposo et al., 2013
Gyrodinium impudicum	Dinoflagellata	43	10.3	2.9	Yim et al., 2004
Flintiella Sanguinaria	Rhodophyta	80	0.6	14	Gaignard et al., 2018

such as *Chlorella, Dunaliella, Kirchneriella aperta,* etc., contain methylated or sulfated glucose, galactose, arabinose, fucose, xylose, mannose, glucuronic acid, and galacturonic acid residues, with mannose being the key sugar to complex with metal ions (Raposo et al., 2013). Exopolysaccharides of *Prophyridium* contain xylose, galactose, glucose, and glucuronic acid in the molar ratios of 1.5/1.3/0.6/0.5 and have a molar mass of $2-7 \times 10^6$ Da (Balti et al., 2018)

9.4 EXTRACTION AND PURIFICATION OF POLYSACCHARIDES

The extraction of polysaccharides from algae is multiple time-consuming processes that involve pre-treatment, extraction, filtration/removal of impurities, purification (drying, milling, and blending), and finally, quality assessment. The quality and quantity of the polysaccharide depend upon various factors, such as reaction time, temperature, algal source, harvest season, etc. (Hernandez-Carmona et al., 2013). Most of the polysaccharides are water-soluble; hence, hot water used as a solvent for extraction is a conventional method, except for hydrophobic alginate extraction, which involves extraction using hot alkali. Though it is a conventional method, hot water treatment usually results in the end product with lower gel-forming (rheological) properties or reduced biological activity due to changes in the molecular structure of the polysaccharide in addition to low extraction efficiency (Khalil et al., 2018). For industrial level pre-treatment (Table 9.4) or extraction, alkali, acid, or nontoxic organic solvent is preferred to get high-quality polysaccharides with optimal yield. During extraction of hydrocolloids, such as carrageenans or agar, pre-treatment with hot alkali solution will cause the conversion of polygalactose chain (non-gelling portion) with sulfate groups into 3,6-anhydrogalactose, which is the gel-forming portion of algal polysaccharides.

Moreover, impurities like phycoerythrine, a major pigment in *Gracilaria,* can be eliminated by this treatment during agar extraction (Zhao et al., 2020). In the case of alginate, the alkali treatment is mandatory to cause the conversion of hydrophobic alginic acid to hydrophilic alginate salts, such as sodium alginate to remove insoluble impurities in the solution. Later, calcium salts are added to form insoluble alginate to remove soluble impurities and again converted into soluble sodium alginate with alcohol and sodium carbonate. Laminarin, which is a non-gelling or thickening polysaccharide with important potential bioactivities, is extracted in two steps. First, the cold-water-soluble laminarin is extracted with ethanol/organic solvents, and later cold water-insoluble laminarin is extracted with acids such as hydrochloric acids and sulfuric acids. Carrageenans are extracted as either refined or semi-refined carrageenans to retain their functional properties. Refined carrageenans are processed like other polysaccharides, while the semi-refined type is retained in seaweed while eliminating other major compounds of the seaweed by treating with KOH ≥5% at 80°C followed by ethanol wash (Hernandez-Carmona et al., 2013). After extraction, modification of the polymer is carried through the addition of functional groups, irradiation, and oxidation/reduction to enhance the activity of the molecule to perform the desired activity (Sedayu et al., 2019; Sukwong et al., 2018).

TABLE 9.4

For Bioethanol Synthesis, Several Examples of Algal Polysaccharides

Source	Pre-treatment Method	Fermentation	Ethanol Yield (%)	References
Gelidium sesquipedale and *Ulva lactuca*	(Mechanical pre-treatment and enzymatic) Rinsed and dried in the air. First, the material was cut and then centrifugally milled at 12,000 rpm. Control: Milled to less than 2 mm with a 5% algal biomass load fermented using the Haliatase enzyme (–glucanase, carrageenase, agarase).	Simultaneous saccharification, fermentation (SSF):*Saccharomyces cerevisiae*	68.7 and 64.3	Amamou et al., 2018
Chaetomorpha linum	(Thermal pre-treatment) Washed, dried (at 40 °C), then ground.Autoclaved at 200°C for 10 min at 1 bar with 4%t (w/v) of seaweed, 12-bar O$_2$ for wet oxidation (WO), 35% DW steam explosion, 1.9 MPa, 200°C, 5 min. Plasma assisted: 2.5 g, 1% O$_3$, 1 h, Ball milling 10% pre-treated algal biomass that has been pre-hydrolyzed and injected with cellulose enzymes.	*Saccharomyces cerevisiae*	Thermal: 39 WO: 44 Steam: 38 Plasma: 41 Ball milling: 44	Schultz-Jensen et al., 2013
Gelidium amansii	(Thermal/acidic pre-treatment) 12% (w/v) slurry content, 358.3 mM H$_2$SO$_4$, and temperature of 142.6°C for 11 min.	*Pichia stipitis*	50	Sukwong et al., 2018

(Continued)

TABLE 9.4 (Continued)
For Bioethanol Synthesis, Several Examples of Algal Polysaccharides

Source	Pre-treatment Method	Fermentation	Ethanol Yield (%)	References
Laminaria digitata	(Acid/enzymatic pre-treatment) 1.5 N H$_2$SO$_4$, 24 min, 121°C, 25%(biomass/reactant) solids loading.	*Saccharomyces cerevisiae*	94	Kostas et al., 2017
Ulva rigida	(Thermal/acid pre-treatment) 4% (v/v) H$_2$SO$_4$, 10% (w/v) biomass loading, and 1 h autoclaved.	*Pachysolen tannophilus*	50	El Harchi et al., 2018
Eucheuma denticulatum	(Thermal acid/enzymatic) 13% (w/v) algal slurry, 180 mM H$_2$SO$_4$ at 121° C for 60 min, enzyme mixtures, Celluclast 1.5 L, and Viscozyme L (5:5 ratio).	*Scheffersomyces stipitis*	48	Ra et al., 2017
Ulva fasciata	(Thermal+ enzymatic) Pre-heat treatment of biomass in aqueous medium at 120 °C for 1 h followed by incubation in 2% (v/v) enzyme (cellulase)for 36 h at 45°C.	*Saccharomyces cerevisiae*	45	Trivedi et al., 2013

If polysaccharides for industrial purposes can be obtained without the need for a chemical treatment step, it is a welcoming approach from both an ecological and an economic point of view. Green method approaches are novel extraction techniques, including enzyme or ultrasonic or microwave-assisted extractions, superficial fluid extraction, and ionic liquid extraction. Their advantages are short extraction time, low energy, low cost, eco-friendliness, and highly efficient extraction. Though there is a need for mild chemical treatments in these approaches, they can still significantly reduce the pollution level in large-scale processing units (Rajak et al., 2020). The microwave-assisted extraction method uses electromagnetic energy in the range of 300 to 300,00 MHz to extract polysaccharides in a short duration. The major drawback is that the high amount of energy can cause degradation or denature of polysaccharides. Ultrasonic-assisted extraction became more convenient since it uses ultrasound waves to form a cavitation phenomenon to break the cell wall and cause the release of water-soluble polysaccharides of the cells into extraction solution at a reasonable temperature of about 66°C within an hour of extraction (Hernandez-Carmona et al., 2013). Hence, this method doesn't alter the nature of the polysaccharide to an appreciable level with a high yield in a short duration. Enzyme-assisted extraction of polysaccharides is a less explored, yet significant potential green approach. The main advantages of this approach are the high yield of quality polymer, rapid process, low energy consumption, high repeatability, and easy manipulation. Enzymes can effectively disrupt the cell wall at an optimal temperature and thus aid in the release of cell polysaccharides into the extraction buffer without affecting its nature (Rhein-Knudsen et al., 2015; Vanegas et al., 2015; Trivedi et al., 2013). Sustainable living, food insecurity, growing population, demand for functional ingredients, etc., are demanding more efficient, green techniques in all industrial sectors.

9.5 POTENTIAL APPLICATIONS OF ALGAL POLYSACCHARIDES

Algal polysaccharides have a wide range of applications in agriculture, medicine, pharmacology, food, and biotechnology, with properties such as antioxidant, anticancer, anticoagulation, immune response modulation, anti-inflammatory, antiviral, antimicrobial in health and nutrition, pharmaceuticals, functional foods, cosmetics (Rajak et al., 2020) (Figure 9.13).

9.5.1 BIOLOGICAL ACTIVITY OF ALGAL POLYSACCHARIDES FOR PHARMACEUTICAL APPLICATIONS

Algal polysaccharides (APS) have different biological capabilities depending on their structural characteristics; some of these positive effects are described in the following sections:

9.5.1.1 Anti-inflammatory and Immunomodulatory Activities

The anti-inflammatory properties of polysaccharides depend on their source and type of inflammation. Inflammation is reduced by the heterofucan from *Dictyota menstrualis*, which binds directly to leukocyte cell surface and prevents leukocyte

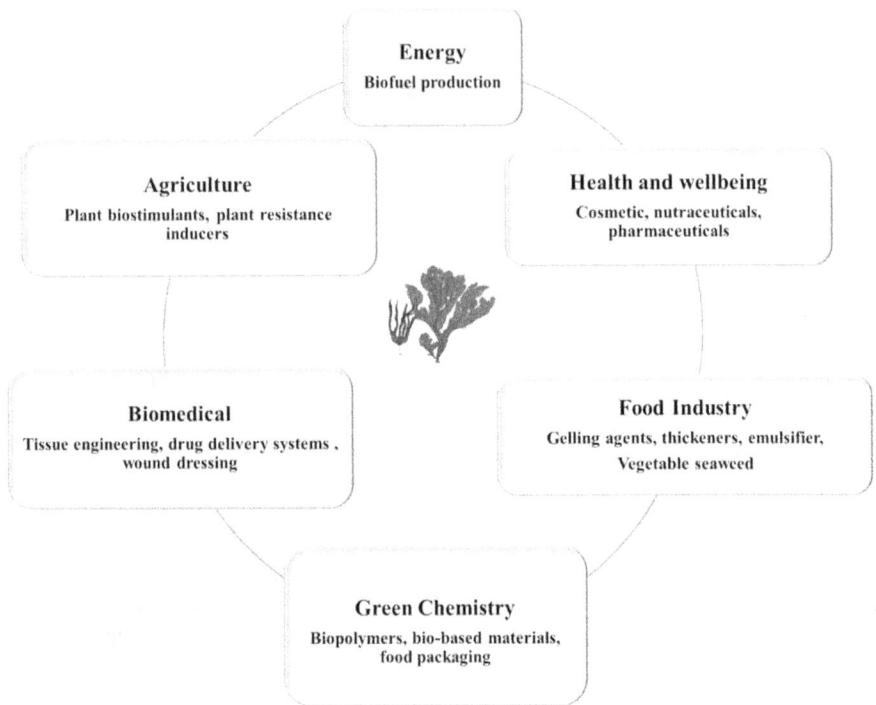

FIGURE 9.13 Applications of algal polysaccharides.

migration into the peritoneal cavity of mice in wounded tissue following simulated pain and inflammation, with no generation of pro-inflammatory cytokines (Albuquerque et al., 2013). The immunostimulating effect of APS mostly depends on macrophages modulation. APS are biological immunomodulators; they maintain homeostasis by regulating NK cells, complement systems, and macrophages, or T/B lymphocytes (Huang et al., 2019).

9.5.1.2   Antioxidant Activity

The antioxidant activity of APS is mostly determined by its functional groups, notably the carboxylic groups of the d-ManpA and l-GulpA units in alginates and the sulphates groups in fucoidans. It's followed by sodium alginates made from brown algae *Cystoseira compressa*, which is also a potent antioxidative (Hentati et al., 2018). Low molecular weight sulfated polysaccharides (SPs) have shown higher antioxidant activity than high molecular weight SPs. Low molecular weight fucoidan derived from *Laminaria japonica* up regulates the activity of superoxide dismutase (SOD) and catalase (CAT) through series of signaling pathways (Zheng et al., 2018). SPs are remarkable candidates in the search for effective, non-toxic substances with potential antioxidant activity.

9.5.1.3   Anticancer and Antitumor Activities

Fucoidans derived from *Sargassum polycystum* have antiproliferative properties and induce programmed cell death in a breast cancer cell line via activating caspase-8

(Palanisamy et al., 2017). Fucoidans from the brown seaweeds *Sargassum duplicatum, Sargassum feldmannii*, and their derivatives have been shown to inhibit colony formation of colon cancer cells (Usoltseva et al., 2019). Apoptosis-inducing properties of fucoidans from *Fucus vesiculosus* have been demonstrated in MC3 human mucoepidermoid carcinoma cells via caspase-dependent apoptosis signaling cascade (Lee et al., 2014).

9.5.1.4 Antiviral Activity

Polysaccharides derived from the red alga *Gelidium robustum* protect embryonic eggs against the mumps virus and influenza B (Gerber et al., 1958). Sulfated polysaccharides (SPs) have been found to hinder the reproduction of enveloped viruses. Before infection, viruses must make contact with cell membrane glycosaminoglycan receptors (GAG), such as heparin sulfate (HS), to enter their host cell. SPs are negatively charged polymers and are chemically identical to HS. SPs hinder viral particle attachment by forming virus-algal PS complex by imitating GAG, eventually protecting the cell from infection (De Jesus Raposo et al., 2015). Carrageenan nasal spray exhibits considerable antiviral activity against three virus subgroups, HRV, human coronavirus, and influenza A virus, with the most significant efficacy reported in patients infected with the human coronavirus (Koenighofer et al., 2014). Polyguluronate sulfate (PGS), a low molecular weight sulfated brown algal polysaccharide, can block the production and secretion of hepatitis B surface antigens HBsAg and HBeAg (Chen et al., 2020). Sulfated polysaccharides from algae might be used as antiviral therapeutics against SARS-CoV-2 (Pereira and Critchley, 2020).

9.5.1.5 Antibacterial Activity

SPs have antibacterial potential against *E. coli* and *Staphylococcus species* (Kraan 2012). APS helps restore the stomach mucosa, eradicate *Helicobacter pylori* colonies, and regenerate biocenosis in the intestines. Fucoidan fraction-2 (Fu-F2) derived from *Sargassum polycystum* has the maximum antibacterial activity against *Pseudomonas aeruginosa* at a dose of 50 g/ml in vitro (Palanisamy et al., 2019). Oxidized κ-carrageenan has broad-spectrum antibacterial action by damaging the bacterial cell wall and cytoplasmic membrane and suppressing the development of gram-positive (*Staphylococcus aureus*) and gram-negative (*Escherichia coli)* bacteria (Zhu, M., et al., 2017).

9.5.1.6 Anticoagulant and Antithrombotic Activities

In blood coagulation, fucoidan fraction from *Nemacystus decipiens* can increase the amount of plasminogen activator inhibitor: tissue plasminogen activator, which has fibrinolytic activity and can be utilized as an antithrombotic medication. The administration of low molecular weight fucoidan decreased the development of arterial thrombosis (Cui et al., 2018). Fucoidans isolated from *Ecklonia cava* were investigated for their anticoagulant effects by measuring prothrombin time, thrombin time, and activated partial thromboplastin time (Athukorala et al., 2006). The anticoagulant activity of a sulfated ramified polysaccharide isolated from *Codium divaricatum* was shown to be dose-dependent (Li et al., 2015). The synthesis and distribution of sulphate groups in the structure of sulfated galactan impact venous antithrombotic and anticoagulant properties. There are also higher anticoagulant effects with higher sulfate-content carrageenans, such as λ-carrageenans, than with lower sulfate-content carrageenans like κ-carrageenans (Necas and Bartosikova, 2013).

9.5.1.7 Antilipidemic and Antihepatotoxic Activities

Algae-derived SPs efficiently impede of human pancreatic cholesterol esterase, an enzyme that increases intestinal intake of cholesterol. Fucoidan produced from *Cladosiphon okamuranus Tokida* effectively suppresses LDL-C peroxidation and hepatic steatosis by stimulating plasma lipoprotein lipase activity, therefore relieving dyslipidemia and atherosclerosis in the animals. Fucoidan reduces ROS formation *in vivo* via oxidative stress-related signaling pathways and then alleviates oxidative damage via several routes (Yokota et al., 2016). Hepatoprotective effects of *Monostroma nitidum* SPs were shown through the increased production of liver detoxicate enzymes (Charles et al., 2007). SPs derived from the green algae *Enteromorpha prolifera* had hypolipidaemic effects, regulated the lipidic profile in the liver and plasma, and raised HDL-cholesterol (Teng et al., 2013).

9.5.1.8 Neuroprotective Activity

Fucoidans extracted from *Sargassum fusiform* has been shown to cure cognitive impairment and enhance cognitive performance (Hu et al., 2016). SPs are useful in treating Alzheimer's disease patients by delaying or halting the progression of the disease. *Splachnidium rugosum* and *Undaria pinnatifida* sulfated fucan extracts inhibited the accumulation of Herpes simplex virus type 1 induced beta-amyloid and AD-like tau in HSV1 cells infected with the Vero virus (Wozniak et al., 2015). Fucoidan, derived from *L. japonica*, was found to have antioxidant properties in Parkinson's disease. Based on this research information, algal-derived polysaccharides can be recommended for their neuroprotective properties and therapeutic application (Hannan et al., 2020).

9.5.2 Cosmetic Applications

APS serves a variety of cosmetic purposes, such as rheology modifiers, suspending agents, hair conditioners, and wound-healing agents, as well as moisturize, hydrate, emulsify, and emolliate. Polysaccharides derived from algae have shown potential inhibitory effects on the matrix metalloproteinases (MMPs) enzymes that contribute to skin degeneration due to UVB irradiation, inflammatory cytokines, some hormones, and pharmacological agents (Wang et al., 2015). In comparison to hyaluronic acid, low molecular polysaccharides isolated from brown macroalgae had the greatest potential for moisture absorption and retention (Wang et al., 2013). Alginate has been used to develop nanofiber-cosmetic patches prepared from polycaprolactone containing *Spirulina* extract, which has moisturizing and adhesive properties (Byeon et al., 2017). *Chnoospora minima* and *Sargassum polycystum* derived fucoidans have broad spectrum skin cosmatizing activity. It is useful in preventing and improving skin aging disorders, such as blotches, freckles, and wrinkles. It can also suppress UV-B-induced MMPs production and tyrosinase, thus protecting connective tissues (Vo and Kim, 2000; Mizutani et al., 2010; Priyan Shanura et al., 2019). Carrageenans are found in various beauty products due to their unique property. Using oil and water, it immediately emulsified; they may also be washed from the skin rapidly. Porphyran has been proven to have potential skin whitening and anti-inflammatory properties (Yun et al., 2013). Products containing 3,6-anhydro-L-galactose are very attractive for manufacturing skin whitening products. A patent addresses rhamnose and fucose's synergistic skin protective and bioactive actions against skin aging (Gesztesi et al., 2006).

9.5.3 Biomedical Applications

APS has potential biomedical uses, including wound care items, medication delivery systems, tissue engineering, medical fibers, and bio textiles. Alginates have been widely employed in various applications, including cell encapsulation, controlled drug release, scaffolding in ligaments, tissue engineering and regeneration of nearly all tissues in the human body, and even mold creation in dentistry (Draget and Taylor, 2011; Kong and Mooney 2005). In a live system, APS is coupled with other chemicals to encapsulate and transport Langerhans islets, ovarian follicles, and stem cells (de Jesus Raposo et al., 2015).

Recent years have seen an increase in the use of algal-derived polysaccharide hydrogels in wound dressing design and tissue engineering due to their high biocompatibility, low immunogenicity and cytotoxicity, and ease of application. Hydrogels' three-dimensional reticulate structure mimics the microarchitectonics of biological tissue's extracellular matrix, functioning as a physical barrier against microorganisms and creating perfect conditions for cell survival *in vivo* (Andryukov et al., 2020).

The alginate therapy Kaltostat® is available on the market. Alginate has been utilized to establish a capillary bed in freshly regenerated tissues in the form of porous scaffolds, and in the form of electrospin nanofibrous scaffolds to produce vascular substitutes comprising endothelial cells and smooth muscle cells (Hajiali et al., 2011).

Due to their abundance and low cost, as well as biocompatibility and degradability, APS is employed in drug delivery systems. They also exhibit a wide range of biological activities. In addition, SPs can be used as specific recognition signals to target immune system cells and increase medication absorption into the inflamed intestine since they can bind cell surface receptors such as TLR4, CD14, and protein kinase receptor, and alter the impact of signaling molecules in cells (Cunha and Grenha, 2016). Using algal polysaccharides as nanoparticles for medication delivery, these systems might be useful in treating inflammatory bowel disease. In combination with chitosan, carrageenan and fucoidan are the most often referenced matrix materials in creating nanoparticles.

Interactions between the amino group of chitosan and the sulphate group of polysaccharides facilitate the creation of nanoparticles and restrict drug release. The pH-responsive profile of fucoidan–chitosan nanoparticles avoids breakdown in the gastrointestinal tract under acidic circumstances, allowing medication absorption in the gut (Da Silva et al., 2017; Cunha and Grenha, 2016). Intestinal epithelial cell defects can be treated with chitosan/fucoidan nanoparticles coated with eggshell protein membranes. Due to the presence of the fucose receptor on intestinal epithelial cells, fucoidan can assist nanoparticles in targeting intestinal epithelial cells. Chitosan and fucoidan nanoparticles have been discovered as potential medication delivery carriers that decrease inflammation of the intestinal epithelium (Lee and Huang, 2019). Inflammation can be reduced by administering zinc and 5-aminosalicylic acid in combination with alginate-chitosan nanoparticles (Duan et al., 2017). Selenium nanoparticles with a polysaccharide coating from the green alga *Ulva lactuca* have low toxicity and anti-inflammatory properties (Zhu, C., et al., 2017).

9.5.4 APPLICATIONS IN FOOD BIOTECHNOLOGY

The use of algae polysaccharides as a food ingredient is widespread and includes dairy products, water dessert gels, meat products, confections, drinks, dressings, and other culinary items. Polysaccharide-based membranes are used in the food industry for packaging and edible coatings of products due to their excellent barrier against oxygen and carbon dioxide (at low or moderate relative humidity) (Galgano et al., 2015). Hydrocolloids, such as agar, alginates, and carrageenan, are highly valued for their gelling, thickening, and stabilizing properties. As a result, food products have a more appealing texture (Scieszka and Klewicka, 2019). Polysaccharides produced from algae that have been used in food include:

Alginate – It is an effective stabilizing and thickening agent in produced goods (Khalil et al., 2018). It improves the appearance of dairy goods and canned food, promotes water retention, improves the appearance of baked items, and provides a smooth texture and uniform dehydration of frozen food. Alginate is also used as a froth stabilizer in beer (Hernandez-Carmona et al., 2013). It is often used in sweets, beverages, ice cream, jellies, syrups, flavor sauces, fruit juices, bakery items, and milkshakes (Brewer et al., 2012). Cellulose nanofibers generated from brown algae (BA) waste following alginate extraction may be utilized as milk thickeners (Gao et al., 2018). The salt of alginic acid determines the food additive code: sodium alginate E-401, potassium alginate E-402, ammonium alginate E-403, calcium alginate E-404, and propylene glycol alginate E-405 (Sartal et al., 2011).

Carrageenans – The efficient water-binding property of carrageenans is likely what defines it as a hydrocolloid with exceptional functional properties. In the dairy, baking, and food processing industries, carrageenans are extensively used to make puddings, milkshakes, nutritional milk drinks, tofu, frozen yogurt, chocolate milk, vegan alternatives to gelatin, pastries, creams, organic product juices, brew, dry food powders such as instant soups, sauce mixes, and flavors, jams, and preserves, canning food enhancer, and pet food (Krempel et al., 2019; Scieszka and Klewicka, 2019). Carrageenans derived from red algae of the *Eucheuma* and *Kappaphycus* genera are utilized as fat substitutes to produce healthier meat products to improve moisture retention and tenderness (Hotchkiss et al., 2016). They are used in the production of edible packaging, film coatings, and mixes. This packaging material avoids discoloration, retains moisture content, maintains food texture, and aids in scent product encapsulation. They are used to wrap meat and poultry delicacies (Sedayu et al., 2019). The food additive code is E-407.

Agar – Agar, sometimes known as "vegetable gelatin", is generally regarded as safe (GRAS), which permits it to be used as a safer ingredient in food items. It can avoid crystallization, which is why it is used to make icing and baking glazes. Even at low concentrations, this substance, which is used to create jellies and fruit candies, produces a rigid gel in water at room temperature without potassium and calcium salts. Furthermore, its melting point (85–95°C) can keep its consistency even at high temperatures, making it a highly appreciated product for food applications (Wang and Rhim, 2015). It is utilized in the production of several food goods, including canned meat products. It serves as a thickening in pie fillings and ice creams and a stabilizer in dairy goods. Biodegradable food packaging is the focus of agar research and development (Salehi et al., 2019). The food additive code is E-406.

9.5.5 APPLICATIONS IN AGRICULTURE

In agricultural applications, algal polysaccharide (APS) extract is more beneficial than chemical fertilizer because of its bio-decomposable, nontoxic, and eco-friendly properties. APS impacts crops by promoting plant growth, seedling growth, and the production of both root hair and secondary roots. It can also enhance nutrient assimilation, fruit setting, resistance properties against pests and diseases, and stress management (drought, salinity, and temperature). Polysaccharides derived from *Lessonia nigrescens* can improve wheat seedling salt stress tolerance. Under stress, polysaccharides significantly enhanced shoot and root lengths and dry and fresh matter of wheat. Polysaccharide treatments preserved the osmotic condition of stressed wheat seedlings by increasing their sugar and proline content and controlling the Na^+/K^+ ratio (Zou et al., 2019).

To enhance plant defense against diseases, algal polysaccharide-based bio-elicitors are a viable option. Due to their structural resemblance to pathogen-derived chemicals, certain bioactive elicitors identified in different APS extract trigger pathogen-associated molecular patterns (PAMPs). This can be achieved effectively and efficiently by priming or evoking the defensive responses of the induced systemic resistance (ISR) and systemic acquired resistance (SAR) pathways. It is interesting to note that different extracts from APS have strikingly different mechanisms of action (Farid et al., 2019):

Alginate – Phenylalanine Ammonium Lyase (PAL) and polyphenol metabolism were modulated by alginate-based elicitors produced from *Bifurcaria bifurcata* and *Fucus spiralis* in date palm roots (Bouissil et al., 2020). Sodium alginate applied to tomato leaves induced SAR by increasing H_2O_2 production in response to pathogen infection and inhibiting the development of *Alternaria solani*, which causes early blight (Dey et al., 2019).

Carrageenan – Plants are protected from viral infections by carrageenans, restricting viral particle attachment to receptors and internalisation into host cells. Sulfated polysaccharide, produced from the red alga *Hypnea musciformis* and containing κ-carrageenan, activates plant defense mechanisms against viral infection by affecting gene expression in salicylic acid (SA) and jasmonic acid/ethylene

(JA/ET) signaling pathways (Ghannam et al., 2013). By boosting the expression of SA-dependent defensive response genes, rubber tree leaves sprayed with carrageenan extracted from the *Acanthophora spicifera* improved defenses against *Phytophthora palmivora* infection (Pettongkhao et al., 2019).

Laminarin – The laminaran and its oligomers are effective elicitors of plant defenses in dicots, such as green bean and tomato, and monocots, such as wheat and rice, providing a valuable tool for disease management in agriculture; also, its application is recommended for integrated management associated with different disease control methods. Laminarins isolated from *Laminaria digitata* elicited a dose-dependent defensive response in tobacco cells. In tobacco plants, laminarin infiltration reduced the progress of soft-rot disease by activating the plant's inherent defensive systems. Laminarins isolated from *Laminaria digitata* protected tea plants from the tea green leafhopper by activating early defense-response genes and boosting the synthesis of SA and abscisic acid (ABA), while also creating oxidative bursts (Klarzynski et al., 2003; Xin et al., 2019).

Ulvan – Ulvans have been widely used in plants to control active defence systems against a range of illnesses in recent years. The use of ulvans as a foliar treatment enhanced bean plant resistance to rust caused by *Uromyces appendiculatus* and angular leaf spot caused by *Pseudocercospora griseola* (Jaulneau et al., 2010). Adding ulvans to an olive branch increased its phenolic metabolism, improving its resistance to *Verticillium wilt* of olive disease. Ulvan treatments changed the expression of genes involved in phytohormone metabolism, according to researchers. There is evidence that Ulvans can control the cross-talk between elicitors and gibberellin responses via inducing transcription factors (Ghannam et al., 2013).

9.5.6 APPLICATIONS IN BIOENERGY

Faced with the environmental consequences of fossil fuel usage, algae, a significant third-generation (3 G) biomass component, has considerable promise for bioenergy development due to its species-abundant, renewable, and carbohydrate-rich features. APS might be used to generate bioethanol or biogas, as shown in Figure 9.14, depicts the metabolic processes involved in creating biogas or bioethanol (Maneein et al., 2018). Hydrolysis of the algae's major polymers yields glucose and galactose from which ethanol is made. For example, altering the bioavailability of polysaccharides before their conversion to sugars might increase the pace and yield of biogas or ethanol, which would result in more biofuel being produced over time. Tables 9.4 and 9.5 summarise ethanol and biogas productivities reported by various groups and studies (Hong and Wu, 2020; Rajak et al., 2020).

9.6 FUTURE PROSPECTS

- Algal polysaccharides have a distinct antiviral mechanism. They interfere with the virus's life cycle or indirectly exhibit antiviral activity by boosting the body's immunity. In anti-virus applications, polysaccharide-based vaccine adjuvants, nanomaterials, and drug-delivery technologies will be crucial.

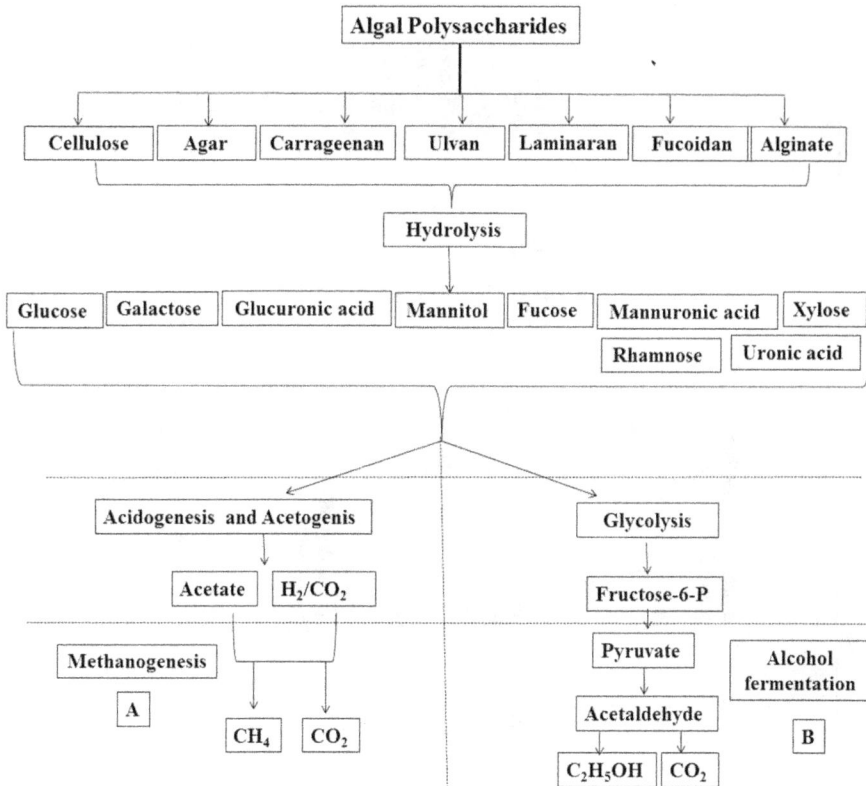

FIGURE 9.14 Biogas generation (A) and ethanol production (B) from algal polysaccharides.

- Sulfated polysaccharide-based natural dressings need the creation of a scientifically sound approach to wound care, as well as the formulation of suitable practical recommendations.
- AP-based biostimulants in agriculture need to be re-evaluated, and products that are helpful and economically feasible in real-world situations need to be developed.
- Future research should concentrate on the partial blending or substitution of APS in food formulations over conventional synthetic or animal-derived compounds, as well as the promotion of environmentally friendly and nutritious processed food products.
- Algal polysaccharide hydrolysis needs more enzymatic combinations than plant polysaccharide hydrolysis. However, one of the primary obstacles in the context of algal polysaccharide enzymatic hydrolysis is identifying hydrolytic enzymes capable of breaking down algal-specific sugars. This necessitates the discovery of novel marine microbes that live on algae and identify their enzyme systems using genomes, transcriptomics, and proteomics methods.

TABLE 9.5

For Methane Generation, Several Examples of Algal Polysaccharides

Source	Pre-Treatment Method	Inoculum/Substrate Ratio; Source	Biochemical Methane Potential (BMP) Method	CH$_4$ Yield (mL g^{-1} VS)	References
Ascophyllum nodosum	(Mechanical pre-treatment) Hollander beater, beating time (15 min)	1:1 in 400 ml anaerobic digestion bioreactors consisted of borosilicate glass flasks of 500 ml in capacity.	Reactors were sealed with borosilicate glass adapters equipped with controlled gas opening valves and incubation time 14 days.	169	Montingelli et al., 2017
Laminaria sp..	(Mechanical pre-treatment) Ball milled, dried at 80°C for 24 h. Particle size: 1–2 mm.	1:1.33 in 400 mL; wastewater treatment plant (WWT)	Bottle sealed with adaptor attached to gas measuring device (GMD), incubated at 38°C for 25 days.	1 mm: 241 2 mm: 260	Montingelli et al., 2016
Laminaria sp.	(Mechanical pre-treatment) Hollander beater, cut without washing, and beaten for 10 min.	1:1.33 in 400 mL; WWT.	GMD connected to bottle with adapter, 38 C, 25 days, manually shook.	335	Montingelli et al., 2016
Laminaria sp.	(Thermal pre-treatment) Cutting and washing the seaweed then heating it in the microwave for 30 sec until the water boils and holding it for 30 sec.	1:1.33 in 400 mL; WWT.	Bottles were sealed and incubated at 38°C for 25 days with the adapter attached GMD.	244	Montingelli et al., 2016
Laminaria digitata	(Thermochemical pre-treatment) Washed with cold water for 3 min, cutting and maceration, hot water at 40°C for 3 min.	2:1,lab-scale continuous stirred-tank reactors, total volume of 650 ml with a working volume of 400 ml.	Bottles were sealed with rubber corks, 30 rpm, and were kept at 37°C.	282	Tabassum et al., 2017

(Continued)

TABLE 9.5 (Continued)

For Methane Generation, Several Examples of Algal Polysaccharides

Source	Pre-Treatment Method	Inoculum/Substrate Ratio; Source	Biochemical Methane Potential (BMP) Method	CH_4 Yield (mL g^{-1} VS)	References
Laminaria digitata	(Enzymatic pre-treatment) The seaweed 20% (w/v) was washed in water, dried at 75°C for 24 hours, and then ground, incubated 300 rpm for 24 h at 37°C with Cellulase, Alginate lyase, and Celluclast.	I/S ratio: 2:1 in 30 mL (pH 7.3–7.5)Pretreated algal biomass was transferred into 120 ml serum bottles.	The headspace in the bottles was flushed with a mix of N_2/CO_2, sealed with rubber stoppers, and capped with aluminium crimps and incubated at 35°C for 32 days.	Cellulase: 243 Alginate lyase: 225 Celluclast: 72	Vanegas et al., 2015
Ulva sp.	(Biological pre-treatment) Inoculated with *A. fumigatus* SL1 *conidia* suspension and incubated for eight days at 50°C after being washed, sun dried, and 35% algae in Mandels' salt solution.	Sugar wastewater industry as a source. In 400 mL with buffer and nutrients, use a 2:1 I/S ratio.	Vials containing vacuum-sealed rubber stoppers and aluminium clamps, 35°C until gas production halts.	153	Yahmed et al., 2017
Ulva lactuca Ascophyllum nodosum Laminaria digitate Saccorhiza polyschides Saccharina latissima	(Ensiling method) *U. lactuca* was cut by hand to a particle size of 10–20mm; other algae were cut using a mincer particle size 5 mm. Biomasses were filled into 1 L glass jars. Silos were equipped with a 3 cm layer of silica-based gravel. Silos were closed airtight, stored at 20°C, storage periods of 2, 4, 7, 14 and 90 days.	2:1, total volume of 400 mL in 500 mL glass bottles, 1 anaerobic digester.	Sealed the reactor system, incubated in a water bath at 37°C, slowly rotating agitator at 30 rpm, operating for 60 s at a time interval of 60 s. 30 days.	255 185 340 358 329	Herrmann et al., 2015

9.7 CONCLUSION

Many biological properties of algal polysaccharides and their structural variations may make them a valuable tool for new therapeutics and industrial applications, such as cosmetics, nutraceuticals, pharmaceuticals, and functional foods, but the discovery of unexpected biological characteristics has made these polymers a highly fascinating study subject. Using algal polysaccharides in medicine is expected to grow in the future, including wound dressings based on alginates and fucoidans. For regulated medication delivery, cell immobilization, and tissue bioengineering, these polymers are useful. To have edible water bubbles made with alginate would be a huge step forward for plastic-free planets. Algal polysaccharides generate value-added goods in addition to bioenergy production, giving resilience to the industry and increasing their economics and usability. For algal polysaccharides, the development of novel extraction and purification techniques may represent the tipping point toward their widespread industrial use.

REFERENCES

Albuquerque, I.R.L., Cordeiro, S.L., Gomes, D.L., Dreyfuss, J.L., Filgueira, L.G.A., Leite, E.L., Rocha, H.A.O., 2013. Evaluation of anti-nociceptive and anti-inflammatory activities of a heterofucan from *Dictyota menstrualis*. *Mar. Drugs.* 11(8), 2722–2740.

Amamou, S., Sambusiti, C., Monlau, F., Dubreucq, E., Barakat, A., 2018. Mechano-enzymatic deconstruction with a new enzymatic cocktail to enhance enzymatic hydrolysis and bioethanol fermentation of two macroalgae species. *Molecules.* 23(1), 174.

Andryukov, B.G., Besednova, N.N., Kuznetsova, T.A., Zaporozhets, T.S., Ermakova, S.P., Zvyagintseva, T.N., Smolina, T.P., 2020. Sulfated polysaccharides from marine algae as a basis of modern biotechnologies for creating wound dressings: Current achievements and future prospects. *Biomedicines.* 8(9), 301.

Armisen, R., Gaiatas, F., 2009. Agar, in: Phillips, G. and Williams, P. (Eds.), *Handbook of Hydrocolloids*. Woodhead Publishing, UK, pp. 82–107.

Arnosti, C., Wietz, M., Brinkhoff, T., Hehemann, J.H., Probandt, D., Zeugner, L., Amann, R., 2021. The biogeochemistry of marine polysaccharides: sources, inventories, and bacterial drivers of the carbohydrate cycle. *Annu. Rev. Mar. Sci.* 13, 81–108.

Asadova, A., Masimov, E.A., Imamaliyev, A.R., Asadova, A.H., 2020. Spectrophotometric investigation of gel formation in water solution of agar. *Mod. Phys. Lett. B.* 34(14), 2050147.

Athukorala, Y., Jung, W.K., Vasanthan, T., Jeon, Y.J., 2006. An anticoagulative polysaccharide from an enzymatic hydrolysate of Ecklonia cava. *Carbohydr. Polym.* 66(2), 184–191.

Balti, R., Le Balc'h, R., Brodu, N., Gilbert, M., Le Gouic, B., Le Gall, S., Massé, A., 2018. Concentration and purification of Porphyridium cruentum exopolysaccharides by membrane filtration at various cross-flow velocities. *Process Biochem.* 74, 175–184.

Bobin-Dubigeon, C., Lahaye, M., Guillon, F., Barry, J.L., Gallant, D.J., 1997. Factors limiting the biodegradation of Ulva sp cell-wall polysaccharides. *J. Sci. Food Agric.* 75(3), 341–351.

Bouissil, S., Alaoui-Talibi, E., Pierre, G., Michaud, P., El Modafar, C., Delattre, C., 2020. Use of alginate extracted from Moroccan brown algae to stimulate natural defense in date palm roots. *Molecules.* 25(3), 720.

Brewer, M.S., 2012. Reducing the fat content in ground beef without sacrificing quality: A review. *Meat Sci.* 91(4), 385–395.

Byeon, S.Y., Cho, M.K., Shim, K.H., Kim, H.J., Song, H.G., Shin, H.S., 2017. Development of a spirulina extract/alginate-imbedded pcl nanofibrous cosmetic patch. *J. Microbiol. Biotechnol.* 27(9), 1657–1663.

Campbell, R., Hotchkiss, S., 2017. Carrageenan industry market overview, in: *Tropical Seaweed Farming Trends, Problems and Opportunities.* Springer, Cham. pp. 193–205.

Charles, A.L., Chang, C.K., Wu, M.L., Huang, T.C., 2007. Studies on the expression of liver detoxifying enzymes in rats fed seaweed (*Monostroma nitidum*). *Food Chem. Toxicol.* 45(12), 2390–2396.

Chee, S.Y., Wong, P.K., Wong, C.L. 2011. Extraction and characterisation of alginate from brown seaweeds (*Fucales*, Phaeophyceae) collected from Port Dickson, Peninsular Malaysia. *J. Appl. Phycol.* 23(2), 191–196.

Chen, X., Han, W., Wang, G., Zhao, X., 2020. Application prospect of polysaccharides in the development of anti-novel coronavirus drugs and vaccines. *Int. J. Biol. Macromol.* 164, 331–343.

Chevolot, L., Mulloy, B., Ratiskol, J., Foucault, A., Colliec-Jouault, S., 2001. A disaccharide repeat unit is the major structure in fucoidans from two species of brown algae. *Carbohydr. Res.* 330(4), 529–535.

Chiovitti, A., Harper, R.E., Willis, A., Bacic, A., Mulvaney, P., Wetherbee, R., 2005. Variations in the substituted 3-linked mannans closely associated with the silicified walls of diatoms. *J. Phycol.* 41(6), 1154–1161.

Chizhov, A.O., Dell, A., Morris, H.R., Haslam, S.M., McDowell, R.A., Shashkov, A.S., Usov, A.I., 1999. A study of fucoidan from the brown seaweed Chorda filum. *Carbohydr. Res.* 320(1-2), 108–119.

Cui, K., Tai, W., Shan, X., Hao, J., Li, G., Yu, G., 2018. Structural characterization and anti-thrombotic properties of fucoidan from *Nemacystus decipiens. Int. J. Biol. Macromol.* 120, 1817–1822.

Cui, S.W. (Ed.)., 2005. *Food Carbohydrates: Chemistry, Physical Properties, and Applications.* CRC Press, Boca Raton.

Cui, Y., Zhu, L., Li, Y., Jiang, S., Sun, Q., Xie, E., Dong, C., 2021. Structure of a laminarin-type β-(1→ 3)-glucan from brown algae *Sargassum henslowianum* and its potential on regulating gut microbiota. *Carbohydr. Polym.* 255, 117389.

Cunha, L., Grenha, A., 2016. Sulfated seaweed polysaccharides as multifunctional materials in drug delivery applications. *Mar. Drugs.* 14(3), 42.

Danyal, A., Mubeen, U., Malik, K.A., 2013. Investigating two native algal species to determine antibiotic susceptibility against some pathogens. *Curr. Res. J. Biol. Sci.* 5(2), 70–74.

Da Silva, T.L., Vidart, J.M.M., da Silva, M.G.C., Gimenes, M.L., Vieira, M.G.A., 2017. Alginate and sericin: environmental and pharmaceutical applications, in: Shalaby, E. (Ed.), *Biological Activities and Application of Marine Polysaccharides.* , InTech, Croatia, pp. 57–86.

De Jesus Raposo, M.F., De Morais, A.M.B., De Morais, R.M.S.C., 2015. Marine polysaccharides from algae with potential biomedical applications. *Mar. Drugs.* 13(5), 2967–3028.

Dey, P., Ramanujam, R., Venkatesan, G., Nagarathnam, R., 2019. Sodium alginate potentiates antioxidant defense and PR proteins against early blight disease caused by *Alternaria solani* in *Solanum lycopersicum* Linn. *PloS one.* 14(9), e0223216.

Dore, C.M.P.G., Alves, M.G.D.C.F., Will, L.S.E.P., Costa, T.G., Sabry, D.A., de Souza Rêgo, L.A.R., Leite, E.L., 2013. A sulfated polysaccharide, fucans, isolated from brown algae Sargassum vulgare with anticoagulant, antithrombotic, antioxidant and anti-inflammatory effects. *Carbohydr. Polym.* 91(1), 467–475.

Draget, K.I., Taylor, C., 2011. Chemical, physical and biological properties of alginates and their biomedical implications. *Food Hydrocoll.* 25(2), 251–256.

Duan, H., Lü, S., Qin, H., Gao, C., Bai, X., Wei, Y., Liu, Z., 2017. Co-delivery of zinc and 5-aminosalicylic acid from alginate/N-succinyl-chitosan blend microspheres for synergistic therapy of colitis. *Int. J. Pharm.* 516(1–2), 214–224.

Durkin, C.A., Mock, T., Armbrust, E.V., 2009. Chitin in diatoms and its association with the cell wall. *Eukaryot. Cell.* 8(7), 1038–1050.

El Harchi, M., Kachkach, F.F., El Mtili, N., 2018. Optimization of thermal acid hydrolysis for bioethanol production from *Ulva rigida* with yeast *Pachysolen tannophilus*. *S. Afr. J. Bot.* 115, 161–169.

Ermakova, S., Men'shova, R., Vishchuk, O., Kim, S.M., Um, B.H., Isakov, V., Zvyagintseva, T., 2013. Water-soluble polysaccharides from the brown alga *Eisenia bicyclis*: Structural characteristics and antitumor activity. *Algal Res.* 2(1), 51–58.

Farid, R., Mutale-Joan, C., Redouane, B., Najib, E.M., Abderahime, A., Laila, S., Hicham, E.A., 2019. Effect of microalgae polysaccharides on biochemical and metabolomics pathways related to plant defense in *Solanum lycopersicum*. *Appl. Biochem. Biotechnol.* 188(1), 225–240.

Ferdouse, F., Holdt, S.L., Smith, R., Murúa, P., Yang, Z., 2018. The global status of seaweed production, trade and utilization. *Globfish Res. Programme.* 124, I.

Fleita, D., El-Sayed, M., Rifaat, D., 2015. Evaluation of the antioxidant activity of enzymatically-hydrolyzed sulfated polysaccharides extracted from red algae; *Pterocladia capillacea*. *LWT.* 63(2), 1236–1244

Ford, C.W., Percival, E., 1965. 1299. Carbohydrates of *Phaeodactylum tricornutum*. Part II. A sulphated glucuronomannan. *J. Am. Chem. Soc.* 1, 7042–7046.

Gaignard, C., Macao, V., Gardarin, C., Rihouey, C., Picton, L., Michaud, P., Laroche, C., 2018. The red microalga *Flintiella sanguinaria* as a new exopolysaccharide producer. *J. Appl. Phycol.* 30(5), 2803–2814.

Galgano, F., Condeli, N., Favati, F., Di Bianco, V., Peretti, G., Caruso, M.C., 2015. Biodegradable packaging and edible coating for fresh-cut fruits and vegetables. *Ital. J. Food Saf.* 27(1), 1–20.

Gao, H., Duan, B., Lu, A., Deng, H., Du, Y., Shi, X., Zhang, L., 2018. Fabrication of cellulose nanofibers from waste brown algae and their potential application as milk thickeners. *Food Hydrocoll.* 79, 473–481.

Gerber, P., Dutcher, J.D., Adams, E.V., Sherman, J.H., 1958. Protective effect of seaweed extracts for chicken embryos infected with influenza B or mumps virus. *Proc. Soc. Exp. Biol. Med.* 99(3), 590–593.

Gesztesi, J.L., Silva, L.V.N., Robert, L., Robert, A., 2006. U.S. Patent Application No. 10/552,375.

Ghannam, A., Abbas, A., Alek, H., Al-Waari, Z., Al-Ktaifani, M., 2013. Enhancement of local plant immunity against tobacco mosaic virus infection after treatment with sulphated-carrageenan from red alga (*Hypnea musciformis*). *Physiol. Mol. Plant Pathol.* 84, 19–27.

Guiry, M.D., 2012. How many species of algae are there?. *J. Phycol.* 48(5), 1057–1063.

Hajiali, H., Shahgasempour, S., Naimi-Jamal, M.R., Peirovi, H., 2011. Electrospun PGA/gelatin nanofibrous scaffolds and their potential application in vascular tissue engineering. *Int. J. Nanomed.* 6, 2133.

Hannan, M., Dash, R., Haque, M., Mohibbullah, M., Sohag, A.A.M., Rahman, M., Moon, I.S., 2020. Neuroprotective potentials of marine algae and their bioactive metabolites: Pharmacological insights and therapeutic advances. *Mar. Drugs.* 18(7), 347.

Hasnain, M.S., Jameel, E., Mohanta, B., Dhara, A.K., Alkahtani, S., Nayak, A.K., 2020. Alginates: sources, structure, and properties, in: Nayak, A.K. and Hasnain, M.S. (Eds.), *Alginates in Drug Delivery*. Academic Press, UK, pp. 1–17.

Hentati, F., Delattre, C., Ursu, A.V., Desbrières, J., Le Cerf, D., Gardarin, C., Pierre, G., 2018. Structural characterization and antioxidant activity of water-soluble polysaccharides from the Tunisian brown seaweed Cystoseira compressa. *Carbohydr. Polym.* 198, 589–600.

Hernandez-Carmona, G., Freile-Pelegrín, Y., Hernández-Garibay, E., 2013. Conventional and alternative technologies for the extraction of algal polysaccharides, in: Dominguez, H. (Ed.), *Functional Ingredients from Algae for Foods and Nutraceuticals*. Woodhead Publishing, UK, pp. 475–516

Herrmann, C., FitzGerald, J., O'Shea, R., Xia, A., O'Kiely, P., Murphy, J.D., 2015. Ensiling of seaweed for a seaweed biofuel industry. *Bioresour. Technol.* 196, 301–313.

Hong, Y., Wu, Y.R., 2020. Acidolysis as a biorefinery approach to producing advanced bioenergy from macroalgal biomass: A state-of-the-art review. *Bioresour. Technol.* 318, 124080.

Hotchkiss, S., Brooks, M., Campbell, R., Philp, K., Trius, A., 2016. The use of carrageenan in food, in: Pereira, L. (Ed.), *Carrageenans: Sources and Extraction Methods, Molecular Structure, Bioactive Properties and Health Effects*. Nova Science Publishers, Inc., Ireland, pp. 229–243.

Hu, P., Li, Z., Chen, M., Sun, Z., Ling, Y., Jiang, J., Huang, C., 2016. Structural elucidation and protective role of a polysaccharide from *Sargassum fusiforme* on ameliorating learning and memory deficiencies in mice. *Carbohydr. Polym.* 139, 150–158.

Huang, L., Shen, M., Morris, G.A., Xie, J., 2019. Sulfated polysaccharides: Immunomodulation and signaling mechanisms. *Trends Food Sci. Technol.* 92, 1–11.

Jaulneau, V., Lafitte, C., Jacquet, C., Fournier, S., Salamagne, S., Briand, X., Dumas, B., 2010. Ulvan, a sulfated polysaccharide from green algae, activates plant immunity through the jasmonic acid signaling pathway. *J Biomed. Biotechnol.* 2010.

Kadam, S.U., O'Donnell, C.P., Rai, D.K., Hossain, M.B., Burgess, C.M., Walsh, D., Tiwari, B.K., 2015. Laminarin from Irish brown seaweeds Ascophyllum nodosum and Laminaria hyperborea: Ultrasound assisted extraction, characterization and bioactivity. *Mar. Drugs* 13(7), 4270–4280.

Khalil, H.P.S., Lai, T.K., Tye, Y.Y., Rizal, S., Chong, E.W.N., Yap, S.W., Paridah, M.T., 2018. A review of extractions of seaweed hydrocolloids: Properties and applications. *Express Polym. Lett.* 12(4).

Kidgell, J.T., Magnusson, M., de Nys, R., Glasson, C.R., 2019. Ulvan: A systematic review of extraction, composition and function. *Algal Res.* 39, 101422.

Klarzynski, O., Descamps, V., Plesse, B., Yvin, J.C., Kloareg, B., Fritig, B., 2003. Sulfated fucan oligosaccharides elicit defense responses in tobacco and local and systemic resistance against tobacco mosaic virus. *Mol. Plant Microbe Interact.* 16(2), 115–122.

Koenighofer, M., Lion, T., Bodenteich, A., Prieschl-Grassauer, E., Grassauer, A., Unger, H., Fazekas, T., 2014. Carrageenan nasal spray in virus confirmed common cold: individual patient data analysis of two randomized controlled trials. *Multidiscip. Respir. Med.* 9(1), 1–12.

Kong, H., Mooney, D., 2005. Polysaccharide-based hydrogels in tissue engineering. Polysacharides, in: Dumitriu, S. (Ed.), *Structural Diversity and Functional Versatility*, 2nd ed. CRC Press, Boca Raton, USA, pp. 817–837

Kostas, E.T., White, D.A., Cook, D.J., 2017. Development of a bio-refinery process for the production of speciality chemical, biofuel and bioactive compounds from *Laminaria digitata*. *Algal Res.* 28, 211–219.

Krempel, M., Griffin, K., Khouryieh, H., 2019. Hydrocolloids as emulsifiers and stabilizers in beverage preservation, in: Grumezescu, A.M. and Holban, A.M. (Eds.), *Preservatives and Preservation Approaches in Beverages*. Academic Press, UK, pp. 427–465.

Kraan, S., 2012. *Algal Polysaccharides, Novel Applications and Outlook*. IntechOpen, Rijeka.

Lahaye, M., 2001. Chemistry and physico-chemistry of phycocolloids. *Cah. Biol. Mar.* 42(1–2), 137–157.

Lahaye, M., Robic, A., 2007. Structure and functional properties of ulvan, a polysaccharide from green seaweeds. *Biomacromolecules.* 8(6), 1765–1774.

Lahaye, M., Rochas, C., 1991. Chemical structure and physico-chemical properties of agar, in: *Int. Workshop Gelidium.* Springer, Dordrecht, pp. 137–148

Lee, H.E., Choi, E.S., Shin, J., Lee, S.O., Park, K.S., Cho, N.P., Cho, S.D., 2014. Fucoidan induces caspase-dependent apoptosis in MC3 human mucoepidermoid carcinoma cells. *Exp. Ther. Med.* 7(1), 228–232.

Lee, K.Y., Mooney, D.J., 2012. Alginate: Properties and biomedical applications. *Prog. Polym. Sci.* 37(1), 106–126.

Lee, M.C., Huang, Y.C., 2019. Soluble eggshell membrane protein-loaded chitosan/fucoidan nanoparticles for treatment of defective intestinal epithelial cells. *Int. J. Biol. Macromol.* 131, 949–958.

Lee, W.K., Lim, Y.Y., Leow, A.T.C., Namasivayam, P., Abdullah, J.O., Ho, C.L., 2017. Biosynthesis of agar in red seaweeds: A review. *Carbohydr. Polym.* 164, 23–30

LeRoux, M.A., Guilak, F., Setton, L.A., 1999. Compressive and shear properties of alginate gel: effects of sodium ions and alginate concentration. *J. Biomed. Mater. Res.* 47(1), 46–53.

Li, N., Mao, W., Yan, M., Liu, X., Xia, Z., Wang, S., Xiao, B., Chen, C., Zhang, L. and Cao, S., 2015. Structural characterization and anticoagulant activity of a sulfated polysaccharide from the green alga *Codium divaricatum. Carbohydr. Polym.* 121, 175–182.

Lim, S.J., Aida, W.M.W., Schiehser, S., Rosenau, T., Böhmdorfer, S., 2019. Structural elucidation of fucoidan from *Cladosiphon okamuranus* (Okinawa mozuku). *Food Chem.* 272, 222–226.

Maneein, S., Milledge, J.J., Nielsen, B.V., Harvey, P.J., 2018. A review of seaweed pretreatment methods for enhanced biofuel production by anaerobic digestion or fermentation. *Fermentation* 4(4), 100.

Mitsuiki, M., Mizuno, A., Motoki, M., 1999. Determination of molecular weight of agars and effect of the molecular weight on the glass transition. *J. Agr. Food Chem.* 47(2), 473–478.

Mizutani, S., Deguchi, S., Kobayashi, E., Nishiyama, E., Sagawa, H., Kato, I., 2010. U.S. Patent No. 7,678,368. U.S. Patent and Trademark Office, Washington, DC.

Mohammed, M.H., Williams, P.A., Tverezovskaya, O., 2013. Extraction of chitin from prawn shells and conversion to low molecular mass chitosan. *Food Hydrocoll.* 31(2), 166–171.

Montingelli, M.E., Benyounis, K.Y., Quilty, B., Stokes, J., Olabi, A.G., 2017. Influence of mechanical pretreatment and organic concentration of Irish brown seaweed for methane production. *Energy* 118, 1079–1089.

Montingelli, M.E., Benyounis, K.Y., Stokes, J., Olabi, A.G., 2016. Pretreatment of macroalgal biomass for biogas production. *Energ. Convers. Manage.* 108, 202–209.

Necas, J., Bartosikova, L., 2013. Carrageenan: A review. *Vet. Med.* 58(4), 187–205.

Nishinari, K., & Fang, Y., 2017. Relation between structure and rheological/thermal properties of agar. A mini-review on the effect of alkali treatment and the role of agaropectin. *Food Struct.* 13, 24–34.

Nishino, T., Nishioka, C., Ura, H., Nagumo, T., 1994. Isolation and partial characterization of a noval amino sugar-containing fucan sulfate from commercial Fucus vesiculosus fucoidan. *Carbohydr. Res.* 255, 213–224.

Palanisamy, S., Vinosha, M., Marudhupandi, T., Rajasekar, P., Prabhu, N.M., 2017. Isolation of fucoidan from Sargassum polycystum brown algae: Structural characterization, in vitro antioxidant and anticancer activity. *Int. J. Biol. Macromol.* 102, 405–412.

Palanisamy, S., Vinosha, M., Rajasekar, P., Anjali, R., Sathiyaraj, G., Marudhupandi, T., Selvam, S., Prabhu, N.M. You, S., 2019. Antibacterial efficacy of a fucoidan fraction (Fu-F2) extracted from Sargassum polycystum. *Int. J. Biol. Macromol.* 125, 485–495.

Pathak, T.S., San Kim, J., Lee, S.J., Baek, D.J., Paeng, K. J., 2008. Preparation of alginic acid and metal alginate from algae and their comparative study. *J. Polym Environ.* 16(3), 198–204

Pereira, L., Critchley, A.T., 2020. The COVID 19 novel coronavirus pandemic 2020: seaweeds to the rescue? Why does substantial, supporting research about the antiviral properties of seaweed polysaccharides seem to go unrecognized by the pharmaceutical community in these desperate times?. *J. Appl. Phycol.* 32, 1875–1877.

Pereira, S., Zille, A., Micheletti, E., Moradas-Ferreira, P., De Philippis, R., Tamagnini, P., 2009. Complexity of cyanobacterial exopolysaccharides: composition, structures, inducing factors and putative genes involved in their biosynthesis and assembly. *FEMS Microbiol. Rev.* 33(5), 917–941

Pettongkhao, S., Bilanglod, A., Khompatara, K., Churngchow, N., 2019. Sulphated polysaccharide from Acanthophora spicifera induced Hevea brasiliensis defense responses against Phytophthora palmivora infection. *Plants.* 8(3), 73.

Priyan Shanura Fernando, I., Kim, K.N., Kim, D., Jeon, Y.J., 2019. Algal polysaccharides: Potential bioactive substances for cosmeceutical applications. *Crit. Rev. Biotechnol.* 39(1), 99–113

Ra, C.H., Kim, M.J., Jeong, G.T., Kim, S.K., 2017. Efficient utilization of *Eucheuma denticulatum* hydrolysates using an activated carbon adsorption process for ethanol production in a 5-L fermentor. *Bioprocess Biosyst. Eng.* 40(3), 373–381.

Rajak, R.C., Jacob, S., Kim, B.S., 2020. A holistic zero waste biorefinery approach for macroalgal biomass utilization: A review. *Sci. Total Environ.* 716, 137067.

Raposo, M.F.D.J., De Morais, R.M.S.C., Bernardo de Morais, A.M.M., 2013. Bioactivity and applications of sulphated polysaccharides from marine microalgae. *Mar. Drugs.* 11(1), 233–252.

Read, S.M., Currie, G., Bacic, A., 1996. Analysis of the structural heterogeneity of laminarin by electrospray-ionisation-mass spectrometry. *Carbohydr. Res.* 281(2), 187–201

Rhein-Knudsen, N., Ale, M.T., Meyer, A.S., 2015. Seaweed hydrocolloid production: an update on enzyme assisted extraction and modification technologies. *Mar. Drugs.* 13(6), 3340–3359.

Rioux, L.E., Turgeon, S.L., Beaulieu, M., 2007. Characterization of polysaccharides extracted from brown seaweeds. *Carbohydr. Polym.* 69(3), 530–537

Rodríguez, M.C., Matulewicz, M.C., Noseda, M.D., Ducatti, D.R.B., Leonardi, P.I., 2009. Agar from *Gracilaria gracilis* (Gracilariales, Rhodophyta) of the Patagonic coast of Argentina–Content, structure and physical properties. *Bioresour. Technol.* 100(3), 1435–1441.

Salehi, B., Sharifi-Rad, J., Seca, A.M., Pinto, D.C., Michalak, I., Trincone, A., Martins, N., 2019. Current trends on seaweeds: Looking at chemical composition, phytopharmacology, and cosmetic applications. *Molecules.* 24(22), 4182.

Sanjeewa, K.A., Lee, J.S., Kim, W.S., Jeon, Y.J., 2017. The potential of brown-algae polysaccharides for the development of anticancer agents: An update on anticancer effects reported for fucoidan and laminaran. *Carbohydr. Polym.* 177, 451–459.

Sartal, C.G., Alonso, M.C.B., Bermejo Barrera, P., 2011. Application of seaweeds in the food industry, in: Kim, S.-K. (Ed.), *Handbook of Marine Macroalgae: Biotechnology and Applied Phycology*. John Wiley and Sons, UK, pp. 522–531.

Schultz-Jensen, N., Thygesen, A., Leipold, F., Thomsen, S.T., Roslander, C., Lilholt, H., Bjerre, A.B., 2013. Pretreatment of the macroalgae Chaetomorpha linum for the production of bioethanol–comparison of five pretreatment technologies. *Bioresour. Technol.* 140, 36–42.

Scieszka, S., Klewicka, E., 2019. Algae in food: A general review. *Crit. Rev. Food Sci. Nutr.* 59(21), 3538–3547.

Sedayu, B.B., Cran, M.J., Bigger, S.W., 2019. A review of property enhancement techniques for carrageenan-based films and coatings. *Carbohydr. Polym.* 216, 287–302.

Sellimi, S., Maalej, H., Rekik, D.M., Benslima, A., Ksouda, G., Hamdi, M., Hajji, M., 2018. Antioxidant, antibacterial and in vivo wound healing properties of laminaran purified from *Cystoseira barbata* seaweed. *Int. J. Biol. Macromol.* 119, 633–644.

Sukwong, P., Ra, C.H., Sunwoo, I.Y., Tantratian, S., Jeong, G.T., Kim, S.K., 2018. Improved fermentation performance to produce bioethanol from *Gelidium amansii* using *Pichia stipitis* adapted to galactose. *Bioprocess Biosyst Eng.* 41(7), 953–960.

Synytsya, A., Choi, D.J., Pohl, R., Na, Y.S., Capek, P., Lattová, E., Park, Y.I., 2015. Structural features and anticoagulant activity of the sulphated polysaccharide SPS-CF from a green alga *Capsosiphon fulvescens*. *Mar. Biotechnol.* 17(6), 718–735.

Tabassum, M.R., Xia, A., Murphy, J.D., 2017. Comparison of pre-treatments to reduce salinity and enhance biomethane yields of Laminaria digitata harvested in different seasons. *Energy.* 140, 546–551

Teng, Z., Qian, L., Zhou, Y., 2013. Hypolipidemic activity of the polysaccharides from *Enteromorpha prolifera*. *Int. J. Biol. Macromol.* 62, 254–256.

Trabelsi, L., M'sakni, N.H., Ouada, H.B., Bacha, H., Roudesli, S. 2009. Partial characterization of extracellular polysaccharides produced by cyanobacterium *Arthrospira platensis*. *Biotechnol. Bioprocess Eng.* 14(1), 27–31.

Trivedi, N., Gupta, V., Reddy, C.R.K., Jha, B., 2013. Enzymatic hydrolysis and production of bioethanol from common macrophytic green alga *Ulva fasciata Delile*. *Bioresour. Technol.* 150, 106–112.

Tuvikene, R., 2021. Carrageenans, in: Phillips, G.O. and Williams, P.A. (Eds.), *Handbook of Hydrocolloids*. Woodhead Publishing, Estonia, pp. 767–804.

Usoltseva, R.V., Anastyuk, S.D., Surits, V.V., Shevchenko, N.M., Thinh, P.D., Zadorozhny, P.A., Ermakova, S.P., 2019. Comparison of structure and in vitro anticancer activity of native and modified fucoidans from *Sargassum feldmannii* and *S. duplicatum*. *Int. J. Biol. Macromol.* 124, 220–228.

Vanegas, C.H., Hernon, A., Bartlett, J., 2015. Enzymatic and organic acid pretreatment of seaweed: effect on reducing sugars production and on biogas inhibition. *Int. J. Ambient Energy.* 36(1), 2–7.

Verma, P., Arun, A., Sahoo, D., 2015. Brown Algae, in: *The Algae World*. Springer, Dordrecht, pp. 177–204

Villay, A., Laroche, C., Roriz, D., El Alaoui, H., Delbac, F., Michaud, P., 2013. Optimisation of culture parameters for exopolysaccharides production by the microalga *Rhodella violacea*. *Bioresour. Technol.* 146, 732–735.

Vo, T.S., Kim, S.K., 2000. Cosmeceutical compounds from marine sources, in: Ley, C. (Ed.), *Kirk-Othmer Encyclopedia of Chemical Technology*. John Wiley & Sons, Inc., Germany, pp. 1–17.

Walsby, A.E., Xypolyta, A. 1977. The form resistance of chitan fibres attached to the cells of *Thalassiosira fluviatilis* Hustedt. *Br. Phycol. J.* 12(3), 215–223.

Wang, H.M.D., Chen, C.C., Huynh, P., Chang, J.S., 2015. Exploring the potential of using algae in cosmetics. *Bioresour. Technol.* 184, 355–362.

Wang, J., Jin, W., Hou, Y., Niu, X., Zhang, H., Zhang, Q., 2013. Chemical composition and moisture-absorption/retention ability of polysaccharides extracted from five algae. *Int. J. Biol. Macromol.* 57, 26–29.

Wang, L.F., Rhim, J.W., 2015. Preparation and application of agar/alginate/collagen ternary blend functional food packaging films. *Int. J. Biol. Macromol.* 80, 460–468.

Wernberg, T., Krumhansl, K., Filbee-Dexter, K., Pedersen, M.F., 2019. Status and trends for the world's kelp forests, in: Sheppard, C. (Ed.), *World Seas: An Environmental Evaluation*. Academic Press, Elsevier, UK, pp. 57–78

Wozniak, M., Bell, T., Dénes, Á., Falshaw, R., Itzhaki, R., 2015. Anti-HSV1 activity of brown algal polysaccharides and possible relevance to the treatment of Alzheimer's disease. *Int. J. Biol. Macromol.* 74, 530–540.

Xin, Z., Cai, X., Chen, S., Luo, Z., Bian, L., Li, Z., Chen, Z., 2019. A disease resistance elicitor laminarin enhances tea defense against a piercing herbivore *Empoasca* (*Matsumurasca*) *onukii Matsuda*. *Sci. Rep.* 9(1), 1–13.

Yahmed, N.B., Carrere, H., Marzouki, M.N., Smaali, I. 2017. Enhancement of biogas production from *Ulva sp.* by using solid-state fermentation as biological pretreatment. *Algal Res.* 27, 206–214.

Yalpani, M., 2013. *Polysaccharides: Syntheses, Modifications and Structure/property Relations.* Elsevier, USA.

Yim, J.H., Kim, S.J., Ahn, S.H., Lee, C.K., Rhie, K.T., Lee, H.K., 2004. Antiviral effects of sulfated exopolysaccharide from the marine microalga *Gyrodinium impudicum* strain KG03. *Mar. Biotechnol.* 6(1), 17–KG25.

Yokota, T., Nomura, K., Nagashima, M., Kamimura, N., 2016. Fucoidan alleviates high-fat diet-induced dyslipidemia and atherosclerosis in ApoEshl mice deficient in apolipoprotein E expression. *J. Nutr. Biochem.* 32, 46–54.

Youssouf, L., Lallemand, L., Giraud, P., Soulé, F., Bhaw-Luximon, A., Meilhac, O., Couprie, J., 2017. Ultrasound-assisted extraction and structural characterization by NMR of alginates and carrageenans from seaweeds. *Carbohydr. Polym.* 166, 55–63

Yu, G., Hu, Y., Yang, B., Zhao, X., Wang, P., Ji, G., Guan, H., 2010. Extraction, isolation and structural characterization of polysaccharides from a red alga *Gloiopeltis furcata*. *J. Ocean Univ.* 9(2), 193–197.

Yun, E.J., Lee, S., Kim, J.H., Kim, B.B., Kim, H.T., Lee, S.H., Kim, K.H., 2013. Enzymatic production of 3, 6-anhydro-L-galactose from agarose and its purification and in vitro skin whitening and anti-inflammatory activities. *Appl. Microbiol. Biotechnol.* 97(7), 2961–2970.

Zhang, Z., Wang, F., Wang, X., Liu, X., Hou, Y., Zhang, Q., 2010. Extraction of the polysaccharides from five algae and their potential antioxidant activity in vitro. *Carbohydr. Polym.* 82(1), 118–121.

Zhao, P., Wang, X., Niu, J., He, L., Gu, W., Xie, X., Wang, G., 2020. Agar extraction and purification of R-phycoerythrin from Gracilaria tenuistipitata, and subsequent wastewater treatment by *Ulva prolifera*. *Algal Res.* 47, 101862.

Zheng, Y., Liu, T., Wang, Z., Xu, Y., Zhang, Q., Luo, D., 2018. Low molecular weight fucoidan attenuates liver injury via SIRT1/AMPK/PGC1α axis in db/db mice. *Int. J. Biol. Macromol.* 112, 929–936.

Zhou, K., Hu, Y., Zhang, L., Yang, K., Lin, D., 2016. The role of exopolymeric substances in the bioaccumulation and toxicity of Ag nanoparticles to algae. *Sci. Rep.* 6(1), 1–11.

Zhu, C., Zhang, S., Song, C., Zhang, Y., Ling, Q., Hoffmann, P.R., Huang, Z., 2017. Selenium nanoparticles decorated with *Ulva lactuca* polysaccharide potentially attenuate colitis by inhibiting NF-κB mediated hyper inflammation. *J. Nanobiotechnology.* 15(1), 1–15.

Zhu, M., Ge, L., Lyu, Y., Zi, Y., Li, X., Li, D., Mu, C., 2017. Preparation, characterization and antibacterial activity of oxidized κ-carrageenan. *Carbohydr. Polym.* 174, 1051–1058.

Zou, P., Lu, X., Zhao, H., Yuan, Y., Meng, L., Zhang, C., Li, Y., 2019. Polysaccharides derived from the brown algae Lessonia nigrescens enhance salt stress tolerance to wheat seedlings by enhancing the antioxidant system and modulating intracellular ion concentration. *Front. Plant Sci.* 10, 48.

10 Algae as Food and Nutraceuticals

Chetan Aware, Rahul Jadhav, and Jyoti Jadhav
Department of Biotechnology, Shivaji University,
Vidyanagar, Kolhapur, Maharashtra, India

Virdhaval Nalavade
Department of Biotechnology, Yashavantrao Chavan
Institute of Science, Satara, Maharashtra, India

*Shashi Kant Bhatia, Yung-Hun Yang, and
Ranjit Gurav*
Department of Biological Engineering, College of
Engineering, Konkuk University, Seoul, South Korea

CONTENTS

10.1 Introduction 296
10.2 Microalgal Diversity and Variability 297
10.3 Digestion and Bioavailability 299
10.4 Microalgae: Uses as Nutraceuticals and Food 303
    10.4.1 Why Should Microalgae Be Promoted as a Nutraceutical? ... 304
10.5 Algal Carbohydrates and Polysaccharides 308
    10.5.1 Ulvan 309
    10.5.2 Fucoidan 309
    10.5.3 Alginates 309
    10.5.4 Carrageenan 310
    10.5.5 Laminarin 310
    10.5.6 Agar 310
    10.5.7 Nutraceutical Advantages of Seaweed Polysaccharides 310
10.6 Algal Lipids, Fatty Acids, and Sterols 311
10.7 Algal Proteins, Peptides, and Amino Acids 312
    10.7.1 Factors Affecting Amino Acid and Protein Synthesis
        in Algae 313
    10.7.2 The Amino Acid Profile of Seaweed 313
10.8 Vitamins 314
    10.8.1 Composition of Vitamins in Seaweed 315

DOI: 10.1201/9781003195405-10

        10.8.2   Factors Influencing Seaweed Vitamins 316
10.9   Phlorotannin..316
10.10   Pigments and Minor Compounds in Algae...317
        10.10.1   Chlorophylls 318
        10.10.2   Carotenoids...318
        10.10.3   Beta-carotene 318
        10.10.4   Fucoxanthin 319
        10.10.5   Phycobiliproteins 320
10.11   Algal Antioxidants 320
10.12   Design of Healthier Foods and Beverages Containing Whole Algae....322
        10.12.1   Use of Algal Components in Foods and Beverages as
                  Healthy Ingredients 323
        10.12.2   Nutritional Quality Standard, Regulations, and Challenges
                  for Algal Products 325
10.13   Prebiotic and Probiotic Use of Algae and Algae-Supplemented
        Products 325
10.14   Future Perspectives of Algal Derived Metabolites 327
10.15   Conclusion 327
Acknowledgment 327
References..328

10.1 INTRODUCTION

The growing demand of consumers for functional food ingredients that prevent the problem of various chronic diseases or enhance health benefits led to extensive research efforts for their identification and characterization. The utilization of microbial systems has the yet-untapped potential to sustainably contribute to the supply of high value-added compounds of commercial value and other environmental applications (Bhatia et al., 2021a; Gurav et al., 2021; Park et al., 2021; Choi et al., 2020). Microbes have a role as promising organisms for producing both basic primary metabolites in human nutrition and unique secondary bioactive compounds with pharmaceutical and food-related applications (Holdt and Kraan, 2011; Mehariya et al., 2021a; Patel et al., 2021; Lee et al., 2021). Currently, products obtained from botanical origin constitute the majority of stake in nutraceutical products marketplace, algae-based phytochemicals, nutraceuticals, and bioactive substances against the traditional role of basic nutrition (e.g., antibiotics, diabetic and obesity control, anti-inflammatories, prevention of other diseases, etc., are in focus due to exceptional features, absent in other plants' terrestrial biomass-derived products (Hafting et al., 2012; Bhatia et al., 2021b; Kumar et al., 2021). Among them, presently marketed as dietary supplements, a major stake compromise carotenoids and omega-3 algal oil. Based on archaeological evidence in many civilizations and cultures around the world (Chile,1400; China, 300 A.D; Ireland, 600 A.D.), algae for thousands of years constituted an essential part of their diet. The annual per capita human consumption of sole algal and algal added products in Japan tops the list; consumption started at 9.6 (2010) up to 11.0 (2014) grams of seaweed per day (MHLW, 2014). On a global basis, the overall increasing trend is seen toward nutritional demand for algal products, mainly focused on health-related products with broad use as food additives. Algae act as a vital source of biologically active phytochemicals and secondary metabolites of natural origin with added advantages like worldwide availability in diverse environmental conditions, ample biomass supply, diversity, and high productivity. In the aquatic ecosystem, algae with its complex composition of primary and secondary metabolites in the food chain support as a primary feedstock, in a similar manner as plants from in terrestrial ecosystem (Bhatia et al., 2019; Gurav et al., 2019; Gurav et al., 2020; Kim et al., 2020). On this basis similarly, as plants act as raw materials to produce food and food additives at industrial-scale algal biomass involving their environmentally friendly and sustainable technologies approach fitting into the concept of biorefinery approach. Further insight and research focus on the composition, functionality, and bioactivity of various algal species are crucial for future sustainable development in versatile, multi-step processes to facilitate rational utilization of algal biomass (Figure 10.1).

Microalgal processes are identical to conventional bioprocesses, but they entail special management (Hafting et al., 2012). The production of photobioreactors for functional food in space, as well as continuous culture processing, present new issues, and challenges. Eventually, despite the difficulties encountered in developing huge photobioreactors, it is worth noting that, while photobioreactors may be insufficient for certain products, such as biofuels, they may be adequate for the generation of products, such as nutraceuticals. Furthermore, utilizing recombinant microalgae, closed photobioreactors support the development of nutritional supplements. This chapter represents a general overview of the concepts particularly in relation to microalgae as active ingredients for the improvement of functional foods and nutraceuticals.

10.2 MICROALGAL DIVERSITY AND VARIABILITY

The name "algae" refers to a diverse and huge variety of polyphyletic, typically photosynthetic organisms, from different origins, evolutionary linage, and biochemical composition. The members of the algal group are currently divided into four kingdoms: bacteria, plantae, chromista, and protozoa. The occurrence of algae is in diverse organization, forms, and sizes (unicellular or colonial or multicellular), which include microalgae and macroalgae (macrophytes or seaweed). If we consider the plant kingdom, microalgae are the most primitive and simple members, which are found in diverse habitats ranging from the fresh and marine aquatic environment, also in a terrestrial ecosystem (soil, snow, mountain crust, tree surfaces, etc.) and present in air microflora. Microalgae have been traditionally classified considering the cytological composition, morphological aspects, type of reserve food metabolites (polyunsaturated fatty acids, oils, proteins, starch), constituents of the cell wall (complex polysaccharides like agar-agar, alginic acids, carrageenans, etc.), and their specific pigment composition (e.g., chlorophyll *a, b, c*, carotenoids, xanthophylls, anthocyanin). If we consider the total living space on Earth, water constitutes 99% space, so there is sufficient space and time for algae to

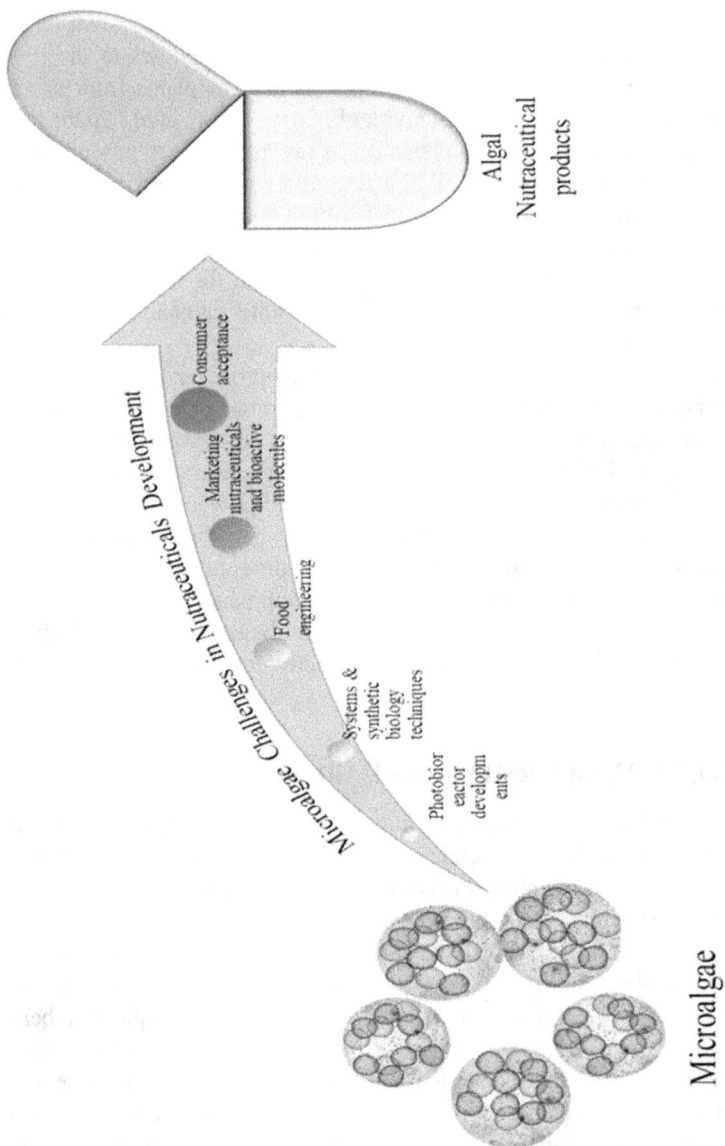

FIGURE 10.1　Various challenges in algal nutraceutical developments.

grow and to develop. The blue-green algae are the oldest members of an algal family with ancient geological history – 3550 million years in relationship to merely 400 million years for higher plants (Sepkoski, 1995). Hence, if we try to answer how many algae are there, we cannot yet estimate the correct number of species. Algal identification is notoriously complicated as they exhibit few morphological characters (although advances in the electron microscope and molecular biology have vastly increased their correct identification), and they display substantial morphological plasticity induced due to environmental conditions and genetic shifts in laboratory culture. Also, it is difficult due to their shorter generation times and evolution proceeds, which place them into numerous sibling species and led to uncertainty in the total number of taxa, reduction by synonyms, doubtful entities, and poor and incomplete descriptions of many taxa (Norton et al., 1996). Currently, morphological characters with additional data generated using advanced micro-scopy and molecular biology methods, have shown highly precise impact on the taxonomical classification of various taxa and species in algae. Insights into a deep study of evolutionary relationships of different algal taxa and their correlation with those algae showing potential sources of bioactive lead to important taxonomic rearrangements (Rindi et al., 2012). Currently, the overall algae recorded, according to the Algae Base database (2021), was conservatively estimated to be about 161,152 species. The predictable identified and non-identified phyla and classes of algal number are shown in Figure 10.2 (Guiry, 2012). As various algal bioactive compounds of interest are yet to be harnessed at their potential commercial level, the full utilization of diverse algal species requires in detail and inerter-related knowledge of environmental impacts, understanding of biochemical and biological variability for assured and stable high-quality products. For commercial utilization of bioactive molecules, the natural variability in the content should not be con-sidered an obstacle to and is dependent on (i) ecological adaptations, (ii) taxonomic or phylogenetic relationships, and (iii) chemical diversity (Stengel et al., 2011).

10.3   DIGESTION AND BIOAVAILABILITY

The term "bioavailability", as stated by Carbonell-Capella et al. (2014) is a combined process of bioactivity and bio-accessibility. Here, bio-accessibility is described as the amount or some portion released from ingested food components in the GI tract, which is finally utilized in various physiological functions of that organism through the absorption process, whereas bioactivity means uptake of metabolites assimilated in tissues, their metabolism, and physio-chemical effects in that organism (Guerra et al., 2012). Hence, bioavailability includes the entire process that follows con-sumption of food elements, their digestibility and solubility food fractions in the gastrointestinal tract, uptake into tissues, metabolism of the digested food material into the circulatory system via the intestinal epithelial cells, and their final utilization in the target site, as shown in Figure 10.3 (Ekmekcioglu, 2002). There are different practical and ethical issues for measurement of bioactivity, the accurate concentration of a given compound or its generated metabolite that finally, absorbed assimilated can be measured as bio-accessible, also it cannot inevitably be considered as bioactive (Holst and Williamson, 2008). The majority of earlier published studies and

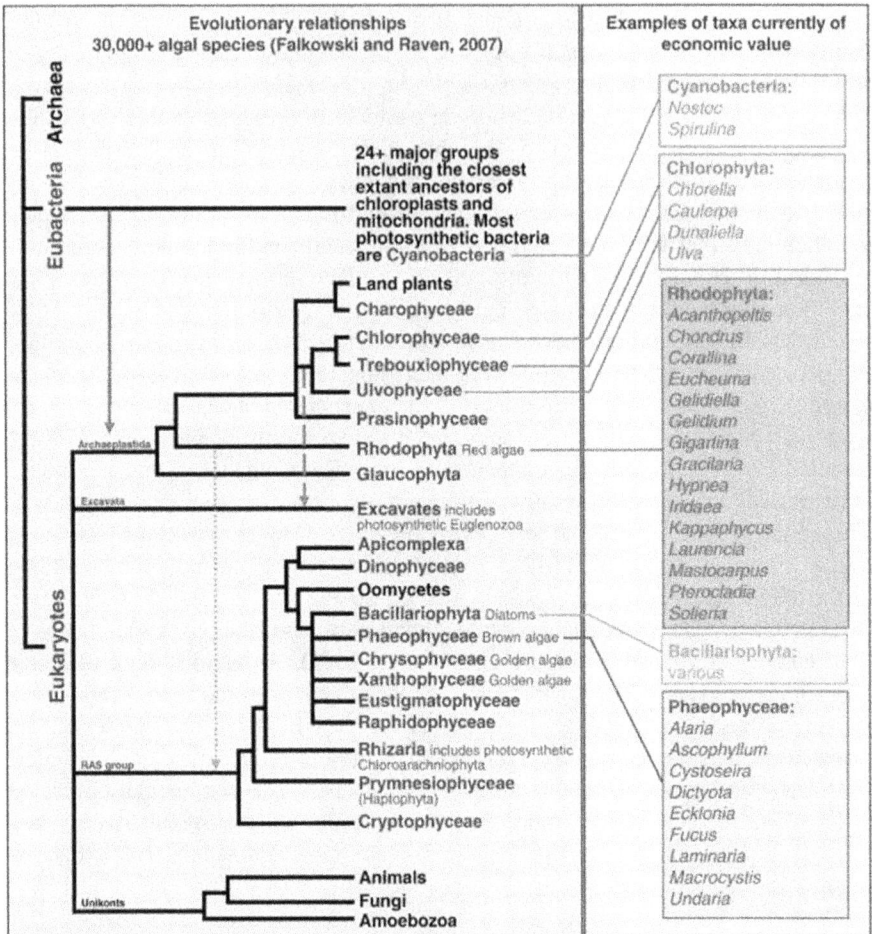

Evolutionary relationships 30,000+ algal species (Falkowski and Raven, 2007)	Examples of taxa currently of economic value

Archaea

Eubacteria

24+ major groups
including the closest
extant ancestors of
chloroplasts and
mitochondria. Most
photosynthetic bacteria
are Cyanobacteria

Eukaryotes

Archaeplastida

- Land plants
- Charophyceae
- Chlorophyceae
- Trebouxiophyceae
- Ulvophyceae
- Prasinophyceae
- Rhodophyta Red algae
- Glaucophyta

Excavata

- Excavates includes
  photosynthetic Euglenozoa
- Apicomplexa
- Dinophyceae
- Oomycetes
- Bacillariophyta Diatoms
- Phaeophyceae Brown algae
- Chrysophyceae Golden algae
- Xanthophyceae Golden algae
- Eustigmatophyceae
- Raphidophyceae

RAS group

- Rhizaria includes photosynthetic
  Chloroarachniophyta
- Prymnesiophyceae
  (Haptophyta)
- Cryptophyceae

Unikonts

- Animals
- Fungi
- Amoebozoa

Cyanobacteria:
Nostoc
Spirulina

Chlorophyta:
Chlorella
Caulerpa
Dunaliella
Ulva

Rhodophyta:
Acanthopeltis
Chondrus
Corallina
Eucheuma
Gelidiella
Gelidium
Gigartina
Gracilaria
Hypnea
Iridaea
Kappaphycus
Laurencia
Mastocarpus
Pterocladia
Solieria

Bacillariophyta:
various

Phaeophyceae:
Alaria
Ascophyllum
Cystoseira
Dictyota
Ecklonia
Fucus
Laminaria
Macrocystis
Undaria

FIGURE 10.2 Evolutionary relationships between the three major lineages of life are Archaea, Eubacteria, and Eukaryotes in consideration of algal taxonomy and diversity. (Adapted from Stengel et al., 2011; Order no: 061780338068).

evaluations for bioactivity of algal foods used only data generated from short-term *in vitro* test models utilizing algal metabolites or extract. These algal products lacked defined qualitative and quantitative biochemical composition; hence, their apparent understanding of food efficacy in test organisms is highly constrained, particularly less understanding of the performance of food components containing algae in the gut environment and their effect on the intestinal microbiome. There is a need to understand the mechanism for digestion and assimilation of these metabolites in humans. Digestion begins in the mouth with chewing consumed food, which causes reduction of food particle size and proper mixture with saliva and is concluded by absorption into the body (Lovegrove et al., 2017).

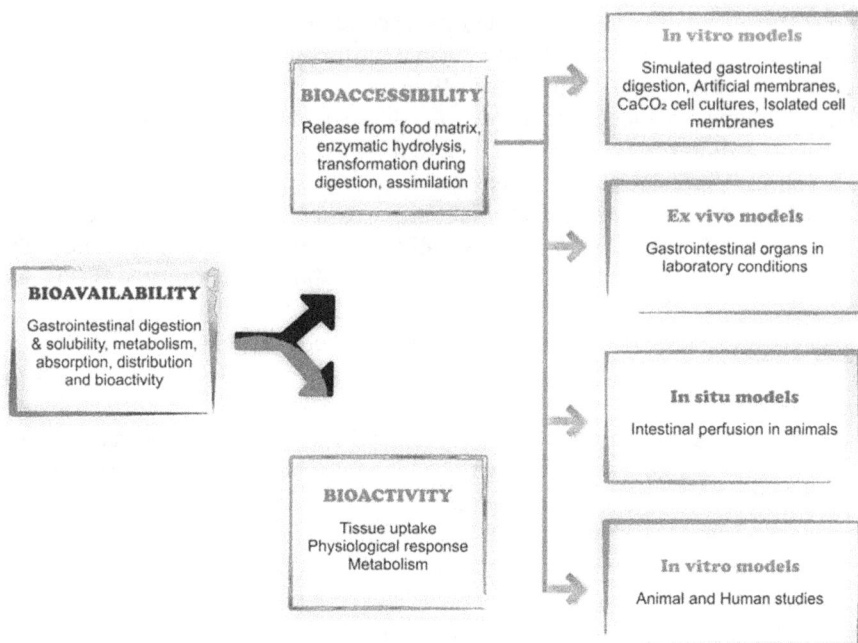

FIGURE 10.3 Schematic representation of digestion and methods that are used to determine bioavailability, bio-accessibility, and bioactivity of proteins and other nutrients (Carbonell-Capella et al., 2014).

The human saliva enzymes mainly consist of alpha (α)-amylase, which is more active in comparison to other primates. Previous data surveys showed a lack of studies on the effect of saliva from the human on algae and particularly polysaccharide from algae-like starch. During protein digestion in the stomach, the enzyme pepsin and the pepsinogens in the presence of acid cause denaturation of proteins and covert their simpler peptides, while lipase secretion in the mouth and stomach activates and catalyzes digestion of triacylglycerols. Apart from this, pancreatic enzymes (lipase, trypsin, α-amylase, carboxypeptidases, etc.) discharged from the small intestine help in further digestion; this system is used as a study of human *in vitro* systems. In the small intestine, various enzymes secretion help the absorption of digested and solubilized primary and secondary metabolites. Humans lack many enzymes that can digest complex polysaccharides (cellulose, lignin, hemicelluloses, pectin, xyloglucan, etc.) and other indigestible materials, which is known as dietary fiber. Hence, the digestion of these dietary fibers in the large intestine (colon) is catalyzed using bacterial-dependent enzymatic processes. The gut microbiomes help in chemical conditioning of the colon, stimulating an immune response, and provide nutritional or functional benefits through absorption and transportation of digested metabolites via the bloodstream (Cian et al., 2015).

The *in vivo* study for the effect of seaweed fiber of *P. tenera* and *U. pinnatifida* on protein digestion in Wistar rats showed dietary fiber has an adverse impact on food efficiency and protein intake and digestion of seaweed (Urbano and Goñi,

2002). The macroalgae *L. japonica* in rats reported decreased protein digestibility, although after three weeks adaptation of rats to the high fiber diet was found in regard to the control diet (Suzuki et al., 1993). The negative impact of digestibility of algal proteins is thought to be due to phlorotannin and high polysaccharide content, as well as resistant cell wall in algae (Wong and Cheung, 2001). Similarly, earlier *in vivo* protein quality studies of bio-accessibility revealed that untreated seaweed proteins reduced protein digestibility in relation to other sources. The seaweed proteins after processing have alike *in vitro* digestibility to commonly consumed plants, legumes (72–92%), grains (69–84%), vegetables (68–80%), and fruits (72–92%) (Tibbetts et al., 2016). The earlier digestibility studies reveal that maximum data is available on seaweed while poorly examined in microalgae. However, microalgae (*Scenedesmus* and *Spirulina*) showed similar digestibility to that of seaweed, having 88.0%, 77.6%, and 76.6% of digestibility coefficient values, respectively, while for casein and egg it was 95.1% and 94.2%. (Becker, 2007). Consequently, to improve algal *in vivo* digestibility, there is a need for disruption of the cell wall structure for improved nutrient accessibility. The algal cell walls disruption can be achieved using the following process: alkaline, acid, and organic solvents (chemical), cellulases (enzymatic), and microfluidics and ultrasonication (mechanical), bead milling, high-pressure homogenization, thermal treatments, and freeze-drying (physical) (Tibbetts et al., 2016).

The study in humans showed undigested algal proteins and carbohydrate factions were found to be helpful via stimulation of immune response in some way via the promoting the gut microbiomes stimuli (Cian et al., 2015). Due to unequal intake of algal products in various regions around the world, the human gut microbiomes had unequal competencies for algal polysaccharides and other metabolites. If we consider the human population from different regions, algal polysaccharide fermentation differs due to unequal competencies of human gut microbiomes. In Japanese people, the common gut bacterium (*Bacteroides plebeius*) exhibited polysaccharide degrading enzymes but was absent in Americans. Further insight showed horizontal gene transfer (HGT) was the reason; it was from a marine bacterium (*Zobellia galactanivorans*), which constitutes the surface microflora of seaweed *Porphyra* used in the preparation of nori sheets and consumed widely in Japan (Hehemann et al., 2012). HGT clarifies the occurrence of A gene cluster seen in *Bacteroides* sp. present in the gut of Japanese people facilitates alginate fermentation present in macroalgae (Thomas et al., 2012). These prominent differences add further complications for the study of the various benefits of foods containing algae as emphasized interactions between food dietary history, customs, and diverse gut microbiomes. Currently, rapidly evolving areas of research include simulated gastrointestinal digestion, colonic flora dietary modulation, xenobiotic animal models, and study the effect of bacterial fermentation byproducts on human health, which will be helpful in assessment for algal-derived foods products (Wells et al., 2017). Despite highly precise and accurate knowledge of different algal product compositions, overall know-how for digestion and bioavailability are not known. Hence, use of suitable model systems and thorough experimental design is necessary to verify the bioavailability of various components present in algal foods products.

10.4 MICROALGAE: USES AS NUTRACEUTICALS AND FOOD

For centuries, marine and freshwater products have been utilized in modern civilization and incorporated in daily dietary ingredients, which are known as "nutraceuticals" and recognized by scientific research for their remarkable health benefits. Nutraceutical is a combination of the two concepts: food and pharmaceutical. Nutraceutical is described as health-related food or food material, which provides health benefits by preventing diseases and required treatments (Kim, 2013). Crucial food items are considered nutraceuticals when they have promising applications in an organism's life and prevent disease induction (Wildman and Taylor, 2006). The word "nutraceutical" has several interpretations and is used as a synonym for functional foods, dietary supplements, or medicinal food (González-Sarrías et al., 2013).

Stephen De Felice invented the word "nutraceuticals" in 1989 by combining the terms "nutrition" and "pharmaceutical" (Vidanarachchi et al., 2012). These food items are being controlled by the U.S. FDA with laws related to food, drugs, and cosmetics, etc., although they are not explicitly specified by law. The term "nutraceutical" is described by the American Nutraceutical Association: "A nutraceutical is any food product considered as food, or food component, that gives medicinal or health advantages, along with disease management."

In comparison to pharmaceutical products, nutritional supplements are not subject to drug regulations and monitoring. These confer them their widespread application. To combat specific and potentially chronic conditions, the growing global population necessitates sufficient and adequate nutrition with high nutritional efficiency (Bishop and Zubeck, 2012). Over more than 470 nutraceuticals are available in the global market that promote significant health attributes. The present nutraceutical trend is to promote not only health but also contradict the potential of neuro-vegetative impairments, including Alzheimer's disease and Parkinson's disease as well as other dysfunctions like heart disease, cancer, and many others. Furthermore, in current decades the requirement for nutraceuticals has intensified due to high healthcare expenses, prolonged life expectancy, and people's willingness to boost their health appetite (Plaza et al., 2010).

Algae are known for being a repository of various vital nutritive elements as well as quick and fast growth characteristics. Microalgae are autotrophs organisms with large family members and have the fast-growing capability by using visible light, trap the atmospheric CO_2, and generating higher productivity per surface area than any other natural resource (Mehariya et al., 2021b, 2021c). Microalgae have the capability to synthesize a wide range of molecules that confer nutraceutical competency and promote further commercial insights (Bishop and Zubeck, 2012). Microalgae are autotrophic organisms, have nutraceutical characteristics, and are advantageous to strengthen the immune system to counteract several health impairments (Bishop and Zubeck, 2012). Because of the primitive immune response of shellfish, nutraceuticals are used in aquaculture to promote shellfish growth under rigorous and regulated environments. *Chlamydomonas, Chlorella, Synechococcus, Scenedesmus, Haematoccoccus, Dunaliella,* and *Porphyridium* are some of the microalgae being utilized in present nutraceuticals products. Furthermore, some microalgae have the proficiency to accumulate crucial macro and microelements, including potassium, selenium, zinc, iron, iodine, manganese, magnesium, copper, cobalt, phosphorus, nitrogen, molybdenum, calcium, and sulfur, which is promoted further in functional food.

Secondary carotenoids, including canthaxanthin and astaxanthin, are bio-synthesized by a carotenogenic group of microalgae and incorporated in the lipid globules of plastids. Some crucial amino acids and unsaturated fatty acids, such as arachidonic acid, docosahexaenoic acid, etc., are abundant in algae (Simoons, 1991) (Table 10.1). Microalgae utilize natural light to convert polluted water, CO_2, and nutrients or growth factors to nutrient-rich biomass, which exhibits an excellent source of proteins, sugars, lipids, and other vital organic components. Inorganic materials are converted into colored products by these microorganisms (Batista et al., 2013). These microorganisms are bestowed with a wide range of essential ailments that benefit human and livestock wellness. Microalgae can be found in a variety of habitats around the globe, and they have a plethora of valued nu-traceuticals, respectively. Their characteristics of adaptations have benefited various creatures in the food chain. For example, the specific properties of several micro-algae, consisting of carotenoids, phycobilins, amino acids, micronutrients, phy-coerythrin, and carbohydrates, have contributed to the development of various organisms with imparting health benefits (Table 10.1). However, considering the features and advantages outlined above of algal compounds, their growth is re-maining in the early stages of development, although there are currently several products available on the market. Numerous microalgal compounds or extracts, including *Dunaliella, Spirulina, Chlorella,* and *Haematococcus,* are being reported in dietary supplements and cosmetics in this decade.

10.4.1 WHY SHOULD MICROALGAE BE PROMOTED AS A NUTRACEUTICAL?

There are many other features and strengths of microalgae that are advantageous for their commercial exploration as a nutraceutical.

- Microalgae require simplistic growth demand, which contains supplies of phosphate and nitrogen, trace metals, sunlight, CO_2, and water.
- Chelf Brown and Wyman (1993) reported that increased productivity can be achieved with relatively limited water consumption. It must be noted that land leaching, fertilized soil erosion, and sewage contribute to the eutrophication of phosphates, nitrates, and other nutrients and minerals, which are released in rivers, ponds, and marine water bodies, where algae absorb them – hence promoting growth in those regions.
- Optimum nutritional concentrations are easily achieved in enriched aqu-eous resources for algal growth compared to the complex soil matrices.
- Microalgae do not support non-photosynthetic systems such as root, stem, and fruits; these are unicellular or colonial organisms and do not expend energy for the storage of complex biomolecules, such as starch in the cellular compartment; however, the whole biomass is functional and en-riched with vital nutrients. In comparison to terrestrial vegetation, which

TABLE 10.1
Overview of Microalgae-derived Products and Their Nutraceutical Applications

Products	Microalgae	Applications	References
Polysaccharides			
Agar	*Gelidium* and *Gracilaria* sp.	Food ingredient, fruit preserves, clarifying brewing agent, hydrocolloids, and paper industry	Vassilev and Vassileva, 2016
Alginate	*Ascophyllum nodosum* and *Laminaria* sp.	Food additive, pharmaceutical, paper and textile printing, fertilizer, and cosmetic industries	Deviram et al., 2020
Carrageenan	*Kappaphycus alvarezii* and *Chondrus crispus*	Pet food, food additives, gels, toothpaste	Deviram et al., 2020
Ulvan	*Ulva* and *Enteromorpha*	Cardiovascular protective effects, antihyperlipidemic influence	Larsen et al., 2011
Fucoidan	*Fucaceae* and *Laminariaceae*	Antiviral efficacy prevents lipid deposition	Park et al., 2011
Laminarin	*Laminaria* sp.	Antimicrobial and anti-inflammatory activity	Bouwhuis et al., 2017
PUFAs			
Eicosapentaenoic acid (EPA)	*Phaeodactylum tricornutum, P. cruentum, Monodus subterraneus, Nannochloropsis, Schizochytrium, Chaetocero scalcitrans*	Anti-inflammatory activity, cardiovascular protective effects	Batista et al., 2013; Pereira et al., 2012
Docosahexaenoic acid (DHA)	*Crypthecodinium cohnii, Pavlova salina, Isochrysis galbana, Schizochytrium*	Important in brain and retina function	Simoons, 1991
Linoleic acid	*D. salina*	Maintains cell membrane fluidity	
γ-linolenic acid	*Spirulina*	Useful in diabetic neuropathy	Koru Edis, 2012
Sterols			
Brassica sterol, sitosterol, and stigmasterol	*D. salina, D. tertiolecta*	Inhibits cardiovascular diseases, reduces cholesterol level	Luo et al., 2015

(Continued)

TABLE 10.1 (Continued)
Overview of Microalgae-derived Products and Their Nutraceutical Applications

Products	Microalgae	Applications	References
Vitamins			
Vitamin B12	*Chlorella, Spirulina*	Helpful in brain function and red blood cell formation	Begum et al., 2016
Vitamin E	*Porphyridum cruentum*	Powerful antioxidant	Sun et al., 2016
β-carotene	*D. salina, Haematococcus pluvialis, Synechococcus, Nanocloropsis gaditana*	Has antioxidants and reduces the risk of cancer development	Lindqvist and Andersson, 2002
Carotenoids, Chlorophyll, Phycobiliproteins			
α-carotene	*Chlorella*	Coloring agent for food, and antioxidant, prevents cervical cancer, a precursor for vitamin A and a supplement for vitamin C	Priyadarshani and Rath, 2012
Astaxanthin	*D. salina, H. pluvialis, Chlorella*	Food supplement as food dye additive and antioxidant	Bhalamurugan et al., 2018 Mehariya et al., 2020
Lutein	*Murellopsis sp., Chlorella, Scenedesmus almeriensis, Auxenochlorella protothecoides*	Prevents age-related macular degeneration (AMD)	García Sartal et al., 2012
Zeaxanthin	*D. salina, Chlorella, Synechococcus, N. gaditana*	Lowers the risk of ophthalmological diseases and cataracts formation	Hosokawa et al., 1999
Canthaxanthin	*Scenedesmus komarekii, D. salina, Chlorella, H. pluvialis*	Prevents erythropoietic protoporphyria (EPP), anti–inflammatory and antioxidant properties, inhibits liver neoplasms	Raposo et al., 2015; Molnár et al., 2010
Fucoxanthin	*Phaeodactyllum tricornutum*	Strong antioxidant, anti-angiogenic	Ganesan et al., 2013
Phytoene	*D. salina*	Strong anti-carcinogenic activity	Nishino et al., 2002

(Continued)

TABLE 10.1 (Continued)
Overview of Microalgae-derived Products and Their Nutraceutical Applications

Products	Microalgae	Applications	References
Phytofluene	*D. salina*	Protects against UV-light-induced erythema	Aust et al., 2005
Violaxanthin	*Chlorella ellipsodea, Synechococcus, N. gaditana*	Anti-inflammatory activity	Cha et al., 2008
Antheraxanthin	*Chlorella ellipsodea*	Anti-inflammatory activity	Soontornchaiboon et al., 2012
Echinenone	*Botrycoccus braunii*	Anti-oxidant property Prevents liver fibrosis	Tonegawa et al., 1998
Cryptoxanthin	*D. salina*	Reduces the risk of type 2 diabetes mellitus (T2DM)	Hozawa et al., 2006
Chlorophyll			
Chlorophyll A	*A. flosaquae*	Content high level of heme and helps in RBC production	Bishop and Zubeck 2012
Phycobiliproteins			
Phycocyanin	*Spirulina*	Food coloring agent	Sonani, 2016
Phycoerythrin	*Porphyridium*	Food and cosmetics coloring agent	Sekar and Chandramohan, 2008
Allophycocyanin	*Spirulina platensis*	Used in immunoassays for cell culture and antiviral compound	Lobban and Harrison, 2009
Phenolic compounds and other volatiles			
Phenolic compounds	*Chlorella, Nostoc, Anabaena, Tolypothrix, Chlamydomonas*	Antioxidant, anti-inflammatory, antimicrobial activities	Plaza et al., 2010
Animal feed	*Chlorella, Spirulina, Tetraselmis, Isochrysis, Pavlova, Phaeodactylum, Chaetoceros, Nannochloropsis, Skeletonema, Thalassiosira*	Used as animal feed for livestock	Yaakob et al., 2014

needs comprehensive up and down systems such as roots and stems to sustain their reproductive constituents, whereas the whole system of seaweed can be used as productive tissues. Algae possess little leaf-like morphology and produce less uneatable material.

- Competency to grow indefinitely: Whereas certain microalgae species go through sexual cycles under some circumstances, they usually propagate asexually through a cellular division called cell fission. To retain optimum culture density, continuous harvesting frequencies can be regulated. This is especially important in continuous cultivation systems like raceway ponds and bioreactors, in which harvesting activities are adjusted to balance productivity. In contrast to terrestrial plants, which have a seasonal plantation, growing, harvesting, and further processing cycles, in a continuous production system, constant processing allows uninterrupted biomass production.
- Microalgae are easy for transporting, storing, and harvesting. Pumping microalgae is simple, but working with terrestrial plants is more complicated and frequently involves the purchase of expensive crop-specific equipment. These characteristics make microalgal crops suitable for extensive industrial exploitation, including nutraceuticals and animal feed.
- Algal culture fulfills the requirement of aquaculture feed. Algal cultivation has the potential to sustain more beneficial species, such as shellfish and several fishes. This attribute can be achieved specifically or through bulk cultivation, e.g., in interconnected mixed aquaculture. In this system, algae utilize animal waste to produce aquaculture nutrition for the cultivation of various fish species.
- Algae can flourish in conditions that are undesirable for several other plants; such environmental circumstances include non-arable land, ponds, and seafloor waters.
- Microalgae's fast doubling time helps them to advance research and innovation at higher levels of magnitude than domestic crops. Besides this, several scientific evidence revealed the pilot-scale and low-cost projects can be effectively transformed to industrial-level food processing.
- Food processing from algae has a specialized societal attribute. Conventional microalgae and seaweed cultivation and processing have benefited societies and specific communities within those communities, including the coastline fisherwomen mentioned in the article by Periyasamy et al. (2014).

10.5   ALGAL CARBOHYDRATES AND POLYSACCHARIDES

Polysaccharides make up most of the seaweed cell wall, which accounts for around half of the dry mass of the algae (Stiger-Pouvreau et al., 2016). These biochemical constituents vary by seaweed community and are affected by a variety of biological and environmental parameters (Rioux and Turgeon, 2015). Cultivation time, algal species, and harvesting protocol have a major influence on polysaccharide functional characteristics and also important structural attributes, including molecular mass, nature of structural elements, sulphate group material, their orientation, form of glycosidic bond, and molecule configuration. Cell-wall polysaccharides and storage polysaccharides are the two main types of polysaccharides found in seaweed. Only brown seaweed, especially the Laminariales and Fucales families, contain laminarins, which are storage polysaccharides. Sulfate groups are present in certain polysaccharides, which are referred to as sulfated polysaccharides (SP). Seaweed, especially green algae, comprise ulvan; red algae have carrageenan; brown seaweed consist of fucoidans and major polysaccharides (Table 10.1). Because of their biocompatibility and availability, these polysaccharides are commonly used in biological and biomedical applications (Venkatesan et al., 2015), though their use as dietary supplements or active food components is yet to be fully exposed.

10.5.1 Ulvan

The green seaweed species, including *Ulva* and *Enteromorpha,* contains a water-soluble sulfated polysaccharide called ulvan. The cell wall of these species is comprised of D-glucose, D-xylose, glucuronic acid, and L-rhamnose (Rioux and Turgeon, 2015) with approximately 10–30% of the dry mass of ulvan polysaccharide, respectively (Lahaye and Robic, 2007). Further, ulvan separation and extraction can be achieved by water and ammonium oxalate (80–90°C) and then precipitated with ethyl alcohol (Lahaye and Robic, 2007).

10.5.2 Fucoidan

Brown seaweed contains sulfated polysaccharides called fucoidans, which are mostly made up of L-fucose. These sulfated polysaccharides (2–10% dry cell mass) are found primarily in the cell walls of brown algal species belonging to the Laminariaceae and Fucaceae families. Fucoidan's antiproliferative, antiangiogenic, and anticancer properties have been documented in recent findings, and its significance as an algal-derived anticancer drug *in vivo* and *in vitro* research was previously reported by Kwak (2014). Furthermore, in separated 3T3-L1 adipocytes, fucoidan prevents lipid deposition. Fucoidan can be promoted as one of the valuable substrates in bone-tissue engineering in the biomedicine field (Lowe et al., 2016).

10.5.3 Alginates

The cell walls of macroalgae like *Laminaria* and *Ascophyllum nodosum* sp. contain alginates as one of the important polysaccharides (Rioux and Turgeon 2015). This is comprised of guluronic acids, hexuronic acids, and mannuronic (Andriamanantoanina and Rinaudo, 2010). Brownlee et al. (2009) state that alginates are being utilized in the food processing industries as a gelling component, emulsifier, food surface coating, and encapsulation. Furthermore, alginates have been shown to have a health-promoting influence on healthy colonic microflora (Brownlee et al., 2009). In bone tissue, engineering alginate-based structural biomaterials are used (Venkatesan et al., 2015). Furthermore, alginates are be-coming popular in the food processing sector.

10.5.4 CARRAGEENAN

Carrageenan is a type of sulfated linear polysaccharide that exists in nature in red algae. It is divided into three categories: κ, ι, and λ type. Carrageenan is typically isolated through the cell walls of *Kappaphycusalvarezii* and *Chondrus crispus*, which comprise 35–85% (dry mass) of carrageenan. The polysaccharide compo-sition of carrageenan is ammonium, potassium, magnesium, calcium, and sodium sulfated esters of units. Carrageenan has initially been examined for its biomedical applications, chemical modifications, and structural evaluation (Campo et al., 2009). The strongest rheological activities were obtained in a fermented food called "tofu," suggesting that carrageenan can act as an important food additive in changing food qualities (Shen and Kuo, 2017).

10.5.5 LAMINARIN

The $(1,3)$-β-D-glucan with presence of β $(1,6)$ branching makes up laminarin structure (Rioux et al., 2007). Laminarin has a molecular mass of about 5000 Da. When administered maternally or exclusively to the piglets, antimicrobial and anti-inflammatory effects are thought to be present in laminarin (Bouwhuis et al., 2017).

10.5.6 AGAR

This seaweed polysaccharide is commonly isolated from *Gelidium* and *Gracilaria* species and can be found in red algae. The cell walls of the red seaweed usually constitutes 20% agar. In certain regions of the world, *Gelidiella* and *Pterocladia* species are also employed to produce agar. Na, Mg, Ca, and K are major compo-nents of agar. In candy and confectionery products, agar would be a key substitute to gelatin (which is derived from animals).

10.5.7 NUTRACEUTICAL ADVANTAGES OF SEAWEED POLYSACCHARIDES

Seaweed polysaccharides were exclusively recorded for their exceptional anti-oxidant and anticoagulant properties (Tang et al., 2017). Certain types of poly-saccharides have anti-cancer properties and may be used to develop an anti-proliferative functional food. Indeed, polysaccharide ulvan was also demonstrated to have substantial cytotoxic effects on several cancerous cell lines, including HepG2, MCF-7, and HeLa (Thanh et al., 2016). *In vivo* in rodents' studies reported that an elevated level of sulfated ulvan has a stronger antihyperlipidemic influence than regular ulvan (Qi et al.,2012). Furthermore, it lowers the levels of triglycerides and low-density lipoprotein cholesterol in a treated animal experimental model. Moreover, feeding polysaccharides from *Ulva* sp. inhibits hamsters from losing their antioxidant weaponry and induction of atherosclerosis (Godard et al., 2009). It has already been stated that the medical advantages of marine algae mainly relate to cardiovascular protective effects (Larsen et al., 2011).

Brown seaweed fucoidans have been exhibited to exert antiviral efficacy against HIV, cytomegalovirus, and herpes simplex virus types 1 and 2. Furthermore, carrageenan prevents the transmission of viruses such as herpes and HIV in cells of the organism, indicating their use as antiviral components in functional foods. Numerous polysaccharides from algal origin are being used in the food sector in recent decades, like agar-agar, carrageenan, and alginates extracted from seaweed, and are commonly exploited as emulsifiers, stabilizers, and solidifiers in food. Fucoidan isolated from *Cladosiphon* (Brown algae) shields the gastric mucosa from the acidic stomach. As a result, fucoidan may be used as an antiulcer component in dietary supplements. Moreover, agar, alginates, and carrageenan are exceptional film-forming products that could be used in the food processing industry as edible coatings (Tavassoli- Kafrani et al., 2016). Furthermore, the addition of fucoidan and laminarin present in Laminaria to meat items showed increased consistency and lifespan of the flesh (Moroney et al., 2013). These reports indicate algal polysaccharides have the potential to be utilized in the food processing industry as potential medicinal and nutritional food items as anticoagulants, antioxidants, anticancer, and antihyperlipidemic weapons.

10.6   ALGAL LIPIDS, FATTY ACIDS, AND STEROLS

Polyunsaturated fatty acids (PUFA) extracted from the algal world are predominantly being used for nutritional applications (Pereira et al., 2012). In human beings, seafood is the principal source of PUFA, which is obtained directly from fish oil (e.g., salmon and mackerel). However, due to diminishing fish populations and rising consumer requirements, the long-term viability of fish supplies for PUFA exploitation is in doubt (FAO, 2010). As a result, renewable alternative supplies of PUFA for feed and food purposes are needed. Algae may be called lipid metabolism managers. As mentioned earlier several microalgal species have significant concentrations of various lipids, which have sparked attention in them as sustainable energy hydrocarbons for biodiesel as well as other bioenergy. Fortunately, several algae synthesize fatty acids and lipids, which are essential to human beings for their health (Farzaneh-Far et al., 2010).

PUFAs are a necessary component of human health in a variety of ways. Omega-3 fatty acids have currently received focus due to their valuable connection to health. The source of such compounds are seafood and several fish, but safety complications occur as fish can consume mercury or other pollutants and make them unsafe for ingestion (Spolaore et al., 2006). Food derived from the fish are not a feasible resource of PUFAs due to an increasing global population with the increasing demand for fish food. Indeed, algae are the main source in marine ecosystems of essential fatty acids (Venugopal, 2009). Algal PUFAs now have the benefit of being vegetarian. PUFAs are extracted from algae using solvent extraction. In certain cases, supercritical carbon dioxide is being used due to chemical preservation concerns (Venugopal, 2009).

The ingestion of PUFAs, specifically omega-3s, has been associated with several potential health benefits (Table 10.1). Consuming these PUFAs is important to inhibit atherosclerosis, arrhythmias, and chronic obstructive pulmonary disease, reduce the risk of stroke, reduce asthma attacks, combat manic-dependent disease, relieve cystic fibrosis symptoms, reduce relapses in Crohn's disease patients, and impede different cancers (Venugopal, 2009). The PUFAs, such as omega-3 Docosahexaenoic Acid (DHA), are a significant element of the brain, brainstem, and eye tissues (Venugopal, 2009). DHA is therefore essential for the growth of children and for developing kids. Some reports indicate improvements in children's behaviors, including children with attention deficit hyperactivity disorder (ADHD), whose diets were balanced with fatty acids (Venugopal, 2009).

PUFAs extracted from marine seaweed like DHA and eicosapentaenoic acid (EPA) are also displayed as a functional food in many health foods shops and pharmacies. *Crypthecodiniumcohnii*, a microalga, is a major heterotrophic DHA producer. DHA is found in high concentrations in the lipids of *C. cohnii*, *Schizochytrium limacinum*, an alga, has already been identified as a good DHA synthesizer (Chi et al., 2007). Research involving 485 people above the age of 55 who were experiencing age-related cognitive disorder (ARCD) was undertaken to see whether DHA could help ARCD. The research revealed that supplementing with DHA derived from *Schizochytrium* sp. strengthened episodic function of the brain in healthy older adults showing moderate memory issues (Yurko-Mauro et al., 2010). Chi et al. provide an example of how commercializing algae can go hand in hand with the development of sustainable fuels. Crude glycerol, an impure material of low commercial benefit, is a significant byproduct of biodiesel processing. Using *Schizochytrium* sp., glycerol was revealed to be an efficient alternative agent for producing DHA. The production efficiency of PUFA in algae can be improved by several different methods. Higher fatty acid synthesis can be achieved by lowering nitrogen levels. By introducing fatty acids and fatty acid precursor chemicals to the culture medium, the level of gamma-linolenic acid in spirulina has been improved (Kim et al., 2012). The use of the two-stage culture of light and dark was also helpful.

10.7   ALGAL PROTEINS, PEPTIDES, AND AMINO ACIDS

Proteins are an important dietary ingredient for livestock and individual survival and a significant carbon source that is around 16%. Proteins are essential for many biochemical processes in the animal body and represent a vital source of energy. Protein recycling is a mechanism in which proteins are constantly produced and processed in cells. The amino acid source is obtained either from food or cellular metabolism and is accumulated in the cell and circulatory system, ready to be integrated into proteins as well as various components or used for energy. For several decades, seaweed has been grown and consumed, especially in China, Japan, and the Republic of Korea. Seaweed has increasingly been produced and incorporated in the cuisines of South America, North America, Europe, and several other countries (FAO, 2003). It has emerged as a great resource of nutritious food because of its low energy level and considerable levels of dietary fiber, minerals, vitamins, and percentage of nitrogen substances such as amino acids and proteins (Fleurence, 1999; Ito and Hori, 1989).

10.7.1 FACTORS AFFECTING AMINO ACID AND PROTEIN SYNTHESIS IN ALGAE

Algal species such as *Porphyra* (nori), *Undaria* (wakame), and *Laminaria* (kombu) are all popular algae that are eaten by humans (FAO, 2002). According to USDA (2010), the mature seaweed *Undaria* sp. represents nearly 3.0% protein and has approximately 15.2% protein of dry weight based on an 80% moisture level. Protein level varies from 5% to 47% depending on the organism, environmental factors, ecosystem, harvesting period, and biochemical method used to determine protein and amino acid (AA) concentrations. Almost all AA are reported in seaweed proteins, which are different depending on the season. The maximum protein content was observed throughout the winter and early spring season, and was low over the summer and beginning of the autumn season (Galland-Irmouli et al., 1999). As per Denis et al. (2010), the content of protein differed in red algae *Grateloupia turuturu* on a seasonal basis. The degradation of phycobiliproteins can be associated with the lowest protein content in the summer. Proteins, in contrast to other polymers and compounds that are made up of a large number of amino acids, were subjected to significant changes over a year. Furthermore, various protein concentrations are said to be dependent on location; for instance, according to Yaich et al. (2011), species from Tunisia's littoral region hold approximately 50% more protein than species found in the Philippines. Renaud and Luong-Van (2006) reported an optimum level of protein was observed in red algae obtained within the summer season (4.8–12.8%), whereas it is considerably lower in the winter season.

Protein concentrations differed significantly between species, often even in the same species (McDermid and Stuercke, 2003). Brown algae have a poor protein content in common. It has also been estimated that it is less than 15%. (Fleurence, 1999). Red seaweed, on the other hand, has high levels of protein, which is mostly similar to that of other commodities like soybeans and eggs. The protein content showed some variations in red and green algae. According to Qasim (1991), the range of 21–28% was calculated in brown algae, 14–26% in green microalgae, and 11–24% in red seaweed. Green algae contain at least 1.5% protein. *Gracilaria changii*, a red alga, contains a moderately large quantity of protein, about 30–40% of dry mass (Norziah and Ching, 2000), which would be equivalent to the protein content in green peas (USDA, 2010). However, protein level varies greatly between species; for instance, the seaweed *Corallina officinalis* has a protein concentration of only about 7%. (Marsham et al., 2007). In general, the protein level of seaweed reduces in the following trend: red > green > brown.

10.7.2 THE AMINO ACID PROFILE OF SEAWEED

All essential amino acids (EAAs) can be found in seaweed (Matanjun et al., 2009). Algae (red and green) appear to be capable of contributing to reasonable levels of complete EAAs in the FAO/WHO (1991) model (Wong and Cheung, 2000). As per Matanjun et al. (2009), green seaweed had elevated levels of an amino acid (AA) whereas red and brown algae exhibited lower levels. However, several studies found that EAAs made up nearly half of complete AAs in red algae, implying a proportion of EAA to AA of 0.4–0.5. Gressler et al. (2011) found that the ratio of EAAs to NEAAs was between 0.7 and 0.9. These findings corroborated those found that *Palmaria palmata* comprised 26–50% EAAs of overall AA and had a protein content similar to those of egg protein. EAA levels in red and green microalgae were found to be 42–48% by Wong and Cheung (2000). Red seaweed had high levels of methionine and cysteine in comparison to green and brown algae (Qasim, 1991); however, red algae had low levels, below 0.3% and 0.1% of protein level (Gressler et al., 2011). The amino acid phenylalanine was found to be higher in three classes of algae (Matanjun et al., 2009). In contrast to Qasim's findings, alanine, glycine, proline, arginine, aspartic acids, and glutamic made up a significant portion in AAs fraction, while AAs methionine, tyrosine, and cysteine were found in smaller amounts (Gressler et al., 2011). Ortiz et al. (2006), on another side, observed that the brown alga *Durvillaeaantartica* had a significant amount of histidine and valine. Glutamic and aspartic acids are found in varying quantities in microalgae, ranging from 15% to 44%. On the other hand, in the case of NEAAs, aspartic and glutamic acid made up the majority of the AA. Glutamic acid is essential for the brain's correct functioning and metacognition. Aspartic acid serves as a source of energy for most essential metabolic pathways, including urea and Krebs (Braverman et al., 2003). However, it is important to note that researchers are unaware of total AAs, including cysteine and tryptophan, resulting in variations in the total AAs (Gressler et al., 2011). In addition, all classes of algae – grey, brown, and red – have the same level of NEAA (Matanjun et al., 2009), whereas nonprotein nitrogenous compounds found in seaweed include free nitrate, AAs, ammonium ions, chlorophyll, and nitrite, nitrogen, and nucleic acids. The AAs, such as aspartic acid and glutamic (Yaich et al., 2011), as well as alanine and glycine (Norziah and Ching, 2000), are likely to be associated with flavors and tastes.

10.8 VITAMINS

Vitamins are organic functional components that are required in very low concentration to the organism for the range of biochemical and physiological activities. Vitamins are usually divided into water- and fat-soluble ones based on their solvation. Individuals may eat foods supplemented with vitamins, such as functional ingredients, to maintain sufficient vitamin consumption in the daily diet, although vitamins are needed in very limited amounts. Furthermore, some vitamins derived from natural resources, mostly seaweed, possess antioxidant potential as well as many other medical benefits, including lowering blood pressure, preventing cardiovascular disease, and lowering cancer risk.

As already mentioned, vitamins are basic compounds that individuals cannot produce or can only synthesize in small amounts, so they must be derived from the diet. Vitamin depletion can be influenced by a lot of factors, including inadequate food consumption, intensified demand by some groups of people (e.g., individuals on dietary restrictions, smokers), incomplete absorption, and inadequate usage. Certain B group vitamins (B1, B2, B12) are commonly found in seaweed

(Table 10.1). Other B-complex vitamins are also available but in small or trace quantities (niacin, B6, biotin, folates). Vitamins C and E, as well as carotenoids, are abundant in some seaweed.

10.8.1 Composition of Vitamins in Seaweed

Just a few articles have been reported on the vitamin level and its bioavailability found in dietary microalgae. Algae usually comprise both water as well as fat, whereas some of them are only present in trace amounts. According to Ortiz et al. (2006), 100 g of marine algae contains greater than twice the daily requirement. The majority of red algae such as *Palmaria* and *Porphyra* exhibit a significant level of provitamin A as well as vitamins belonging to B-complex, which are often found in green algae. Brown seaweed (*Undaria, Laminaria*) possesses a lower vitamin concentration than green seaweed (*Undaria, Laminaria*), but it has a higher vitamin C level (Mabeau and Fleurence, 1993). Vegetarians can get enough vitamin B12 from certain seaweed (such as *Porphyra*).

Vitamins B1 and B2 are abundant in microalgae, particularly brown and red algae. Wakame and kombu had the greatest level of both vitamins between 0.3 and 0.24 mg B1/100 g of dry weight; 1.35 and 0.85 mg B_2/100 g of dry weight, respectively (MacArtain et al., 2007). In wakame, for example, the French Institut de Phytonutrition (FIP) reports a large value of these vitamins—5 mg B1/100 g of dry mass or 11.7 mg B2/100 g dry weight (MacArtain et al., 2007). In strictly vegetarian and vegan diets, vitamin B12 consumption is generally minimal; as a result, deficiency in this vitamin is common. When vitamin B12 levels in the diet are low, it can lead to decreased DNA methylation or increased homocysteine levels, both of which are risk factors for cardiovascular disease (Hernandez et al., 2003). Seaweed, products supplemented with it, or seaweed extracts are especially high in vitamin B12, making them an excellent source of the vitamin for vegetarians. As a result, vegans can avoid B12 depletion by consuming certain seaweed (nori) (Hernandez et al., 2003). Red *Porphyra* sp. (nori) has the largest percentage of vitamin B12 in algae, with 133.8 mg B12 per 100 g dry weight, in the form that is effective in humans (Miyamoto et al., 2009). Takenaka et al. (2003) has found that the B12 level of this algae ranges from 12.02 to 68.8 mg/100 g DW. Green laver *Enteromorpha* sp. has the highest B12 material, followed by dulse, and small amounts in *Ulva* sp., hijiki, and wakame (MacArtain et al., 2007).

Vitamin C is abundant in green and brown algae, but not so much in red algae. *Enteromorpha flexuosa* and *Ulva fasciata* have been reported to have the maximum amount of vitamin C (McDermid and Stuercke, 2003). In wakame, higher levels of vitamin C (184.75 mg/100 g DW), sea lettuce, and red laver are seen, according to FIP (MacArtain et al., 2007). The group of Chan et al. (1997) discovered that freeze-dried algae *Sargassum hemiphylum* has a high amount of vitamin c of 150 mg/100 g dry mass, compared to the lowest level in heat-dried seaweed. Although algae, for example, don't possess intrinsic vitamin A, they do produce provitamins including b-carotene. The red macroalgae *Gracilariachanggi* and *K. alvarezzi*, as well as the brown algae kombu, had high levels of b-carotene with vitamin A concentration (5.2, 5.26, and 2.99 mg/100 g dry weight). Reasonable vitamin A concentrations have been evaluated in sea grapes, arame, wakame, and sea salmon (Hernández-Carmona et al., 2009).

10.8.2 Factors Influencing Seaweed Vitamins

The content and thus vitamin profiles of microalgae differ and are influenced by algal organisms, algal growth phase, geographical region and salinity, the weather of every year, light intensity, and water temperature (Norziah and Ching, 2000). Vitamin B12 level in seaweed, for example, differs widely between the same species (Norziah and Ching, 2000). Since light is a critical regulator of vitamin biosynthetic pathways in many instances, plants grown in intense light correspond to greater ascorbic acid levels (Smith et al., 2007). Furthermore, algae collected from depths of 9–18 m have a greater level of vitamin C than algae obtained from the littoral region (Smith et al., 2007). Certain environmental influences, like certain components of sea, can also influence the occurrence of vitamins in algae (Smith et al., 2007).

Vitamin degradation can be caused by storage and processing conditions, such as sunlight and oxygen exposure. Furthermore, mechanical processing like drying and sterilization, along with culinary methods, such as frying, baking, and roasting, may have a negative impact on vitamin contents due to water extrusion and elevated temperatures applied during such processes. This has been found in extremely unstable ascorbic acid, for example (Smith et al., 2007). As per Hernández-Carmona et al. (2009) vitamin assessment, significant variations in vitamin level occurred due to seasonal patterns, in Eisenia arborea, for example. The high concentration of certain vitamins such as A, B1, B 2, and vitamin C was reported in spring, while the lower concentration of vitamin E was seen during this season. Summer had the highest carotenoid quality, while winter had the minimum (Lovstad Holdt and Kraan, 2010).

10.9 PHLOROTANNIN

Phlorotannin (1,3,5-trihydroxy benzene) belongs to the family of polyphenolic bioactive molecules and is found in brown seaweed, contributing around 6–12% of the dry weight (Shibata et al., 2004). Polymerization of phloroglucinol oligomers with the acetate-malonate (polyketide) pathway will lead to the formation of phlorotannin. In coastal areas of Japan and Korea, *Ecklonia cava, Eisenia bicyclis,* and *Ecklonia kurome* are popular. *Eisenia bicyclis* contain 2–5% of the total dry weight of the algae in the form of phlorotannin (Table 10.1). According to HPLC findings, *E. bicyclis* contains phloroglucinol (0.9%), eckol (7.5%), phloroglucinol tetramer (4.5%), dieckol (23.4%), phlorofucofuroeckol A (21.8%), and 8,80-bieckol (24.7%), as well as 17.3% of unidentified polyphenols (Shibata et al., 2004). Individual organisms, geographic areas, and isolation strategies can all affect phlorotannin levels in marine algae (Jégou et al., 2015). Further, fucodiphlorethol G has been initially extracted from *E. cava* and acetylated to generate multiple forms (Young et al., 2007).

Phlorotannins are often extracted by traditional extraction of organic solvents and chromatographic techniques for the purification of the molecules from algae, especially brown algae (Kim et al., 2016). In addition, NMR spectroscopy is commonly used to identify the compound structure. Saravana et al. (2017) has utilized supercritical carbon dioxide to isolate phlorotannin. Melanson and Mackinnon (2015) have reported the most popular tool for characterizing phlorotannin is HPLC-HRMS (high-resolution mass spectrometer). The Folin–Ciocalteu assay is a widely employed method for determining the amount of phlorotannin in a specific population. To obtain polyphenol substances from Irish macroalgae *Pelvetiacanaliculata, Ascophyllum nodosum, Ulva intestinalis,* and *Fucus spiralis,* Tierney et al. (2014) established a pressurized liquid extraction method, compared to the conventional solid-liquid extraction methodologies with respect to their antioxidants. The findings show that the pressurized liquid extraction obtained higher polyphenols than conventional processing approaches (Tierney et al., 2014).

Because of the structural features of a phlorotannin, beneficial pharmacological properties are present such as antioxidants, anti-HIV anti-proliferative, radio-protective, antidiabetic, and anti-allergenic (Tierney et al., 2014). Recent research has found that phlorotannin isolated from marine brown algae plays a significant role as bioactive compounds in human nutrition and health. Designing and developing food ingredients and pharmaceuticals to assist in the prevention or regulation of diet-related chronic malfunctions is a hopeful prospect as a result of their essential biological roles and health-promoting properties. Phlorotannin has a strong capacity as a functional component in the nutritional supplement, cosmeceutical, and healthcare commodities.

10.10 PIGMENTS AND MINOR COMPOUNDS IN ALGAE

Green seaweed (*Chlorophyceae*), red seaweed (*Rhodophyceae*), and brown seaweed (*Phaeophyceae*) are the three main classes of seaweed, divided on the basis of their coloration due to their pigment composition (Holdt and Kraan, 2011). Chlorophylls, carotenoids, and phycobiliproteins are the three fundamental groups of pigments present in algal species, respectively. Several pigments have now been identified in seaweed under such groups, each with unique biological activities and medical implications (Holdt and Kraan, 2011). Chl a and b, β-carotene, 9-cis-carotene, trans carotene violaxanthin, antheraxanthin, zeaxanthin, astaxanthin, lutein, and cryptoxanthin have all been found in the green seaweed *U. lactuca* (L.) in decreasing order (Abd El-Baky et al., 2008) (Table 10.1). Algal pigments are most often utilized as food colorants and in dietary additives in the industrial sector in 2009, the worldwide demand for food colors was projected to be over $1.5 billion (Leatherhead Food Research, 2010). Natural derived food colorants are presently used in around 50,000 tones around the globe. The world market valuation is anticipated to hit $1.6 billion at the end of this decade, up 10% from current levels. Natural colors and coloring fruits and vegetables accounted for nearly a third of the global demand for natural colors between 2005 and 2009, with natural colors and coloring food products anticipated to account for almost all the future expansion. Food coloring accounts for roughly 67% of the market share, led by beverages

(28%) and alcoholic drinks (13%). Because of its bright and attractive color, β-carotene is highly demanded in various food processing sectors as a precursor of pro-vitamin A and for its health benefits.

10.10.1 CHLOROPHYLLS

Chlorophylls are photosynthesis green pigments found in both higher plants and algae, along with cyanobacteria. Chlorophyll a is required for photosynthesis in the respiratory chain of the thylakoid (García Sartal et al., 2012). Chlorophyll is believed to be transformed in prepared vegetable foods into pheophytin, pyrophytin, and pheophorbide and human intake. These derivatives have an antimutagenic activity and may be useful in cancer reduction. Pheophorbide had higher cellular absorption and suppression of myeloma cell number than pheophytin. However, dependent on the quantity of chlorophyll derivative interacting with cells, pheophytin was found to be slightly cytotoxic compared to pheophorbide (Chernomorsky et al., 1999). Chlorophyll intensity showed three times greater in algae grown in harbour zones than in an open sea location, according to initial research (Larsen et al., 2011). The brown species recorded a chlorophyll A composition of 0.5–2 g/kg on a dry matter. Chlorophylls, for example, one of the ingredients in the production of jam, jelly, toffees, and ice cream, documented in the European Union under the E-number E-140 (García Sartal et al., 2012).

10.10.2 CAROTENOIDS

Carotenoids are the most common natural pigments, and they form an important component of pigments in algae, terrestrial plants, and photosynthetic bacteria. In yellow, orange, and red wavelengths, they reflect photosynthesis pigments. Carotenes are natural isoprene pigments that form normal, strongly conjugated, 40-carbon structures and are polymerized enzymatically with 15 conjugated double bonds. Carotenoid biosynthesis in algal species is through a mevalonic acid-based metabolic pathway. Species of green marine algae contain lutein, β-carotene, neoxanthin, violaxanthin, and zeaxanthin, whereas red marine algae comprise α and β-carotene, zeaxanthin, and lutein (Mehariya et al., 2019). Species of brown seaweed include β-carotene, violaxanthin, and fucoxanthin (Haugan and Liaaen-Jensen, 1994).

10.10.3 BETA-CAROTENE

Beta-carotene is a common food ingredient that both acts as colorant and has medical benefits. E160a is the European Union's authorized additive (García Sartal et al., 2012). Special carotenoids often serve as precursor intermediates for vitamins, which are identified based on their biological function rather than their structure integrity, (Mayne, 1996). Vitamins have a wide range of biochemical roles, including functioning as hormones, antioxidants, cell signaling mediators, and growth and differentiation moderators. The gastrointestinal enzyme β-carotene 15,15'-monooxygenase catalyzes the conversion of provitamin A carotenoids to retinal (Lindqvist and Andersson, 2002). On a dry matter basis, the β-carotene concentration varied from 40 to 4500 mg/kg (or ppm), with *Porphyra agardh* holds the greatest level at 456 ppm, and *Palmaria* gets the subsequent highest at 455 ppm. *Palmaria* has seasonal fluctuations in carotenes, with the maximum level in the mid-summer (450 ppm) and the smallest in the winter (40 ppm), respectively (Yuan, Roubos et al., 2008). Algal carotenoids' antioxidant potential has been shown to support the prevention of oxidative load (Yuan, Roubos et al., 2008). In addition, research suggests that carotene, which has provitamin A involvement, can help to prevent cancer, particularly respiratory cancer (Hosokawa et al., 1999). Recent research has found a connection between eating a carotenoids-rich diet and a lower threat of cardiovascular disease, cancer, and ophthalmological disorders (Hosokawa et al., 1999). Several pieces of evidence have linked daily consumption of carotenoids to a reduced risk of certain diseases (Aust et al., 2005). Carotenoids effectively protect the body from photooxidation caused by UV radiation (Aust et al., 2005). Astaxanthin has shown positive evidence in the avoidance of several human clinical manifestations, including photo-oxidation of the skin due to UV, inflammation, mammary and prostate carcinogenesis, *Helicobacter pylori* infection lead ulcers, and several age-related disorders (Guerrin et al., 2003). Several of the optimistic medical and nutritional studies have suggested potential antioxidant capacity of carotenoids may be a critical aspect in lowering the prevalence of many diseases, particularly those that are thought to be induced by light (Astley et al., 2004). Although there are substantial epidemiological data that connect a higher intake of carotenoids from foods to lower chances of certain cancers, consumption timelines with artificial carotenoids do not appear to be beneficial for the health of people (Astley et al., 2004).

10.10.4 Fucoxanthin

Fucoxanthin is a xanthophyll that formed by an unusual allenic bond and 5,6-monoepoxide within the structure. Amongst the most common carotenoids present in plants and algae is fucoxanthin (Matsuno, 2001). The composition of seaweed differs depending on the environmental conditions and development process. Besides destruction in the drying method and preservation at the usual temperature, it is very stable over the stress of organic components. Fucoxanthin is prone to oxidation in its natural form (Haugan and Liaaen-Jensen, 1994). Fucus serratus has a gross carotenoid level of about 0.08% of the dried harvested samples, with fucoxanthin accounting lasting for approximately 70% (Haugan and Liaaen-Jensen, 1994). Fucoxanthin abundance in brown algal species varies from 170–750 mg/kg dry mass, with the greatest concentration in F. serratus L. (Tsukui et al., 2007). However, values of 3700 mg/kg have been recorded in *Sargassum horneri* (Turner) C.Agardh (Tsukui et al., 2007). When provided in drinkable water, fucoxanthin from *Undaria* predominantly decreases the proliferation of prostate cancer in humans, as well as the proportion of malignant cells in animal cancer models and the total number of tumors per animal (Miyashita and Hosokawa, 2008). Recent investigations have shown anticancer influence, comprising prevention of human leukemia cell line multiplication (HL-60), and have led to apoptotic cell death

(Miyashita and Hosokawa, 2008). The decrease of white adipose tissue in rodents and obese diabetic mice model was illustrated by Maeda et al. (2005), once fed with fucoxanthin from brown algae. Fucoxanthin-rich diets have reduced total cholesterol in mice and increased the expression of a thermoregulatory protein in adipose tissues. Extracted fucoxanthin prevented cellular lipid uptake in 3T3-L1 adipocytes. This indicates that fucoxanthin is a useful organic component for preventing obesity (Miyashita and Hosokawa, 2008). Antioxidant function, anti-inflammatory potential, neuroprotective activities, antiangiogenic properties, and skin defensive influence are among the certain biological functions attributed to fucoxanthin (Miyashita et al., 2012).

10.10.5 PHYCOBILIPROTEINS

Phycobiliproteins are soluble in water, in contrast to chlorophylls and carotenoids. They form clusters called phycobilisomes on the thylakoid surface instead of trapped inside the membrane. Various configurations in the two main phycobilins, phycocyanobilin (blue) and phycoerythrobilin (red), can absorb multiple wavelength ranges, leading to distinct spectroscopic analysis (Lobban and Harrison, 2009). R- or B-phycoerythrin, allophycocyanin, and phycocyanin are common names for the isolated pigments. Phycobiliproteins play a significant function in the photosynthetic phase of the algal group *viz.* Rhodophyta, Cyptophyta, and Cyanophyta, which are included inside phycobilisomes (Aneiros and Garateix, 2004). Phycoerythrin constitutes a significant portion of red algal protein molecules, with concentrations as high as 1–1.5% and 0.5% for *Gracilaria tikvahiae* McLachlan and *P. palmata*, respectively, on a dry weight basis. These pigments are commonly incorporated as natural organic colorants in chewing gum and milk products as well as in cosmetics, including eyeliners and lipsticks (Sekar and Chandramohan, 2008). A Japan-based company known as Dainippon Ink & Chemical Inc. exports natural food coloring phycobiliprotein phycocyanine, and India-based Parry Nutraceuticals uses their brand Parry Blue (Parry Nutraceuticals, 2012). Distinct phycobiliproteins were proven to be antioxidant, neuroprotective, anti-inflammatory, hepatoprotective, anti-obesity, antiviral, hepatic, antitumor, atherosclerosis, lipase inhibition, and serum lipid reduction (Sekar and Chandramohan, 2008).

10.11 ALGAL ANTIOXIDANTS

Unsurprisingly, there is significant evidence of research on algal species as reservoirs of natural antioxidants for human nutrition. Continuous redox disequilibrium is needed for photosynthetic energy acquisition and transformations, resulting in the development of reactive species, which can reduce life expectancy, well-being, and fitness. Both algal forms, micro and macro, similar to other living things, have biological substances and biocatalysts that protect them from oxidative stress caused by reduced reactive free radicals, such as superoxide radical ionic species, hydrogen peroxide, hydroxyl ion, and singlet oxygen (Halliwell and Gutteridge, 1996). Although the antioxidant advantages of many terrestrial plant sources have been identified, little is learned concerning the antioxidant properties of algal products. Despite the immense variety of natural antioxidants found in seaweed, their active oxygen species degradation is complicated (Cornish and Garbary, 2010); however, the implication of advantageous outcomes from human ingestion of such substances is much less clear. The development of enzymatic antioxidants and the generation of metabolites that function as sacrificial foragers of free radicals are both examples of antioxidant potential. In mice brown algae, polyphenolic compounds prevented UVB radiation-caused skin cancer progression (Hwang et al., 2006), and although the biocompatibility has yet to be evaluated in humans, it does confirm that algal products are important functional foods. Superoxide dismutase causes elimination of anions of superoxide radicals, whereas catalases and peroxidases transform H_2O_2 to water; they are the main key enzymes in minimizing oxidative damage in a green world. Peroxidases utilize a reducing agent to transform hydrogen peroxide to water, whereas catalase has a Fe-containing heme cofactor. Cu and Zn, especially Fe, are being used as enzyme cofactors in a variety of human biochemical activities. The influence of consumed antioxidant enzymes in the animal model system is by absorption of metal cofactors throughout the gastrointestinal tract, as they are absorbed in the intestine. The effects of any indigestible enzyme or extracted metal cofactors on the gastrointestinal microbiota have still not been reported. In metazoans like mammals, there seems to be a closer connection between selenium in nutrition and antioxidants strength. Selenium (Se) is required for the synthesis of Se-requiring glutathione peroxidase, which is then employed to metabolize lipid hydroperoxides and hydrogen peroxide in eukaryotic organisms and also some algae (Gobler et al., 2011). Se is rarely included in evaluations of the elementary components of microalgae and macro-algae (Quigg et al., 2011), although it exists in both cases. Algae accumulate Se rapidly (Cases et al., 2001), and Se deficiency in rats can be treated with the incorporation of Se-rich *Arthrospira*, as shown by enhanced expression and activity of glutathione peroxidase (Se-containing) in the liver and kidneys (Cases et al., 2001). Moreover, in another study in rats, the enzyme catalyst improved higher than those of the same dose of Se-rich cyanobacterium, given selenomethionine or selenite, due to a decreased biological availability of the cyanobacterial Se. The mechanisms that control the amount of Se in algal nutrition and its accessibility are exciting research topics. *In vitro* and *in vivo*, algae produce a wide variety of molecules that can scavenge free radicals. Phytochemicals, such as phlorotannin (Shibata et al., 2007) and halophenols, as well as sulfated polysaccharides and alginate (Zhao et al., 2012), excavate free radicals. Nwosu et al. (2011) validated and expanded earlier studies indicating the antioxidant properties of polyphenolic products from four dietary marine seaweed species in preventing cell growth and -glucosidase function in Caco-2 colon cancer. In comparison to the various algae, *U. lactuca* had less abundance of polyphenols, but brown and red algal phenolic extracts served antioxidants. Bromophenols derived from the red seaweed *Vertebrata lanosa* substantially suppressed oxidative stress and lipid peroxidation in the fetal lung (MTC-5) and hepatocellular liver carcinoma cultures from humans, according to Olsen et al. (2013). Bromophenol was demonstrated to be able to penetrate cells and thus theoretically pass from the gastrointestinal tract into circulation in this situation. Overall, further research into the *in vivo* evaluation of the antioxidant powers of phenols and certain algal food components in humans and other mammals is needed. The phlorotannin present in brown algae has remarkable but unpredictable antioxidant potential (Wang et al., 2014), The majority of research was done *in vitro*, with the majority of it focusing on the phlorotannin impact on carbohydrate hydrolyzing enzymes. Kawamura-Konishi et al. (2012) have observed that four *Sargassum* species of phlorotannin extracts effectively suppress the α-amylase in saliva *in vitro,* and rat pancreas pulmonary α-glucosidase action with amylopectin has been blocked by the novel phlorotannin from Sargassum patens. Oral dosing of extracts prevented lipid peroxidation in the liver, red blood cells, plasma, and KK-Ay mice kidney, suggesting that phlorotannin's antioxidant function has positive health benefits in minimizing oxidative stress diabetes (Iwai, 2008). The effects of food-grade phlorotannin extracted from *Ascophyllum nodosum* in clinical findings revealed phlorotannin intestinal alteration *in vitro*, the presence of phlorotannin compounds in plasma and urine, and a substantial improvement in the cytokine IL-8 (Corona et al., 2016). At all levels, from identification between species to impact on intestinal microbiome and transportation throughout the gastrointestinal tract to impacts on human physiology, the efficacy of algal foods is well established. Over the next decade, this will be an important field of emerging science.

10.12 DESIGN OF HEALTHIER FOODS AND BEVERAGES CONTAINING WHOLE ALGAE

Healthier products can be developed by following a range of potential strategies by initiating qualitative and/or quantitative alterations in foods and beverages. Compared to other modifications, product processing strategies constitute an important step and are particularly promising. The baseline strategy is to boost the concentration of compounds having favorable physiological properties and, on the other hand, reduce the quantity of others having unfavorable health benefits. There is scope for use of algae in currently marketed processed food products due to their functional effects provided by various bioactive compounds and their technological prospects in product reformulation for various food products (Wells et al., 2017). The algae species diversity provides nutritional components like high-quality polysaccharide fractions, protein, different vitamins, a high percentage of necessary unsaturated fatty acids, microelements (Ca, Na, Mg, P, K, Fe, and Zn), polyphenols, carotenoids, tocopherols, and dietary fiber (DF) (Bocanegra et al., 2009). From a nutritional, perspective, algae are low-energy foods, and possibilities to alter their composition for a desired bioactive substance are affected by a number of factors (habitat, species, growth conditions, maturity, environmental changes, or handling or pretreatment methods) (Holdt and Kraan, 2011). From a medicinal and pharmaceutical perspective, algae has great diversity of physiologically active substances, such as polysaccharides with anticoagulant, antitumor, and antiviral effects, and algal DF showing antihypertensive, hypocholesterolemic, and antioxidant effects. Also, algal proteins exert physiological effects in human cells as anti-inflammatory, metastasis-induced cancer cell apoptosis and differentiation, mineral binding, antibiotic, lipase activity inhibition, anti-HIV activity, inhibition of platelet aggregation, antioxidant action, etc. Much of the chronic disease risk reduction is derived from epidemiological evidence relating to Asian and Western populations. There is a low risk of cardiovascular or different diet-related cancers due to habitual and higher proportion of algae in their diet. This has led to increased interest in research and development of functional foods from marine algae (Yuan et al., 2011). One of the disadvantages of an increase in marine algal consumption involves the probable risk associated with higher heavy metals concentrations (Bocanegra et al., 2009).

10.12.1 Use of Algal Components in Foods and Beverages as Healthy Ingredients

Apart from traditional Asian food recipes, which mainly constitute seaweed like *Porphyra* sp., *Undaria sp., Laminaria sp., Enteromorpha sp.*, etc., a variety of products containing algae are being developed and can be seen on shelves worldwide (Kim et al., 2004). In meat and fish products, various algae like *Undaria pinnatifida, Himanthalia elongata, Porphyra umbilicalis, Laminaria japonica, Sargassum thunbergii, Gelidium amansii,* and *Kappaphycus alverezii* are used (López-López et al., 2011; Cofrades et al., 2012). Algae in meat and fish products act as functional ingredients by providing antioxidant, low-salt, low-fat, low cholesterol, and gel/emulsion characteristics (Bocanegra et al., 2009).

Another nutrient-rich food that supplies a considerable level of several nutrients in the diet is various dairy products, which provide calcium, potassium, magnesium, phosphorus, zinc, amino acids, protein, vitamins like A, D, B_{12} and riboflavin. In dairy products like cheese, calcium is locked due to casein; hence, reformulating and incorporating algae improves various essential nutritional qualities. In other cases, the addition of algae helps individuals who lack enzymes related to casein-degradation and who are unable to reabsorb calcium show hypocalcaemic type (Anderson and Sjöberg, 2001). Dairy products incorporated with brown seaweed (*Laminaria sp.*) showed iodine enrichment, showed protection against radiation, and helped in dealing with endemic goiter. The addition of *Chlorella* (around 1%) in cheese showed some differences in texture and color in comparison to control (Jeon, 2006). Cheese added with *Chlorella* powder showed an increase in lactic acid bacterial counts than in the control cheese.

In many countries, cereal-based products (noodles, bread, pasta, etc.) with other bakery foodstuffs are their staples and broadly accepted due to their lower cost, easy preparation, sensory traits, versatility, high nutrition quality, and extended shelf-life. Many of these food products are developed and used for improvement in the nutrition and health benefits of these products. Macroalgae like *U. pinnatifida, Sargassum marginatum, Monostromanitidum, Ulva lactuca, Laminaria,* and*Enteromorpha compressa,* and microalgae like *Chlorella* and *Spirulina* are added in various cereal-based products, as shown in Figures 10.4 and 10.5 (Prabhasankar et al., 2009; Gade, 2018).

If we consider the food beverages industry, there is increasing demand for algal and other natural ingredients to improve nutritional deficits, nutrient content, health, texture, color, and appearance, and this is seen as the fastest-developing segment in

FIGURE 10.4 Cookies prepared by addition of *Spirulina* dry powder.

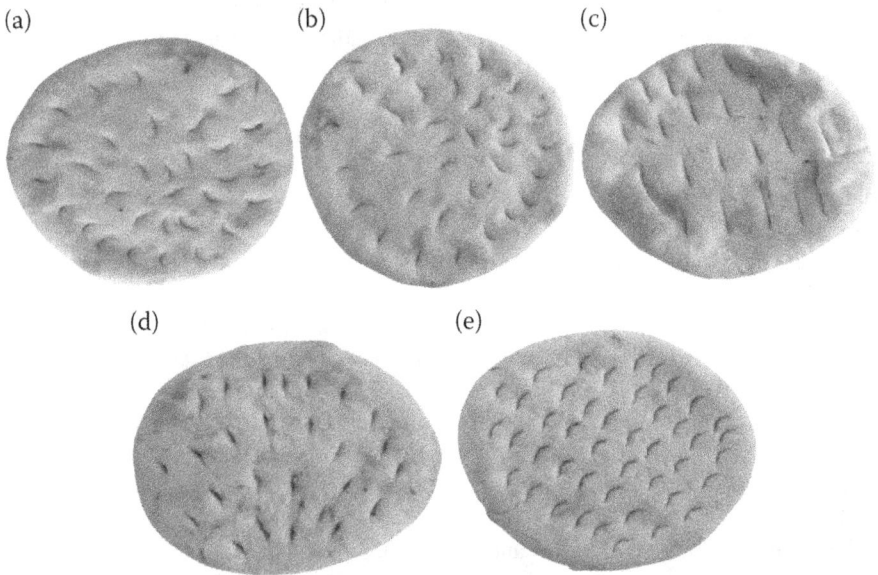

FIGURE 10.5 Pizza bases prepared by enrichment with *Spirulina* powder at a various concentration [(a) 0, (b) 0.5, (c) 1, (d) 1.5, and (e) 2%] (Gade, 2018).

the functional food market. Currently, if we consider the drinks and beverage market, the incorporation of algae shows varied challenges. Also, numerous reports have been found in the literature, and many patents are registered for the development of algae-added beverages. Seaweed like sea mustard, hijiki, sea tangle, wakame, sea trumpet, and sea lettuce are being added to various beverages, which provides nutritional and health benefits (Cofrades et al., 2013). In another study using *Gracilaria fisheri* and *Lactobacillus plantarum* DW3, Prachyakij et al. (2008) used four different fermentation processes and made a fermented beverage. These lactic acid fermented algae beverages contain more microelements, such as iron, copper, and zinc, antioxidants, polyphenols, etc., and have no changes in the organoleptic characters of the drink. For extracting the cellular liquid from fresh algae, different mechanical processes (centrifugation, cryo-grinding and ultra-filtration) were employed. These algal beverages were effective in tackling the problems faced due to lifestyle-related diseases (cardiovascular, cancer and dia-betes) and had an antihypertensive effect (Ścieszka and Klewicka, 2019).

Durmaz et al. (2020) incorporated three different spray-dried microalgae (*Nannochloropsis, Porphyridium,* and *Diacronema*) to ice cream formulation (0.1, 0.2, and 0.3% concentrations). The various parameters considered were color, flow behavior, melting sensory, and functional properties. Results showed that flow behavior, sensory characteristics, and ice color were markedly affected by micro-algal species with their used concentration. Also, a high concentration of micro-algae leads to phenolic contents increase of the ice cream samples.

10.12.2 NUTRITIONAL QUALITY STANDARD, REGULATIONS, AND CHALLENGES FOR ALGAL PRODUCTS

The microalgae-based products developed or proposed for human or animal diet and nutrition are subjected to a range of regulations and standards. Presently, for unconventional foodstuffs, there are various international testing programs with different standards, which are prerequisites before consumption. Out of various microalgae used, only for *Spirulina* stipulated legislative standards have been formed by a few countries so far. Before approval of algae biomass or developed products for human and animal consumption, Becker (2007) suggested several steps needed to be evaluated. It seems helpful to make available the following specifi-cations: i) The proximate chemical composition for nutritional factors like quality and quantity of protein and EAAs; ii) analysis of toxic biogenic (phycotoxins, and other toxicants) and non-biogenic (heavy microelements) and substances and toxic compounds, iii) the sanitary analyses (microbial contamination), and iv) the study of physiological, toxicological, and safety evaluations with model experimental animals.

10.13 PREBIOTIC AND PROBIOTIC USE OF ALGAE AND ALGAE-SUPPLEMENTED PRODUCTS

Prebiotics are defined as "microbial digested specific fermented substances that alter specific alterations in the gastrointestinal microflora either in the composition and/or activity that bestow benefits to host well-being" (Roberfroid, 2007). These selectively fermented substances, known as prebiotics, include increased fermentable poly-saccharides content or glycoprotein. They act as a growth stimulant on bifidobacteria and some probiotic gram-positive bacteria (*Lactobacillus* sp., *Streptococcus* sp.,

Bifidobacterium sp., *Lactococcus* sp., and *Saccharomyces* sp.). The commonly used prebiotics in the human and animal diet are lactulose, galacto-oligosaccharides (GOS), arabinose, fructooligosaccharides (FOS), mannan-oligosaccharides, raffinose, inulin, xylooligosaccharides, lactosucrose, malto-oligosaccharides (MOS), and resistant starch (RS), present in onion, garlic, chicory, asparagus, bananas, tomatoes, artichoke, leek, etc. (Gupta, 2017).

As macroalgae are rich in polysaccharides (25–75% of the dry weight), most of these polysaccharides act as dietary fibers due to the inability of the human gastrointestinal tract to digest them, which make them an attractive choice as prebiotics. Similarly, microalgal polysaccharides, which are polymeric carbohydrates (CHO), are resistant to being hydrolyzed in the small intestine (SI) of humans by the endogenous enzymes; hence, they also can act as efficient prebiotics. Also, microalgae constitute approximately 50–85% water-soluble fiber (Michaud, 2018). There is scientific evidence that these algal polysaccharides and other substances alter and affect the microbiota balance in the human digestive tract (Seema and Arun, 2011). These prebiotic substances and their microbial fermented derivative compounds have a positive impact on health like antimicrobial effects on pathogens, improved mineral absorption; anti-inflammatory, immune system stimulation; antiallergenic, etc. (Rodrigues et al., 2010). In macroalgal polysaccharide alginic acid found in members of Phaeophytahad, a positive impact was found on an increase in populations of bifidobacteria and decreasing levels of pathogenic bacteria was found in fecal matter. Yuan et al. (2011) reported in red seaweed *Kappaphycus striatum* on the chemical modification of kappa-carrageenan created sulphated, acetylated, and phosphorylated derivatives, which were the most effective against tumors. In green seaweed-like *Ulva* and other members of Chlorophyta, the polysaccharide ulvan is present upon fermentation by bifidobacteria in the gut, which act as a growth promotor involved in the growth of intestinal epithelial and repair mechanism of wounds. Nwosu et al. (2011) reported that inhibition of amylase and glucosidase enzymes due to extracts from *Ascophyllum, Palmaria,* and *Alaria*, which can be useful in controlling diabetes as a result of action on carbohydrate digesting enzymes. In the caecum of pigs, laminarin and fucoidan present in seaweed increased the counts of lactobacilli, by increasing the production of butyric acid in the caecum and colon and decreasing the count in the faeces for *Salmonella typhimurium*. The extracts of *U. pinnatifida* supplementation in broiler chicks showed an increased nutrient digestibility, and in the ileum, decreased population dynamics of coliform bacteria were seen (Sweeney et al., 2012).

On the other hand, microalgal oligosaccharides from *Chlorella* sp. and *Spirulina* sp. have confirmed antimicrobial of pathogenic bacteria and effects the growth-promoting activity in *Bifidobacterium animalis* and *Lactobacillus casei*, as well as induced generation of short-chain fatty acids (lactic and acetic acids) in these microorganisms (Sauer Leal et al., 2017). Currently, the research is focused on the development of combinations of new prebiotics and probiotics (synbiotics) that include the diet (Gourbeyre et al., 2011). However, there is a need for innovation in processing techniques, e.g., enzymatic treatment) and lower production costs to industrially produce prebiotics, which can be tackled using novel prebiotic resources like microalgae and seaweed (Michaud, 2018).

10.14 FUTURE PERSPECTIVES OF ALGAL DERIVED METABOLITES

For a long time, it has been known that seaweed produces biologically and functionally unique compounds. Seaweed-produced bioactive compounds have a wide range of biocompatibility, including antioxidant, anti-cancer, anti-inflammatory, anti-HIV and antidiabetic activities (Manilal et al., 2009). As a result, microalgae may be thought of as highly intriguing natural resources, including novel constituents with an extensive range of biological properties that might be utilized as functional components in a variety of sectors, including the food supplement, pharmaceuticals, and cosmetic industries (Plaza et al., 2008). Therefore, investigating secondary metabolites originating from seaweed, a distinct source of biological products, has shown to be a promising domain of functional constituent exploration. The reality is that the market for natural products is booming, but it now faces a new obstacle as producing algae on a massive scale without substantially damaging the marine ecosystem (Robic et al., 2009). The exploitation of scientific advancement in algal communities could surmount the adverse challenges and flourish the algal-derived compounds to compete with the other natural resources.

10.15 CONCLUSION

The algal biomass and feedstock are considered vast renewable and sustainable feedstock with enormous potential for mass-scale production of bioactive compounds used in functional food and nutraceuticals. The positive attributes of algae are: (i) they are an undiscovered species with ecological and chemical diversity; (ii) they have algal ability to adapt in diverse ecological niches and environmental conditions; and (iii) they have a less complicated reproductive cycle, for mass-scale production of biomass. Further insights are needed for the study of bio-accessibility of algal products considering their complex function bioactive metabolites, the processing techniques used for the preparation of the alga as food, digestibility of the specific algal matrix individually or in combination with a consortium of bacteria, hydrolysis due to their competent enzymatic activity, and the occurrence of anti-nutritional components that may obstruct or boost gastrointestinal absorption. Advancement in know-how for the bioavailability of algal foods can be accurately done with precise analytical methods and routine analysis of studies encompassing all these above-mentioned factors for the development of new functional algal foods and nutraceutical applications.

ACKNOWLEDGMENT

Mr. Chetan B. Aware and Prof. Jyoti P. Jadhav offer thanks to the Department of Biotechnology, Government of India for Interdisciplinary Program for Life Sciences (IPLS) (Reference no.: BT/ PR4572/INF/22/147/2012) for giving fellowship and research facility. Mr. Chetan B. Aware wishes to thank the Council of Scientific & Industrial Research (CSIR), New Delhi, India, for a Senior Research Fellowship award (Reference no.: 09/816(0046)/2019-EMR-I).

REFERENCES

Abd El-Baky, H.H., El Baz, F.K., El-Baroty, G.S., 2008. Evaluation of marine alga *Ulva Lactuca* L. as a source of natural preservative ingredient. *Electron. J. Environ. Agric. Food Chem.* 7, 3353–3367.

Anderson, J.J.B., Sjöberg, H.E., 2001. Dietary calcium and bone health in the elderly: Uncertainties about recommendations. *Nutr. Res.* 21, 263–268.

Andriamanantoanina, H., Rinaudo, M., 2010. Characterization of the alginates from five madagascan brown algae. *Carbohydr. Polym.* 82, 555–560.

Aneiros, A., Garateix, A., 2004. Bioactive peptides from marine sources: Pharmacological properties and isolation procedures. *J. Chromatogr. B Anal. Technol. Biomed. Life Sci.* 803, 41–53.

Astley, S.B., Hughes, D.A., Wright, A.J.A., Elliott, R.M., Southon, S., 2004. DNA damage and susceptibility to oxidative damage in lymphocytes: Effects of carotenoids in vitro and in vivo. *Br. J. Nutr.* 91, 53–61.

Aust, O., Stahl, W., Sies, H., Tronnier, H., Heinrich, U., 2005. Supplementation with tomato-based products increases lycopene, phytofluene, and phytoene levels in human serum and protects against UV-light-induced erythema. *Int. J. Vitam. Nutr. Res.* 75, 54–60.

Batista, A.P., Gouveia, L., Bandarra, N.M., Franco, J.M., Raymundo, A., 2013. Comparison of microalgal biomass profiles as novel functional ingredient for food products. *Algal Res.* 2, 164–173.

Becker, E.W., 2007. Micro-algae as a source of protein. *Biotechnol. Adv.* 25, 207–210.

Begum, H., Yusoff, F.M.D., Banerjee, S., Khatoon, H., Shariff, M., 2016. Availability and Utilization of Pigments from Microalgae. *Crit. Rev. Food Sci. Nutr.* 56, 2209–2222.

Bhalamurugan, G.L., Valerie, O., Mark, L., 2018. Valuable bioproducts obtained from microalgal biomass and their commercial applications: A review. *Environ. Eng. Res.* 23, 229–241.

Bhatia, S.K., Gurav, R., Choi, T.-R., Han, Y.H., Park, Y.-L., Park, J.Y., Jung, H.-R., Yang, S.-Y., Song, H.-S., Kim, S.-H., Choi, K.-Y., Yang, Y.-H. 2019. Bioconversion of barley straw lignin into biodiesel using *Rhodococcus* sp. YHY01. *Bioresour. Technol.* 289, 121704.

Bhatia, S.K., Gurav, R., Choi, Y.-K., Choi, T.-R., Kim, H.-j., Song, H.-S., Mi Lee, S., Lee Park, S., Soo Lee, H., Kim, Y.-G., Ahn, J., Yang, Y.-H. 2021a. Bioprospecting of exopolysaccharide from marine *Sphingobium yanoikuyae* BBL01: Production, characterization, and metal chelation activity. *Bioresour. Technol.* 324, 124674.

Bhatia, S.K., Mehariya, S., Bhatia, R.K., Kumar, M., Pugazhendhi, A., Awasthi, M.K., Atabani, A.E., Kumar, G., Kim, W., Seo, S.-O., Yang, Y.-H. 2021b. Wastewater based microalgal biorefinery for bioenergy production: Progress and challenges. *Sci. Total Environ.* 751, 141599.

Bishop, W.M., Zubeck, H.M., 2012. Evaluation of microalgae for use as nutraceuticals and nutritional supplements. *J. Nutr. Food Sci.* 2, 5–10.

Bocanegra, A., Bastida, S., Benedí, J., Ródenas, S., Sánchez-Muniz, F.J., 2009. Characteristics and nutritional and cardiovascular-health properties of seaweeds. *J. Med. Food* 12, 236–258.

Bouwhuis, M.A., Sweeney, T., Mukhopadhya, A., McDonnell, M.J., O'Doherty, J.V., 2017. Maternal laminarin supplementation decreases Salmonella Typhimurium shedding and improves intestinal health in piglets following an experimental challenge with S. Typhimurium post-weaning. *Anim. Feed Sci. Technol.* 223, 156–168.

Braverman, E.R., Pfeiffer, C.C., Blum, K., Smayda, R., 2003, Glutamate amino acids, in: Hirsch, C. (Ed.), *The Healing Nutrients Within: Facts, Findings, and New Research on Amino Acids.* Basic Health Publications, inc., Laguna Beach, pp. 165–200.

Brownlee, I.A., Seal, C.J., Wilcox, M., Dettmar, P.W., Pearson, J.P., 2009. Applications of Alginates in Food. *Alginates: Biology and Applications*, pp. 211–228.

Campo, V.L., Kawano, D.F., Silva, D.B. Da, Carvalho, I., 2009. Carrageenans: Biological properties, chemical modifications and structural analysis - A review. *Carbohydr. Polym.* 77, 167–180.

Carbonell-Capella, J.M., Buniowska, M., Barba, F.J., Esteve, M.J., Frígola, A., 2014. Analytical methods for determining bioavailability and bioaccessibility of bioactive compounds from fruits and vegetables: A review. *Compr. Rev. Food Sci. Food Saf.* 13, 155–171.

Cases, J., Vacchina, V., Napolitano, A., Caporiccio, B., Besançon, P., Lobinski, R., Rouanet, J.M., 2001. Selenium from selenium-rich Spirulina is less bioavailable than selenium from sodium selenite and selenomethionine in selenium-deficient rats. *J. Nutr.* 131, 2343–2350.

Cha, K.H., Koo, S.Y., Lee, D.U., 2008. Antiproliferative effects of carotenoids extracted from Chlorella ellipsoidea and Chlorella vulgaris on human colon cancer cell. *J. Agric. Food Chem.* 56, 10521–10526.

Chan, J.C.C., Cheung, P.C.K., Ang, P.O., 1997. Comparative Studies on the Effect of Three Drying Methods on the Nutritional Composition of Seaweed Sargassum hemiphyllum (Turn.) C. Ag. *J. Agric. Food Chem.* 45, 3056–3059.

Chelf, P., Brown, L.M., Wyman, C.E., 1993. Aquatic biomass resources and carbon dioxide trapping. *Biomass Bioenerg.* 4, 175–183.

Chernomorsky, S., Segelman, A., Poretz, R.D., 1999. Effect of dietary chlorophyll derivatives on mutagenesis and tumor cell growth. *Teratog. Carcinog. Mutagen.* 19, 313–322.

Chi, Z., Ma, C., Wang, P., Li, H.F., 2007. Optmization of medium and cultivation conditions for alkaline protease production by the marine yeast Aureobasidium pullulans. *Bioresour. Technol.* 08, 534–538.

Choi, Y.-K., Choi, T.-R., Gurav, R., Bhatia, S.K., Park, Y.-L., Kim, H.J., Kan, E., Yang, Y.-H., 2020. Adsorption behavior of tetracycline onto *Spirulina* sp. (microalgae)-derived biochars produced at different temperatures. *Sci. Total Environ.* 710, 136282.

Cian, R.E., Drago, S.R., De Medina, F.S., Martínez-Augustin, O., 2015. Proteins and carbohydrates from red seaweeds: Evidence for beneficial effects on gut function and microbiota. *Mar. Drugs.* 13, 5358–5383.

Cofrades, S., Lopez-Lopez, I., Jimenez-Colmenero, F., 2012. Applications of seaweed in meat-based functional foods, in: Kim, S.K. (Ed.), *Handbook of Marine Macroalgae: Biotechnology and Applied Phycology*. Wiley. pp. 491–499

Cofrades, S., Serdaroğlu, M., Jiménez-Colmenero, F., 2013. Design of healthier foods and beverages containing whole algae, in: Dominguez H. (Ed.), *Functional Ingredients from Algae for Foods and Nutraceuticals*. Woodhead Publishing Series, Burlington, pp. 609–633

Cornish, M.L., Garbary, D.J., 2010. Antioxidants from macroalgae: Potential applications in human health and nutrition. *Algae.* 25, 155–171.

Corona, G., Ji, Y., Anegboonlap, P., Hotchkiss, S., Gill, C., Yaqoob, P., Spencer, J.P.E., Rowland, I., 2016. Gastrointestinal modifications and bioavailability of brown seaweed phlorotannins and effects on inflammatory markers. *Br. J. Nutr.* 115, 1240–1253.

Denis, C., Morançais, M., Li, M., Deniaud, E., Gaudin, P., Wielgosz-Collin, G., Barnathan, G., Jaouen, P., Fleurence, J., 2010. Study of the chemical composition of edible red macroalgae Grateloupia turuturu from Brittany (France). *Food Chem.* 119, 913–917.

Deviram, G., Mathimani, T., Anto, S., Ahamed, T.S., Ananth, D.A., Pugazhendhi, A., 2020. Applications of microalgal and cyanobacterial biomass on a way to safe, cleaner and a sustainable environment. *J. Clean. Prod.* 253, 119770.

Durmaz, Y., Kilicli, M., Toker, O.S., Konar, N., Palabiyik, I., Tamtürk, F., 2020. Using spray-dried microalgae in ice cream formulation as a natural colorant: Effect on physicochemical and functional properties. *Algal Res.* 47.

Ekmekcioglu, C., 2002. A physiological approach for preparing and conducting intestinal bioavailability studies using experimental systems. *Food Chem.* 76, 225–230.

FAO, 2002. *Prospects for Seaweed Production in Developing Countries. FAO Fisheries Circular, 968*. FAO, Rome.

FAO, 2003. *A guide to the Seaweed Industry*. FAO Fisheries Technical Paper 441. FAO, Rome.

FAO, Fisheries and Aquaculture Department, 2010. *The State of World Fisheries and Aquaculture 2010*. Food and Agriculture Organization of the United Nations, Rome, p. 197.

FAO/WHO, 1991. *Report of a Joint FAO/WHO/UNU Expert Consultation*. FAO Food and Nutritional Paper 51. Protein Quality Evaluation. WHO Press, Rome

Farzaneh-Far, R., Lin, J., Epel, E.S., Harris, W.S., Blackburn, E.H., Whooley, M.A., 2010. Association of marine omega-3 fatty acid levels with telomeric aging in patients with coronary heart disease. *JAMA - J. Am. Med. Assoc.* 303, 250–257.

Fleurence, J., 1999. *Seaweed proteins: Biochemical, nutritional aspects and potential uses.* Trends Food Sci. Technol. 10, 25–28.

Gade, S.J., 2018. *Nutritional enhancement of pizza base by enrichment with Spirulina platensis*. Bachelors' degree diss, Yashwantrao Chavan Institute of Science, Satara, India.

Galland-Irmouli, A.V., Fleurence, J., Lamghari, R., Luçon, M., Rouxel, C., Barbaroux, O., Bronowicki, J.P., Villaume, C., Guéant, J.L., 1999. Nutritional value of proteins from edible seaweed Palmaria palmata (Dulse). *J. Nutr. Biochem.* 10, 353–359.

Ganesan, P., Matsubara, K., Sugawara, T., Hirata, T., 2013. Marine algal carotenoids inhibit angiogenesis by down-regulating FGF-2-mediated intracellular signals in vascular endothelial cells. *Mol. Cell. Biochem.* 380, 1–9.

García Sartal, C., Barciela Alonso, M.C., Barmejo Barrera, P., 2012. Application of seaweed in the food industry, in: Kim, Se-Kwon (Ed.), *Handbook of Marine Macroalgae: Biotechnology and Applied Phycology*, 1st ed. John Wiley and Sons Ltd. pp. 522–531.

Gobler, C.J., Berry, D.L., Dyhrman, S.T., Wilhelm, S.W., Salamov, A., Lobanov, A. V., Zhang, Y., Collier, J.L., Wurch, L.L., Kustka, A.B., Dill, B.D., Shah, M., VerBerkmoes, N.C., Kuo, A., Terry, A., Pangilinan, J., Lindquist, E. a., Lucas, S., Paulsen, I.T., Hattenrath-Lehmann, T.K., Talmage, S.C., Walker, E. a., Koch, F., Burson, A.M., Marcoval, M.A., Tang, Y.Z., LeCleir, G.R., Coyne, K.J., Berg, G.M., Bertrand, E.M., Saito, M. a., Gladyshev, V.N., Grigoriev, I. V., 2011. Niche of harmful alga Aureococcus anophagefferens revealed through ecogenomics. *Proc. Natl. Acad. Sci. U.S.A.* 108, 4352–4357.

Godard, M., Décordé, K., Ventura, E., Soteras, G., Baccou, J.C., Cristol, J.P., Rouanet, J.M., 2009. Polysaccharides from the green alga Ulva rigida improve the antioxidant status and prevent fatty streak lesions in the high cholesterol fed hamster, an animal model of nutritionally-induced atherosclerosis. *Food Chem.* 115, 176–180.

González-Sarrías, A., Larrosa, M., García-Conesa, M.T., Tomás-Barberán, F.A., Espín, J.C., 2013. Nutraceuticals for older people: Facts, fictions and gaps in knowledge. *Maturitas.* 75, 313–334.

Gourbeyre, P., Denery, S., Bodinier, M., 2011. Probiotics, prebiotics, and synbiotics: impact on the gut immune system and allergic reactions. *J. Leukoc. Biol.* 89, 685–695.

Gressler, V., Fujii, M.T., Martins, A.P., Colepicolo, P., Mancini-Filho, J., Pinto, E., 2011. Biochemical composition of two red seaweed species grown on the Brazilian coast. *J. Sci. Food Agric.* 91, 1687–1692.

Guerra, A., Etienne-Mesmin, L., Livrelli, V., Denis, S., Blanquet-Diot, S., Alric, M., 2012. Relevance and challenges in modeling human gastric and small intestinal digestion. *Trends Biotechnol.* 30, 591–600.

Guerrin, M., Huntley, M.E., Olaizola, M., 2003. *Haematococcus* astaxanthin: applications for human health and nutrition. *Trends Biotechnol.* 21, 210–216.

Guiry, M.D., 2012. How many species of algae are there? *J. Phycol.* 48, 1057–1063.

Gupta, C., 2017. Prebiotic efficiency of Blue Green Algae on probiotics microorganisms. *J. Microbiol. Exp.* 4, 11–12.

Gurav, R., Bhatia, S.K., Choi, T.-R., Choi, Y.-K., Kim, H.J., Song, H.-S., Lee, S.M., Lee Park, S., Lee, H.S., Koh, J., Jeon, J.-M., Yoon, J.-J., Yang, Y.-H., 2021. Application of macroalgal biomass derived biochar and bioelectrochemical system with *Shewanella* for the adsorptive removal and biodegradation of toxic azo dye. *Chemosphere.* 264, 128539.

Gurav, R., Bhatia, S.K., Choi, T.-R., Kim, H.J., Song, H.-S., Park, S.-L., Lee, S.-M., Lee, H.-S., Kim, S.-H., Yoon, J.-J., Yang, Y.-H. 2020. Utilization of different lignocellulosic hydrolysates as carbon source for electricity generation using novel *Shewanella marisflavi* BBL25. *J. Clean. Prod.* 277, 124084.

Gurav, R., Bhatia, S.K., Moon, Y.-M., Choi, T.-R., Jung, H.-R., Yang, S.-Y., Song, H.-S., Jeon, J.-M., Yoon, J.-J., Kim, Y.-G., Yang, Y.-H. 2019. One-pot exploitation of chitin biomass for simultaneous production of electricity, n-acetylglucosamine and poly-hydroxyalkanoates in microbial fuel cell using novel marine bacterium *Arenibacter palladensis* YHY2. *J. Clean. Prod.* 209, 324–332.

Hafting, J.T., Critchley, A.T., Cornish, M.L., Hubley, S.A., Archibald, A.F., 2012. On-land cultivation of functional seaweed products for human usage. *J. Appl. Phycol.* 24, 385–392.

Halliwell, B., Gutteridge, J.M.C., 1996. Book reviews. *Free Radic. Res.* 24, 495–496.

Haugan, J.A., Liaaen-Jensen, S., 1994. Algal carotenoids 54. Carotenoids of brown algae (Phaeophyceae). *Biochem. Syst. Ecol.* 22, 31–41.

Hehemann, J.H., Kelly, A.G., Pudlo, N. a., Martens, E.C., Boraston, A.B., 2012. Bacteria of the human gut microbiome catabolize red seaweed glycans with carbohydrate-active enzyme updates from extrinsic microbes. *Proc. Natl. Acad. Sci. U. S. A..* 109, 19786–19791.

Hernandez, B.Y., McDuffie, K., Wilkens, L.R., Kamemoto, L., Goodman, M.T., 2003. Diet and premalignant lesions of the cervix: Evidence of a protective role for folate, ribo-flavin, thiamin, and vitamin B12. *Cancer Causes Control* 14, 859–870.

Hernández-Carmona, G., Carrillo-Domínguez, S., Arvizu-Higuera, D.L., Rodríguez-Montesinos, Y.E., Murillo-Álvarez, J.I., Muñoz-Ochoa, M., Castillo-Domínguez, R.M., 2009. Monthly variation in the chemical composition of Eisenia arborea J.E. Areschoug. *J. Appl. Phycol.* 21, 607–616.

Holdt, S.L., Kraan, S., 2011. Bioactive compounds in seaweed: Functional food applications and legislation. *J. Appl. Phycol.* 23, 543–597.

Holst, B., Williamson, G., 2008. Nutrients and phytochemicals: from bioavailability to bioefficacy beyond antioxidants. *Curr. Opin. Biotechnol.* 19, 73–82.

Hosokawa. M., Wanezaki, S., Miyauchi, K., Kunihara, H., Kohno, H., Kawabata, J., 1999. Apoptosis-inducing effect of fucoxanthin on human leukemia cell line HIL- 60. *Food Sci. Technol. Res.* 5 (3), 243–246.

Hozawa, A., Jacobs, D.R., Steffes, M.W., Gross, M.D., Steffen, L.M., Lee, D.H., 2006. Associations of serum carotenoid concentrations with the development of diabetes and with insulin concentration: Interaction with smoking: The Coronary Artery Risk Development in Young Adults (CARDIA) Study. *Am. J. Epidemiol.* 163, 929–937.

Hwang, H., Chen, T., Nines, R.G., Shin, H.C., Stoner, G.D., 2006. Photochemoprevention of UVB-induced skin carcinogenesis in SKH-1 mice by brown algae polyphenols. *Int. J. Cancer* 119, 2742–2749.

Ito, K., Hori, K., 1989. Seaweed: Chemical composmon and potential food uses. *Food Rev. Int.* 5, 101–144.

Iwai, K., 2008. Antidiabetic and antioxidant effects of polyphenols in brown alga Ecklonia stolonifera in genetically diabetic KK-Ay mice. *Plant Foods Hum. Nutr.* 63, 163–169.

Jégou, C., Kervarec, N., Cérantola, S., Bihannic, I., Stiger-Pouvreau, V., 2015. NMR use to quantify phlorotannins: The case of Cystoseira tamariscifolia, a phloroglucinol-producing brown macroalga in Brittany (France). *Talanta* 135, 1–6.

Jeon, H., 2006. Effect of *Chlorella* addition on the quality of processed cheese. *J. Kor. Soc. Food Sci. Nutri.* 35, 373–377.

Kawamura-Konishi, Y., Watanabe, N., Saito, M., Nakajima, N., Sakaki, T., Katayama, T., Enomoto, T., 2012. Isolation of a new phlorotannin, a potent inhibitor of carbohydrate-hydrolyzing enzymes, from the brown alga Sargassum patens. *J. Agric. Food Chem.* 60, 5565–5570.

Kim, E.K., Choi, G.G., Kim, H.S., Ahn, C.Y., Oh, H.M., 2012. Increasing γ-linolenic acid content in Spirulina platensis using fatty acid supplement and light-dark illumination. *J. Appl. Phycol.* 24, 743–750.

Kim, J., Um, M., Yang, H., Kim, I., Lee, C., Kim, Y., Yoon, M., Kim, Y., Kim, J., Cho, S., 2016. Method development and validation for dieckol in the standardization of phlorotannin preparations. *Fish. Aquat. Sci.* 19, 1–6.

Kim, J.E., Bhatia, S.K., Song, H.J., Yoo, E., Jeon, H.J., Yoon, J.-Y., Yang, Y., Gurav, R., Yang, Y.-H., Kim, H.J., Choi, Y.-K. 2020. Adsorptive removal of tetracycline from aqueous solution by maple leaf-derived biochar. *Bioresour. Technol.* 306, 123092

Kim, S.J., Moon, J.S., Park, J.W., Park, J. I., Kim, J. B., Rhim, M. W., Jung, S. T., Kang, S. G., 2004. Quality of soybean paste (doenjang) prepared with sweet tangle, sea mustard and anchovy powder. *Kor. J. Food Sci. Nutr.* 33, 875–879.

Kim, S.K., 2013. *Marine Nutraceuticals: Prospects and Perspectives.* CRC Press, Boca Raton, FL.

Koru, Edis, 2012. Earth Food Spirulina (Arthrospira): Production and Quality Standarts. *Food Additive.* 10.5772/31848.

Kumar, M.D., Kavitha, S., Tyagi, V.K., Rajkumar, M., Bhatia, S.K., Kumar, G., Banu, J.R. 2021. Macroalgae-derived biohydrogen production: Biorefinery and circular bioeconomy. *Biomass Conver. Bioref.*

Kwak, J.Y., 2014. Fucoidan as a marine anticancer agent in preclinical development. *Mar. Drugs* 12, 851–870.

Lahaye, M., Robic, A., 2007. Structure and function properties of *Ulvan* a polysaccharide from green seaweeds. *Biomacromolecules* 8, 1765–1774.

Larsen, R., Eilertsen, K.E., Elvevoll, E.O., 2011. Health benefits of marine foods and ingredients. *Biotechnol. Adv.* 29, 508–518.

Leatherhead Food Research, 2010. Market report, *The Global Market for Good Colours.* http://www.marketresearch.com/Leatherhead-Food-Research-v1435.

Lee, S.M., Lee, H.-J., Kim, S.H., Suh, M.J., Cho, J.Y., Ham, S., Song, H.-S., Bhatia, S.K., Gurav, R., Jeon, J.-M., Yoon, J.-J., Choi, K.-Y., Kim, J.-S., Lee, S.H., Yang, Y.-H., 2021. Engineering of *Shewanella marisflavi* BBL25 for biomass-based poly-hydroxybutyrate production and evaluation of its performance in electricity production. *Inter. J. Biol. Macromol.* 183, 1669–1675.

Lindqvist, A., Andersson, S., 2002. Biochemical properties of purifi ed recombinant human β-carotene 15,15′-monooxygenase. *J. Biol. Chem.* 277, 23942–23948

Lobban, C.S., Harrison, P.J., 2009. Morphology, life histories, and morphogenesis. *Seaweed Ecol. Physiol.* 1–68.

López-López, I., Cofrades, S., Cañeque, V., Díaz, M.T., López, O., Jiménez-Colmenero, F., 2011. Effect of cooking on the chemical composition of low-salt, low-fat Wakame/olive oil added beef patties with special reference to fatty acid content. *Meat Sci.* 89, 27–34.

Lovegrove, a., Edwards, C.H., De Noni, I., Patel, H., El, S.N., Grassby, T., Zielke, C., Ulmius, M., Nilsson, L., Butterworth, P.J., Ellis, P.R., Shewry, P.R., 2017. Role of polysaccharides in food, digestion, and health. *Crit. Rev. Food Sci. Nutr.* 57, 237–253.

Lovstad Holdt, S., Kraan, S., 2010. Bioactive compounds in seaweed: Functional food applications and legislation. *J. Appl. Phycol.* 23, 543–597

Lowe, B., Venkatesan, J., Anil, S., Shim, M.S., Kim, S.K., 2016. Preparation and characterization of chitosan-natural nano hydroxyapatite-fucoidan nanocomposites for bone tissue engineering. *Int. J. Biol. Macromol.* 93, 1479–1487.

Luo, X., Su, P., Zhang, W., 2015. Advances in microalgae-derived phytosterols for functional food and pharmaceutical applications. *Mar. Drugs.* 13, 4231–4254.

Mabeau, S., Fleurence, J., 1993. Seaweed in food products: biochemical and nutritional aspects. *Trends Food Sci. Technol.* 4, 103–107.

MacArtain, P., Gill, C.I.R., Brooks, M., Campbell, R., Rowland, I.R., 2007. Nutritional value of edible seaweeds. *Nutr. Rev.* 65, 535–543.

Maeda, H., Hosokawa, M., Sashima, T., Funayama, K., Miyashita, K., 2005. Fucoxanthin from edible seaweed, Undaria pinnatifida, shows antiobesity effect through UCP1 expression in white adipose tissues. *Biochem. Biophys. Res. Commun.* 332, 392–397.

Manilal, A., Sujith, S., Kiran, G.S., Selvin, J., Shakir, C., Gandhimathi, R., Panikkar, M.V.N., 2009. Biopotentials of seaweeds collected from southwest coast of India. *J. Mar. Sci. Technol.* 17, 67–73.

Marsham, S., Scott, G.W., Tobin, M.L., 2007. Comparison of nutritive chemistry of a range of temperate seaweeds. *Food Chem.* 100, 1331–1336.

Matanjun, P., Mohamed, S., Mustapha, N.M., Muhammad, K., 2009. Nutrient content of tropical edible seaweeds, Eucheuma cottonii, Caulerpa lentillifera and Sargassum polycystum. *J. Appl. Phycol.* 21, 75–80.

Mayne, S.T., 1996. β -Carotene, carotenoids and disease prevention in humans. *FASEB J.* 10, 690–701.

McDermid, K. J., Stuercke, B., 2003. Nutritional composition of edible Hawaiian seaweeds. *J. Appl. Phycol.* 15, 513–524.

Mehariya, S., Fratini, F., Lavecchia, R., Zuorro, A., 2021a. Green extraction of value-added compounds form microalgae: A short review on natural deep eutectic solvents (NaDES) and related pre-treatments. J. Environ. Chem. *Engineer.* 9(5), 105989.

Mehariya, S., Goswami, R.K., Karthikeysan, O.P., Verma, P., 2021b. Microalgae for high-value products: A way towards green nutraceutical and pharmaceutical compounds. *Chemosphere* 280, 130553.

Mehariya, S., Goswami, R.K., Verma, P., Lavecchia, R., Zuorro, A., 2021c. Integrated Approach for Wastewater Treatment and Biofuel Production in Microalgae Biorefineries. *Energies.* 14, 2282. 10.3390/en14082282

Mehariya, S., Iovine, A., Di Sanzo, G., Larocca, V., Martino, M., Leone, G.P., Casella, P., Karatza, D., Marino, T., Musmarra, D., Molino, A. 2019. Supercritical fluid extraction of lutein from *Scenedesmus almeriensis. Molecules.* 24(7), 1324.

Mehariya, S., Sharma, N., Iovine, A., Casella, P., Marino, T., Larocca, V., Molino, A., Musmarra, D. 2020. An integrated strategy for nutraceuticals from *Haematoccus pluvialis*: From cultivation to extraction. *Antioxidants.* 9(9), 825.

Melanson, J.E., Mackinnon, S.L., 2015. Characterization of phlorotannins from brown algae by LC-HRMS. *Methods Mol. Biol.* 1308: 253–266.

MHLW. 2014. The National Health and Nutrition Survey in Japan, 2004–2014. *The Ministry of Health, Labour and Welfare.* http://www.mhlw.go.jp/bunya/kenkou/kenkou_eiyou_chousa.html.

Michaud, P., 2018. Polysaccharides from microalgae, what's future? *Adv. Biotechnol. Microbiol.* 8, 1–2.

Miyamoto, E., Yabuta, Y., Kwak, C.S., Enomoto, T., Watanabe, F., 2009. Characterization of vitamin B12 compounds from Korean purple laver (Porphyra sp.) products. *J. Agric. Food Chem.* 57, 2793–2796.

Miyashita, H., Hosokawa, M., 2008. Beneficial health effects of seaweed carotenoid, fucoxanthin, in: Barrow, C., Shahidi, F. (Eds.), *Marine Nutraceuticals and Functional Foods*. CRC Press, Boca Raton, FL, pp. 297–320.

Miyashita, K., Narayan, B., Tsukui, T., Kamogawa, H., Abe, M., Hosokawa, M., 2012. Brown seaweed lipids as potential source of omega-3 PUFA in biological systems, in: Kin, S.-K. (Ed.), *Handbook of Marine Macroalgae: Biotechnology and Applied Phycology*, 1st ed. John Wiley and Sons, pp. 329–335.

Molnár, P., Deli, J., Tanaka, T., Kann, Y., Tani, S., Gyémánt, N., Molnár, J., Kawases, M., 2010. Carotenoids with anti-Helicobacter pylori activity from Golden delicious apple. *Phytother. Res.* 24, 644–648.

Moroney, N.C., O'Grady, M.N., O'Doherty, J. V., Kerry, J.P., 2013. Effect of a brown seaweed (Laminaria digitata) extract containing laminarin and fucoidan on the quality and shelf-life of fresh and cooked minced pork patties. *Meat Sci.* 94, 304–311.

Nishino, H., Murakoshi, M., Ii, T., Takemura, M., Kuchide, M., Kanazawa, M., Yang Mou, X., Wada, S., Masuda, M., Ohsaka, Y., Yogosawa, S., Satomi, Y., Jinno, K., 2002. Carotenoids in cancer chemoprevention. *Cancer Metastasis Rev.* 21, 257–264.

Norton, T.A., Melkonian, M., Andersen, R.A., 1996. Algal biodiversity. *Phycologia* 35, 308–326.

Norziah, M.H., Ching, C.Y., 2000. Nutritional composition of edible seaweed Gracilaria changgi. *Food Chem.* 68, 69–76.

Nwosu, F., Morris, J., Lund, V.A., Stewart, D., Ross, H.A., McDougall, G.J., 2011. Antiproliferative and potential anti-diabetic effects of phenolic-rich extracts from edible marine algae. *Food Chem.* 126, 1006–1012.

Olsen, E.K., Hansen, E., Isaksson, J., Andersen, J.H., 2013. Cellular antioxidant effect of four bromophenols from the red algae, vertebrata lanosa. *Mar. Drugs* 11, 2769–2784.

Ortiz, J., Romero, N., Robert, P., Araya, J., Lopez-Hernandez, J., Bozzo, C., Navarrete, E., Osorio, A., Rios, A., 2006. Dietary fiber, amino acid, fatty acid and tocopherol contents of the edible seaweeds *Ulva lactuca* and *Durvillaea antarctica*. *Food Chem.* 99, 98–104.

Park, M.K., Jung, U., Roh, C., 2011. Fucoidan from marine brown algae inhibits lipid accumulation. *Mar. Drugs.* 9, 1359–1367.

Park, Y.-L., Song, H.-S., Choi, T.-R., Lee, S.M., Park, S.L., Lee, H.S., Kim, H.-J., Bhatia, S.K., Gurav, R., Park, K., Yang, Y.-H. 2021. Revealing of sugar utilization systems in *Halomonas* sp. YLGW01 and application for poly(3-hydroxybutyrate) production with low-cost medium and easy recovery. *Int. J. Biol. Macromol.* 167, 151–159.

Pereira, H., Barreira, L., Figueiredo, F., Custódio, L., Vizetto-Duarte, C., Polo, C., Rešek, E., Aschwin, E., Varela, J., 2012. Polyunsaturated fatty acids of marine macroalgae: Potential for nutritional and pharmaceutical applications. *Mar. Drugs* 10, 1920–1935.

Periyasamy, C., Anantharaman, P., Balasubramanian, T., 2014. Social upliftment of coastal fisher women through seaweed (Kappaphycus alvarezii (Doty) Doty) farming in Tamil Nadu, India. *J. Appl. Phycol.* 26, 775–781.

Plaza, M., Cifuentes, A., Ibanez, E., 2008. In the search of new functional food ingredients from algae. *Trend. Food Sci. Technol.* 19, 31–39.

Plaza, M., Santoyo, S., Jaime, L., 2010. Screening for bioactive compounds from algae. *J. Pharm. Biomed. Anal.* 51, 450–455.

Prabhasankar, P., Ganesan, P., Bhaskar, N., Hirose, A., Stephen, N., Gowda, L.R., Hosokawa, M., Miyashita, K., 2009. Edible Japanese seaweed, wakame (Undaria pinnatifida) as an ingredient in pasta: Chemical, functional and structural evaluation. *Food Chem.* 115, 501–508.

Prachyakij, P., Charernjiratrakul, W., Kantachote, D., 2008. Improvement in the quality of a fermented seaweed beverage using an antiyeast starter of Lactobacillus plantarum DW3 and partial sterilization. *World J. Microbiol. Biotechnol.* 24, 1713–1720.

Priyadarshani, I., Rath, B., 2012. Commercial and industrial applications of micro algae – A review. *J. Algal Biomass Util.* 3, 89–100.

Qasim, R., 1991. Amino acids composition of some common seaweeds. *Pak. J. Pharm. Sci.* 4, 49–54.

Qi, H., Huang, L., Liu, X., Liu, D., Zhang, Q., Liu, S., 2012. Antihyperlipidemic activity of high sulfate content derivative of polysaccharide extracted from *Ulva pertusa* (Chlorophyta). *Carbohyd. Polym.* 87 (2), 1637–1640.

Quigg, A., Irwin, A.J., Finkel, Z. V., 2011. Evolutionary inheritance of elemental stoichiometry in phytoplankton. *Proc. R. Soc. B Biol. Sci.* 278, 526–534.

Raposo, M.F., De Morais, A.M.M.B., De Morais, R.M.S.C., 2015. Carotenoids from marine microalgae: A valuable natural source for the prevention of chronic diseases. *Mar. Drugs* 13, 5128–5155.

Renaud, S.M., Luong-Van, J.T., 2006. Seasonal variation in the chemical composition of tropical Australian marine macroalgae. *J. Appl. Phycol.* 18, 381–387.

Rindi, F., Soler-Vila, A., Guiry, M. D., 2012. Taxonomy of marine macroalgae used as sources of bioactive compounds, in: Hayes, M (Ed.), *Marine Bioactive Compounds: Sources, Characterization and Applications.* Springer Science & Business Media, New York, pp. 1–54

Rioux, L.E., Turgeon, S.L., 2015. Seaweed carbohydrates, in: Tiwari, B.K., Troy, D. (Eds.), *Seaweed Sustainability.* Elsevier, San Diego, pp. 141–192.

Rioux, L. E., Turgeon, S. L., Beaulieu, M., 2007. Characterization of polysaccharides extracted from brown seaweeds. *Carbohydr. Polym.* 69(3), 530–537.

Roberfroid, M., 2007. Prebiotics: The concept revisited. *J. Nutr.* 137.

Robic, a., Rondeau-Mouro, C., Sassi, J.F., Lerat, Y., Lahaye, M., 2009. Structure and interactions of ulvan in the cell wall of the marine green algae Ulva rotundata (Ulvales, Chlorophyceae). *Carbohydr. Polym.* 77, 206–2106.

Rodrigues, J.A.G., Torres, V.M., de Alencar, D.B., Sampaio, A.H., Farias, W.R.L., 2010. Natural heparinoids isolated from Halymenia sp. (Rhodophyceaes) delivery on the Ceará coast. *Acta Sci. Biol. Sci.* 32, 235–242.

Saravana, P.S., Getachew, A.T., Cho, Y.J., Choi, J.H., Park, Y.B., Woo, H.C., Chun, B.S., 2017. Influence of co-solvents on fucoxanthin and phlorotannin recovery from brown seaweed using supercritical CO2. *J. Supercrit. Fluids* 120, 295–303.

Sauer Leal, B.E., Real, P.M., Grzybowski, A., Tiboni, M., Koop, H. S., Blitzkow, S.L., Cardoso, S.A., Takamatsu, A.A., Farfus dos Santos, A., Ferreira C.V., Fontana, J. D., 2017. Potential prebiotic oligosaccharides from aqueous thermopressurized phosphoric acid hydrolysates of microalgae used in treatment of gaseous steakhouse waste. *Alg. Res.* 24, 138–147.

Ścieszka, S., Klewicka, E., 2019. Algae in food: A general review. *Crit. Rev. Food Sci. Nutr.* 59, 3538–3547.

Seema, P., Arun, G., 2011. Functional oligosaccharides: Production, properties and applications. *World J. Microbiol. Biotechnol.* 27, 1119–1128.

Sekar, S., Chandramohan, M., 2008. Phycobiliproteins as a commodity: Trends in applied research, patents and commercialization. *J. Appl. Phycol.* 20, 113–136.

Sepkoski, J.J., 1995. Large scale history of biodiversity, in: *Global Biodiversity Assessment.* United Nations Environmental Programme. Cambridge University Press, pp. 202–212.

Shen, Y.R., Kuo, M.I., 2017. Effects of different carrageenan types on the rheological and water-holding properties of tofu. *LWT – Food Sci. Technol.* 78, 122–128.

Shibata, T., Ishimaru, K., Kawaguchi, S., Yoshikawa, H., Hama, Y., 2007. Antioxidant activities of phlorotannins isolated from Japanese Laminariaceae. *J. Appl. Phycol.* 20: 705–711.

Shibata, T., Kawaguchi, S., Hama, Y., Inagaki, M., Yamaguchi, K., Nakamura, T., 2004. Local and chemical distribution of phlorotannins in brown algae. *J. Appl. Phycol.* 16, 291–296.

Simoons, F.J., 1991. *Food in China: A Cultural and Historical Inquiry*. CRC Press, Boca Raton, FL, USA.

Smith, A.G., Croft, M.T., Moulin, M., Webb, M.E., 2007. Plants need their vitamins too. *Curr. Opin. Plant Biol.* 10, 266–275.

Sonani, R.R., 2016. Recent advances in production, purification and applications of phyco-biliproteins. *World J. Biol. Chem.* 7, 100.

Soontornchaiboon, W., Joo, S.S., Kim, S.M., 2012. Anti-inflammatory effects of violaxanthin isolated from microalga Chlorella ellipsoidea in RAW 264.7 macrophages. *Biol. Pharm. Bull.* 35, 1137–1144.

Spolaore, P., Joannis-Cassan, C., Duran, E., Isambert, A., 2006. Commercial applications of microalgae. *J. Biosci. Bioeng.* 101, 87–96.

Stengel, D.B., Connan, S., Popper, Z.A., 2011. Algal chemodiversity and bioactivity: sources of natural variability and implications for commercial application. *Biotechnol. Adv.* 29, 483–501.

Stiger-Pouvreau, V., Bourgougnon, N., Deslandes, E., 2016. Carbohydrates from Seaweeds. *Seaweed in Health and Disease Prevention*. Academic Press. 223–274.

Sun, S., Wang, H., Xie, J., Su, Y., 2016. Simultaneous determination of rhamnose, xylitol, arabitol, fructose, glucose, inositol, sucrose, maltose in jujube (Zizyphus jujube Mill.) extract: Comparison of HPLC-ELSD, *LC-ESI-MS/MS and GC-MS*. *Chem. Cent. J.* 10, 1–9.

Suzuki, T., Nakai, K., Yoshie, Y., Shirai, T., Hirano, T., 1993. Seasonal variation in the dietary fiber content and molecular weight of soluble dietary fiber in brown alga, Hijiki. *Nippon Suisan Gakkaishi* 59, 1633–1633.

Sweeney, T., Collins, C.B., Reilly, P., Pierce, K.M., Ryan, M., O'Doherty, J. V., 2012. Effect of purified β-glucans derived from Laminaria digitata, Laminaria hyperborea and Saccharomyces cerevisiae on piglet performance, selected bacterial populations, volatile fatty acids and pro-inflammatory cytokines in the gastrointestinal tract of pig. *Br. J. Nutr.* 108, 1226–1234.

Takenaka, S., Takubo, K., Watanabe, F., Tanno, T., Tsuyama, S., Nanano, Y., Tamura, Y., 2003. Occurrence of coenzyme forms of vitamin B12 in a cultured purple laver (Porphyla yezoensis). *Biosci. Biotechnol. Biochem.* 67, 2480–2482.

Tang, L., Chen, Y., Jiang, Z., Zhong, S., Chen, W., Zheng, F., Shi, G., 2017. Purification, partial characterization and bioactivity of sulfated polysaccharides from Grateloupia livida. *Int. J. Biol. Macromol.* 94, 642–652.

Tavassoli-Kafrani, E., Shekarchizadeh, H., Masoudpour-Behabadi, M., 2016. Development of edible films and coatings from alginates and carrageenans. *Carbohydr. Polym.* 137, 360–374.

Thanh, T.T.T., Quach, T.M.T., Nguyen, T.N., Vu Luong, D., Bui, M.L., Tran, T.T. Van, 2016. Structure and cytotoxic activity of ulvan extracted from green seaweed Ulva lactuca. *Int. J. Biol. Macromol.* 93, 695–702.

Thomas, F., Barbeyron, T., Tonon, T., Génicot, S., Czjzek, M., Michel, G., 2012. Characterization of the first alginolytic operons in a marine bacterium: From their emergence in marine Flavobacteriia to their independent transfers to marine Proteobacteria and human gut Bacteroides. *Environ. Microbiol.* 14, 2379–2394.

Tibbetts, S.M., Milley, J.E., Lall, S.P., 2016. Nutritional quality of some wild and cultivated seaweeds: Nutrient composition, total phenolic content and in vitro digestibility. *J. Appl. Phycol.* 28, 3575–3585.

Tierney, M.S., Soler-Vila, A., Rai, D.K., Croft, A.K., Brunton, N.P., Smyth, T.J., 2014. UPLC-MS profiling of low molecular weight phlorotannin polymers in Ascophyllum nodosum, Pelvetia canaliculata and Fucus spiralis. *Metabolomics* 10, 524–535.

Tonegawa, I., Okada, S., Murakami, M., Yamaguchi, K., 1998. Pigment Composition of the Green Microalga Botryococcus braunii Kawaguchi-1. *Fish. Sci.* 64, 305–308.

Tsukui, T., Konno, K., Hosokawa, M., Maeda, H., Sashima, T., Miyashita, K., 2007. Fucoxanthin and fucoxanthinol enhance the amount of docosahexaenoic acid in the liver of KKAy obese/diabetic mice. *J. Agric. Food Chem.* 55, 5025–5029.

Urbano, M. G., Goñi, I., 2002. Bioavailability of nutrients in rats fed on edible seaweeds, nori (*Porphyra tenera*) and wakame (*Undaria pinnatifida*), as a source of dietary fibre. *Food Chem.* 76, 281–286.

USDA, 2010. *Composition of Foods Raw, Processed, Prepared. USDA National Nutrient Database for Standard Reference. Release 23. U.S.* Department of Agriculture, Beltsville.

Vassilev, S. V., Vassileva, C.G., 2016. Composition, properties and challenges of algae biomass for biofuel application: An overview. *Fuel.* 181, 1–33.

Venkatesan, J., Bhatnagar, I., Manivasagan, P., Kang, K.H., Kim, S.K., 2015. Alginate composites for bone tissue engineering: A review. *Int. J. Biol. Macromol.* 72, 269–281.

Venugopal, V., 2009. *Marine Products for Healthcare.* CRC Press, Boca Raton.

Vidanarachchi, J.K., Kurukulasuriya, M.S., Malshani Samaraweera, a., Silva, K.F.S.T., 2012. Applications of Marine Nutraceuticals in Dairy Products. *Adv. Food Nutr. Res.* 65, 457–478.

Wang, D., He, F., Lv, Z., Li, D., 2014. Phytochemical composition, antioxidant activity and HPLC fingerprinting profiles of three Pyrola species from different regions. *PLoS One.* 9, 1–11.

Wells, M.L., Potin, P., Craigie, J.S., Raven, J. a., Merchant, S.S., Helliwell, K.E., Smith, A.G., Camire, M.E., Brawley, S.H., 2017. Algae as nutritional and functional food sources: revisiting our understanding. *J. Appl. Phycol.* 29, 949–982.

Wildman, R., Taylor, C., 2006. *Wallace Handbook of Nutraceuticals and Functional Foods*, second ed., CRC Press, Boca Raton, FL.

Wong, K.H., Cheung, P.C.K., 2000. Nutritional evaluation of some subtropical red and green seaweeds. Part I – Proximate composition, amino acid profiles and some physico-chemical properties. *Food Chem.* 71, 475–482.

Wong, K., Cheung, P.C., 2001. Influence of drying treatment on three *Sargassum* species. *J. Appl. Phycol.* 13, 43–50.

Yaakob, Z., Ali, E., Zainal, A., Mohamad, M., Takriff, M. S., 2014. An overview: Biomolecules from microalgae for animal feed and aquaculture. *J. Biol. Res. (Thessalon)* 21, 6.

Yaich, H., Garna, H., Besbes, S., Paquot, M., Blecker, C., Attia, H., 2011. Chemical composition and functional properties of *Ulva lactuca* seaweed collected in Tunisia. *Food Chem.* 128, 895–901.

Young, M.H., Jong, S.B., Jin, W.H., Nam, H.L., 2007. Isolation of a new phlorotannin, fucodiphlorethol G, from a brown alga *Ecklonia cava. Bul. Kor. Chem. Soc.* 28, 1595–1597.

Yuan, J.H., Xu, R.K., Zhang, H., 2011. The forms of alkalis in the biochar produced from crop residues at different temperatures. *Bioresour. Technol.* 102, 3488–3497.

Yuan, X.L., Roubos, J.A., Van Den Hondel, C.A.M.J.J., Ram, A.F.J., 2008. Identification of InuR, a new Zn(II)2Cys6 transcriptional activator involved in the regulation of inulinolytic genes in Aspergillus niger. *Mol. Genet. Genomics* 279, 11–26.

Yurko-Mauro, K., McCarthy, D., Rom, D., Nelson, E.B., Ryan, A.S., Blackwell, A., Salem, N., Stedman, M., 2010. Beneficial effects of Docosahexaenoic acid on cognition in age-related cognitive decline. *Alzheimer's Dement.* 6, 465–464.

Zhao, X., Li, B., Xue, C., Sun, L., 2012. Effect of molecular weight on the antioxidant property of low molecular weight alginate from Laminaria japonica. *J. Appl. Phycol.* 24, 295–300.

11 Cyanobacterial Phycobiliproteins – Biochemical Strategies to Improve the Production and Its Bio Application

Khushbu Bhayani and Sandhya Mishra
CSIR-Central Salt and Marine Chemicals Research Institute, Council of Scientific and Industrial Research, Bhavnagar, Gujarat, India

Imran Pancha
Department of Biology, SRM University-AP, Amaravati, Andhra Pradesh, India

CONTENTS

11.1 Introduction 340
11.2 Light-Harvesting Bio-molecule – Phycobiliproteins 341
    11.2.1 Types of Phycobiliproteins (PBPs) 341
    11.2.2 Structure of Phycobiliproteins 342
        11.2.2.1 C-Phycoerythrin (CPE) 342
        11.2.2.2 C-Phycocyanin (CPC) 343
        11.2.2.3 Allophycocyanin (APC) 343
    11.2.3 Bilin and Linker Polypeptide 343
11.3 Role of Different Abiotic Stress Conditions in the Production of Light-Harvesting Biomolecules 344
    11.3.1 Production of PBPs 344
    11.3.2 Factors Affecting Accumulation of Phycobiliproteins in Cyanobacteria 345
        11.3.2.1 Effect of Light Stress 345
        11.3.2.2 Effect of Temperature Stress 346

DOI: 10.1201/9781003195405-11

          11.3.2.3   Effect of Nutrient Accessibility.....................................347
11.4   Downstream Processing for Light-Harvesting Biomolecules..................348
       11.4.1   Extraction of PBPs 348
                11.4.1.1   Ultrasonication 350
                11.4.1.2   Freezing-Thawing 350
                11.4.1.3   Osmotic Shock..350
                11.4.1.4   Biological and Chemical Extraction 350
11.5   Purification of PBPs 351
       11.5.1   Different Purification Methods...351
                11.4.2.2   Stability of Phycobiliproteins 351
11.5   Biotechnological Importance...355
       11.5.1   Antioxidant Capacity of Phycobiliproteins.......................355
       11.5.2   Anti-Tumor Activity of Phycobiliproteins.........................355
       11.5.3   Phycobiliproteins as Neuroprotective Agents...........................356
       11.5.4   Phycobiliproteins as Fluorescent Probe.............................357
       11.5.5   Phycobiliproteins as Food Colorants 357
11.6   Conclusion and Future Prospects..357
Acknowledgments...359
References..359

11.1   INTRODUCTION

Cyanobacteria are oxygenic photosynthetic, gram-negative bacteria. It is believed
that they originated 3.5 billion years ago and spread in both aquatic and terrestrial
ecosystems (Fischer, 2008). They are the predominant organism required for primary productivity, aquatic biomass production, and maintaining the tropic energy
dynamics (Stock et al., 2014). They can also fix atmospheric nitrogen, which makes
them essential for plant development and biomass production (Vaishampayan et al.,
2002). During evolution, cyanobacteria faced various stress environments like high
levels of CO_2, extreme temperature, light intensity, and the presence of other gases
like NH_4, H_2S, and CH_4 in the environment. The ability to survive under such
conditions made selective pressure in the genome of cyanobacteria for diversity in
morphology. Due to these variabilities and adaptability toward diverse stress even
today, they are found in various extreme conditions like hot springs, soda lakes,
polar regions, etc. (Puzorjov and McCormick, 2020). Seaweeds and other macroalgae also produce phycobiliproteins (PBPs). Compared to terrestrial plants, microalgae have high photosynthetic ability and growth rates (Mehariya et al., 2021a,
2021b, 2021c); additionally, it is very easy to perform genetic engineering due to
the smaller genome size of cyanobacteria to make desired bioproducts (Bharadwaj
et al., 2020). Cyanobacteria can thus be considered potential bio-factories for
producing various economical and important nutraceutical compounds like carotenoids, fatty acids, vitamins, sterols, polysaccharide, etc. (Parmar et al., 2011a).
Apart from these nutraceutical components, cyanobacteria are also studied to produce biofuels and other industrially important chemicals (Bharadwaj et al., 2020).
Among other cellular metabolites, PBPs comprise almost 60% of cyanobacteria's

total protein content. Cyanobacterial PBP is an excellent antioxidant compound. Due to these properties, it is well explored for the production of pharmaceutical and nutraceutical compounds and use as an anti-aging component (Singh et al., 2015). Apart from this application, PBPs are also used as colorants and fluorescent probes in various medical and industrial applications (Singh et al., 2015). Due to its commercial applications, various research groups tried to optimize the content of PBPs and optimize the downstream processing for harvesting and purification from cyanobacteria (Patel et al., 2005).

As mentioned earlier, cyanobacteria are found in almost all conditions, including extreme environments. Production and composition of particular PBPs is dependent on the cultivation conditions, like the source of nitrogen and carbon in the growth medium, the light intensity, and the pH. Optimization of such cultural parameters and downstream processing like harvesting, extraction, and purification is crucial for large-scale production of these cyanobacterial proteins. In this chapter, we focus on PBPs from cyanobacteria along with their downstream processing and the effect of abiotic stress on them. A detailed discussion based on the biotechnological application of PBPs is also provided.

11.2   LIGHT-HARVESTING BIO-MOLECULE – PHYCOBILIPROTEINS

11.2.1   Types of Phycobiliproteins (PBPs)

Cyanobacteria are the most important prokaryotic photosynthetic organism found in almost all types of environments on earth. Cyanobacterial photosynthetic apparatus is made up of phycobilisome along with two main light-harvesting photosystems, i.e., Photosystem I and Photosystem II (Pagels et al., 2019). Phycobilisome is macromolecular complex structure comprised of pigment-binding PBPs that absorb visible light and transfer it to Chlorophyll a (Sidler, 1994). PBPs can make up 40–50% of the total protein content in cyanobacteria (Biswas, 2011). However, its content varies from species to species, and some factors also affect the growth of the cyanobacteria (Gantt, 1975). PBPs are a water-soluble, brilliantly colored, and fluorescent protein-pigment complex made up of linear tetrapyrroles chromophores known as phycobilins. PBPs are composed of different phycobilins, which absorbs different wavelengths of light due to their small structural difference. Each phycobilin is bound *via* thio-ether linkages to specific component of the PBP (Bhayani et al., 2018). There are three types of PBPs: C-Phycocyanin (CPC; blue color), C-Phycoerythrin (CPE; red color), and C-Allophycocyanin (APC; light blue color) (MacColl, 1998). Table 11.1 represents the different absorbance range of PBPs. Researchers have observed that PBPs from different species also have similar spectral properties. For example, the spectral properties of Phycocyanin from some algae and cyanobacteria are similar to each other. CPC and CPE are major components, while allophycocyanin functions as the connecting pigment between PBPs and the photosynthetic lamella. Allophycocyanin is almost always a minor component, whereas amount of phycocyanin and phycoerythrin depends on the organism and its growth conditions (Gantt, 1975).

TABLE 11.1
Absorbance Range of Different Phycobiliproteins

Phycobiliproteins (PBPs)	Absorbance Range (nm)	Color
C-Phycoerythrin (CPE)	540–570	Orange-red
C-Phycocyanin (CPC)	610–620	Blue
Allophycocyanin (APC)	650–655	Green-blue

11.2.2 STRUCTURE OF PHYCOBILIPROTEINS

PBSs helps the organism to survive in a broad range of environmental conditions, such as different depths of ocean and other dark regions where sunlight is very limited. PBPs are broadly classified in three colorful types according to their absorption characteristics. They are composed of a number of different chromophoric, water-soluble polypeptides called PBP and nonchromophoric polypeptide. Hydrophobic in nature called linker proteins, the principal photoreceptor in blue-green algae (Chaiklahan et al., 2012). Figure 11.1 illustrates the overall structure of PBPs.

11.2.2.1 C-Phycoerythrin (CPE)

Red-colored phycoerythrin, with λ_{max} – 540–570 nm, serves as light-harvesting pigments in cyanobacteria (Schoenleber et al., 1984). CPE absorbs the light in the green region of the visible light spectrum and is located in the distal part of PBPs. According

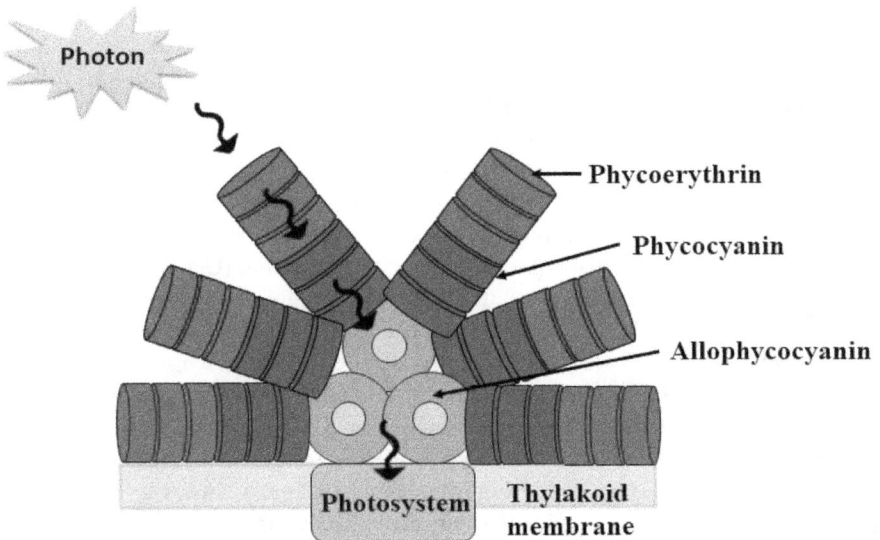

FIGURE 11.1   Structure of Phycobilisome.

to their production in different species, PE can be classified into three classes: CPE (Cyanobacterial PE), B-PE (Bangiophyceae PE), and R-PE (Rhodophyta PE) (Liu et al., 2005). α and β polypeptide of CPE contains two or three chromophores (MacColl, 1998). The most PE-containing distal part of the PBPs rod is composed of two types of PE (PE I and PE II). PE I binds either with two to three covalently attached tetrapyrrole chromophores commonly called phycoerythrobilin (PEB) or to both PEB and PUB (phycourobilin), whereas PE II always binds to both PEB and PUB. Optical properties at 550 nm and 495 nm are because of the presence of PEB and PUB, respectively, in the PBPs. Usually, PUB shifts the absorption range of PBPs closer to shorter wavelengths. Therefore, so many chromophores bind to PBP, and linkers benefit the PBS to harvest a wide range of wavelengths during the process of evolution (Talarico and Maranzana, 2000).

11.2.2.2 C-Phycocyanin (CPC)

Velvet blue color phycocyanin with an absorption maximum of 610–620 nm, serves as the major light harvesting protein in *Spirulina*. CPC extend the light-harvesting capability of the PBPs in the red region of the visible light spectrum, and is located at part of the antenna (rod) of PBS in association with APC. PC is found in complex of hexamers (α3β3)2 and other oligomers along with chromophore. Generally, one chromophore is attached along with α- and two with β- subunit. There are many types of phycocyanin depending on the type of attached chromophores, found in cyanobacteria. CPC contain only phycocyanobilin (PCB) λ_{max} = 620 nm, R-PC II and R-PC III contain both phycoerythrobilin (PEB) (λ_{max} = 540 nm) (Szalontai et al., 1994). Due to excellent fluorescence properties, it has been used in immunodiagnostics as a substitute of synthetic dyes (Paswan et al., 2016, Kronick and Grossman, 1983).

11.2.2.3 Allophycocyanin (APC)

Allophycocyanin (APC), with λ_{max} – 650–655 nm, is the most efficient biliprotien-transferring light energy (Gysi and Zuber, 1979) and is the key pigment in funnelling the photons from the biliproteins to the chlorophyll "a" of the PS-II. APC has an emission maximum of about 657 nm, which allows them to transfer energy efficiently to chlorophyll (MacColl, 1998). APC at neutral pH is found as trimer (αβ)$_3$, three α, and three β polypeptides; each polypeptide has only one bilin and is the important part of phycobilisome. The fluorescence assets of adjacent PC and absorption properties of AP ensure their function as an efficient acceptor.

11.2.3 Bilin and Linker Polypeptide

Phycobilins are tetrapyrrole structures attached to the cysteine residues by thioether linkages and serve as the site of fluorescence origin (Glazer, 1994). Attachment as well as detachment of bilin to PBPs is mediated by different types of lyases (Scheer and Zhao, 2008; Wiethaus et al., 2010), which allow better light harvesting ability in particular regions of the spectra. Allophycocyanin and C-phycocyanin have a bilin called phycocyanobilin (PCB), and CPE has a bilin called phycoerythrobilin (PEB). Two other types of bilin, phycourobilins (PUB) and phycoviolobilin (PVB), are also found on certain cyanobacteria (Samsonoff and MacColl, 2001). Crystallographic and spectroscopic studies on PBSs have identified donor and acceptor bilin in CPC and in other PBPs (Schirmer and Vincent, 1987). Linkers are non-chromophoric basic polypeptides hydrophobic in nature that join the back-to-back PBPs to form PBS (De Marsac and Cohen-Bazire, 1977). It is very important for PBS structure maintenance and the energy conduct mechanism (De Marsac and Cohen-Bazire, 1977; Sidler, 1994). Linkers are involved in closer aggregation of PE, PC, and PEC trimers that are involved in construction of the PBS. The process is mediated by end-to-end attachment of hexameric structure into cylinders of rods and cores (Figure 11.1).

11.3 ROLE OF DIFFERENT ABIOTIC STRESS CONDITIONS IN THE PRODUCTION OF LIGHT-HARVESTING BIOMOLECULES

11.3.1 PRODUCTION OF PBPs

The characterization of new prospective species is essential for cyanobacterial biotechnology with a focus on producing high quantities of PBPs with the selection and identification of their growth conditions. Growth of any cyanobacteria is selective, based on their ability to survive and colonize the environment. Therefore, researchers have developed different media in which these organisms can grow and develop easily (Table 11.2). Their requirement of suitable media depends on the

TABLE 11.2
Common Culture Media Used for Cyanobacteria Growth

Environment	Culture Media	References
Fresh Water	BG-11 Medium	Allen and Stanier, 1986; Rippka et al., 1979
	Chu #10 Medium	Chu, 1942
	Bold's Basal Medium (BBM)	Bold, 1949
	Allen's Blue-Green Algal Medium	Allen, 1986
	D11 Medium	Graham et al., 1982
	Fraquil Medium	Kilham et al., 1998
	D Medium	Waterbury, 2006
	Spirulina Medium	Aiba and Ogawa, 1977
	MA Medium	
	COMBO Medium	Kilham et al., 1998
Marine	ASN-III Medium	Rippka, 1988
	K Medium	Keller et al., 1987
	F/2 Medium	Guillard, 1975; Guillard and Ryther, 1962
	YBC-II Medium	Chen et al., 1996
	SN Medium	Waterbury, 1986
	L1 Medium	Guillard and Hargraves, 1993

chemical environment of the particular habitat and the specific need of nutrients (Fábregas et al., 1998). Thus, several approaches have been applied for the betterment of growth and pigment production. Production of PBPs is influenced by different physicochemical parameters. Besides, introducing a new high-yielding strain, designing of a suitable growth medium is important, which affects product concentration, yield, and downstream processing cost (Kennedy and Krouse, 1999). In general, photoautotrophic production of PBPs can be done either by growing cyanobacteria at large scale in artificially fabricated open ponds or in closed bioreactor underneath sunlight. So many industries around the globe use *Spirulina platensis* for PBP production because it can easily grow in alkaline conditions (pH 10.5) in open ponds with significantly less contamination (Richmond and Grobbelaar, 1986; Radmann et al., 2007) More than 3000 tons of *Spirulina platensis* cell mass is produced worldwide (Vonshak and Richmond, 1988; Spolaore et al., 2006). Closed photobioreactor can be used instead of open raceway pond to avoid poor light availability to the organism, which affects the overall growth and PBP production. Moreover, it provides temperature control and equal distribution of light, CO_2, etc. (Soni et al.,2017). In heterotrophic cultivation, cyanobacteria accumulate available organic compounds (e.g., glucose, acetate) in dark conditions (Mohan et al., 2015). In this type of cultivation, contamination chances are there, so pigments obtained from this kind of cultivation cannot be used in the food industry (Bachchhav et al., 2017). In the current situation, phycocyanin is produced commercially in open raceway ponds by autotrophic cultivation of *Spirulina platensis,* and *Porphyridium spp.* is used for the production of Phycoerythrin (Christaki, et al., 2015; Kathiresan et al., 2007). Due to low production of biomass and pigments, large-scale production of PBPs is still limited (Bachchhav et al., 2017). This can be solved by developing an economically viable process and culture growth optimization. This ability to cope with certain limitations or starvation is a requirement for survival in different conditions.

11.3.2 Factors Affecting Accumulation of Phycobiliproteins in Cyanobacteria

Various abiotic stress (e.g., light, temperature, nutrient accessibility) affects the synthesis of CPC in cells (Paliwal et al., 2017). Cyanobacteria have tremendous acclimatization capacity toward abiotic stress factors and can develop pigments accordingly. Such unique features of cyanobacteria can be used for PBP production in a biorefinery approach. The expression of genes encoding PBPs also changes according to the abiotic stress. The content and composition of PBPs can be affected by environmental factors like temperature, biochemical stress (e.g., phosphorus, carbon dioxide, sulphate concentration, nitrogen depletion, different salts), light intensity and wavelength, and metal ions (Allen and Smith, 1969; Fatma, 2009).

11.3.2.1 Effect of Light Stress

The lower light intensities limit photosynthetic efficiency, while higher intensities damage the photosynthetic apparatus (George et al., 2014). In low light conditions, carotenoids also serve as accessory pigments along with PBPs, so they can prosper under a wide range of environmental conditions (Olaizola and Duerr, 1990). People have reported that UVB irradiation has an adverse effect on several species of cyanobacteria (Rinalducci et al., 2006; Six et al., 2007). UV irradiation damages PBPs' pigmentation and the structure of linker polypeptides in the cell (Six et al. 2007; Kannaujiya and Sinha 2016). UVB exposure hampers the spectral properties of PBPs by inhibiting oxygen evolution in *Synechococcus* sp. (Kulandaivelu et al., 1989). Rajagopal et al. (1998) demonstrated that after UVB irradiation, polypeptide structure of PBPs was altered in *Spirulina platensis*. Higher strength of photosynthetically active radiation (PAR; 400–700 nm) decreases the fluorescence of PBPs (Rastogi and Incharoensakdi, 2015). Other than UV radiation, different light intensities also affect PBP structure in cyanobacteria. Interestingly, PBP productivity is also dependent on the maximum color intensity of light in selected cyanobacteria.

Moreover, the particular wavelength of light is very important because according to it, the organism regulates pigment production in the cell. This mechanism is known as complementary chromatic adaptation of phycobiliproteins (CCA) (Sonani et al., 2016). CCA phenomena only occur in a cyanobacteria producing phycocyanin and phycoerythrin (da Fontoura Prates et al., 2018; Singh et al., 2012; Khatoon et al., 2018). In this type of phenomena, pigmentation changes according to spectral illumination. Cyanobacteria can modify production of CPC and CPE based on CCA. Thus, CPE and CPC will be produced well in green light and red light, accordingly. Mishra et al. (2011) showed that phycoerythrin production was observed in 39.2 mg L^{-1}green light, and phycocyanin production was higher in red light (10.9 mg L^{-1}) when *Pseudanabaena sp.* was grown under different quality of light. Further production of pigments can be enhanced by manipulation of certain environmental parameters (Fatma, 2009). However, more research is recommended to use the CCA for PBP production.

11.3.2.2 Effect of Temperature Stress

Temperature is the most persuasive element for the cells of the organism as it affects cell structure and its overall cellular mechanism. The maximum temperature required for growth and its tolerance level to the extreme values, varies between different species based on its native habitat. Researchers have conducted several experiments to confirm the specific role of temperature for the CPC productivity in cyanobacteria from different climatic conditions (Colla et al., 2007; Chaneva et al., 2007; Kenekar and Deodhar, 2013; Tiwari et al., 2019). Patel at al. (2005) purified CPC with higher purity ratio from three different cyanobacteria; *Phormidium, Lyngbya,* and *spirulina* by growing at 20 ± 2 °C temperature. However, the maximum temperature for CPC production has been found to be 35 °C in *Spirulina platensis and Arthronema africanum* species (Colla et al., 2007; Chaneva et al., 2007). Sudden changes in temperature also cause stress, e.g., in hot water, free oxygen unavailability will adversely affect the growth and pigment production (Brock, 2012). It has been observed that optimal temperature for pigment production in Anabaena NCCU-9 was near 30 °C, although variation in optimal temperature decreases yield of PBPs (Fatma, 2009). Studies have also exerted that optimized growth temperature for *Synechococcus* and *Arthronema africanum* were 37 °C and 36 °C, respectively (Sakamoto et al., 1998; Chaneva et al., 2007). When *Synechococcus sp.* PCC 7942 is exposed to high temperature, it will produce heat shock proteins, which directly inactivate the production of PBPs (Nakamoto and Honma, 2006). Moon et al. (2014) isolated thermostable phycocyanin with 81.93% yield and higher purity ratio from red algae *G. sulphuraria*, which remained stable at high temperatures up to 60°C. Higher temperature does not correspond to higher yield, but the optimization of other parameters is also necessary for fruitful results in each strain.

11.3.2.3 Effect of Nutrient Accessibility

Cyanobacterial growth can be inhibited in both nutrient-enriched as well as nutrient-limited growth media. Therefore, optimized nutrients are key factors for cell growth and pigment content (Hemlata and Fatma, 2009).

In nutrient-deprived media, color changes from bluish green to yellowish green due to CPC degradation in cyanobacteria (Allen and Smith 1969). As nitrogen is an essential element in nucleic acid and protein structure, it is one of the important macronutrients for PBP synthesis and biomass production. Several studies have been reported that nitrogen source affects the PBP content and growth of cyano-bacteria (Luque et al., 1994; Deschoenmaeker et al. 2017). CPC is a part of nitrogen storage compounds (PBPs) in the cyanobacteria in marine environments where nitrogen is in limited amounts (Boussiba & Richmond, 1980). Its production is totally based on cell growth patterns and varies from species to species; for example, in *Spirulina platensis*, higher CPC content was found after the 30 days of culture inoculation, which means after longer growth cycle (Soundarapandian and Vasanthi, 2010), whereas Kenekar and Deodhar (2013) reported maximum CPC content after 15 days of *Geiterinema sulphureum* culture inoculation. Although these tiny organisms are able to balance their CPC content (PBP composition) with respect to various stress signals (Fatma, 2009; Singh et al., 2012; Singh et al., 2013), cyanobacteria utilize nitrate, nitrite, or ammonium as a nitrogen source for their growth. A number of species are also capable of fixing N_2 (Herrero et al., 2001). Cyanobacterial cells absorb nitrate by active transport system, and nitrate reductase enzyme consecutively reduces it to nitrite, and after that, nitrite reductase reduce it to ammonium (Luque et al., 1994). Urea is metabolized by the activity of enzymes like urease and urea starch lyase in cyanobacteria. Ajayan et al. (2012) reported that better growth of *S. platensis* culture was observed when urea was used as a nitrogen source for its growth compared to other nitrogen sources (e.g., KNO_3 and $NaNO_3$). Cyanobacteria, which are not efficient for molecular nitrogen fixing, utilize their own PBPs (Allen, 1986; Collier and Grossman, 1994). There are several reports on phosphorus and nitrogen starvation-induced changes, since these two are the most responsible for the synthesis of various cellular components (Xu et al., 2013). Chen et al. (2013) have observed that further increase in the $NaNO_3$ concentration did not increase the CPC content or productivity. Conversely, nitrogen-limited media stimulates the PBP degradation, since they are nitrogen storage compound (Boussiba and Richmond, 1980). Therefore, a sufficient amount of nitrogen is required for PBP production in cyanobacteria.

Sulphur plays an important role for regulation of cyanobacterial growth and PBP production in it. Research has been shown that, when *Synechococcus* sp. was grown in sulphur-limiting media, PBP production decreased in the cell (Collier and Grossman, 1994; Richaud et al., 2001).

Phosphorus is an essential element for cyanobacterial cells and their physiological mechanism (Ritchie et al., 1997). It is supplied to cyanobacteria in its orthophosphate (PO_4^{2-}) form, and its uptake is energy dependent. It is one of the most potent growth-limiting factors (Richmond, 2008). Many cyanobacteria are capable of accumulating small molecule for osmoregulation in response to increases in salinity. Increased salinity may increase the total lipids and carotenoids content, but decrease total PBPs in freshwater cyanobacteria (Richmond, 2008; Fatma, 2009). Appropriate salt concentration is also required for metabolism and proper osmotic balance; the effect of salt on cellular mechanism was studied in salt-tolerant cyanobacteria *Synechocystis* PCC 6803 and *Euhalothece* sp. (Huang et al., 2007). The amount of CPC decreases in salt stress, and thus the energy exchange rate of PBPs is interrupted in the photosystem of cyanobacteria (Vonshak et al., 2000). Fatma (2009) reported that when Anabaena NCCU-9 culture was grown in lower salt concentration media compared to normal ones, PBP production increased.

11.4 DOWNSTREAM PROCESSING FOR LIGHT-HARVESTING BIOMOLECULES

Extraction and purification of these beautiful natural coloring molecules involves combined steps with several norms and standardization. One important step is to select a suitable technique that extracts PBPs without harming their structures and functional ability. It means a higher amount of ruptured cells will confirm higher yields of PBPs. Therefore, it is essential to take care of each step involved in the whole process. Algae cell walls are composed of a diverse array of fibrillar, matrix, and other polymers with complex composition (Gerken et al., 2013). These complex compositions are barriers, and thus they are required to break down the cell wall for the extraction of these high value pigments. Various researchers have reported the successful segregation of PBPs (Viskari and Colyer, 2003; Cuellar-Bermudez et al., 2015; Pan-utai and Iamtham, 2019). Such methods are time consuming and require multiple steps, which increases cost of downstream process, thus limiting its widespread application. If the number of steps increases, it results in a massive loss of product.

11.4.1 EXTRACTION OF PBPs

People have provided so many methods for pigment's characterization but not a single method exists that can practically be applied for technology-based downstreaming in a real manner. Common methods for cell disruption include physical (grinding, high-pressure homogenization, ultrasonication, repeated freezing and thawing, osmotic shock), chemical (Ionic liquids), and biological (enzymatic hydrolysis) extraction methods (Sekar and Chandramohan, 2008; Gerken et al, 2013) (Table 11.3).

TABLE 11.3
Extraction of Phycobiliproteins Using Different Protocol

Phycobiliproteins	Species	Extraction Protocol	Yield after Purification	Reference
CPC	*Spirulina platensis*	Homogenized aqueous two-phase extraction	79%	(Chethana et al., 2015)
CPC	*Spirulina platensis*	Freezing and thawing	14%	(Kumar et al., 2014)
CPC	*Limnotrhrix sp*	Freezing and thawing	8%	(Gantar et al., 2012a)
CPC	*Synechocystis aquatilis*	Osmotic shock	69%	(Ramos et al., 2011)
CPC	*Spirulina platensis*	Homogenized	77.3%	(Silveira et al., 2008)
CPC and APC	*Arthronema africanum*	Freeze–thawed	55% and 35%	(Minkova et al., 2007)
CPC	*Spirulina platensis*	*Klebsiella neumoniae* treatment	91%	(Zhu et al., 2007)
CPC	*Spirulina platensis; Phormidium sp.; Lyngbya sp.*	Freezing and thawing, Sonication	45.6%; 35.2%; 36.8%	(Patel et al., 2005)
CPCCPC	*Calothrix sp.*	EDTA and lysozyme treatment	80%	(Santiago-Santos et al.,2004)
APC	*Spirulina platensis*	Freezing and thawing	43%	(Su et al., 2010)
CPC	*Galdieria sulphuraria*	Homogenized	39%	(Sørensen et al., 2013)

11.4.1.1 Ultrasonication

This method ruptures cells by compressing and decompressing sound waves, which creates fragmentation, shear stress, and destruction of the cell wall (Chemat et al., 2017). Proportion of this mechanism differs according to the biomass types and process parameters. Generally, this method in labor saving in lab-scale experiments; thus, it is applied for disrupting cells of cyanobacteria with greater proficiency (Le Guillard et al., 2015; Mittal et al., 2017; Horváth et al., 2013). However, in some procedures heat production due to ultrasound waves denatures the structure of the protein molecule (Fleurence, 1999).

11.4.1.2 Freezing-Thawing

Freezing-thawing is a conventional method among all, which is very fruitful for PBP extraction form cyanobacteria. In this treatment, the cell wall structure is ruptured and thus pigments are extracted (Horváth et al., 2013; Lawrenz et al., 2011). Normally, at −20 °C or 0 °C cyanobacteria are frozen for several hours and, after that, thawed at 4 °C or room temperature. A repeated cycle of freezing-thawing will definitely give better results (Patel et al., 2005).

11.4.1.3 Osmotic Shock

A specific pH buffer of distilled water fresh biomass of cyanobacteria could be mixed and kept for several hours as is. In this mixture, due to the hypotonic effect, cell wall lyses and pigment oozing will occurs. Researchers used freeze-dried biomass of cyanobacteria prior to mix it with distilled water or buffer to enhance the extraction of PBPs (Patel et al., 2005; Kissoudi et al., 2018).

11.4.1.4 Biological and Chemical Extraction

Cell wall degradation by applying enzymatic hydrolysis is the lucrative stratagem with high biological specificity and without harming the protein structure. Phycocyanin was extracted by using 2 mg of lysozyme/g wet biomass, from *Calothrix* sp. (Santiago-Santos et al., 2004).

In the current scenario, ionic liquids (ILs) are widely used in various biological fields for the purpose of selective extraction (Tang et al., 2012). ILs are basically different salts made up of organic cations and various anions with melting point below 100 °C. ILs have negligible vapor pressure, which makes them very desirable solvents in contrast to organic solvents. Apart from this, ILs can be synthesized based on choice of cations and anions or can be mixed together for possible synergistic effects to be chemically stable, and soluble. Due to their magnificent properties, ILs are widely used in extraction (Pfruender et al., 2006; Huddleston et al., 2001; Marsh et al., 2002), dissolving cellulose (Swatloski et al., 2002), dissolution and extraction of protein (Lee et al., 2017), and extraction of essential oils (Liu et al., 2016). The interesting part of IL is that the properties of IL also applied to IL aqueous solutions. Since the extraction of bioactive compounds using IL is becoming more convenient and economical, various extraction processes are being developed to replace organic solvents with ILs. Such selection of proper extraction methods depends on cyanobacterial species.

11.5 PURIFICATION OF PBPS

11.5.1 Different Purification Methods

In a view of marvellous biotechnological application and commercial optimization, pigment's purity makes a major contribution; hence, a preferable purification method is needed. After extraction, PBP concentration is moderately low, ergo the purification step must ameliorate the purity of PBPs. The purification protocol is made up of different steps. Routine methods contain a combination of ammonium sulphate precipitation and chromatography technique (Table 11.4). Ammonium sulphate precipitation involves two steps; in the first step, it precipitates impurities up to 25–30% (W/V) ammonium sulphate, and in the second step, it precipitates impurities up to 60–70% (w/v). It removes cell debris and other impurities, so this step is necessary before chromatographic purification of PBPs. For further purification of PBPs, researchers have developed different column chromatography techniques, e.g., ion exchange chromatography, gel filtration chromatography, expanded bed adsorption chromatography, hydrophobic interaction chromatography, and hydroxyapatite chromatography (Abalde et al., 1998; Soni et al., 2006; Patil et al., 2006; Johnson et al., 2014; Lauceri et al., 2018). Gel filtration chromatography (size-exclusion chromatography) separates proteins according to their size, and ion exchange chromatography separates proteins according to their affinity toward ions. These two are common methods people are applying in lab scale. Sonani et al. (2014) have separated CPC, APC, and CPE with greater purity ratio by using a combination of ammonium sulphate precipitation, ion exchange, and size exclusion chromatography. Gradients of ionic strength help to purify PBPs while using ion exchange chromatography, but PBPs elution with different pH gradients is a more fruitful method (Su et al., 2010; Yan et al., 2011; Kumar et al., 2014). In expanded bed adsorption chromatography, protein separates directly from crude extract; thus, it eliminates a few steps from the purification protocol, which is a work-saving protocol (Niu et al., 2010). Aqueous two-phase extraction is globally used for purification of colourful PBPs. It can be applied for large-scale setups, but compared to this method, chromatographic methods are more proficient.

The column purification method requires changes and optimization to made it economically feasible for large-scale purification of PBPs. In the current scenario, tangential filtration and ultrafiltration protocols attracting people for separation of proteins, which are commonly used in biomedical and biotechnological application. Still, these techniques have not optimized for commercialization due to membrane-fouling phenomena. It can be used as a primary step before purification by chromatography to concentrate PBPs from other proteins in the solution.

11.4.2.2 Stability of Phycobiliproteins

Being proteinic in nature, stability of PBPs is a major factor that hampers the purification production and preservation; pH, light, temperature and ionic strength are the main parameters that should not be avoided throughout the extraction purification of PBPs. (Jespersen et al., 2005; Berns and MacColl, 1989; Sarada et al., 1999; Wu et al., 2016). Fukui et al. (2004) have concluded that remodeling of

TABLE 11.4

Purification of Phycobiliproteins

PBPs	Species	Purification Methods	Purity	Reference
CPC	Synechochoccus sp.	Hydrophobic interaction chromatography; ion exchange chromatography	4.85	(Abalde et al., 1998)
PE	Palmaria palmate	Preparative "native" PAGE	3.2	(Galland-Irmouli et al., 2000)
CPC	S. fusiformis	Rivanol treatment; ammonum sulphate precipitation; gel filtration chromatography; ammonum sulphate precipitation	4.3, 45.7	(Minkova et al., 2003)
CPC	O.quadripunctulata	Ammonum sulphate precipitation; gel filtration chromatography; anion exchange chromatography	3.31	(Soni et al., 2006)
CPC	S. platensis	Chitosan adsorption; two-phase aqueous extraction; ion exchange chromatography	6.69	(Patil et al., 2006)
CPC	P. fragile	Ammonum sulphate fractionation; hydrophobic interaction chromatography	4.52	(Soni et al., 2008)
CPC and APC	S. platensis	Ammonum sulphate precipitation; ion-exchange chromatography	5.59 and 5.19	(Yan et al., 2011)
PE	Heterosiphonia japonica	Ammonum sulphate precipitation; gel filtration chromatography; ion exchange chromatography	4.89	(Sun et al., 2009)
APC	Geitlerinema sp. A28DM	"Ethodin" precipitation; gel filtration chromatography	3.2	(Parmar et al., 2010)
PE	Porphyra yezoensis	Expanded bed absorption; ion-exchange chromatography	>4	(Niu et al., 2010)
APC	S. platensis	Ammonum sulphate precipitation; hydroxyapatite chromatography; ion-exchange chromatography	5	(Su et al., 2010)
PE	Phormidium sp. A27DM Lyngbya sp. A09DM,Halomicronema sp. A32DM	Ammonum sulphate precipitation; gel filtration chromatography	> 3.5	(Parmar et al., 2011b)

(Continued)

TABLE 11.4 (Continued)
Purification of Phycobiliproteins

PBPs	Species	Purification Methods	Purity	Reference
PE	*Pseudanabaena* sp.	Ammonum sulphate precipitation; gel filtration chromatography; ion exchange chromatography	6.86	(Mishra et al., 2011)
CPC	*Nostoc* sp.	Ion exchange chromatography; two-phase aqueous extraction	3.55	(Johnson et al., 2014)
PE, CPC nd APC	*Lyngbya* sp. A09DM	Triton X-100 mediated ammonium sulphate precipitation; ion exchange chromatography; gel filtration chromatography	6.75, 5.53 and 5.43	(Sonani et al., 2014)
PE	*S. platensis*	Purified by ammonium sulphate precipitation, followed by anion exchange chromatography using DEAE cellulose	4.74.	(Kamble et al., 2018)
CPC	*S. platensis*	Hydrophobic interaction membrane chromatography (HIMC)	4.20	(Lauceri et al., 2018)
CPCAPCPE	*S. platensis* AICB49*Synechocystis* sp. AICB51*Fremyella sp.* UTEX481, *Coelomaron pussilum* AICB1012	Aqueous two-phase system	>4.2>2.95.24.5	(Porav et al., 2020)

protein conformation ameliorates its stability. Sodium azide (NaN_3) and dithio-threitol (DTT) are used to avoid microbial contamination. However, they are toxic and cannot be used for edible purposes; so, they need to be replaced with natural or organic preservatives without affecting the compound's stability (Kannaujiya & Sinha, 2016). Natural or consumable preservatives can (e.g., citric acid, sucrose, sodium chloride, ascorbic acid, calcium chloride, and benzoic acid) improve the stability of purified CPC when incorporated in liquid media (Mishra et al., 2010; Chaiklahan et al., 2012). In aqueous solution, pH is a very important feature, which influences PBP structure association and dissociation. Hexameric structures stay stable at pH close by 7.0; thus, they avoid denaturation. At pH much higher or lower than pH 7, such structures dissociate. In pH 5.5 to 7, hexameric form of CPC was more stable, but at acidic pH it was precipitated (Chaiklahan et al., 2012). It is recommended to store and handle PBPs at lower temperature, because it degrades at higher temperature. At higher temperature, the amount of alpha helix in PBP structure decreases, which results in degradation and instability (Munier et al., 2014). For PBP preservation, suggested optimal temperature is around 4 °C; although it can stay stable until 40–45 °C; still, deliberate degradation happens. So, it is not good to preserve it in higher temperatures as it is more susceptible to microorganism con-tamination (Chaiklahan et al., 2012). Reports have been observed that CPC and CPE are light-sensitive pigments (Munier et al., 2014; Wu et al., 2016). It was observed that CPC degradation occurs after exposing it to 100 μ mol m^{-2} s^{-1} light intensity compared to lower intensity (Wu et al., 2016). When PBPs were exposed to light for longer terms, their chromophores started to degrade; thus, decoloration and loss in stability of PBPs occurs (Munier et al., 2014).

However, such factors consequently affect the stability of PBPs; still, there are some options to improve their stability, e.g., encapsulation, use of preservatives, nanofibers, and micelles, intramolecular crosslinking. The most easy and coherent option to increase PBPs' stability is use of additives, which ameliorates the thermal stability and does not require expensive instrumentation (Martelli et al., 2014; Gonzalez-Ramirez et al., 2014; Braga et al., 2016). Chaiklahan et al. (2012) re-ported that glucose, sucrose, and sodium chloride remarkably increased stability of CPC compared to sodium azide and sorbic acid. Higher sugar concentration also improves the stability of PBPs (Martelli et al., 2014). These salts work as protein stabilizing agents. Addition of sugar raises the surface tension of water, and thus the thermo stability of pigments increases (Chaiklahan et al., 2012; Braga et al., 2016). Use of nanofibers for stability purposes can reduce the denaturation rate by in-tegration with PBP structures, thus restraining conformation of protein in un-favorable conditions (Kannaujiya and Sinha, 2016; Braga et al., 2016). According to one study, intramolecular crosslinking between silver nanoparticles and protein molecules increased the stability of phycoerythrin (Bekasova et al., 2016; Selig et al., 2018). Stabilization of phycocyanin molecules was also done by covalent crosslinking between α and β subunits of CPC by formaldehyde; it prohibits the dissociation of subunits (Sun et al., 2006). In the microencapsulation approach, the protein structure is insulated by coating material (Yan et al., 2014). Micelles are amphiphilic molecules that comprise of hydrophilic and hydrophobic fragments; it also helps PBPs to retain their structure. Sodium dodecyl sulfate (SDS) micelles confines the phycocyanin molecule inside and stabilizes it via hydrophobic inter-actions. Thus, the blue color of CPC is maintained and the formation of the pro-tonated (green) form is prohibited (Falkeborg et al., 2018). As pigments, stability is the major issue that hampers PBPs' application in various areas. Different research groups are working on it. However, further research is needed to compare these methods.

11.5 BIOTECHNOLOGICAL IMPORTANCE

Recently, PBPs were considered to be essential biomolecules for various pharma-ceutical, nutraceutical, and cosmetic applications. They are considered in such in-dustrial applications, mainly due to their non-toxic and non-carcinogenic effects, making them an excellent agent for various food and pharmaceutical procedures. The following are some of the major applications explored by various research groups for suitable pharmaceutical and nutraceuticals from cyanobacterial PBPs.

11.5.1 Antioxidant Capacity of Phycobiliproteins

In the last few decades, growing interest in natural compounds for the production of antioxidants is increasing due to its non-toxic and human health-promoting activity. Among such natural compounds of cyanobacteria, PBPs are the promising candi-dates due to their high antioxidant activity (Patel et al., 2005). Due to various conditions like pathogen attacks, metabolic disorder cells generate a high amount of toxic reactive oxygen species (ROS) (Gill and Tuteja, 2010). These ROS are toxic to the cells and damage the cells' DNA, proteins, and lipids. The structure of phycobilin is similar to human biliverdin and bilirubin, which serve as a natural antioxidant for the body (Strasky et al., 2013). A recent study by Sonani et al. showed that among all the PBP extracted from cyanobacterium *Lyngbya* sp. A09DM, CPE exhibits the highest antioxidant activity (Sonani et al., 2014). Further, they mentioned that the difference in PBPs' antioxidant activity may be due to a difference in amino acid composition in the side chain (Sonani et al., 2014). Another study indicates selenium-rich phycocyanin obtained from selenium-cultivated *Spirulina platensis* scavenge superoxide, hydrogen peroxide, and 2,2-diphenyl-1-picryl-hydrazine (DPPH) very efficiently (Huang et al., 2007). A study with recombinant APC expressed in *E. coli* indicates that the recombinant APC can scavenge hydroxyl ions. Further, it is also indicated that the independent α and β subunits show higher activity compared to their combination (Ge et al., 2006).

11.5.2 Anti-Tumor Activity of Phycobiliproteins

Cancer is among the most prevalent diseases worldwide and results in the deaths of many people. Globally, various research organizations and NGOs are funding heavily to find the cure of various cancers. Recently, extracts from plants, algae, and other organisms showed the potential to use anti-cancer agents due to their tumor-suppressive activity (Simmons et al., 2005). PBPs work as a potential anti-tumor compound because they are non-toxic to the cells. However, at the same time, they will kill the cancerous cells due to their antioxidative and anti-tumor activity. Various studies conducted using different types of PBPs (PC, PE, and APC) indicated potential *in vitro* and *in vivo* activity of these cyanobacterial PBPs. An *in vivo* study with mouse skin tumors indicated that the CPC-treated cells showed a reduction in tumor-specific markers like ornithine decarboxylase, cyclooxygenase-2, and interleukin 6 and other cells transducers, which otherwise have a very high amount in control. Additionally, the authors also mentioned that CPC's tumor-suppressive effect is dose-dependent (Gupta and Gupta, 2012). *In vivo* studies also indicate that CPC can inhibit the growth of HT-29 (colon cancer) and A549 (lung cancer). The anti-cancer activity is due to the effect of CPC in inhibition of DNA replication (Thangam et al., 2013). How PBPs inhibit cancerous cells' growth is not known fully until recently to date, and various research groups are working on this to identify the molecular mechanism behind this process. It might be possible that PBPs inhibit the cell cycle regulation in such cancerous cells or might be involved in regulations of apoptosis or hyperplasia, which is mainly responsible for cancer. It has also been possible that PBPs affect caspase-9 and caspase-3 enzymes involved in cancer (Gantar et al., 2012b). Apart from various *in vitro* and *in vivo* anti-tumor activity, PBPs are also an excellent photosensitizer for photodynamic therapy for cancer treatment. Phycocyanin from cyanobacteria *Microcystis* showed the potential to use as photodynamic therapy. HepG2 cells when incubated with phycocyanin from *Microcystis* and when illuminated with laser radiation. The MTT indicates that 200 µg/mL phycocyanin dose is most effective in inhibiting the growth of HepG2 (Wang et al., 2012). Another study also indicates that when phycocyanin is incubated with breast cancer cells MDA-MB-231 and irradiated, it generates reactive oxygen species and kills the cancerous cells, indicating these PBPs have the potential to develop as agents for photodynamic therapy (Bharathiraja et al., 2016).

11.5.3 Phycobiliproteins as Neuroprotective Agents

In cells, usually reactive oxygen species are generated through the mitochondrial respiratory system and neutralize through the cell's antioxidative systems. However, during the lateral part of life, this balance of ROS production and counterattack by one's own antioxidative system is impaired. These conditions result in the generation of various age-related disorders like atherosclerosis, arthritis, cancer, diabetes, and neurodegeneration (Pan et al., 2011). It has been reported that aging is due to the accumulation of a large amount of ROS in the cells or due to the downstream effects of the insulin/insulin-like growth factor pathway (Blagosklonny, 2008). As PBPs are excellent antioxidants, supplementation of PBPs may work as a neuroprotective agent and improve the organisms' life span. A study by Sonani et al. indicates that feeding of purified PE to the model organism *Caenorhabditis elegans* improves the organisms' thermal and oxidative stress tolerance (Sonani et al., 2014). Additionally, they have also reported that feeding PE to *C. elegans* also inhibits αβ plaque formation. However, they do not find that the ILS pathway's involvement in this process indicates that PE might use different pathways and mechanisms for anti-aging activity (Sonani et al., 2014). Another PBP allophycocyanin is isolated from *Phormidium* sp. A09DM also showed anti-aging activity (Chaubey et al., 2020). 100 µg/mL APC treatment improves the lifespan of *C.elegans* from 16 to 21 days and improves the various physiological functions like pharyngeal pumping. Further, the study also indicates that an increase in the life span involved DAF-16 dependent and SKN-1 dependent (Chaubey et al., 2020). Crystal structure analysis and molecular modeling study of phycocyanin indicate it interacts with enzyme β-secretase, which is involved in amyloid precursor protein's proteolysis from plaques. The study also indicates that phycocyanin interaction is energetically more favorable than all known inhibitors of β-secretase (Singh and Montgomery, 2013). These results indicate that PBPs might be novel therapeutics for the treatment of various age-related disorders.

11.5.4 PHYCOBILIPROTEINS AS FLUORESCENT PROBE

PBPs have a high molar extinction coefficient and fluorescence quantum yield, which leads to high fluorescent activity in PBPs. Due to the fluorescent activity of PBPs, they are also used as a fluorescent probe in various applications like flow cytometers, fluorescence-activated cell sorting, immunodiagnostics, etc. PBPs are generally conjugated with immunoglobulins, protein A, biotin, and avidin and are used in various applications (Oi et al., 1982). Among various PBPs, PE is one of the best for development as a fluorescent probe. A PE-conjugated detection system for the Hendra virus will take only 20 min, while the conventional detection system will take almost 2–3 hr to detect the virus. Such a study indicates the development of PBPs as a fluorescent probe will reduce the time and give more accuracy to the detection system (Gao et al., 2015).

11.5.5 PHYCOBILIPROTEINS AS FOOD COLORANTS

Color is essential in various food and cosmetic industries. Colors give an attractive look as well as help to distinguish products from different manufacturers. Generally, artificial dye is used for such purposes; however, as these artificial colors are synthesized using a chemical reaction, sometimes they are toxic and cause cancer and other diseases. On the other hand, natural color extracted from plants, algae, and cyanobacteria is a non-toxic and natural compound that may be replacing these toxic synthetic colors. Cyanobacterial PBPs are used as natural colorants in various food industries; for example, phycocyanin from Spirulina is used as colorants for ice creams, yogurts, confectionery, and jellies (Jespersen et al., 2005). One of the major problems in using PBPs in such industries is their reduced stability at operational parameters, such as temperature, pH, salinity, etc. Improvement in the stability of PBPs by adding certain chemicals or exploration of extremophilic strain will be required to commercialize PBPs as natural colorants.

11.6 CONCLUSION AND FUTURE PROSPECTS

Microalgae and cyanobacteria are diverse photosynthetic organisms that are considered to be one of the best bioresources for the production of biofuels and industrially relevant compounds. However, few cyanobacteria also have the ability to

obtain organic nutrients from various types of wastewaters, which means further
reduction in environmental pollution load and production of sustainable compounds. As they are photosynthetic organisms, in order to perform efficient photosynthesis and to produce different products, they are evolved with different
pigment compositions, such as chlorophyll-a, chlorophyll-b, carotenoids, and PBPs.
Among various light-harvesting pigments, PBPs are considered to be one of the
very important nutraceutical compounds due to their high antioxidative and other
characteristics. Various industrial and medicinal applications require high-quantity
and high-purity PBPs from cyanobacteria. Naturally occurring cyanobacteria generally contain a lot PBP content. Various cultural parameters like source of nitrogen, carbon, light intensity, pH, etc. significantly affect the growth of
cyanobacteria and the production potential of PBPs. Optimization of these cultural
parameters through one factor at a time as well as using various statistical tools is
among the best methods to produce cyanobacterial biomass with a high concentration of desired PBPs. As cyanobacteria are cosmopolitan organisms, they
have the ability to utilize various amounts and spectrums of light, so among various
parameters, light is one of the best parameters that affects the quality and quantity
of PBPs in the cyanobacteria. Apart from light cyanobacteria, also present in the
environment with high salt concentration, high light intensity, etc., further understanding of such parameters and how they affect the PBP composition in cyanobacteria is very important for large-scale development. Most of the time,
optimization of growth and cultural parameters will not give much enhancement in
the production of desired end products, including PBPs in the cells, due to various
metabolic constraints. In such conditions, metabolic engineering is the best strategy
to improve the growth and production of PBPs in cyanobacteria. Due to availability
of complete genome sequences of various model cyanobacteria, such as
Synechococcus elongatus PCC 7942, *Synechocystis* sp. PCC 6803, etc., as well as
availability of various molecular genetic tools, it is very easy in recent times to
generate the desired engineered strain with high growth and PBP content. Apart
from this approach, utilization of diazotrophic cyanobacteria, which have the ability
to obtain nitrogen from the atmosphere and directly incorporate it into the metabolism for production of various compounds, including PBPs, is one of the novel
subjects of research. Further exploitation of such an approach will help our understanding of PBP synthesis and its further industrial utilization. After optimization, the next biggest challenge in commercialization of PBP-based products is
extraction and purification of these pigments. As PBPs are a type of protein, any
extreme physical, chemical, or mechanical process can damage the structure and
activity of these proteins. As they do not have any cell wall, most of the time the
normal freeze-thaw cycle is sufficient for extraction of intracellular components,
including PBP, from the cells. However, at large or commercial scale, such an
approach is not good, so use of ultrasonication or identification of a cheap source of
enzyme that can be used is a good alternative. Apart from such challenges, identification of cyanobacteria from an extreme environment is one of the important
areas in recent times because it is believed that the PBPs extracted from such
extremophilic cyanobacteria have high stability at various industrial processes
like high temperature, salinity, pH, etc. Isolation of such cyanobacteria and

characterization and structural elucidation of their PBPs is very important for commercialization of these algal products. Additionally, these compounds have known antioxidant and anticancer activity; however, further identification of molecular mechanism behind this phenomenon using model systems like *C. elegans* is essential before its application to humans. Overall, PBPs are an excellent phytomedicine for various neurodegenerative diseases, cancers, and other disorders. Additionally, they are used as colorants for the food industry.

ACKNOWLEDGMENTS

KB and SM would like to thank the Council of Scientific and Industrial Research for financial support. IP would like to acknowledge financial support from DST-SERB (SRG/2020/000165).

REFERENCES

Aiba, S., Ogawa, T., 1977. Assessment of growth yield of a blue—green alga, *Spirulina platensis*, in axenic and continuous culture. *Microbiology*. 102, 179–182.

Ajayan, K. V., Selvaraju, M., Thirugnanamoorthy, K., 2012. Enrichment of chlorophyll and phycobiliproteins in *Spirulina platensis* by the use of reflector light and nitrogen sources: An in-vitro study. *Biomass Bioenerg*. 47, 436–441.

Allen, M.M., Smith, A.J., 1969. Nitrogen chlorosis in blue-green algae. *Archiv für Mikrobiologie*. 69, 114–120.

Allen, M.M., 1986. Simple conditions for growth of unicellular blue-green algae on plates 1, 2. *J. Phycol*. 4, 1–4.

Allen, M.M., Stanier R.Y., 1986. Growth and division of some unicellular blue-green algae. *Microbiology*. 51, 199–202.

Bachchhav, M., Bhanudas, M.V.K., Ingale, A.G., 2017. Enhanced phycocyanin production from *Spirulina platensis* using light emitting diode. *J. Inst. Eng. (India): E*. 98, 41–45.

Bekasova, O.D., Borzova, V.A., Shubin, V.V., Kovalyov, L.I., Stein-Margolina, V.A., Kurganov, B.I., 2016. An increase in the resistance of R-phycoerythrin to thermal aggregation by silver nanoparticles synthesized in nanochannels of the pigment. *Appl. Biochem. Microbiol*. 52, 98–104.

Berman-Frank, I., Lundgren, P., Falkowski, P., 2003. Nitrogen fixation and photosynthetic oxygen evolution in cyanobacteria. *Res. Microbiol*. 154, 157–164.

Berns, D.S., MacColl, R., 1989. Phycocyanin in physical chemical studies. *Chem. Rev*. 89, 807–825.

Bharadwaj, S.V., Vamsi, S.R., Pancha, I., Mishra, S., 2020. Recent trends in strain improvement for production of biofuels from microalgae, in: *Microalgae Cultivation for Biofuels Production*. Academic Press, pp. 211–225.

Bharathiraja, S., Seo, H., Manivasagan P., Moorthy M.S., Park, S., Oh J., 2016. In vitro photodynamic effect of phycocyanin against breast cancer cells. *Molecules*. 21, 1470.

Bhayani, K., Paliwal, C., Ghosh, T., Mishra, S., 2018. Nutra-cosmeceutical potential of pigments from microalgae, in: Rastogi, R.P. (Ed.), *Sunscreens: Source, Formulations, Efficacy and Recommendations*. NovaScience Publishers, Inc., New York, pp. 29–52.

Biswas, A., 2011. Identification and characterization of enzymes involved in the biosynthesis of different phycobiliproteins in cyanobacteria. Thesis.

Blagosklonny, M.V., 2008. Aging: Ros or tor. *Cell Cycle*. 7, 3344–3354.

Bold, Harold C., 1949. The morphology of Chlamydomonas chlamydogama, sp. nov. *Bull. Torrey Bot. Club*. 76, 101–108.

Boussiba, S., Richmond, A.E., 1980. C-phycocyanin as a storage protein in the blue-green alga *Spirulina platensis*. *Arch. Microbiol.* 125, 143–147.

Braga, A.R.C., Figueira F.S., Silveira J.T., Morais M.G., Costa J.A.V., Kalil S.J., 2016. Improvement of thermal stability of c-phycocyanin by nanofiber and preservative agents. *J. Food Process. Preserv.* 40, 1264–1269.

Brock, T.D., 2012. *Thermophilic microorganisms and life at high temperatures*. Springer Science & Business Media.

Chaiklahan, R., Chirasuwan, N., Bunnag, B., 2012. Stability of phycocyanin extracted from *Spirulina* sp.: Influence of temperature, pH and preservatives. *Process Biochem.* 47, 659–664.

Chaneva, G., Furnadzhieva, S., Minkova, K., Lukavsky, J., 2007. Effect of light and temperature on the cyanobacterium *Arthronema africanum* – A prospective phycobiliprotein-producing strain. *J. Appl. Phycol.* 19, 537–544.

Chaubey, M.G., Patel, S.N., Rastogi, R.P., Madamwar, D., Singh, N.K., 2020. Cyanobacterial pigment protein allophycocyanin exhibits longevity and reduces Aβ-mediated paralysis in *C. elegans*: Complicity of FOXO and NRF2 ortholog DAF-16 and SKN-1. *3 Biotech*. 10, 1–11.

Chen, C.-Y., Kao, P.-C., Tsai, C.-J., Lee, D.-J., Chang, J.-S., 2013. Engineering strategies for simultaneous enhancement of C-phycocyanin production and CO_2 fixation with *Spirulina platensis*. *Bioresour. Technol.* 145, 307–312.

Chen, Y.-B., Zehr, J.P., Mellon, M., 1996. Growth and nitrogen fixation of the diazotrophic filamentous nonheterocystous cyanobacterium *Trichodesmium* sp. Ims 101 in defined media: Evidence for a circadian rhythm. *J. Phycol.* 32, 916–923.

Chethana, S., Nayak, C.A., Madhusudhan, M.C., Raghavarao, K.S.M.S., 2015. Single step aqueous two-phase extraction for downstream processing of C-phycocyanin from *Spirulina platensis*. *J. Food Sci. Technol.* 52, 2415–2421.

Christaki, Efterpi, Bonos, E., Florou-Paneri, P., 2015. Innovative microalgae pigments as functional ingredients in nutrition, in: Kim, Se-Kwon (Ed.), *Handbook of Marine Microalgae*. Academic Press, pp. 233–243.

Chu, S. P., 1942. The influence of the mineral composition of the medium on the growth of planktonic algae: Part I. Methods and culture media. *J. Ecol.*, 284–325.

Colla, L.M., Reinehr, C.O., Reichert, C., Costa, J.A.V., 2007. Production of biomass and nutraceutical compounds by *Spirulina platensis* under different temperature and nitrogen regimes. *Bioresour. Technol.* 98, 1489–1493.

Collier, J.L., Grossman, A.R., 1994. A small polypeptide triggers complete degradation of light-harvesting phycobiliproteins in nutrient-deprived cyanobacteria. *EMBO J.*. 13, 1039–1047.

Cuellar-Bermudez, S.P., Aguilar-Hernandez, I., Cardenas-Chavez, D.L., Ornelas-Soto, N., Romero-Ogawa, M.A., Parra-Saldivar, R., 2015. Extraction and purification of high-value metabolites from microalgae: Essential lipids, astaxanthin and phycobili-proteins. *Microb. Biotechnol.* 8, 190–209.

da Fontoura Prates, D., Radmann, E.M., Duarte, J.H., Morais, M.G., Costa, J.A.V., 2018. *Spirulina* cultivated under different light emitting diodes: Enhanced cell growth and phycocyanin production. *Bioresour. Technol.* 256, 38–43.

De Marsac, N.T., Cohen-Bazire, G., 1977. Molecular composition of cyanobacterial phycobilisomes. *Proc. Natl. Acad. Sci.* 74, 1635–1639.

Deschoenmaeker, F., Bayon-Vicente, G., Sachdeva, N., Depraetere, O., Pino, J.C.C., Leroy, B.,... Wattiez, R., 2017. Impact of different nitrogen sources on the growth of *Arthrospira* sp. PCC 8005 under batch and continuous cultivation – A biochemical, transcriptomic and proteomic profile. *Bioresour. Technol.* 237, 78–88.

Fábregas, J., García, D., Morales, E., Domínguez, A., Otero, A., 1998. Renewal rate of semicontinuous cultures of the microalga *Porphyridium cruentum* modifies phycoerythrin, exopolysaccharide and fatty acid productivity. *J. Ferment. Bioeng.* 86, 477–481.

Falkeborg, M.F., Roda-Serrat, M.C., Burnæs, K.L., Nielsen, A.L.D., 2018. Stabilising phycocyanin by anionic micelles. *Food Chem.* 239, 771–780.

Fatma, T., 2009. Screening of cyanobacteria for phycobiliproteins and effect of different environmental stress on its yield. *Bull. Environ. Contam. Toxicol.* 83, 509.

Fischer, W.W., 2008. Life before the rise of oxygen. *Nature.* 455, 1051–1052.

Fleurence, J., 1999. Seaweed proteins: Biochemical, nutritional aspects and potential uses. *Trends Food Sci. Technol.* 10, 25–28.

Fukui, K., Saito, T., Noguchi, Y., Kodera, Y., Matsushima, A., Nishimura, H., Inada, Y., 2004. Relationship between color development and protein conformation in the phycocyanin molecule. *Dyes Pigm.* 63, 89–94.

Galland-Irmouli, A.V., Pons, L., Lucon, M., Villaume, C., Mrabet, N.T., Guéant, J. L., Fleurence, J., 2000. One-step purification of R-phycoerythrin from the red macroalga *Palmaria palmata* using preparative polyacrylamide gel electrophoresis. *J. Chromatogr. B Biomed. Sci. Appl.* 739, 117–123.

Gantar, M., Dhandayuthapani, S., Rathinavelu, A., 2012b. Phycocyanin induces apoptosis and enhances the effect of topotecan on prostate cell line LNCaP. *J. Med. Food.* 15, 1091–1095.

Gantar, M., Simović, D., Djilas, S., Gonzalez, W.W., Miksovska, J., 2012a. Isolation, characterization and antioxidative activity of C-phycocyanin from *Limnothrix* sp. strain 37-2-1. *J. Biotechnol.* 159, 21–26.

Gantt, E., 1975. Phycobilisomes: Light-harvesting pigment complexes. *Bioscience.* 25, 781–788.

Gao, Y., Pallister, J., Lapierre, F., Crameri, G., Wang, L.-F., Zhu, Y., 2015. A rapid assay for Hendra virus IgG antibody detection and its titre estimation using magnetic nanoparticles and phycoerythrin. *J. Virol. Methods.* 222, 170–177.

Ge, B., Qin, S., Han, L., Lin, F., Ren Y., 2006. Antioxidant properties of recombinant allophycocyanin expressed in *Escherichia coli*. *J. Photochem. Photobiol. B: Biol.* 84, 175–180.

George, B., Pancha, I., Desai, C., Chokshi, K., Paliwal, C., Ghosh, T., Mishra, S., 2014. Effects of different media composition, light intensity and photoperiod on morphology and physiology of freshwater microalgae *Ankistrodesmus falcatus* – A potential strain for bio-fuel production. *Bioresour. Technol.* 171, 367–374.

Gerken, H.G., Donohoe, B., Knoshaug, E.P., 2013. Enzymatic cell wall degradation of *Chlorella vulgaris* and other microalgae for biofuels production. *Planta,* 237, 239–253.

Gill, S.S., Tuteja N., 2010. Reactive oxygen species and antioxidant machinery in abiotic stress tolerance in crop plants. *Plant Physiol. Biochem.* 48, 909–930.

Glazer, A.N., 1994. Phycobiliproteins – A family of valuable, widely used fluorophores. *J. Appl. Phycol.* 6, 105–112.

Gonzalez-Ramirez, E., Andujar-Sanchez, M., Ortiz-Salmeron, E., Bacarizo, J., Cuadri, C., Mazzuca-Sobczuk, T., Ibáñez, M.J., Camara-Artigas, A., Martinez-Rodriguez, S., 2014. Thermal and pH stability of the B-Phycoerythrin from the Red Algae *Porphyridium cruentum*. *Food Biophys.* 9, 184–192.

Graham, J.M., Auer, M.T., Canale, R.P., Hoffmann, J.P., 1982. Ecological studies and mathematical modeling of *Cladophora* in Lake Huron: 4. Photosynthesis and respiration as functions of light and temperature. *J. Great Lakes Res.* 8, 100–111.

Guillard, R.R.L.,1975. Culture of phytoplankton for feeding marine invertebrates, in: *Culture of Marine Invertebrate Animals*. Springer, Boston, MA, pp. 29–60.

Guillard, R.R.L., Hargraves, P.E., 1993. Stichochrysis immobilis is a diatom, not a chrysophyte. *Phycologia.* 32, 234–236.

Guillard, R.R.L., Ryther, J.H., 1962. Studies of marine planktonic diatoms: I. *Cyclotella nana* Hustedt, and *Detonula confervacea* (Cleve) Gran. *Can. J. Microbiol.* 8, 229–239.

Gupta, N.K., Gupta, K.P., 2012. Effects of C-Phycocyanin on the representative genes of tumor development in mouse skin exposed to 12-O-tetradecanoyl-phorbol-13-acetate. *Environ. Toxicol. Pharmacol.* 34, 941–948.

Gysi, J.R., Zuber, H., 1979. Properties of allophycocyanin II and its α-and β-subunits from the thermophilic blue–green alga *Mastigocladus laminosus. Biochem. J.* 181, 577–583.

Herrero, A., Muro-Pastor, A.M., Flores, E., 2001. Nitrogen control in cyanobacteria. *J. Bacteriol.* 183, 411–425.

Horváth, H., Kovács, A.W., Riddick, C., Présing, M., 2013. Extraction methods for phycocyanin determination in freshwater filamentous cyanobacteria and their application in a shallow lake. *Eur. J. Phycol.* 48, 278–286.

Huang, Z., Guo, B.J., Wong, R.N.S., Jiang, Y., 2007.Characterization and antioxidant activity of selenium-containing phycocyanin isolated from *Spirulina platensis. Food Chem.* 100, 1137–1143.

Huddleston, J.G., Visser, A.E., Matthew Reichert, W., Willauer, H.D., Broker, G.A., Rogers, R.D., 2001. Characterization and comparison of hydrophilic and hydrophobic room temperature ionic liquids incorporating the imidazolium cation. *Green Chem.* 3, 156–164.

Jespersen, L., Strømdahl, L.D., Olsen, K., Skibsted, L.H., 2005. Heat and light stability of three natural blue colorants for use in confectionery and beverages. *Eur. Food Res. Technol.* 220. 261–266.

Jian-Feng, N.I.U., Guang-Ce, W.A.N.G., Lin, X., Zhou, B.-C., 2007. Large-scale recovery of C-phycocyanin from *Spirulina platensis* using expanded bed adsorption chromatography. *J. Chromatogr. B.* 850, 267–276.

Johnson, E.M., Kumar, K., Das, D., 2014. Physicochemical parameters optimization, and purification of phycobiliproteins from the isolated *Nostoc* sp. *Bioresour. Technol.* 166, 541–547.

Kamble, S.P., Vikhe, G.P., Chamle, D.R., 2018. Extraction and purification of phycoerythrin—A natural colouring agent from *Spirulina platensis. J. Pharm. Chem. Biol. Sci.* 6, 78–84.

Kannaujiya, V.K., Sinha, R.P., 2016. Thermokinetic stability of phycocyanin and phycoerythrin in food-grade preservatives. *J. Appl. Phycol.* 28, 1063–1070.

Kathiresan, S., Sarada, R., Bhattacharya, S., Ravishankar, G.A., 2007. Culture media optimization for growth and phycoerythrin production from *Porphyridium purpureum. Biotechnol. Bioeng.* 96, 456–463.

Keller, M.D., Selvin, R.C., Claus, W., Guillard, R.R.L., 1987. Media for the culture of oceanic ultraphytoplankton. *J. Phycol.* 23, 633–638.

Kenekar, A.A., Deodhar, M.A., 2013. Effect of varying physicochemical parameters on the productivity and phycobiliprotein content of indigenous isolate *Geitlerinema sulphureum. Biotechnology.* 12, 146–154.

Kennedy, M., Krouse, D., 1999. Strategies for improving fermentation medium performance: A review. *J. Ind. Microbiol. Biotechnol.* 23, 456–475.

Khatoon, H., Leong, L.K., Rahman, N.A., Mian, S., Begum, H., Banerjee, S., Endut, A., 2018. Effects of different light source and media on growth and production of phycobiliprotein from freshwater cyanobacteria. *Bioresour. Technol.* 249, 652–658.

Kilham, S.S., Kreeger, D.A., Lynn, S.G., Goulden, C.E., Herrera, L., 1998. COMBO: A defined freshwater culture medium for algae and zooplankton. *Hydrobiologia.* 377, 147–159.

Kissoudi, M., Sarakatsianos, I., Samanidou, V., 2018. Isolation and purification of food-grade C-phycocyanin from *Arthrospira platensis* and its determination in confectionery by HPLC with diode array detection. *J. Sep. Sci.* 41, 975–981.

Kronick, M.N., Grossman, P.D., 1983. Immunoassay techniques with fluorescent phycobiliprotein conjugates. *Clin. Chem.* 29, 1582–1586.

Kulandaivelu, G., Gheetha, V., Periyanan, S., 1989. Inhibition of energy transfer reactions in cyanobacteria by different ultraviolet radiation, in: *Photosynthesis*. Springer, Berlin, Heidelberg. pp. 305–313.

Kumar, D., Dhar, D.W., Pabbi, S., Kumar, N., Walia, S., 2014. Extraction and purification of C-phycocyanin from *Spirulina platensis* (CCC540). *Indian J. Plant Physiol.* 19, 184–188.

Lauceri, R., Zittelli, G.C., Maserti, B., Torzillo, G., 2018. Purification of phycocyanin from *Arthrospira platensis* by hydrophobic interaction membrane chromatography. *Algal Res.* 35, 333–340.

Lawrenz, E., Fedewa, E.J., Richardson, T.L., 2011. Extraction protocols for the quantification of phycobilins in aqueous phytoplankton extracts. *J. Appl. Phycol.* 23, 865–871.

Lee, S. Ying, I.K., Ooi, C.W., Ling, T.C., Show, P.L., 2017. Recent advances in protein extraction using ionic liquid-based aqueous two-phase systems. *Sep. Purif. Rev.* 46, 291–304.

Liu, L.N., Chen, X.L., Zhang, Y.Z., Zhou, B.C., 2005. Characterization, structure and function of linker polypeptides in phycobilisomes of cyanobacteria and red algae: An overview. *Biochim. Biophys. Acta (BBA)-Bioenerg.* 1708, 133–142.

Liu, Z., Chen, Z., Han, F., Kang, X., Gu, H., Yang, L., 2016. Microwave-assisted method for simultaneous hydrolysis and extraction in obtaining ellagic acid, gallic acid and essential oil from *Eucalyptus globulus* leaves using Brönsted acidic ionic liquid [HO3S (CH2) 4mim] HSO4. *Ind. Crops Prod.* 81, 152–161.

Luque, I., Flores, E., Herrero, A., 1994. Molecular mechanism for the operation of nitrogen control in cyanobacteria. *EMBO J.* 13, 2862–2869.

MacColl, R., 1998. Cyanobacterial phycobilisomes. *J. Struct. Biol.* 124, 311–334.

Marsh, K.N., Deev, A., Wu, A.C.T., Tran, E., Klamt, A., 2002. Room temperature ionic liquids as replacements for conventional solvents – A review. *Korean J. Chem. Eng.* 19, 357–362.

Martelli, G., Folli, C., Visai, L., Daglia, M., Ferrari, D., 2014. Thermal stability improvement of blue colorant C-Phycocyanin from *Spirulina platensis* for food industry applications. *Process Biochem.* 49, 154–159.

Mehariya, S., Fratini, F., Lavecchia, R., Zuorro, A., 2021a. Green extraction of value-added compounds form microalgae: A short review on natural deep eutectic solvents (NaDES) and related pre-treatments. *J. Environ. Chem. Eng.* 9, 105989.

Mehariya, S., Goswami, R.K., Karthikeysan, O.P., Verma, P., 2021b. Microalgae for high-value products: A way towards green nutraceutical and pharmaceutical compounds. *Chemosphere.* 130553.

Mehariya, S., Goswami, R.K., Verma, P., Lavecchia, R., Zuorro, A., 2021c. Integrated approach for wastewater treatment and biofuel production in microalgae biorefineries. *Energies.* 14, 2282.

Minkova, K.M., Tchernov, A.A., Tchorbadjieva, M.I., Fournadjieva, S.T., Antova, R.E.,Busheva, M.Ch., 2003. Purification of C-phycocyanin from *Spirulina (Arthrospira) fusiformis*. *J. Biotechnol.* 102, 55–59.

Minkova, K., Tchorbadjieva, M., Tchernov, A., Stojanova, M., Gigova, L., Busheva, M., 2007. Improved procedure for separation and purification of *Arthronema africanum* phycobiliproteins. *Biotechnol. Lett.* 29, 647–651.

Mishra, S.K., Shrivastav, A., Mishra, S., 2011. Preparation of highly purified C-phycoerythrin from marine cyanobacterium *Pseudanabaena* sp. *Protein Expr. Purif.* 80, 234–238.

Mohan, S.V., Rohit, M.V., Chiranjeevi, P., Chandra, R., Navaneeth, B., 2015. Heterotrophic microalgae cultivation to synergize biodiesel production with waste remediation: progress and perspectives. *Bioresour. Technol.* 184, 169–178.

Moon, M., Mishra, S.K., Kim, C.W., Suh, W.I., Park, M.S., Yang, J.-W., 2014. Isolation and characterization of thermostable phycocyanin from *Galdieria sulphuraria*. *Korean J. Chem. Eng.* 31, 490–495.

Munier, M., Jubeau, S., Wijaya, A., Morancais, M., Dumay, J., Marchal, L., Jaouen, P., Fleurence, J., 2014. Physicochemical factors affecting the stability of two pigments: R-phycoerythrin of *Grateloupia turuturu* and B-phycoerythrin of *Porphyridium cruentum*. *Food Chem.* 150, 400–407.

Nakamoto, H., Honma, D., 2006. Interaction of a small heat shock protein with light-harvesting cyanobacterial phycocyanins under stress conditions. *FEBS Lett.* 580, 3029–3034.

Niu, J.-F., Chen, Z.-F., Wang, G.-C., Zhou, B.-C., 2010. Purification of phycoerythrin from *Porphyra yezoensis* Ueda (Bangiales, Rhodophyta) using expanded bed absorption. *J. Appl. Phycol.* 22, 25–31.

Oi, V.T., Glazer, A.N., Stryer, L., 1982. Fluorescent phycobiliprotein conjugates for analyses of cells and molecules. *J. Cell Biol.* 93, 981–986.

Olaizola, M., Duerr, E.O., 1990. Effects of light intensity and quality on the growth rate and photosynthetic pigment content of *Spirulina platensis*. *J. Appl. Phycol.* 2, 97–104.

Pagels, F., Catarina Guedes, A., Amaro, H.M., Kijjoa, A., Vasconcelos, V., 2019. Phycobiliproteins from cyanobacteria: Chemistry and biotechnological applications. *Biotechnol. Adv.* 37, 422–443.

Paliwal, C., Mitra, M., Bhayani, K., Vamsi Bharadwaj, S.V., Ghosh, T., Dubey, S., Mishra, S., 2017. Abiotic stresses as tools for metabolites in microalgae. *Bioresour. Technol.* 244, 1216–1226.

Pan, M., Jiang, T.S., Pan, J.L., 2011. Antioxidant activities of rapeseed protein hydrolysates. *Food Bioproc. Technol.* 4, 1144–1152.

Pan-Utai, W., Iamtham, S., 2019. Extraction, purification and antioxidant activity of phycobiliprotein from *Arthrospira platensis*. *Process Biochem.* 82, 189–198.

Parmar, A., Singh, N.K., Kaushal, A., Sonawala S., Madamwar, D., 2011b. Purification, characterization and comparison of phycoerythrins from three different marine cyanobacterial cultures. *Bioresour. Technol.* 102, 1795–1802.

Parmar, A., Singh, N.K., Madamwar, D., 2010. Allophycocyanin from a local isolate *Geitlerinema* sp. A28DM (Cyanobacteria): A simple and efficient purification process. *J. Phycol.* 46, 285–289.

Parmar, A., Singh, N.K., Pandey, A., Gnansounou, E., Madamwar, D. , 2011a. Cyanobacteria and microalgae: A positive prospect for biofuels. *Bioresour. Technol.* 102, 10163–10172.

Paswan, M.B., Chudasama, M.M., Mitra, M., Bhayani, K., George, B., Chatterjee, S., Mishra, S. , 2016. Fluorescence quenching property of C-phycocyanin from Spirulina platensis and its binding efficacy with viable cell components. *J. Fluoresc.* 26, 577–583.

Patel, A., Mishra, S., Pawar, R., Ghosh, P.K., 2005. Purification and characterization of C-Phycocyanin from cyanobacterial species of marine and freshwater habitat. *Protein Expr. Purif.* 40, 248–255.

Patil, G., Chethana, S., Sridevi, A.S., Raghavarao, K.S.M.S., 2006. Method to obtain C-phycocyanin of high purity. *J. Chromatogr. A.* 1127, 76–81.

Pfruender, H., Jones, R., Weuster-Botz, D., 2006. Water immiscible ionic liquids as solvents for whole cell biocatalysis. *J. Biotechnol.* 124, 182–190.

Porav, A.S., Bocăneală, M., Fălămaş, A., Bogdan, D.F., Barbu-Tudoran, L., Hegeduş, A., Dragoş, N., 2020. Sequential aqueous two-phase system for simultaneous purification of cyanobacterial phycobiliproteins. *Bioresour. Technol.* 315, 123794.

Puzorjov, A., McCormick, A.J., 2020. Phycobiliproteins from extreme environments and their potential applications. *J. Exp. Bot.* 71, 3827–3842.

Radmann, E., Martha, C.O.R., Costa, J.A.V., 2007. Optimization of the repeated batch cultivation of microalga *Spirulina platensis* in open raceway ponds. *Aquaculture.* 265, 118–126.

Rajagopal, S., Jha, I.B., Murthy, S.D.S., Mohanty, P., 1998. Ultraviolet-B Effects on *Spirulina platensis* Cells: Modification of chromophore–Protein interaction and energy transfer characteristics of phycobilisomes. *Biochem. Biophys. Res. Commun.* 249, 172–177.

Ramos, A., Gabriel Acien, F., Fernandez-Sevilla, J.M., Gonzalez, C.V., Bermejo, R., 2011. Development of a process for large-scale purification of C-phycocyanin from *Synechocystis aquatilis* using expanded bed adsorption chromatography. *J. Chromatogr. B.* 879, 511–519.

Rastogi, R.P., Incharoensakdi, A., 2015. Occurrence and induction of a ultraviolet-absorbing substance in the cyanobacterium *Fischerella muscicola* TISTR8215. *Phycological Res.* 63, 51–55.

Richaud, C., Zabulon, G., Joder, A., Thomas, J.-C., 2001. Nitrogen or sulfur starvation differentially affects phycobilisome degradation and expression of the nblA gene in *Synechocystis* strain PCC 6803. *J. Bacteriol.* 183, 2989–2994.

Richmond, A., (Ed.). 2008. *Handbook of Microalgal Culture: Biotechnology and Applied Phycology.* John Wiley & Sons.

Richmond, A., Grobbelaar J.U., 1986. Factors affecting the output rate of *Spirulina platensis* with reference to mass cultivation. *Biomass.* 10, 253–264.

Rinalducci, S., Hideg, E., Vass, I., Zolla, L., 2006. Effect of moderate UV-B irradiation on Synechocystis PCC 6803 biliproteins. *Biochem. Biophys. Res. Commun.* 341, 1105–1112.

Rippka, R., Deruelles, J., Waterbury, J.B., Herdman, M., Stanier, R.Y., 1979. Generic assignments, strain histories and properties of pure cultures of cyanobacteria. *Microbiology.* 111, 1–61.

Rippka, R., 1988. Isolation and purification of cyanobacteria. *Methods Enzymol.* 167, 3–27.

Ritchie, R.J., Trautman, D.A., Larkum, A.W.D., 1997. Phosphate uptake in the cyanobacterium *Synechococcus* R-2 PCC 7942. *Plant Cell Physiol.* 38, 1232–1241.

Sakamoto, T., Delgaizo, V.B., Bryant, D.A., 1998. Growth on urea can trigger death and peroxidation of the cyanobacterium *Synechococcus* sp. strain PCC 7002. *Appl. Environ. Microbiol.* 64, 2361–2366.

Samsonoff, W.A., MacColl, R., 2001. Biliproteins and phycobilisomes from cyanobacteria and red algae at the extremes of habitat. *Arch. Microbiol.* 176, 400–405.

Santiago-Santos, M.C., Ponce-Noyola, T., Olvera-Ramírez, R., Ortega-López, J., Cañizares-Villanueva, R.O., 2004. Extraction and purification of phycocyanin from *Calothrix* sp. *Process Biochem.* 39, 2047–2052.

Sarada, R.M.G.P., Pillai M.G., Ravishankar, G.A., 1999. Phycocyanin from *Spirulina* sp: Influence of processing of biomass on phycocyanin yield, analysis of efficacy of extraction methods and stability studies on phycocyanin. *Process Biochem.* 34, 795–801.

Scheer, H., Zhao, K.H., 2008. Biliprotein maturation: The chromophore attachment. *Mol. Microbiol.* 68, 263–276.

Schirmer, T., Vincent, M.G., 1987. Polarized absorption and fluorescence spectra of single crystals of C-phycocyanin. *Biochim. Biophys. Acta (BBA)-Bioenerg.* 893, 379–385.

Schoenleber, R.W., Lundell, D.J., Glazer, A.N., Rapoport, H., 1984. Bilin attachment sites in the alpha and beta subunits of B-phycoerythrin. Structural studies on the singly linked phycoerythrobilins. *J. Biol. Chem.* 259, 5485–5489.

Sekar, S., Chandramohan, M., 2008. Phycobiliproteins as a commodity: Trends in applied research, patents and commercialization. *J. Appl. Phycol.* 20, 113–136.

Selig, M.J., Malchione, N.M., Gamaleldin, S., Padilla-Zakour, O.I., Abbaspourrad, A., 2018. Protection of blue color in a spirulina derived phycocyanin extract from proteolytic and thermal degradation via complexation with beet-pectin. *Food Hydrocoll.* 74, 46–52.

Sidler, W.A., 1994. Phycobilisome and phycobiliprotein structures, in: *The Molecular Biology of Cyanobacteria.* Springer, Dordrecht, pp. 139–216.

Silveira, S., Terra, L.K.M., Quines, C.A.V.B., Kalil, S.J., 2008. Separation of phycocyanin from *Spirulina platensis* using ion exchange chromatography. *Bioprocess Biosyst. Eng.* 31, 477–482.

Simmons, T.L., Andrianasolo, E., McPhail, K., Flatt, P., Gerwick, W.H., 2005. Marine natural products as anticancer drugs. *Mol. Cancer Ther.* 4, 333–342.

Singh, N.K., Parmar, A., Sonani, R.R., Madamwar, D., 2012. Isolation, identification and characterization of novel thermotolerant *Oscillatoria* sp. N9DM: Change in pigmentation profile in response to temperature. *Process Biochem.* 47, 2472–2479.

Singh, N.K., Sonani, R.R., Rastogi, R.P., Madamwar, D., 2015. The phycobilisomes: An early requisite for efficient photosynthesis in cyanobacteria. *EXCLI J.* 14, 268.

Singh, S.P., Montgomery, B.L., 2013. Salinity impacts photosynthetic pigmentation and cellular morphology changes by distinct mechanisms in *Fremyella diplosiphon*. *Biochem. Biophys. Res. Commun.* 433, 84–89.

Six, C., Thomas, J.C., Garczarek, L., Ostrowski, M., Dufresne, A., Blot, N., Partensky, F., 2007. Diversity and evolution of phycobilisomes in marine *Synechococcus* spp.: A comparative genomics study. *Genome Biol.* 8, 1–22.

Sonani, R.R., Rastogi, R.P., Patel, R., Madamwar, D., 2016. Recent advances in production, purification and applications of phycobiliproteins. *World J. Biol. Chem.* 7, 100.

Sonani, R.R., Singh, N.K., Awasthi, A., Prasad, B., Kumar, J., Madamwar, D., 2014. Phycoerythrin extends life span and health span of *Caenorhabditis elegans*. *Age.* 36, 1–14.

Sonani, R.R., Singh, N.K., Kumar, J., Thakar, D., Madamwar, D., 2014. Concurrent purification and antioxidant activity of phycobiliproteins from *Lyngbya* sp. A09DM: An antioxidant and anti-aging potential of phycoerythrin in *Caenorhabditis elegans*. *Process Biochem.* 49, 1757–1766.

Soni, B., Kalavadia, B., Trivedi, U., Madamwar, D., 2006. Extraction, purification and characterization of phycocyanin from *Oscillatoria quadripunctulata*—Isolated from the rocky shores of Bet-Dwarka, Gujarat, India. *Process Biochem.* 41, 2017–2023.

Soni, B., Trivedi, U., Madamwar, D., 2008. A novel method of single step hydrophobic interaction chromatography for the purification of phycocyanin from *Phormidium fragile* and its characterization for antioxidant property. *Bioresour. Technol.* 99, 188–194.

Soni, R.A., Sudhakar, K., Rana, R.S., 2017. Spirulina–From growth to nutritional product: A review. *Trends Food Sci. Technol.* 69, 157–171.

Sørensen, L., Hantke, A., Eriksen, N.T., 2013. Purification of the photosynthetic pigment C-phycocyanin from heterotrophic *Galdieria sulphuraria*. *J. Sci. Food Agric.* 93, 2933–2938.

Soundarapandian, P., Vasanthi, B., 2010. Effects of chemical parameters on *Spirulina platensis* biomass production: Optimized method for phycocyanin extraction. *Int. J. Zool. Res.* 6, 293–303.

Spolaore, P., Joannis-Cassan, C., Duran, E., Isambert, A., 2006. Commercial applications of microalgae. *J. Biosci. Bioeng.* 101, 87–96.

Stock, C.A., Dunne, J.P., John, J.G., 2014. Global-scale carbon and energy flows through the marine planktonic food web: An analysis with a coupled physical–biological model. *Prog. Oceanogr.* 120, 1–28.

Strasky, Z., Zemankova, L., Nemeckova, I., Rathouska, J., Wong, R.J., Muchova, L., Subhanova, I., et al. 2013. *Spirulina platensis* and phycocyanobilin activate atheroprotective heme oxygenase-1: A possible implication for atherogenesis. *Food Funct.* 4, 1586–1594.

Su, H.-N., Xie, B.-B., Chen, X.-L., Wang, J.-X., Zhang, X.-Y., Zhou, B.-C., Zhang, Y.-Z., 2010. Efficient separation and purification of allophycocyanin from *Spirulina (Arthrospira) platensis. J. Appl. Phycol.* 22, 65–70.

Sun, Li, Wang, S., Gong, X., Zhao, M., Fu, X., Wang, L., 2009. Isolation, purification and characteristics of R-phycoerythrin from a marine macroalga *Heterosiphonia japonica. Protein Expr. Purif.* 64, 146–154.

Sun, Li, Wang, S., Qiao, Z., 2006. Chemical stabilization of the phycocyanin from cyanobacterium *Spirulina platensis. J. Biotechnol.* 121, 563–569.

Swatloski, R.P., Spear, S.K., Holbrey, J.D., Rogers, R.D., 2002. Dissolution of cellose with ionic liquids. *J. Am. Chem. Soc.* 124, 4974–4975.

Szalontai, B., Gombos, Z., Csizmadia, V., Bagyinka, C., Lutz, M., 1994. Structure and interactions of phycocyanobilin chromophores in phycocyanin and allophycocyanin from ananalysis of their resonance Raman spectra. *Biochemistry* 33, 11823–11832.

Talarico, L., Maranzana, G., 2000. Light and adaptive responses in red macroalgae: An overview. *J. Photochem. Photobiol. B: Biol.* 56, 1–11.

Tang, B., Bi, W., Tian, M., Row, K.H., 2012. Application of ionic liquid for extraction and separation of bioactive compounds from plants. *J. Chromatogr. B.* 904, 1–21.

Thangam, R., Suresh, V., Asenath Princy, W., Rajkumar, M., Senthilkumar, N., Gunasekaran, P., Rengasamy, R., Anbazhagan, C., Kaveri, K., Kannan, S., 2013. C-Phycocyanin from *Oscillatoria tenuis* exhibited an antioxidant and in vitro antiproliferative activity through induction of apoptosis and G0/G1 cell cycle arrest. *Food Chem.* 140, 262–272.

Tiwari, O.N., Bhunia, B., Chakraborty, S., Goswami, S., Devi, I., 2019. Strategies for improved production of phycobiliproteins (PBPs) by *Oscillatoria* sp. BTA170 and evaluation of its thermodynamic and kinetic stability. *Biochem. Eng. J.* 145, 153–161.

Vaishampayan, A., Sinha, R.P., Hader, D-P., Dey, T., Gupta, A.K., Bhan, U., Rao, A.L., 2002. Cyanobacterial biofertilizers in rice agriculture. *Bot. Rev.* 67, 453–516.

Viskari, P.J., Colyer, C.L., 2003. Rapid extraction of phycobiliproteins from cultured cyanobacteria samples. *Anal. Biochem.* 319, 263–271.

Vonshak, A., Cheung, S.M., Chen, F., 2000. Mixotrophic growth modifies the response of *Spirulina (Arthrospira) platensis* (Cyanobacteria) cells to light. *J. Phycol.* 36, 675–679.

Vonshak, A., Richmond, A., 1988. Mass production of the blue-green alga Spirulina: An overview. *Biomass.* 15, 233–247.

Wang, C., Wang, X., Wang, Y., Zhou, T., Bai ,Y., Li, Y., Huang, B., 2012. Photosensitization of phycocyanin extracted from Microcystis in human hepatocellular carcinoma cells: implication of mitochondria-dependent apoptosis. *J. Photochem. Photobiol. B: Biol.* 117, 70–79.

Waterbury, J.B., 1986. Biological and ecological characterization of the marine unicellular cyanobacterium Synechococcus. *Photosyn. Picoplankton,* 71–120.

Waterbury, J.B., 2006. The cyanobacteria—Isolation, purification and identification. *The Prokaryotes.* 4, 1053–1073.

Wiethaus, J., Busch, A.W., Dammeyer, T., Frankenberg-Dinkel, N., 2010. Phycobiliproteins in *Prochlorococcus marinus*: Biosynthesis of pigments and their assembly into proteins. *Eur. J. Cell Biol.* 89, 1005–1010.

Wu, H.-L., Wang, G.-H., Xiang, W.-Z., Li, T., He, H., 2016. Stability and antioxidant activity of food-grade phycocyanin isolated from *Spirulina platensis*. *Int. J. Food Prop.* 19, 2349–2362.

Xu, X., Thornton, P.E., Post, W.M., 2013. A global analysis of soil microbial biomass carbon, nitrogen and phosphorus in terrestrial ecosystems. *Glob. Ecol. Biogeogr.* 22, 737–749.

Yan, M., Liu, B., Jiao, X., Qin, S., 2014. Preparation of phycocyanin microcapsules and its properties. *Food Bioprod. Process.* 92, 89–97.

Yan, S.-G., Zhu, L.-P., Su, H.-N., Zhang, X.-Y., Chen, X.-L., Zhou, B.-C., Zhang, Y.-Z., 2011. Single-step chromatography for simultaneous purification of C-phycocyanin and allophycocyanin with high purity and recovery from *Spirulina (Arthrospira) platensis*. *J. Appl. Phycol.* 23, 1–6.

Zhu, Y., Chen, X.B., Wang, K.B., Li, Y.X., Bai, K.Z., Kuang, T.Y., Ji, H.B., 2007. A simple method for extracting C-phycocyanin from *Spirulina platensis* using *Klebsiella pneumoniae*. *Appl. Microbiol. Biotechnol.* 74, 244–248.

Index

symbols

ι-carrageenan, 281

α-amylase, 86, 317, 338
α-carotene, 68, 158, 322
α-chitin, 283
α-glucosidase, 338
α-l-fucopyranose, 284
α-l-galactose, 36
α-l-guluronic, 45
α-l-iduronopyranosic, 286
α-linolenic, 78
α-l-rhamnose, 286
α-tocopherol, 45, 68

β-carotene, 67, 68, 121, 158, 159, 161, 163, 165,
    166, 171, 175–77, 180, 184, 188, 192,
    194, 203, 210, 223, 226, 322, 333–35
β-chitin, 283
β-d-galactopyranose, 36
β-d-galactose, 36, 285
β-d-glucan, 326
β-d-glucanswith, 284
β-d-glucuronosyluronic, 286
β-d-mannans, 277
β-d-xylopyranose, 286
β-d-xylose, 286
β-glucan, 45, 112, 228, 284
β-secretase, 372
β-subunits, 377

γ-tocopherols, 45
γ-valerolactone, 65

κ-carrageenan, 27, 50, 281, 294, 298, 310

λ-carrageenans, 281

A

abiotic, 46, 155, 186, 199, 245, 354, 356, 359,
    360, 376, 379
acclimatization, 360
acid-catalyzed, 76, 88
acid-enriched, 145
acidification, 123, 124, 282
acidobacteria, 241
acidocalcisomes, 247

adsorbents, 35, 72, 137, 144
adsorption, 42, 91, 136, 137, 143, 145, 308, 345,
    366, 367, 377, 380
aerobic, 186
affinity, 111, 136, 137, 169, 171, 366
agar-extracted, 58, 76, 80
agrobacterium, 164
agrobacterium-mediated, 167
agro-industrial, 108
agronomy, 225, 231
agro-waste, 234
algae-assisted, 102, 104, 105
algae–bacteria, 239, 250
algae-supplemented, 312
algal-bacterial, 129, 240
alginate-based, 298, 325
alginate-chitosan, 297
alkali, 26, 28, 80, 171, 266, 282, 289, 307
amylase, 76, 317, 342
anaerobically, 73, 105
animal-derived, 300
anti-cancer, 45, 177, 184, 193, 197, 203, 204, 326,
    343, 370, 371
antidiabetic, 44, 140, 333, 343, 347
antifungal, 66, 77, 140
anti-genotoxic, 38
anti-inflammatories, 312
antilipidemic, 275, 295
antimicrobial, 38, 45, 57, 64, 68, 69, 78, 82, 88,
    141, 147, 192, 241, 275, 292, 321, 323,
    326, 342
antioxidants, 36, 68, 69, 111, 140, 154, 155, 159,
    177, 180, 186, 193, 194, 312, 322, 327,
    333, 334, 336, 337, 341, 345, 347, 349,
    370, 371
anti-tumor, 45, 68, 177, 178, 355, 370, 371
antiviral, 32, 39, 62, 66, 85, 89, 140, 193, 205,
    275, 292, 294, 299, 308, 310, 321, 323,
    327, 336, 338
apoptosis, 294, 307, 338, 371, 376, 382
aquacultural, 86
aquafeed, 196, 204, 205, 207, 208, 210, 222,
    228, 256
archaea, 191, 316
aromatics, 65
arthronema, 361, 362, 364, 375, 378
aspergillus, 69, 249, 353

astaxanthin, 67, 68, 109, 136, 158, 159, 165, 166,
   169, 171, 174–76, 178–80, 185–92,
   194, 195, 203–5, 210, 223, 226, 228,
   263–65, 269–71, 320, 322, 333, 335,
   346, 375
atmosphere, 74, 116, 154, 230, 233, 267, 373
atmospheric, 73, 119, 131, 233–35, 246, 247, 264,
   265, 319, 355
auxin, 77, 245
azotobacter, 231, 238, 239, 242, 253

B

bacillariophyceae, 197, 288
bacillariophyta, 59, 237
bacterial, 94, 97, 105, 167, 207, 230, 237–41, 249,
   250, 252, 253, 267, 294, 303, 318,
   339, 352
bacterial-dependent, 317
beta-carotene, 68, 147, 312, 334
beverages, 62, 142, 297, 306, 312, 333, 338–41,
   345, 377
bicarbonate, 122, 198
bifidobacteria, 341, 342
biflagellate, 174
bioabsorbent, 35
bioaccumulation, 310
bioactivity, 55, 67, 68, 78, 199, 201, 227, 276,
   306, 308, 313, 315–17, 352
bio-based, 30, 60, 85, 106, 173, 188, 246, 255
biobutanol, 28
biocatalysis, 380
biochar, 27, 35, 37, 41, 42, 48, 55, 58, 72, 347,
   348, 353
biocolorants, 183
biodegradable, 45, 94, 95, 255, 256, 259, 260,
   262, 268, 298, 305
biodiesel, 21, 22, 26, 28, 31, 35, 40, 41, 48, 50, 52,
   54, 55, 58, 61, 69–71, 73, 76, 80–84,
   86–88, 92, 93, 105–7, 173, 176,
   184–86, 190, 194, 228, 271, 327, 328,
   344, 379
bioeconomy, 23, 25, 29, 30, 38, 43, 49, 51, 55, 84,
   89, 103, 108, 146, 153, 182, 183, 222,
   256, 260, 267, 271
bioelectricity, 84, 91, 92, 95–97, 99, 103–9
bioenergy, 26, 30, 33, 35, 36, 43, 47–49, 53, 54,
   72, 80, 83, 85, 87, 92, 93, 95, 102, 103,
   105, 108, 148, 261, 263, 268, 275, 299,
   303, 306, 327, 344
biofactory, 189
biofertilizers, 32, 36, 41, 42, 230–32, 235, 243,
   245, 252, 253, 382
biofilm, 95, 97, 238, 240, 252, 253
biofixation, 154, 155
bioflocculants, 287

biogas, 21, 22, 26, 27, 34, 35, 41, 47, 48, 54, 55,
   58, 65, 71–73, 75, 81, 84, 87, 93, 109,
   299, 300, 307, 309, 310
biohydrogen, 21, 22, 31, 34, 41, 48, 51, 58, 74, 89,
   91, 100, 104, 107, 150, 261, 348
biomacromolecules, 89, 105, 307, 348
biomedical, 32, 36, 46, 62, 260, 262, 275, 296,
   304, 307, 325, 326, 366
biomolecules, 52, 62, 111, 137, 138, 146, 185,
   275, 320, 353–55, 359, 363, 370
biopolymer, 81, 254, 258–60, 262, 267,
   269–71, 275
bioprocess, 55, 86, 87, 103, 146, 151, 189, 193,
   269, 271, 308, 309, 381
bioproducts, 22, 47, 106, 146, 147, 150, 167, 186,
   258, 270, 344, 355
bioreactor, 96, 199, 222, 360
biorefineries, 53, 55, 56, 81, 83, 104–6, 108, 225,
   226, 270, 349, 378
bioremediation, 35, 105, 106, 148, 188, 251, 259,
   267, 272
biostimulant, 38, 247
blue-bioeconomy, 89
botryococcus, 68, 73, 89, 119, 202, 223, 272, 352
bromophenols, 337, 350
brown-algae, 308
butanol, 26, 34, 48, 52, 54, 55
byproducts, 50, 51, 73, 77, 83, 148, 260, 262, 263,
   265, 268, 318

C

calcium, 32, 45, 66, 280, 286, 289, 297, 298, 320,
   326, 339, 344, 369
caloxanthin, 158
cancerous, 326, 371
carbohydrate, 33, 34, 37, 39, 48, 49, 52, 59, 62,
   63, 71, 76, 81, 112, 117–20, 123, 124,
   147, 189, 202, 225, 260, 263, 266, 267,
   271, 273, 277, 279, 280, 288, 303, 318,
   338, 342, 348
cardiovascular, 32, 68, 140, 141, 178, 201, 321,
   327, 330, 331, 335, 339, 341
carotene, 158, 165, 184, 223, 333–35, 348, 349
carotenoid, 45, 112, 159, 161, 163, 165, 167, 171,
   185, 187–89, 191, 193, 195, 203, 228,
   246, 332, 334, 335, 347, 350
carpophyllum, 27, 48, 56
cellulose, 33, 34, 39, 49, 76, 291, 302
chemo-enzymatic, 71, 85
chitin, 62, 95, 274, 278, 282, 283, 305, 307, 347
chitosan, 36, 274, 278, 282, 283, 296, 297,
   307, 367
chlamydomonas, 63, 68, 87, 88, 98, 99, 107, 121,
   123, 136, 145, 151, 167, 180, 181, 186,

189, 192, 202, 239, 241, 252, 261, 271, 319, 323, 374
chlorococcum, 66, 78, 82, 177, 204, 258, 269
chlorophyll-a, 373
chloroplast, 113, 159, 161, 163, 164, 167–69, 186, 187
cholesterol, 75, 206, 295, 321, 326, 336, 339, 346
chronic, 160, 312, 319, 328, 333, 339, 351
clostridium, 33, 34, 52, 208
co-cultivation, 236, 252, 253
co-digestion, 41, 50, 54, 87
coenzyme, 352
colorants, 66, 158, 160, 176, 179, 182, 187, 333, 336, 355, 356, 372, 374, 377
commercialization, 44, 49, 59, 65, 80, 88, 173, 184, 210, 223, 351, 366, 373, 374, 381
commodities, 59, 329, 333
communities, 105, 231, 234–36, 249, 250, 324, 343
copolymer, 269, 277, 282
cosmeceutical, 22, 36, 39, 43, 55, 153, 177, 178, 182, 194, 308, 309, 333
cost-effective, 35, 47, 48, 70, 81, 182, 183, 245, 256, 260, 267
c-phycocyanin, 136, 140, 160, 177, 187, 195, 223, 354, 356–58, 375–80, 383
cryptococcus, 76, 88
cryptoxanthin, 323
cyanobacterium, 147, 188, 247, 248, 269, 309, 337, 370, 375, 379, 380, 382
cysteine, 201, 330, 358
cytoplasmic, 119, 159, 294
cytosol, 112, 161, 163, 164
cytotoxic, 326, 334, 352

D

dairy, 183, 187, 253, 258, 273, 297, 298, 339, 353
dark-heterotrophic, 109
deacetylation, 282
decarboxylase, 371
deficiency, 68, 121, 123, 139, 163, 243, 331, 337
density, 94, 98, 99, 102, 166, 174, 175, 198, 223, 265, 324
de-oiled, 65, 69–71, 73–76, 82, 86, 87
diabetes, 22, 32, 140, 141, 323, 338, 341, 342, 347, 371
diadinoxanthin, 67, 158, 203
diatoms, 59, 67, 197, 203, 237, 282, 287, 304, 305, 377
dietary, 26, 30, 39, 45, 62, 77, 84, 146, 147, 178, 208, 224–26, 228, 312, 317–20, 325, 327, 328, 330, 331, 333, 337, 338, 342, 344, 345, 350, 352, 353
dimorphus, 63, 69, 194, 232, 233, 236, 238, 247
dinoflagellate, 207

disease, 37, 51, 67, 75, 88, 145, 179, 205, 223, 246, 295, 296, 299, 304, 310, 319, 328, 330, 331, 335, 339, 346, 349, 352
docosahexaenoic, 64, 77, 201, 224, 227, 320, 321, 328, 353
downstream, 30, 34, 53, 144, 147, 149, 176, 185, 189, 193, 256, 262, 263, 267, 268, 355, 356, 360, 363, 371, 375
drug, 36, 177–79, 184, 194, 204, 275, 296, 304, 305, 319, 325
drying, 35, 39, 46, 48, 75, 175, 225, 263, 264, 266, 268, 289, 332, 335, 345, 353
dunaliella, 68, 69, 71, 84, 85, 89, 99, 121, 124, 138, 147, 148, 150, 154, 158, 171, 173, 175, 177, 181, 184, 186–89, 192, 195, 199, 202–4, 210, 216, 226, 228, 257, 287, 289, 319, 320
dynamics, 107, 126, 191, 241, 342, 355

E

earthworms, 231
eco-friendly, 35, 41, 43, 65, 230, 231, 243, 245, 267, 298
economically, 59, 61, 65, 70, 74, 102, 131, 144, 162, 174, 180, 198, 209, 256, 268, 300, 360, 366
ecosystem, 22, 64, 75, 86, 234, 235, 239, 242, 243, 245, 246, 252, 276, 282, 313, 329, 343
ecotoxicology, 86
effluent, 23, 46, 100, 101, 105, 253, 257, 258, 273
eicosapentaenoic, 77, 78, 201, 223, 321, 328
electricity, 48, 93, 94, 99, 100, 102–9, 347, 348
electrocoagulation, 249
electro-fermentation, 105
electrons, 92–96, 155, 157
element, 122, 123, 139, 328, 361–63
emissions, 91, 189, 209, 229, 233–35, 251, 252
emulsification, 142
emulsifiers, 62, 275, 306, 327
encapsulation, 36, 190, 228, 260, 296, 298, 325, 369
endoplasmic, 113, 157
engineered, 49, 102, 145, 164, 167, 169, 373
enteromorpha, 279, 295, 309, 321, 325, 331, 339
environment, 29, 33, 44, 59, 64, 75, 79, 91, 95, 96, 104, 105, 115, 137, 152, 153, 166, 183, 194, 231, 244, 245, 247, 260–62, 287, 313, 316, 345, 355, 359, 360, 373
enzymatic, 26–28, 33, 38–40, 46, 47, 50, 53, 55, 71, 76, 83, 87, 111, 119, 127, 128, 130, 131, 135, 144, 146, 148, 151, 237, 261, 263, 266, 290, 291, 300, 302, 303, 309, 310, 317, 318, 337, 342, 343, 363, 365, 376

epidemiological, 335, 339
escherichia, 33, 69, 209, 259, 294, 376
eukaryotic, 58, 92, 112, 153, 187, 197, 202, 230,
    247, 287, 337
eutrophication, 29, 35, 37, 42, 43, 235, 320
evolutionary, 155, 186, 193, 234, 246, 313, 315,
    316, 351
exopolysaccharides, 145, 177, 195, 202, 224, 289,
    303, 308, 309
exoskeleton, 282
extracellular, 67, 83, 84, 112, 122, 177, 180,
    226–28, 242, 251, 262, 287, 288,
    296, 309
extraction, 21–24, 26–28, 30–33, 35–41, 43, 44,
    46–53, 55–57, 59, 60, 62, 64–66,
    69–85, 87, 88, 104, 106, 111, 112, 115,
    124–32, 134, 143–51, 153, 160,
    169–76, 185–90, 192, 224, 226, 227,
    254, 256, 259–71, 273, 275, 276, 285,
    289, 292, 297, 303, 304, 306–8, 310,
    325, 327, 333, 349, 355, 356, 363–68,
    373, 375, 377–82
extremophilic, 273, 372

F

fast-growing, 55, 319
fed-batch, 115, 118, 123, 273
feedstocks, 26, 28, 52, 234, 235
fermentation, 26–28, 31, 33, 34, 39–41, 46–48,
    51–53, 55, 59, 71, 73, 74, 76, 78, 79,
    81, 85–87, 103, 104, 224, 256, 259–61,
    267, 272, 290, 291, 303, 307, 309, 310,
    318, 341, 342, 377
fertilizer, 22, 23, 26, 27, 32, 33, 42, 44, 45, 54, 58,
    59, 62, 72, 75, 76, 80, 86, 229, 231,
    233–35, 243, 245–47, 252, 253,
    298, 321
fiber, 32, 62, 77, 286, 317, 318, 328, 338, 342,
    350, 352
fibrosis, 323, 328
filamentous, 249, 375, 377
filtration, 32, 38, 46, 133, 134, 145, 146, 175, 289,
    303, 366–68
fisheries, 208, 224, 225, 346
fishmeal, 75, 205, 206, 225, 227
fixation, 55, 58, 81, 91, 93, 95, 108, 119, 121, 123,
    147, 151, 155, 189, 225, 230, 231, 233,
    235, 237, 242, 243, 251, 374, 375
flavobacteriia, 352
flavonoids, 109
flavoxanthin, 158
flocculation, 75, 184
fluid, 24, 64, 106, 126, 145, 173, 184, 209, 254,
    265–67, 292, 349

fluorescent, 38, 159, 160, 194, 204, 355, 356, 372,
    378, 379
foaming, 142, 145
food-grade, 39, 338, 377, 383
footprints, 255
formaldehyde, 101, 369
formulation, 67, 121, 158, 183, 185, 204, 216,
    217, 222, 260, 300, 341, 345
fractionation, 33, 54, 60, 62, 77, 144, 145,
    150, 367
freeze-drying, 262, 318
freezing-thawing, 171, 355, 365
freshwater, 22, 58, 64, 81, 92, 139, 145, 197, 203,
    207, 214, 218, 224, 226, 234, 241, 259,
    273, 319, 363, 376, 377, 379
fruit, 191, 236, 238, 249, 297, 298, 321
fucoidans, 62, 274, 277, 278, 283, 293–95, 303,
    304, 309, 325, 327
fucoxanthin, 44, 45, 47, 67, 87, 112, 124, 158,
    178, 203, 312, 322, 334–36, 347, 349,
    351, 353
fuels, 34, 40, 41, 53, 58, 72, 74, 81, 85, 86, 88, 91,
    195, 328
full-scale, 259, 267, 272
fumigatus, 249, 302
fungi, 79, 94, 190, 191, 228, 241, 282

G

galactose, 59, 76, 77, 278, 284–87, 289, 299,
    309
gamma-linolenic, 202, 328
gaseous, 55, 71–73, 351
gasification, 31, 35, 73, 74, 82, 83, 89
gastrointestinal, 296, 315, 318, 334, 337, 338,
    341–43, 345, 352
gelatin, 297, 298, 305, 326
genetic, 29, 33, 35, 49, 52, 63, 88, 102, 153, 163,
    164, 167, 169, 187, 222, 231, 315,
    355, 373
geographical, 23, 36, 55, 332
germination, 42, 46, 190, 235, 236, 238, 239, 241,
    268, 269
gloaguen, 52, 85, 224
glucosamine, 278, 282
glucose, 40, 59, 71, 92, 94, 101, 122, 141, 198,
    207, 261, 278, 282–84, 287, 289, 299,
    352, 360, 369
glucuronic, 30, 33, 278, 287, 289, 325
glutamate, 161, 344
glutamic, 46, 201, 330
glutathione, 64, 69, 337
glycerol, 41, 65, 70, 77, 166, 194, 328
glycerol-galactosides, 52, 77, 85
glycogen, 112, 275

glycolipids, 77
glycoprotein, 261, 341
glycosidic, 45, 275, 276, 278, 280, 282, 287, 325
gram-negative, 141, 294, 355
gram-positive, 141, 294, 341

H

habitat, 69, 153, 159, 338, 360, 361, 379, 380
haematococcus, 67, 68, 78, 83, 86, 87, 119, 134,
        145, 146, 155, 158, 165, 171, 174, 177,
        178, 181, 184–95, 201, 203, 204, 208,
        210, 212, 214, 215, 217, 220, 227, 228,
        256, 258, 261, 263–65, 268–71, 273,
        320, 322, 346
halomonas, 350
haptophyte, 157
harvesting, 49, 65, 66, 69, 70, 75, 80, 84, 104, 153,
        155, 156, 160, 170, 172, 174, 175, 182,
        184, 189, 191, 199, 222, 324, 329, 354,
        356, 358, 375, 379
healthcare, 193, 319, 333, 353
health-promoting, 228, 325, 333, 370
helicobacter, 294, 335
hemicellulose, 70, 95, 112, 260
heterotrophic, 90, 92, 95–97, 105, 121, 166, 188,
        191, 194, 197–99, 201, 207, 208,
        223–26, 228, 239, 248, 249, 328, 360,
        379, 381
high-value, 61, 62, 77, 81, 82, 105, 108, 174, 183,
        184, 190, 223, 228, 240, 246
homogenization, 38, 111, 126, 127, 145, 146, 171,
        262, 318, 363
hydrocarbons, 40, 67, 71, 103, 104, 154, 156,
        159, 327
hydrocolloids, 78, 276, 277, 289, 297, 303, 306,
        309, 321
hydrogels, 280, 296, 306
hydrolysate, 33, 37–40, 71, 141, 151, 193,
        266, 303
hydrophilic, 72, 280, 284, 289, 369, 377
hydrophobic, 72, 111, 113, 136, 137, 260, 289,
        357, 359, 366–70, 377, 378, 381
hydrothermal, 26–28, 31, 42, 48, 82, 84, 89, 129,
        149, 265, 266, 271
hypotonic, 131, 365

I

illumination, 97, 202, 260, 348, 361
immune, 66–68, 178, 205, 207, 208, 210, 223–26,
        292, 296, 317–19, 342, 346
immuno-modulating, 67
industrial-scale, 256, 265, 313
infection, 153, 177, 207, 294, 298, 299, 305,
        308, 335

inflammation, 22, 140, 209, 223, 292, 293, 297,
        310, 335
inhibitory, 35, 50, 69, 78, 141, 161, 166, 180, 295
insulin, 68, 141, 347, 371
intake, 139, 206, 228, 295, 317, 318, 334, 335
interaction, 136, 137, 230, 239–41, 245, 248, 347,
        366–68, 372, 378–81
intracellular, 115, 127, 128, 130, 145, 149, 158,
        170, 242, 262, 310, 346, 373
ion-exchange, 93, 96, 137, 367
iron-based, 37, 42
irradiation, 128, 131, 132, 289, 295, 361, 380
isoelectric, 28, 55, 134, 150
isolation, 32, 46, 49, 64, 71, 80, 82, 83, 88, 115,
        132, 140, 145, 151, 183, 187, 226, 261,
        272, 307, 310, 332, 344, 348, 353, 373,
        376, 378–82

J

japonica, 28, 47, 48, 52, 53, 69, 279, 295, 318,
        339, 353, 367, 382
jellies, 297, 298, 372

K

kidney, 338
kinetics, 94, 115, 145, 209, 272
kitchen, 105, 106, 189
klebsiella, 364, 383
klebsormidium, 238

L

laccases, 143
lactic, 28, 34, 58, 76, 86, 88, 260, 261, 339,
        341, 342
lactob acillus, 76, 86, 208, 209, 341, 342, 350
landfill, 42, 100, 255
leachate, 42, 100
leaching, 248, 320
leaf-derived, 348
leftover, 30, 32, 35, 38, 40–42, 44, 48, 50, 61
leukemia, 335, 347
leukocyte, 292
l-fucose, 278, 283, 325
light-dark, 119, 156, 348
light-harvesting, 112, 155, 156, 158, 159, 161,
        191, 354–59, 363, 373, 376
lignin, 29, 34, 59, 70, 95, 131, 282, 317, 344
lignocellulosic, 80, 95, 150, 272, 347
linkages, 77, 275–77, 280, 283, 287, 356, 358
lipid-extracted, 32, 57, 58, 65, 66, 69, 71, 75, 82,
        85–87, 89
lipidomics, 167, 169
lipophilic, 112, 157
lipoprotein, 203, 295, 326

liquefaction, 31, 71, 72, 82, 83, 85, 89
liver, 178, 295, 304, 310, 322, 323, 337, 338, 353
livestock, 85, 138, 205, 320, 323, 328
l-lactic, 260
long-chain, 35, 45, 71, 205
long-term, 95, 143, 169, 181, 201, 327
loroxanthin, 67, 158
low-cost, 134, 324, 350
lutein, 67, 68, 106, 107, 112, 124, 158, 159, 161,
        165, 166, 180, 185, 186, 194, 203, 322,
        333, 334, 349
lyase, 302, 362
lycopene, 158, 161, 186, 189, 203, 236, 344
lycopersicum, 248, 304, 305
lysozyme, 207, 364, 365

M

macroalgae-based, 22, 43
macroalgae-derived, 53, 348
macroalgae-sustainable, 108
macroalgal, 21–25, 29–38, 41, 42, 46, 47, 49–51,
        55, 56, 85, 274, 277, 306, 308,
        342, 347
macronutrients, 75, 112, 113, 121, 235, 239, 362
magnesium, 32, 75, 156, 186, 280, 286, 320,
        326, 339
magnitude, 96, 136, 324
mannuronic, 45, 277, 278, 325
manufacture, 45, 79, 80
manure, 129, 131, 234, 235
marine-based, 29, 35
mass-cultivation, 52
mass-scale, 343
matrices, 204, 320
maxima, 63, 99, 138, 141, 151, 181, 184, 206, 227
medical, 45, 76, 182, 203, 204, 296, 327, 330,
        333–35, 356
medicinal, 178, 186, 228, 319, 327, 338, 373
medium-chain-length, 259
melanogenesis, 180, 189
membranes, 92, 94, 106, 128, 130, 132–34,
        197, 297
metabolism, 38, 66, 93, 108, 109, 113, 116,
        119–21, 138, 141, 154–56, 163, 167,
        169, 184, 189, 190, 193, 195, 197–99,
        201, 203, 209, 221, 227, 234, 239, 243,
        249, 252, 257, 298, 299, 315, 327, 328,
        363, 373
metabolites, 68, 69, 92, 97, 112, 117, 121–24, 146,
        161, 166, 177, 179, 185, 191, 193, 194,
        197–99, 208, 222, 224, 227, 276, 277,
        305, 312, 313, 315–18, 337, 343, 355,
        375, 379
metabolomics, 169, 305, 352

metal, 35, 42, 43, 104, 142, 249, 289, 308, 337,
        344, 360
methane, 31, 35, 37, 41, 51, 71, 73, 82, 87, 103–5,
        266, 267, 272, 301, 307
microalga, 85, 88, 89, 120, 121, 145–48, 150, 151,
        154, 170, 174, 186, 192–94, 198, 202,
        208, 225–28, 239, 247, 271–73, 305,
        309, 310, 328, 352, 376, 380
microalgae-based, 144, 147, 148, 150, 172, 177,
        178, 180, 181, 185, 191, 210, 211, 213,
        215, 217–19, 221, 222, 270, 341
microbes, 33, 49, 72, 73, 91, 100, 247, 260, 270,
        300, 312, 347
microbiome, 230, 236, 241, 249, 250, 316,
        338, 347
microchloropsis, 122, 148
microcystis, 95, 140, 147, 233, 371, 382
microflora, 208, 225, 226, 231, 313, 318, 325,
        341
microwave, 26, 55, 64, 111, 128, 129, 131–33,
        149, 151, 171, 262, 271, 301
microwave-assisted, 24, 27, 31, 128, 133, 146,
        148, 192, 276, 292, 378
milling, 46, 111, 126, 127, 134, 144, 149, 171,
        262, 289, 290, 318
minerals, 22, 36, 38, 44, 75, 231, 232, 283,
        320, 328
mitigation, 98, 105, 157, 231, 246, 253, 270
mitochondrial, 167, 168, 195, 248, 371
mixotrophic, 70, 96, 109, 121, 148, 174, 190, 197,
        198, 224, 226, 228, 382
modulation, 49, 155, 292, 293, 318
moisture, 34, 43, 232, 235, 295, 297, 329
molar, 135, 179, 258, 284, 289, 372
monomer, 261, 275, 280
morphology, 22, 95, 150, 260, 324, 348, 355, 374,
        376, 381
multicellular, 95, 276, 313
mycosporine-like, 69, 141, 150, 179, 192

N

n-acetyl-d-glucosamine, 278, 282
nannochloropsis, 71, 74–76, 81, 86, 88, 96, 119,
        129, 131, 135, 136, 141, 148, 149, 173,
        177, 180, 181, 188, 193, 202, 206,
        208–12, 216–19, 223, 224, 232, 238,
        261, 272, 321, 323, 341
nanocrystals, 58, 79, 86
nanofibers, 36, 79, 297, 305, 369
nanoparticles, 69, 88, 187, 260, 269, 296, 297,
        307, 310, 369, 374
neoalgae, 210, 220
neoxanthin, 67, 158, 180, 203, 334
neuroprotective, 39, 275, 295, 305, 336, 355, 371

nitrogen-fixing, 233, 234
non-degradable, 43, 255
non-polar, 131, 170, 175, 176
nostoc, 68, 160, 179, 180, 188, 232, 233, 236, 238,
    241, 242, 247, 249, 259, 267, 269, 323,
    368, 377
nutraceutical, 33, 36, 37, 39, 42, 43, 53, 66, 106,
    140, 153, 158, 176–78, 182, 187, 189,
    190, 259, 262, 263, 265, 268, 311, 312,
    314, 319–23, 326, 343, 349, 355, 356,
    370, 373, 375, 378

O

obesity, 22, 37, 45, 50, 141, 312, 336
oil-extracted, 65, 89
oleaginous, 76, 88, 195
oligosaccharides, 45, 55, 202, 207, 240, 275, 306,
    342, 351
omics, 167, 169
orthodontic, 192
osmoregulation, 77, 247, 363
osmotic, 28, 55, 111, 121, 131, 132, 148, 171,
    298, 355, 363–65
over-accumulation, 169, 271
overexpression, 165, 167
oxalate, 32, 325
oxidation, 69, 79, 95, 124, 227, 231, 289, 290, 335

P

pancreatic, 295, 317
pathogens, 69, 86, 207, 304, 342
patients, 294, 295, 328, 346
pectin, 62, 275, 317
peptides, 37, 38, 44, 50, 83, 123, 137, 140, 141,
    144, 146, 147, 151, 168, 266, 311, 317,
    328, 344
pesticides, 138, 139, 252
petroleum, 60, 71, 91, 106, 136
phaeodactylum, 67, 78, 83, 84, 178, 181, 212, 215,
    216, 227, 237, 246, 288, 305, 321, 323
phaeophyceae, 22, 51, 52, 59, 157, 277, 279, 283,
    304, 333, 347
phaeophyta, 43, 50
phagocytosis, 208
pharmaceutical, 21, 22, 36, 37, 39, 43, 45, 46, 53,
    62, 66, 68, 69, 106, 140, 146, 153, 158,
    176–78, 182, 190, 192, 193, 203, 204,
    228, 240, 259, 260, 262, 263, 265, 268,
    275, 292, 304, 308, 312, 319, 321, 338,
    349, 350, 356, 370, 378
pharmacological, 37, 62, 64, 66, 69, 78, 295, 305,
    333, 344
phenolic, 112, 180, 187, 299, 323, 337, 341, 352

phenomena, 128, 141, 171, 172, 233, 361, 366
phospholipids, 70, 77, 112
photoacclimation, 165
photoautotrophic, 99, 105, 109, 154, 162, 174,
    203, 228, 254, 257, 360
photobioreactor, 29, 117, 118, 120, 121, 123, 147,
    155, 191, 259, 360
photochemical, 155, 160
photodynamic, 371, 374
photoinhibition, 117, 119, 165
photooxidation, 158, 335
photoperiod, 117–19, 159, 165, 195, 376
photoprotective, 179, 192
photosynthesis, 22, 91–93, 97, 98, 112, 113, 117,
    119, 121, 122, 155–57, 159, 165, 167,
    185, 189, 190, 193–95, 197, 243, 253,
    258, 334, 373, 376, 378, 381
phototrophic, 91, 109, 191, 194
phycobilins, 112, 156, 159, 320, 336, 356, 358, 378
phycobiliprotein, 66, 204, 336, 377, 379, 381
phycocolloids, 23, 76, 79, 277, 306
phycocyanin, 38, 84, 136, 140, 159, 160, 170, 175,
    177, 178, 183, 184, 186, 187, 189, 192,
    194, 204, 323, 336, 356, 358, 360–62,
    365, 369–72, 374–83
phycoerythrin, 38, 159, 170, 192, 323, 336, 356,
    357, 360, 361, 369, 376, 377, 379, 381
phycology, 69, 82, 189, 190, 196, 223, 308, 345,
    346, 350, 380
phycoremediation, 182, 251
physicochemical, 30, 79, 84, 125, 133, 142, 143,
    146, 148, 172, 345, 360, 377, 379
physiological, 140, 202, 207, 208, 223, 238, 243,
    248, 315, 330, 338, 341, 346, 363, 372
physiology, 140, 151, 165, 189, 205, 228,
    338, 376
phytochemicals, 105, 191, 312, 313, 337, 347
phytohormones, 32, 231
phyto-medicine, 374
phytoplankton, 59, 213, 245, 251, 272, 351,
    376, 378
pigment, 26, 38, 84, 112, 140, 149, 150, 156–58,
    160–67, 171–74, 176, 179, 186, 189,
    191–94, 203–5, 225, 289, 313, 333, 352,
    356, 358, 360–62, 365, 373–76, 379, 381
planktonic, 239, 375, 377, 382
plant-algae, 230, 241
plasmid, 164, 167
plasticizer, 43
plastids, 195, 320
p-nitrophenol, 95, 106
polar, 112, 128, 131, 158, 170, 355
policies, 82, 251, 255
pollutants, 35, 42, 100, 144, 327
polycystum, 293–95, 307, 349

polyhydroxyalkanoates, 254, 259, 269
polymer, 45, 59, 82, 207, 269, 277, 281, 284, 285, 289, 292
polypeptide, 159, 354, 357, 358, 361, 375
polyphenols, 36, 44, 50, 62, 64, 69, 77, 154, 332, 333, 337, 338, 341, 347
polysaccharide, 27, 30, 32, 36, 45, 62, 67, 76, 78, 79, 83–85, 87, 88, 202, 227, 276–80, 284–86, 289, 292, 294, 296–98, 300, 303, 304, 306–10, 317, 318, 324–26, 338, 342, 348, 351, 355
polyunsaturated, 39, 62, 63, 68, 77, 156, 165, 201, 223, 226, 256, 313, 327, 350
porphyridium, 67, 68, 88, 105, 145, 184, 192, 202, 216, 227, 303, 319, 323, 341, 360, 376, 377, 379
poultry, 43, 75, 85, 140, 208, 218, 220, 221, 223, 225, 258, 298
prebiotic, 66, 207, 312, 341, 342, 347, 351
precipitation, 28, 46, 55, 132, 134, 150, 171, 172, 176, 366–68
precursor, 77, 161, 322, 328, 334, 372
predominantly, 156, 280, 285, 287, 327, 335
pretreatment, 26, 33, 34, 41, 44, 46–49, 54, 70, 71, 74, 79, 84, 87–89, 109, 128–30, 145, 146, 149, 150, 248, 254, 262, 263, 266, 268, 271–73, 307–10, 338
prevention, 51, 69, 145, 160, 187, 192, 202, 223, 312, 333, 335, 349, 351, 352
probiotic, 312, 341
prokaryotic, 153, 197, 230, 287, 356
proteases, 130, 143, 273
protective, 64, 86, 140, 160, 251, 296, 305, 306, 321, 327, 347
protein-based, 147, 262
proteobacteria, 94, 241, 352
provitamin, 331, 334, 335
proximate, 148, 341, 353
pseudomonas, 69, 238, 294
pulmonary, 328, 338
pulsed-electric, 171
pyrenoidosa, 63, 68, 74, 95, 98, 99, 109, 118, 120, 130, 132, 142, 146, 151, 177, 194, 233, 253
pyrolysis, 26, 27, 31, 35, 41, 42, 48, 51, 54, 71, 86, 89, 103
pyrophosphate, 161
pyrrhoxanthin, 158

Q

quadricellulare, 119, 151
quantification, 145, 147, 150, 183, 378
quantitative, 316, 338
quantum, 194, 372
quorum-sensing, 247, 248, 253

R

raceway, 116, 189, 228, 324, 360, 380
radiation, 67, 92, 141, 170, 174, 179, 192, 335, 339, 361, 371, 378
radical, 38, 69, 84, 141, 223, 336
reaction, 69, 92, 94–96, 102, 128, 154, 289, 372
reactive, 109, 130, 157, 179, 336, 370, 371, 376
reactors, 101, 116, 175, 199, 270, 301
reclamation, 235
recombinant, 45, 164, 313, 348, 370, 376
recycling, 59, 73, 75, 87, 182, 235, 252, 255, 328
refinery, 226, 254, 255
regulation, 148, 155, 164, 169, 186, 188–90, 192, 201, 203, 248, 249, 252, 333, 353, 363, 371
renewable, 30, 35, 41, 43, 55, 58, 60, 63, 68, 72, 81, 82, 84–86, 91, 93, 98, 100, 104, 105, 107, 108, 149, 182, 231, 260, 270, 273, 299, 327, 343
residual, 23, 30–33, 37–39, 41, 48–50, 57, 58, 62, 66, 69, 70, 76, 85, 169, 234, 242, 251
resistance, 75, 94, 98, 121, 124, 138, 205, 207, 239, 260, 273, 298, 299, 306, 309, 310, 374
respiratory, 68, 96, 334, 335, 371
retinol, 67
rhamnose, 30, 33, 34, 287, 296, 352
rhodophyta, 54, 66, 154, 157–60, 203, 279, 288, 308, 336, 358, 379
riboflavin, 339, 347
rigidity, 144, 171, 261, 268
r-phycoerythrin, 310, 374, 376, 382
rubisco, 126, 143
ruminants, 209, 223

S

saccharification, 26, 27, 33, 39, 40, 51, 54, 59, 71, 85, 87, 88, 148, 290
salinity, 29, 58, 105, 110, 118, 120, 121, 131, 150, 159, 165, 166, 174, 188, 190, 197, 202, 259, 271, 272, 298, 309, 332, 363, 372, 373, 381
salmonella, 69, 342, 344
scaffolds, 36, 296, 305
scale-up, 126, 176, 182, 199, 223
scytonemin, 179, 192
seawater, 23, 81, 92
seaweeds, 33, 51, 52, 54–56, 59, 62, 64, 66, 78, 82, 84, 85, 104, 251, 276, 279, 280, 294, 304, 306–8, 310, 344, 345, 348–53, 355
secretion, 141, 145, 197, 207, 288, 294, 317
seedling, 43, 44, 298
selenium, 42, 297, 310, 320, 337, 345, 370

selenoproteins, 66, 81
semi-continuous, 115, 116
sensor, 101
separation, 60, 91, 111, 129, 132–37, 144–48, 153,
    169–76, 183, 185, 188, 263, 325, 366,
    378, 381, 382
sequestration, 22, 35, 60, 72, 91, 92, 97, 106, 186,
    231, 235
serum, 181, 207, 208, 302, 336, 344, 347
sewage, 41, 189, 258, 320
short-chain, 342
shrimp, 42, 203, 210–17, 256
signalling, 189, 237, 250, 293–96, 299, 306, 334
single-cell, 66
single-chamber, 97, 99, 103
siphonaxanthin, 158
size-exclusion, 366
socio-economy, 49
sodium, 32, 41, 46, 55, 135, 248, 280, 286, 289,
    293, 297, 298, 304, 307, 326, 345, 369
soils, 234, 246–49, 252
solubilisation, 48, 55, 77, 115, 128, 130, 134,
    172, 231
spectrum, 23, 58, 60, 112, 156–59, 185, 295,
    357, 358
spirilloxanthin, 203
spirulina, 63, 66, 68, 72, 96, 119–21, 136, 141–43,
    145–49, 151, 154, 155, 160, 177, 178,
    180, 181, 185–88, 194, 199, 206–10,
    214, 215, 217, 219–21, 223–28, 232,
    233, 247, 253, 259, 269, 288, 295, 304,
    318, 320–23, 328, 339–42, 345, 346,
    348, 358–62, 364, 370, 372, 374, 375,
    377–83
staphylococcus, 69, 83, 141, 294
starch, 36, 71, 79, 83, 86, 95, 100, 112, 141, 151,
    255, 260, 261, 271, 273, 277, 313, 317,
    320, 342, 362
starvation, 122–24, 147, 151, 155, 164, 165, 202,
    225, 259, 261, 273, 360, 380
stream, 50, 51, 116, 262, 267
streptococcus, 341
streptomyces, 33
stress-mediated, 78
sulphate, 132–35, 281–83, 285, 295, 296, 325,
    360, 366–68
sunlight, 59, 95, 102, 157–59, 194, 320, 332,
    357, 360
sunscreen, 179, 192, 194
supercritical, 24, 64, 69, 74, 87, 89, 106, 170, 176,
    186, 254, 265–67, 269, 271, 273, 327,
    333, 349, 351
surfactants, 101, 143
survival, 75, 95, 205, 206, 208, 225, 296, 328, 360
susceptibility, 133, 304, 344

sustainability, 54, 60, 81, 91, 102, 108, 190, 197,
    267, 351
symbiotic, 213, 239, 241, 242, 247, 249
synechococcus, 63, 68, 138, 141, 160, 181, 193,
    259, 319, 322, 323, 361–63, 373,
    380–82

T

taxonomy, 155, 316, 351
techno-economic, 194
techno-functionality, 148
temperature-dependent, 119
terrestrial, 22, 23, 29, 34, 44, 58, 64, 80, 91, 138,
    139, 142, 196, 204, 205, 208, 210, 218,
    219, 221, 243, 246, 258, 261, 268, 312,
    313, 320, 324, 334, 336, 355, 383
tetracycline, 345, 348
tetrapyrrole, 156, 159, 186, 358
textile, 46, 182, 186, 240, 255, 258, 260, 270, 321
therapeutic, 38, 39, 51, 64, 67, 86, 177, 189,
    295, 305
thermo-alkaline, 74, 89
thermochemical, 26, 30, 31, 41, 42, 71, 72, 74, 83,
    85, 89, 301
thermodynamic, 172, 382
thermophilic, 34, 151, 249, 375, 377
thermostable, 362, 379
thylakoid, 159, 165, 334, 336
tisochrysis, 122, 135, 149, 216, 257, 272
tocopherol, 45, 64, 69, 350
tolerance, 120, 166, 205, 225, 257, 298, 310, 361,
    371, 376
tolypothrix, 233, 242, 323
tomato, 232, 236, 238, 246, 248, 249, 253, 298,
    299, 344
toothpaste, 157, 321
toxic, 34, 42, 65, 69, 94, 95, 121, 124, 130, 139,
    176, 341, 347, 369, 370, 372
toxins, 245, 252
transcriptional, 66, 87, 186, 353
transcriptomics, 167, 169, 300
transesterification, 40, 41, 70
transformation, 123, 161, 167–69
treatments, 26, 31, 34, 76, 124, 125, 127, 129,
    159, 171, 247, 262, 292, 298, 299,
    318, 319
triacylglycerol, 191, 192
triglycerides, 40, 70, 203, 326
trimethylene, 255
tropical, 43, 52, 250, 272, 304, 349, 351
turnover, 49, 205
two-phase, 131, 184, 201, 364, 366–68, 375,
    378, 380
two-stage, 51, 74, 127, 166, 184, 190, 328

U

ultrafiltration, 133, 134, 341, 366
ultraphytoplankton, 377
ultrasonication, 111, 125, 127, 171, 262, 318, 355,
    363, 365, 373
ultrasound, 26, 127, 128, 132, 149–51, 271, 292,
    306, 365
ultrasound-assisted, 24, 31, 64, 78, 147, 150, 310
ultraviolet, 156, 159, 179, 183, 378, 380
unicellular, 59, 95, 192, 227, 276, 313, 320,
    374, 382
unsaturated, 67, 320, 338
upstream, 66, 75, 144, 256, 259, 262, 263, 268
uptake, 122, 234, 235, 243, 256, 257, 270, 315,
    336, 363, 380
urea, 121, 124, 164, 257, 330, 362, 380

V

valorization, 23, 51–54, 57, 58, 60, 61, 66, 69, 76,
    78, 80, 81, 86, 149, 150, 194, 273
value-added, 23, 30, 35, 37, 38, 42, 44, 45, 47–49,
    53, 58, 60, 65, 76, 77, 81, 103, 106,
    146, 189, 226, 267, 303, 312, 349, 378
variability, 227, 311, 313, 315, 352
vascular, 241, 242, 296, 305, 346
vegetables, 40, 204, 305, 318, 333, 345
vermiculophylla, 41, 54
versatility, 51, 85, 268, 306, 339
vesiculosus, 278, 283, 284, 294, 307
violaxanthin, 67, 68, 158, 180, 203, 323, 333,
    334, 352
virus-algal, 294
viscosity, 40, 119, 127
vitamin, 45, 68, 69, 158, 204, 236, 239, 248, 322,
    330–32, 347, 349, 352
volatile, 74, 105, 176, 259, 352
voltage, 94, 100, 128

W

wastewater, 35, 42, 50, 53, 60, 81, 83, 87, 91, 92,
    94, 95, 98–104, 106, 107, 120, 122,
    131, 139, 143, 144, 146, 148, 151, 175,
    182, 183, 188, 189, 197, 204, 230, 235,
    236, 238, 248, 249, 251, 253, 256–60,
    267, 269–73, 301, 302, 310, 344,
    349, 378
water-soluble, 62, 67, 87, 112, 113, 159, 171, 179,
    266, 286, 289, 305, 342, 356, 357
wavelengths, 67, 156, 159, 334, 356, 358
wetland, 253
wheat, 64, 139, 149, 232, 235, 236, 238, 241, 242,
    245, 247–53, 298, 299, 310
whole-cell, 177, 265

X

xanthophyceae, 157
xanthophyllomyces, 204
xanthophylls, 67, 158, 159, 170, 192, 194,
    203, 313
xenobiotic, 35, 318
xerophthalmia, 68
x-ray, 24
xylan, 276
xyloglucan, 30, 287, 317
xylose, 30, 278, 283, 284, 287, 289

Y

yeast, 33, 40, 54, 71, 76, 78, 79, 88, 174, 204, 207,
    305, 345
yogurt, 185, 297

Z

zeaxanthin, 67, 158, 203, 322, 333, 334
zebrafish, 206, 223
zero-waste, 30, 38, 59, 60
zofingiensis, 151, 165, 167, 185, 191, 195, 204,
    260, 273
zooplankton, 207, 214, 247, 377
zooxanthellae, 213
zucchini, 251
zygotes, 43

For Product Safety Concerns and Information please contact our EU
representative GPSR@taylorandfrancis.com
Taylor & Francis Verlag GmbH, Kaufingerstraße 24, 80331 München, Germany